3rd International Conference on Measurement Control and Instrumentation (MCAI 2023)

Journal of Physics: Conference Series Volume 2724

Guangzhou, China
24 – 26 November 2023

ISBN: 978-1-7138-9279-3
ISSN: 1742-6588

Printed from e-media with permission by:

Curran Associates, Inc.
57 Morehouse Lane
Red Hook, NY 12571

Some format issues inherent in the e-media version may also appear in this print version.

Institute of Physics
Dirac House, Temple Back
Bristol BS1 6BE UK

Phone: 44 1 17 929 7481
Fax: 44 1 17 920 0979

techtracking@iop.org

Additional copies of this publication are available from:

Curran Associates, Inc.
57 Morehouse Lane
Red Hook, NY 12571 USA
Phone: 845-758-0400
Fax: 845-758-2633
Email: curran@proceedings.com
Web: www.proceedings.com

TABLE OF CONTENTS

Journal of Physics: Conference Series **2724** (2024) 011001 doi:10.1088/1742-6596/2724/1/011001

Preface

The 2023 3rd International Conference on Measurement Control and Instrumentation (MCAI 2023) was grandly held in Guangzhou, China from November 24th to 26th, 2023 (hybrid event). Scholars, experts and enterprise representatives from all over the world gathered together to discuss the latest research results, technological advances and innovative applications in the field of measurement control and instrumentation.

This Conference was organized in a variety of formats, including keynote speeches, oral presentations, poster presentations, Q&A sessions, academic investigation, etc., aiming to provide a platform for participants to present their achievements, have in-depth exchanges, and discuss collaborations. Over 300 representatives attended the conference, including experts with outstanding achievements in the field of measurement control and instrumentation as well as young researchers and graduate students interested in the field. They are actively committed to advancing the field and promoting academic exchanges and collaborations through this international conference. Participants listened to the presentations of outstanding researchers from all over the world and engaged in in-depth discussions and exchanges with them.

The Proceedings of MCAI 2023 contains top academic papers from all over the world, which cover a wide range of topics and cutting-edge researches in the field of measurement control and instrumentation, including but not limited to: Numerical Control Technology, Precision Assessment, Microsensor Technology and Applications, Intelligent Instruments and Machinery, etc. The Editorial Board, a group of experts with rich experience and outstanding achievements in related fields, were responsible for reviewing and selecting the best papers to ensure the high quality and academic impact of the Proceedings.

By sharing the latest research results and technology applications, we are confident that MCAI 2023's Proceedings will be an important literature resource in the field of measurement control and instrumentation, providing researchers with an extensive knowledge base that will help them keep abreast of the latest research development and technology trends.

Finally, we would like to thank all the people and organizations that have participated in and supported MCAI 2023. Sincere thanks to all the authors and editorial board members who have contributed to this Proceedings. Through the publication and dissemination of the Proceedings, we can facilitate the cooperation between academia and industry, promote the translation and application of scientific research results, and contribute to the solution of practical problems.

The Committee of MCAI 2023

Committee Member

Conference General Chair
Prof. Guixiong Liu, South China University of Technology, China
Prof. Zhen Li, South China Agricultural University, China

Publication Chair
Prof. Talha Ali Khan, UE Germany, Campus Innovation Hub, Germany

Technical Program Committee Chair
Prof. Zhenmin Zhu, East China Jiaotong University, China

Organizing Committee
Prof. Jiang Guo, Dalian University of Technology, China
Prof. Vikas Gupta, Technocrats Institute of Technology, Bhopal, M.P., India
Prof. Ning Sun, Nankai University, China
Prof. Weixing Wang, South China Agricultural University, China
Prof. Liming Wu, Guangdong University of Technology, China
Prof. Xue Wang, Tsinghua University, China
Prof. Zhen Li, South China Agricultural University, China
Prof. Yaohua Deng, Guangdong University of Technology, China
Prof. Guojian Huang, Guangdong Mechanical & Electronical College of Technology, China
Assoc. Prof. Jie Jia, Chang'an University, China
Dr. Ravi Panwar, IIIT Jabalpur, India
Dr. Xuefeng Zhou, Guangdong University of Technology, China

Technical Program Committee
Prof. Jiaxiang Xue, South China University of Technology, China
Prof. Shuqiang Huang, Jinan University, China
Prof. Dr. Murat Tolga OZKAN, Gazi University, Turkey
Prof. Tirumala Ramashri, SVUniversity College of Engineering, India
Assoc. Prof. Sivakumar Dhar Malingam, Universiti Teknikal Malaysia Melaka, Malaysia
Assoc. Prof. Md Mozasser Rahman, Universiti Tun Hossein Onn Malaysia, Malaysia
Assoc. Prof. Ghous Bakhsh, NED University, Pakistan
Assoc. Prof. Ankit Sharma, Chitkara University, India
Dr. Ajay, Shree Guru Gobind Singh Tricentenary University, India
Dr. Ranjithkumar Rajamani, Viyen Biotech LLP, India

Journal of Physics: Conference Series **2724** (2024) 011002 doi:10.1088/1742-6596/2724/1/011002

Peer Review Statement

All papers published in this volume have been reviewed through processes administered by the Editors. Reviews were conducted by expert referees to the professional and scientific standards expected of a proceedings journal published by IOP Publishing.

- **Type of peer review:** Single Anonymous
- **Conference submission management system:** Morressier
- **Number of submissions received:** 125
- **Number of submissions sent for review:** 100
- **Number of submissions accepted:** 66
- **Acceptance Rate (Submissions Accepted / Submissions Received × 100):** 52.8
- **Average number of reviews per paper:** 2
- **Total number of reviewers involved:** 13
- **Contact person for queries:**
  Name: Xuexia Ye
  Email: xx.ye@keoaeic.org
  Affiliation: AEIC Academic Exchange Information Centre

Mechanism of Seebeck coefficient variation at the output of NiCr/NiSi thin film thermocouple with different wires

Zhihui Liu [1, 2], Tengda Guo[1], Yongjun Cheng[3], Bi Wang[4], Kai Shen[2], Kai Hu[1], Jiankang Zhou[1], Chuanbing Zhang[1], Zixi Wang[1,5] and Wanyu Ding[2]

[1]State Key Laboratory of Tribology in Advanced Equipment, Tsinghua University, Beijing 100084, China)
[2]College of Materials Science and Engineering, Dalian Jiaotong University, Dalian 116028, China)
[3]Laboratory of Low frequency Electromagnetic Communication Technology with the WMCRT CSSC)
[4]State Key Laboratory of High-end Compressor and System Technology, Hefei, Anhui, 230031, China)

[5]Corresponding author's e-mail: zxwang@tsinghua.edu.cn

Abstracts. In this paper, by using magnetron sputtering to prepare NiCr/NiSi thin film thermocouples, the static calibration method is used for NiCr/NiSi thermocouples with rapid temperature calibration experiments. Different temperature calibration curves are obtained. The Seebeck coefficient of NiCr/NiSi thin film thermocouples connected to NiCr/NiSi wires is significantly higher (41.39 μV/°C) than that of NiCr/NiSi wires (0 μV/°C). The Seebeck coefficient (41.39 μV/°C) of the NiCr/NiSi thin-film thermocouple connected to copper wire is significantly higher than that of the thermocouple connected to copper wire (0 μV/°C). The problem of the Seebeck coefficient of the K-type thermocouple is analyzed by experimental data, which provides the relevant parameter basis for the use of the K-type thermocouple. The method has the advantages of simple equipment, convenient operation, accurate and reliable data, and provides a basis for the sensor to measure the temperature measurement.

1. Introduction

Current research on thin-film thermocouples mainly focuses on improving the preparation methods and expanding the application areas. However, regardless of the preparation method and the application of thin-film thermocouples, their temperature measurement principle is based on the Seebeck effect [1-5]. Therefore, the use of thermocouples to accurately measure transient surface temperatures presupposes the ability to accurately calibrate the Seebeck coefficient of the thin film thermocouple. The static calibration process for ordinary filament thermocouples involves placing them in a metering furnace with a standard platinum resistor, fixing the temperature of the cold end and keeping the measuring end constant at different temperatures, and establishing the thermocouple's thermopotential-temperature relationship by measuring the calibration temperature from the standard filament thermocouple [6-7]. When this method is used to calibrate the tiny-size thin-film thermocouple, the entire sensor is placed in the homogenized temperature field of the metering furnace. The temperature gradient is mainly concentrated in the compensation wire, which is inconsistent with the temperature distribution of thin-film thermocouple applications, and it is not possible to realize effective calibration, which affects the temperature measurement accuracy of thin-film thermocouples in practical applications [8-9]. As the NiCr/NiSi thermocouple connected to a different line Seebeck is also different, the thermoelectric properties of the electric potential difference are still very large. The calibration in this paper analyzes the above problems well. Only by analyzing the static calibration method of thin-film thermocouple in detail can we accurately evaluate its static characteristics so as to further improve the accuracy of temperature measurement.

2 Structural design and preparation

2.1 Thin film thermocouple structure design

When designing the NiCr/NiSi thin film thermocouple temperature, first of all, the structure of the thin film thermocouple temperature sensor is simple, and the preparation process is easy to realize, which can meet the needs of the sensor and calibration. The NiCr/NiSi thin film thermocouple temperature has a sensitivity and fast response speed, which can meet the requirements of temperature testing [10-14]. The connection between the thin film thermocouple temperature thermal electrode material and the compensation wire is stable and reliable, will not fall off easily, and can stably transmit the temperature signal. Based on the above principles, the NiCr/NiSi thin-film thermocouple temperature design uses a glass substrate with a rectangular shape, as shown in Figure 1. For the shape of a thin-film thermocouple, thermal electrodes are prepared in this paper using a mechanical mask.

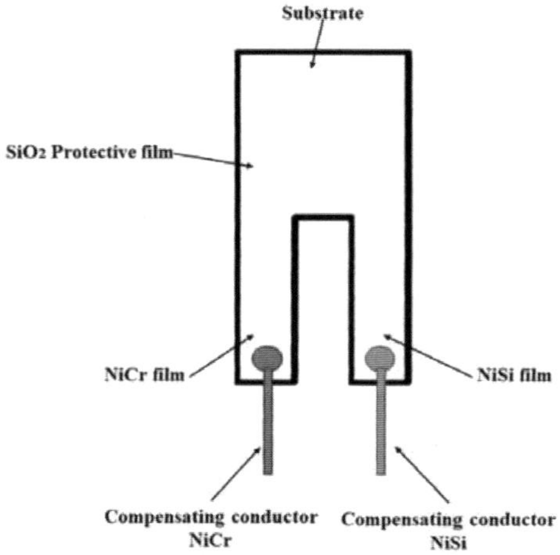

Figure 1. NiCr/NiSi thin-film thermocouple temperature sensor.

2.2 Preparation of NiCr/NiSi thin film thermocouple

NiCr/NiSi thin-film thermocouple was prepared as follows: the polished mask, fixture, and glass substrate were cleaned with acetone, ethanol, and deionized water for 20 minutes, respectively, and blown dry with nitrogen. The mask is positioned and mounted on the fixture and fixed on the sample tray with high-temperature tape. The mask should be mounted carefully to tightly bond the mask to the substrate to avoid diffraction. When the vacuum degree of the vacuum coating chamber reaches 1.0×10^{-3}, the vacuum chamber of the vacuum coating equipment is closed. When the vacuum degree of the vacuum coating chamber reaches 6.0×10^{-3}, the air pressure of the vacuum coating chamber is adjusted to 0.6 Pa. After the pressure in the vacuum chamber is balanced, the power switch is activated, and at the same time, the baffle is turned on for pre-sputtering for 5 minutes in order to eliminate the impurities on the target material and to ensure the stability of the sputtering process. After pre-sputtering, the target baffle was closed, and the sputtering of NiCr functional films was formally initiated. After the sputtering of the NiCr functional film is finished, it is necessary to gradually close the gas valve molecular pump, the insert valve, and the mechanical pump, and then wait for the completion of the venting and cooling, and finally remove the sample. The mask and target material are replaced to sputter NiSi and SiO_2 in turn. The sputtering parameters are shown in Table 1[18], and the operation procedure is the same as that for NiCr.

Table 1. Plating parameters [18].

Parameters	Vacuum	Working gas	Working Pressure /pa	Ar flow /sccm	Sputtering power /W	Sputtering time /min
NiSi	6.5×10^{-3}	Ar	0.6	20	150	20
NiCr	6.5×10^{-3}	Ar	0.6	20	150	20
SiO_2	6.2×10^{-3}	Ar/O_2	0.57	20/8	200	100

2.3 Performance study of NiCr/NiSi thin film thermocouple

Sensor performance testing is an important part of the sensor production and commissioning process [15]. In order to ensure that the developed sensors can be used in industry, scientific research, and other places to obtain accurate, reliable, stable, and rapid measurement results, in sensor manufacturing, packaging must be tested after the performance parameters. Static characteristics are important performance indicators of thin film thermocouple temperature sensors [16]. The NiCr/NiSi thin-film thermocouple temperature sensor developed in this paper is a non-standard sensor. In this paper, the developed NiCr/NiSi thin film thermocouple temperature sensor is calibrated using static calibration of laboratory equipment.

2.3.1 Calibration system design. Temperature sensor static calibration refers to the temperature sensor in the temperature field that reaches the static standard conditions and determines the linear relationship between the sensor input and output [17]. The aim is to obtain the developed sensor's Seebeck coefficient, linearity, repeatability, and other indicators. The temperature sensor used in this paper mainly consists of FLUKE-9144 dry metering furnace, PLANCK-6190A type icemaker, and DMM7510 digital multimeter. The thin film thermocouple temperature static calibration schematic system is shown in Figure 2[19-20].

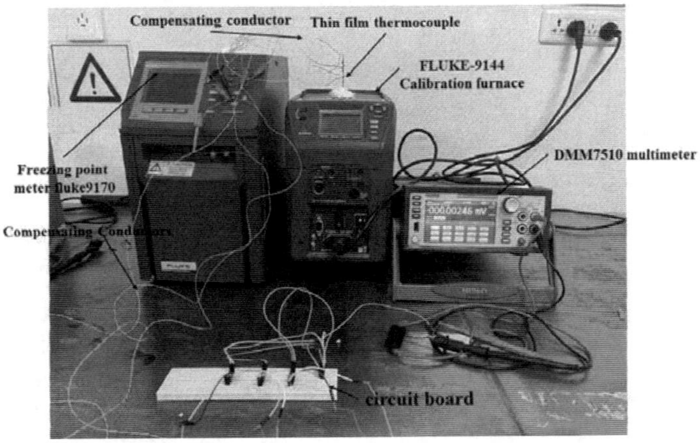

Figure 2. Thin film thermocouple temperature sensor[19-20].

2.3.2 Static calibration of thin-film thermocouples with NiCr/NiSi wires. The NiCr/NiSi thermocouple is placed in the hot end of the FLUKE-9144 dry metering furnace. The compensation wire connected to the NiCr/NiSi wire is placed in the PLANCK-6190A freezing point of the thermostat. 0 °C constant is set. The static calibration temperature range is set as 50 ~ 400 °C. The dry metering furnace temperature rises 10 °C per the heating speed is 10 °C/min and the stepping temperature is 0.5°C. The cold end temperature is provided by Fluke 9170 calibration furnace and kept at 0°C.. The temperature constant is kept for 5 min. A DMM7510 multimeter is used to record each time the temperature constant sensor output thermopotential value. According to the static calibration experiments to derive the thin film thermocouple temperature sensor thermopotential and temperature relationship, the fitted curve is shown in Figure 3.

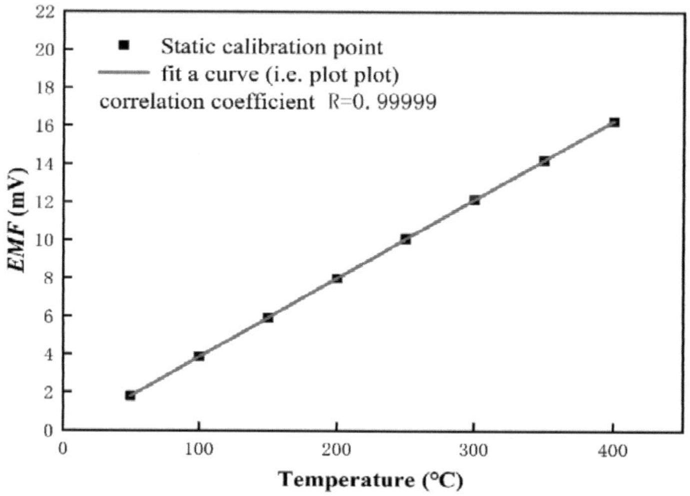

Figure 3. Static calibration curve of thin film sensor.

As can be seen from Figure 3, the equation between the thermopotential E sensor and the temperature θ is: E = 0.04139θ + 0.04139. By analyzing the results of the static calibration experiments, it can be concluded that the developed NiCr/NiSi temperature sensor is connected to the NiCr/NiSi wire with a Seebeck coefficient of 41.39 μV / °C.

2.3.3 Static calibration of thin-film thermocouples with copper leads. In the same way, the developed NiCr/NiSi thin-film sensor is placed in the hot end. The compensation wire connected to the copper wire is placed in a freezing-point constant thermostat. 0 °C constant is set. The static calibration temperature range is set as 50 ~ 400 °C. The dry metering furnace temperature is maintained at every 10 °C rise 5 min constant temperature for calibration to obtain data. The expression obtained by fitting the data is shown in Figure 4:

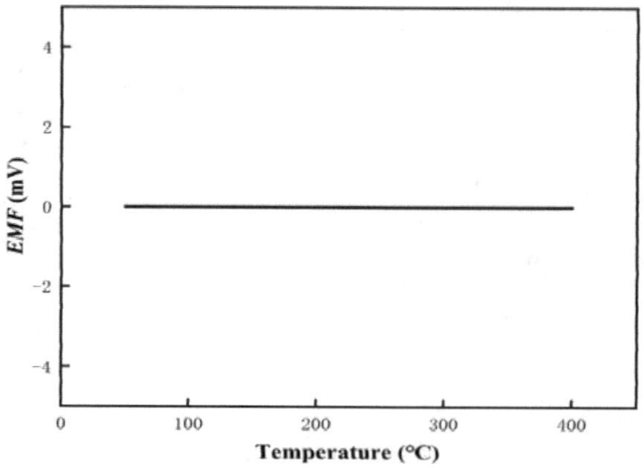

Figure 4. Static calibration curves for thin film thermocouples.

As can be seen from Figure 4, the thin film thermocouple thermoelectric potential E and the temperature θ of the equation between E = 0. By analyzing the results of the static calibration experiments, it can be concluded that the developed NiCr/NiSi thin-film sensor is connected to the copper wire and has a Seebeck coefficient of 0 μV / °C.

3. Experiments and results

Through the previous study, this paper found that the Seebeck of NiCr/NiSi thin film thermocouple connected to NiCr/NiSi wire (41.39 μV/°C) is significantly higher than that of NiCr/NiSi thin film thermocouple connected to copper wire (0 μV/°C). Detailed comparison data are shown in Figure 5.

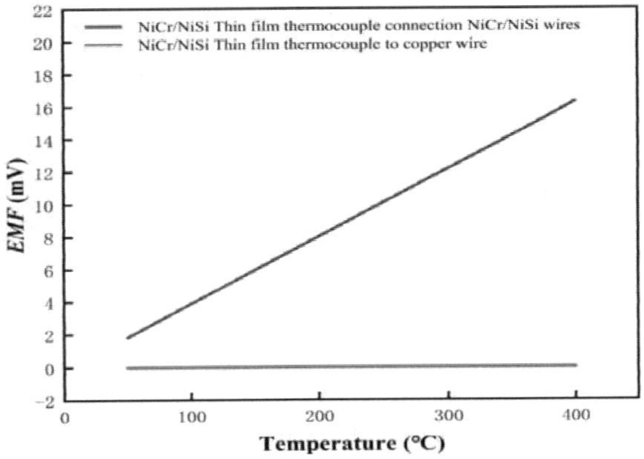

Figure 5 Static calibration curves for wo outlets of a thin-film Sensor.

This is because no electric potential is generated between the copper compensation wires. Therefore, the connection to the electric potential of the copper wires is provided only by the film thermocouple and the hot and cold ends of the film thermocouple are in the same temperature field, so the Seebeck coefficient is 0 μV/°C.

Through the NiCr/NiSi thin film thermocouple connected to the wire is not the same as analyzing the Seebeck coefficient value. It can be reliably applied to temperature measurement, smart tools, smart manufacturing, and other fields.

4. Conclusion

In this paper, a NiCr/NiSi thermocouple is connected to different wires with different Seebeck. Using rapid temperature calibration experiments, different temperature calibration curves are obtained. The study of NiCr/NiSi thin film thermocouple connected to the NiCr/NiSi wire Seebeck coefficient (41.39 μV / °C) is significantly higher than that of NiCr/NiSi film thermocouple connected to copper wire (0 μV/°C). The problem of the Seebeck coefficient of the K-type thermocouple is analyzed by experimental data, which provides the relevant parameter basis for the use of the K-type thermocouple. The method has the advantages of simple equipment, convenient operation, accurate and reliable data, and provides a basis for the sensor to measure the temperature measurement. Using the comparative method of NiCr/NiSi thermocouple connected to a different line, Seebeck was analyzed in detail for accurate measurement of temperature. The study of tool wear, improved machining accuracy, machining efficiency, and tool durability is of great significance.

Acknowledgment

1. This work was supported by the National Key R&D Program of China (No.2021YFB4001500).

2. Supported by the Open Foundation of the Laboratory of Low-Frequency Electromagnetic Communication Technology with the WMCRT CSSC, No. DPJJ-2022-03.

3. Supported by Open Foundation of State Key Laboratory of High-end Compressor and System Technology, No SKL-YSJ202201.

References

[1] Bohara B B, Batra A K. Development of multi-functional nano-paint for energy harvesting applications[J].Progress in Natural Science: Materials International, 2018.

[2] Liang G, Li P.Research on thermoelectrics transducers for harvesting energy from asphalt pavement based on Seebeck effects[J]. 2015.

[3] Zhicun Xong. T-type flexible thin film thermocouple response speed test and its application [J]. Shanghai Measurement Testing, 2023,50(03):19-21+24.

[4] Xiaohui Zhang. Development of a Thermocouple for Thermal Conductivity and Insulation Surface Testing [J]. Metrology and test technology, 2022,49(10):35-37.

[5] Chengbing Fang. Design and testing of a thin film thermocouple calibrator for aerospace applications [D]. Shandong Agricultural University,2023.

[6] ZHAO Yuan-Shen. Preparation of T-type thin-film thermocouple and its performance research[D]. Shanghai University of Technology, 2012.

[7] Tolganbek N, Mentbayeva A, Serik N, et al. Design and preparation of thin film gel polymer electrolyte for 3D Li-ion battery[J].Journal of Power Sources, 2021, 493:229686.

[8] Chen Yang, Fu Min, Liu Bo. Research on detecting train bearing damage by acoustic resonance demodulation technique FJ]. Information Technology, 2004, 8(1):27-29.

[9] Huang X F , Luo D , Wang G Q ,et al.State of the Art in Condition Monitoring in Wind Turbines Based on Embedded FBG Sensors[C]//Asia-pacific Power & Energy Engineering Conference.IEEE, 2010.5448939.

[10] Yidan Wang. Thermal oxidation reliability and structural optimisation of thin film thermocouples [J]. Journal of Beijing University of Aeronautics and Astronautics,2023,49(04):943-948.

[11] Zhang JG. Research on fast response thin film temperature sensor technology[D]. National University of Defense Technology, 2008.

[12] YAO Fei, JIANG Hongchuan, ZHANG Wanli, et al. Preparation and performance of metal-based NiCr-NiSi thin film thermocouples[J]. Electronic Components and Materials, 2010, (9): 3.

[13] Nouri M, Fussell B K. Real-Time Milling Force Control and Tool Wear Monitoring Using Constraint-Based Feedrates[C]//North American Manufacturing Research Conference.2013.

[14] Yeh, Fang-Wei. Development of thin film thermocouples and their applications[J]. Materials Herald. 1995(05): 28-32.

[15] Zhang JG. Research on fast response thin film temperature sensor technology[D]. National University of Defense Technology, 2008.

[16] Keith Riley. Temperature sensor calibration[J]. Control Engineering, 2019, 66(3): 26-27.

[17] Oracle International Corporation; Patent Issued for Statistical Temperature Sensor Calibration Apparatus And Methodology (USPTO 10,768,057)[J]. Computer Weekly News, 2020: 6078.

[18] C. Zhang Can, W. Ding, H. Wang, W. Chai, The influence of O2/Ar flow ratio on the photocatalytic activity of TiO2 films prepared by DC pulse magnetron sputtering, Zhenkong Kexue Yu Jishu Xuebao, 30 (2010) 106-110. (in Chinese)

[19] W. Ding, D. Ju, W. Chai, A simple route to synthesize nanocrystalline TiO2 films at room temperature, Mater. Lett. 64 (2010) 1634-1637. https://doi.org/10.1016/j.matlet.2010.04.019.

[20] W. Ding, H. Wang, D. Ju, W. Chai, Composition and crystal structure of N doped TiO2 film deposited with different O2 flow rates, Acta Phys. Sin. 60 (2011) 028105. https://doi.org/10.7498/aps.60.028105.

Error analysis of IMU pipeline bending strain detection based on full-scale test

Tong Shi[1], Xiaoben Liu[1,4], Yanan Lu[2], Haoning Zhang[1], Rui Li[3], Kuan Fu[3], Qingshan Feng[3] and Qiyu Huang[1]

[1]National Engineering Laboratory for Pipeline Safety/MOE Key Laboratory of Petroleum Engineering/Beijing Key Laboratory of Urban Oil and Gas Distribution Technology, China University of Petroleum-Beijing, China
[2]East Branch, Petrochina Natural Gas Marketing Company, China
[3]National Petroleum and natural gas pipeline network Group Co.,Ltd., China

[4]Corresponding author's e-mail: xiaobenliu@cup.edu.cn

Abstract. The strain identification technology based on inertial mapping unit (IMU) internal detectors is currently a widely used internal detection technology for identifying pipeline bending strain both domestically and internationally. It calculates the bending strain of the entire pipeline through the detected centerline coordinate data. However, there are issues such as abnormal vibration and noise during the actual operation of the inertial navigation internal detector, as well as trajectory deviation during the detector's operation. This issue leads to a certain deviation between the strain detected by IMU and the actual strain of the pipeline. This article utilizes a self-designed full size pipeline bending strain loading device to conduct full size pipeline bending strain tensile tests. Compare the internal detection data obtained from the experiment with the actual bending strain data of the pipeline and conduct error analysis. The research conclusion indicates that as the bending strain increases, the error between the IMU detection strain and the actual strain of the pipeline increases. When there is vibration noise, the maximum error is 0.077% for 0.33% true strain, and the maximum error rate for larger strain segments is 27.1%. The comparison between noise reduction detection data and true strain shows that when the true strain is 0.12%, 0.21%, and 0.33%, the average error is 0.0285%, 0.0315%, and 0.0555%, respectively. The research conclusion provides technical support and guidance for obtaining true strain of pipelines based on IMU internal detection in the future.

1. Introduction

Pipeline is currently the most efficient and common tool for transporting oil and natural gas[1], and is an important part of ensuring energy stability in China. Oil and gas pipelines pass through complex areas, and when settlement or surrounding soil collapse occurs, rock pressure and other factors can cause local deformation and displacement of the pipeline, resulting in bending strain[2][3]. When the strain exceeds the limit of the pipeline material, defects such as cracks and depressions will occur. In severe cases, it can lead to pipeline failure and fracture[4] resulting in pipeline leakage, causing casualties[5] and economic losses.

At present, inertial navigation technology is gradually maturing for internal detection of pipeline bending strain. IMU measures the three-dimensional coordinates of the pipeline and obtains the centerline of the pipeline[6], thereby obtaining the strain situation at each position of the pipeline. China Petroleum Pipeline Company[7] independently developed a high-precision IMU inertial navigation surveying and mapping system to calculate pipeline bending strain. However, due to the complexity of the IMU inertial navigation system during the detection process, there are certain differences between the results and the actual results[8]. Li Rui et al.[9] used the IMU inertial navigation system for pipeline detection in permafrost regions, proposed a method based on IMU data for bending strain and displacement, and corrected the problem of displacement divergence. Feng Qingshan et al.[10] used GPS and mileage wheel data to correct the pipeline position and improve detection accuracy.

In summary, current domestic and foreign researchers have revised and optimized the IMU detection data[11][12], but have not conducted precise error analysis between the IMU detection data and the true strain of the pipeline. The relationship between the measured value of pipeline bending strain and the actual value cannot be determined. To this end, tensile tests were conducted on IMU internal detectors under different bending strain loading conditions for full-scale pipelines, and the error between the IMU internal detection data and the true strain data was ultimately obtained.

2. Calculation method of bending strain of pipeline

2.1. Bending strain calculation based on center line coordinates

The bending curvature of the pipeline can be calculated from the projection of the centerline of the pipeline on the horizontal plane and the projection on the vertical plane, respectively. Its total curvature, horizontal component curvature and vertical component curvature[13] have the following relationship:

$$k=\sqrt{k_v^2+k_h^2} \tag{1}$$

In the formula: $\boldsymbol{k}$ is the total curvature of the pipeline; $\boldsymbol{k_v}$ is the curvature of the pipeline in the horizontal plane; $\boldsymbol{k_h}$ is the curvature of the vertical plane of the pipe.

Let the equation of the pipeline center line curve be: $y=f_{(x)}$, and $f_{(x)}$ has a second derivative.

The curvature $\boldsymbol{k_v}$ and $\boldsymbol{k_h}$ of this point are calculated according to the first derivative $\boldsymbol{y'}$ horizontal, $\boldsymbol{y'}$ vertical and the second derivative $\boldsymbol{y''}$, $\boldsymbol{y''}$ vertical, respectively. The specific calculation method is as follows:

$$k=\frac{|y''|}{(1+y'^2)^{\frac{3}{2}}} \tag{2}$$

$\boldsymbol{k_v}$ and $\boldsymbol{k_h}$ are calculated, and the total strain $\boldsymbol{k}$ is calculated by the formula, and then the bending strain of the pipeline is calculated by the following formula.

$$\varepsilon=\frac{D}{2}k \tag{3}$$

In the formula: $\boldsymbol{D}$ is the outer diameter of the pipeline, mm; $\boldsymbol{\varepsilon}$ is the bending strain of the pipe, %.

2.2. Bending strain solution based on IMU attitude data

Based on the above analytical calculation method for pipeline bending strain, it can be seen that within the elastic deformation range, the bending strain of the pipeline is directly proportional to the change in curvature. Therefore, when calculating pipeline bending strain, the curvature of the pipeline should be calculated first.Among them, the pitch change of the internal detector represents the change in inclination angle of the pipeline relative to the horizontal plane in a fixed observation pipeline, and the heading represents the angle between the direction along the pipeline and the due north direction[14], as follows:

$$k_v=\Delta\varphi/\Delta s \tag{4}$$

$$k_h=\Delta\alpha\cos\varphi/\Delta s \tag{5}$$

Where: $\boldsymbol{k_v}$ is the curvature of the pipeline in the horizontal plane; $\boldsymbol{k_h}$ is the curvature of the vertical plane of the pipeline; $\boldsymbol{\varphi}$ is the pitch Angle of the inertial measurement device along the vertical tangent direction of the pipe, (°); $\boldsymbol{\Delta s}$ is the length of the fixed pipe section, m; $\boldsymbol{\Delta\alpha}$ is the angular change of the inertial measurement device along the horizontal tangent direction of the pipe, (°).

The curvature $\boldsymbol{k_v}$ of the horizontal plane of the pipeline and the curvature $\boldsymbol{k_h}$ of the vertical plane of the pipeline are obtained and the total strain $\boldsymbol{k}$ is obtained by equation (1). The bending strain $\boldsymbol{\varepsilon}$ of the pipeline can be obtained by bringing it into equation (3). The bending strain of the pipeline is related to the pitch Angle of the inertial measurement device along the vertical tangent direction of the pipeline and the Angle along the horizontal tangent direction of the pipeline as follows:

$$\varepsilon = \frac{D}{2}\left((\Delta\varphi / \Delta s)^2 + (\Delta\alpha \cos\varphi / \Delta s)^2\right)^{\frac{1}{2}} \tag{6}$$

3. Traction test study

3.1. Test scheme and device

The IMU full size pipeline pulling test was carried out using 508×10mm full size pipeline, full size pipeline bending strain loading device and IMU internal inertial navigation detector. The full-size pipe bending strain loading device was designed and used to load the load to the preset working condition. The IMU inertial navigation detection data collected by the IMU detector was obtained by the traction device, and then the error analysis between the full-size pipe IMU test detection data and the real pipe strain data was carried out.

Table 1. Full size loading device main technical parameters table.

Pointer object	Technical index
Loading pipe parameter	508*10mm、100m
Maximum output load of hydraulic jack	500kN
Number of hydraulic units	4
Loading pipe length	3m
Reaction frame design load	2000kN

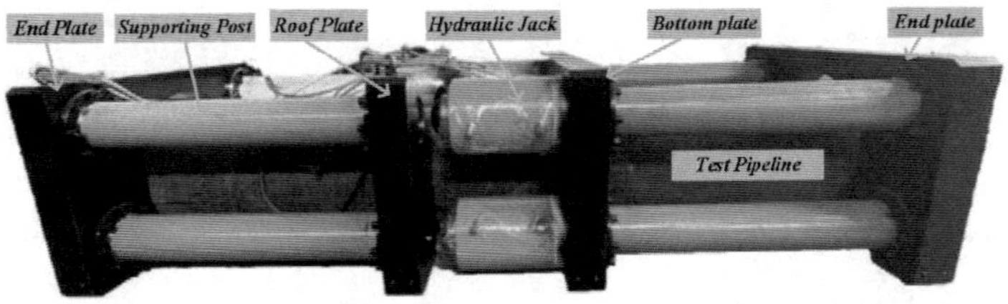

Figure 1. Full size axial loading device assembly drawing.

The length of the stretch test pipe is 100m, and the length of the bending deformation section pipe is 20m. The bending strain loading device of the pipeline comprises the structure of support plate, end plate, column and so on. The output load of a single hydraulic jack of 508mm full-size pipeline bending strain loading device is 500kN. During the bending test loading, two hydraulic jacks above starting can realize vertical upward bending of the pipeline. The overall length of the device is 3m, and the load of the reaction frame is 2000kN. The main technical parameters of the device are shown in table 1, and the assembly drawing of the full-scale axial loading device is shown in Figure 1.

A total of 4 groups of different bending strain conditions of the pipeline were set up in the test. First, the basic working conditions were tested when the pipeline was in a straight line. The design strains of the pipeline strain section from the second to the fourth times were 0.1%, 0.2% and 0.3%, respectively. Considering the errors of the actual field test process and device loading, the actual offset of the field pipeline was determined as the basis for the comparison between the strain detected by IMU and the true strain. The actual offset of the pipeline in each group of field test conditions was 68mm, 124mm and 180mm. The design bending strain and actual offset under different working conditions are shown in table 2.

Table 2. Pull test different preset conditions and actual offset.

Working condition	Design bending strain	Actual bending
1	0%	0
2	0.1%	68
3	0.2%	124
4	0.3%	180

3.2. Test result

For each group of strain conditions, the drawing speed of 1m/s is pulled twice. The bending strain of the pipeline was calculated using the attitude information measured by IMU, and the data collected by the IMU in-inertial navigation detector at the same drawing speed and bending strain of the pipeline were analyzed. When using IMU to calculate the bending strain, the position coordinates of the center line of the pipeline are calculated by using the combined inertial navigation algorithm. After obtaining the exact coordinates of the whole line of the pipeline and the attitude information such as pitch Angle and azimuth Angle, the bending strain is further calculated.

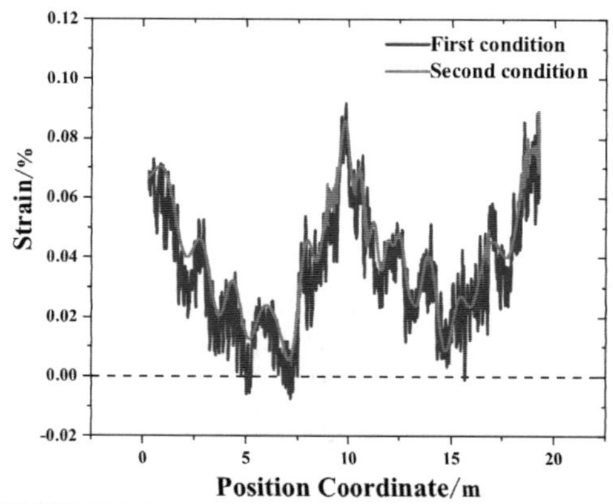

Figure 2. Data before and after noise reduction of the first pull IMU detection in the second working condition.

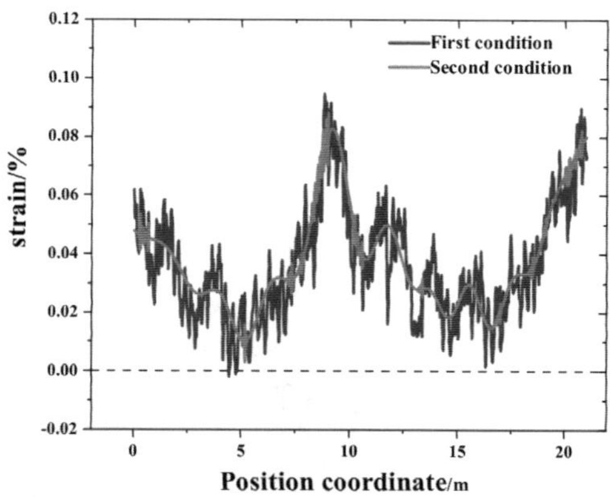

Figure 3. Data before and after noise reduction of the second pull IMU detection in the second working condition.

Select condition one as the basic condition. The aligned data is denoised using wavelet to reduce the interference of fluctuating sawtooth noise in the detection results. Based on the results of the tensile test, it can be seen that the bending strain detected by the inertial navigation detector is basically the same under the same working condition of the strain deformation section. The data before and after noise reduction in IMU for the two groups of IMU pull test under condition 2 are shown in Figure 2-3. The strain results detected in IMU under different test conditions are shown in table 3.

Table 3. Strain measured in IMU under different test conditions.

Working condition	Different groups	Design bending strain	Actual bending	Measured strain
1	A	0%	0	0%
	B			0%
2	A	0.1%	68	0.091%
	B			0.092%
3	A	0.2%	124	0.181%
	B			0.183%
4	A	0.3%	180	0.276%
	B			0.273%

4. Error analysis

4.1. True strain calculation

According to the bending strain calculation method based on center line coordinates, the true strain value of pipeline under different test offset conditions is calculated. The center line data of IMU pipeline without interference and error under different working conditions and the true bending strain of the pipeline under different working conditions are shown in Figure 4-5. The true strain data of the pipeline under different pipeline migration are 0.12%, 0.21% and 0.33% respectively.

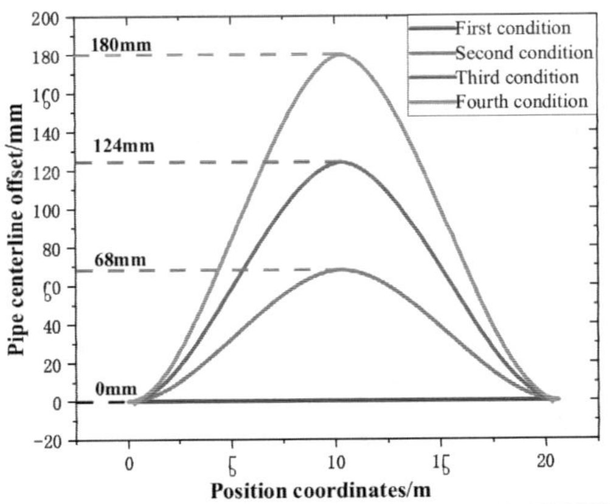

Figure 4. IMU reconstructed model to obtain pipeline center line data with different working conditions.

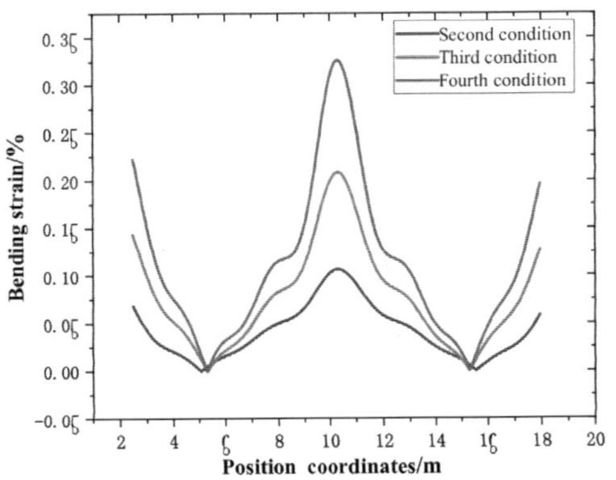

Figure 5. Bending strain of pipeline under different working conditions.

4.2. Inclusion noise error

Error calculation was carried out based on test data under different working conditions and pipeline true strain data, and error results were shown in Figure 6. The error calculation results show that the error increases gradually with the increase of strain. When the true strain is 0.33%, the maximum error reaches 0.0772%. The error rate of IMU detection data and true strain data was calculated (Figure 6-7). It was found that the error rate of the section with large strain, that is, the middle deformation area, was higher, reaching 27.1%, nearly twice that of the two sides. In this paper, the mapping research is carried out for the large strain section, that is, the 5m-15m part of the deformation section of the whole pipeline.

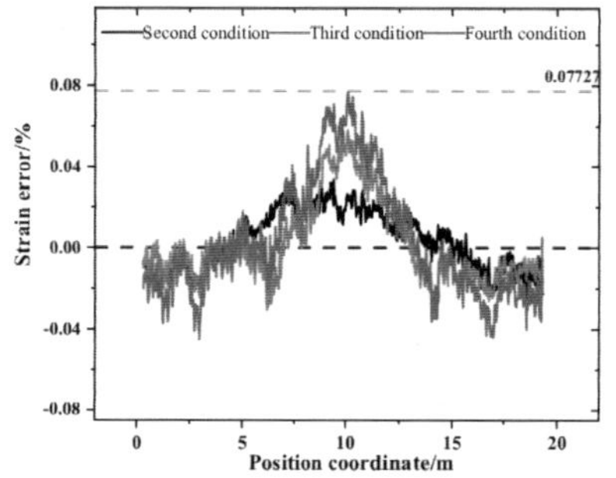

Figure 6. The detector in IMU detects strain data and true.

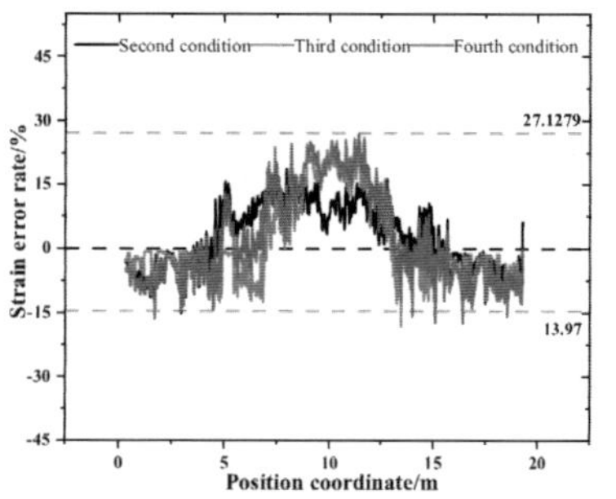

Figure 7. The detector in the IMU detects the strain data and the true strain error rate.

4.3. Error after noise reduction

The attitude data obtained by the detector in IMU was calculated for pipe bending strain. Working condition 1 was selected as the basic working condition, interference of relevant objective factors was excluded, and wavelet noise reduction algorithm was selected for data noise reduction, and the data after noise reduction was obtained. The errors of noise reduction data and true strain data after tests under different working conditions are calculated, and the error results are shown in table 4. When the true strain is 0.33%, the average error reaches 0.0555%. It can be seen from the calculation results that after the abnormal noise such as vibration is removed, there is still a certain error between the strain detected by the detector in IMU and the true bending strain of the pipeline due to the deviation of the inner detector's running track line and other reasons.

Table 4. Detection of noise reduction data and true strain error by IMU under different test conditions.

Working condition	Different groups	Measured strain	True strain	Error	Average error
1	A	0%	0	/	/
	B	0%		/	
2	A	0.091%	0.12%	0.029%	0.0285%
	B	0.092%		0.028%	
3	A	0.181%	0.21%	0.032%	0.0315%
	B	0.183%		0.031%	
4	A	0.276%	0.33%	0.054%	0.0555%
	B	0.273%		0.057%	

5. Conclusion

In this paper, a full-size pipe bending strain loading device is designed and a 508mm diameter pipe is used to carry out the pull test of IMU in-inertial navigation detector. Four groups of operating conditions were carried out, with a total of eight pulling tests twice in each group. Through the error analysis of the strain detected by the detector in IMU and the true strain value of the pipeline, it can be seen that the detection error of the detector in IMU comes from two factors, such as abnormal vibration noise and the track deviation caused by the detector size. The research conclusions are as follows:

(1) In this paper, the full-scale pipeline detector pull test is carried out respectively. When the actual pipeline offset of each group of working conditions is 68mm, 124mm and 180mm in the field test, the detected strain is 0.091% and 0.092%; 0.181%, 0.183%; 0.276%, 0.273%.

(2) As the bending strain of the pipeline gradually increases, the error between the noisy data detected by the detector in IMU and the real data gradually increases. When the true strain is 0.33%, the error reaches 0.077%, and the error rate between the actual strain and the true strain is detected by IMU.

(3) As the bending strain of the pipeline gradually increases, the error caused by the trajectory deviation of the detector in the IMU gradually increases. When the true strain is 0.12%, 0.21% and 0.33%, the average error is 0.0285%, 0.0315% and 0.0555%, respectively.

Acknowledgments
①National Key R&D Program of China (Grant No. 2022YFC3070100), ②Natural Science Foundation of China (Grant No. 52004314), ③Young Elite Scientists Sponsorship Program by Beijing Association for Science and Technology (Grant No. BYESS2023261), ④the Science and Technology Research Project of Pipechina (Grant No. WZXGL202105), ⑤the Science and Technology Research Project of Pipechina (Grant No. WZXGL202104), ⑥Science Foundation of China University of Petroleum, Beijing (No.2462023BJRC005).

References
[1] Gao Zhenyu, ZhangHuiyu, Gao Peng. New progress in China's oil and gas pipeline construction in 2022[J]. International Petroleum Economics,2023,31(03):16-23.
[2] Li Ping. Research and application of geological hazard monitoring system for buried oil and gas pipelines[J]. Technology Innovation and Application,2022,12(02):168-170.DOI:10.19981/j. CN23-1581/G3.2022.02.047.
[3] Liu Peng, Li Yuxing, Zhang Yu, et al. Stress calculation of buried pipelines affected by typical geological hazards[J]. Oil & Gas Storage and Transportation,2021,40(02):157-165.
[4] Zhang Hong, Ji Beilei, Liu Shen, et al. Digital Twin mechanism model for the structural safety of pipelines in geohazards area[J]. Oil & Gas Storage and Transportation,2021,40(10):1099-1104+1130.
[5] Zhang Hong, Ji Beilei， Liu Shen， et al. Digital Twin mechanism model for the structural safety of pipelines in geohazards area[J]. Oil & Gas Storage and Transportation, 2021,40(10):1099-1104+1130.
[6] Li Rui, Zhao Xiaoli, Zhao Lianyu, et al. Central Line and Strain & Displacement Inspection System of Long Distance Pipelines[J]. Petroleum Science and Technology Forum,2016,35(S1):197-200+256-257.
[7] Ding Nan, Chen Hao, Sun Jing, et al. Data Integration Methods of Intelligent Pipeline Construction and Discussion on Problems Existing in Data Integration[J]. Technology Supervision in Petroleum Industry, 2021,37(07):45-48.
[8] Zhang, Penghe , et al. Low-cost IMU and odometer tightly coupled integration with Robust Kalman filter for underground 3-D pipeline mapping[J]. Measurement， 2019， 137: 454-463.
[9] Li Rui, Liu Hezi, Chen Pengchao, et al. Bending strain and displacement monitoring method for pipelines in permafrost zones[J]. Oil & Gas Storage and Transportation,2017,36(04):393-397+408.
[10] Li Rui, Feng Qingshan, Zhang Junzheng, et al. Measurement method of pipeline central line based on inertial navigation[J]. Oil & Gas Storage and Transportation, 2013,32(09):979-981.
[11] Sun Peng, Xie Yuxi, Zhang Yong, et al. Alignment method of pipeline in-line inspection data and its application[J]. Oil & Gas Storage and Transportation,2021,40(02):151-156.
[12] Liu Wei. Research on alignment method of pipeline ontology data and internal inspection data[D]. Xi'an University of Science and Technology,2018.
[13] Feng Qingshan, Li Rui, Zhang Hong. Modeling and Calculation of Dent Based on Pipeline Bending Strain[J].Journal of Sensors, 2016, 2016:1-8.DOI:10.1155/2016/8126214.
[14] Liu, Shucong, et al. A novel algorithm for pipeline displacement and bending strain of in-line inspection based on inertia measurement technology [J].Advances in Mechanical Engineering, 2018, 10(12):168781401881675.DOI:10.1177/1687814018816755.

Research on rapid detection of characteristic parameters of aeroengine blade surface based on laser scanning

Junfang Song [1,4], Bin Sun [2], Tengjiao Wang [1] and Huauyou Deng [3]

[1]School of Information Engineering, Xizang Minzu University, Xianyang 712082, China
[2]School of Optoelectronics Engineering, Xi'an Technological University, Xi'an 710021, China
[3]Luoyang LYC Bearing Co., Ltd., China

[4] Corresponding author's e-mail: 284786635@qq.com

Abstract: How to detect and evaluate the profile quality of aeroengine blades and provide technical support for blade manufacturing has become one of the key technical problems in the field of aviation manufacturing quickly and effectively. This paper studies the evaluation criteria of medium surface quality in the processing and manufacturing process of aeroengine blades, optimizes and innovates the fast extraction algorithms of various geometric parameters such as aeroengine blade section line, middle arc, front/rear edge center, front/rear edge radius, blade chord length, and chord angle, and establishes the evaluation criteria of aeroengine blade surface. According to the geometric relationship of blade chord length and front/rear edge radius in the blade section profile, the parameters are solved one by one through an optimization algorithm, combined with non-contact laser scanning measurement, and the blade detection efficiency is improved by more than 40%.

1. Introduction

The blade is an important component of aeroengine, whose size, shape accuracy, and usage have a direct impact on engine performance and life [1-3]. Manufacturers of aeroengine blades generally require measurement of blade profile parameters closely related to blade performance, such as edge head radius, string line, chord length, chord angle, mid-arc, and maximum thickness, and take this as the basis for judging whether the blade is qualified [4-5].

Aeroengine blades are usually made of high-strength, corrosion-resistant, and heat-resistant titanium alloy materials. From rough to finished products, multiple processes such as cutting, grinding, and polishing are required. To ensure accuracy, multiple processes such as machining, measurement, re-processing, re-measurement, and shaping are usually required. Therefore, the requirements for measurement accuracy and efficiency are becoming increasingly high. The solution of blade feature parameters is a complex engineering problem, which must consider the determination of starting and ending points, as well as the shortest distance between points and lines and the determination of intersection points. Gao et al. [6] proposed a regional sampling method for measuring points in response to the uneven distribution of measurement points on thin-walled blade profiles. When planning measurement points, the distribution of machining errors in the front and rear edges of the blade profile, as well as in the transition areas of the blade basin and blade back, was considered. Zhang et al. [7] proposed a blade measurement sampling method based on contour deviation statistical analysis. This method can improve measurement efficiency and ensure the measurement accuracy of mass-produced blades. Sun et al. [8] elaborated on the 3D reconstruction technology of blade profiles based on laser scanning point cloud and conducted an in-depth analysis of the profile reconstruction error. Liu et al. [9] proposed an iterative nearest point method of improved main region principal component analysis. This method uses the constraint plane constructed on the tenon to define and extract the main region of the blade. Feng et al. [10] proposed an adaptive surface sampling strategy based on the bending moment theory for the variation of curvature in different blade regions. It is an effective method for reconstructing blade curves and surfaces based on a small number of measurement points.

This topic studies and innovates the rapid extraction algorithm of geometrical parameters such as arc, section line, the center of front and rear edges, the radius of front and rear edges, blade chord length, and string angle, and uses point laser displacement sensor to carry out experiments on the four-coordinate measuring system of aero-engine blade.

2. Extraction of blade profile feature parameters

2.1. Mid-arc and maximum thickness

This manuscript is based on an inscribed circle blade mid-arc search and section thickness calculation method, as shown in Figure 1. The algorithm has greatly improved efficiency and accuracy and is suitable for engineering applications. It is assumed that the distance from a point on the back of the leaf to the leaf basin is the minimum value from that point to each point of the leaf basin, and this minimum value can be considered as the diameter of the inscribed circle of the blade profile at that point, and then the midpoint of the minimum distance between the back of the leaf and the corresponding leaf basin is the point on the mid-arc line. The mid-arc of the blade section line can be obtained by calculating the set of minimum distance points between the corresponding leaf back and the corresponding leaf basin. If the minimum distance is the blade thickness of the corresponding point, the largest of all the minimum distances will be the maximum section thickness. If the density of measurement points is insufficient in the calculation, B-spline interpolation can be used to increase the density of measurement points, or the sampling data of the laser displacement sensor can be increased.

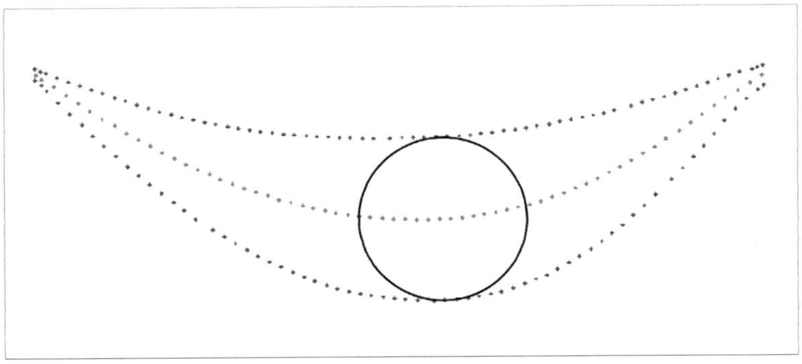

Figure 1. Extraction results of mid-arc and maximum thickness.

If the blade back point in the blade profile is listed as $Q_j(x,y), j=0,1,\cdots n$ and the blade back point is listed as $P_i(x,y), i=0,1,\cdots m$, the mathematical expression for the maximum thickness $D_{\max}$ of the blade profile is as follows:

$$D_{\max} = \max_{i\in(0,m)}(\min_{j\in(0,n)}(|\overrightarrow{Q_iP_j}|)) \tag{1}$$

Through the above analysis, the algorithm for calculating the central arc of the blade cross-section profile and the maximum thickness of the cross-section can be obtained.

According to the algorithm for extracting the mid-arc and maximum thickness of the blade cross-section profile above, the extraction process of the mid-arc and maximum thickness of the blade cross-section using a point laser displacement sensor for measuring the characteristic parameters of the blade cross-section is shown in Figure 2.

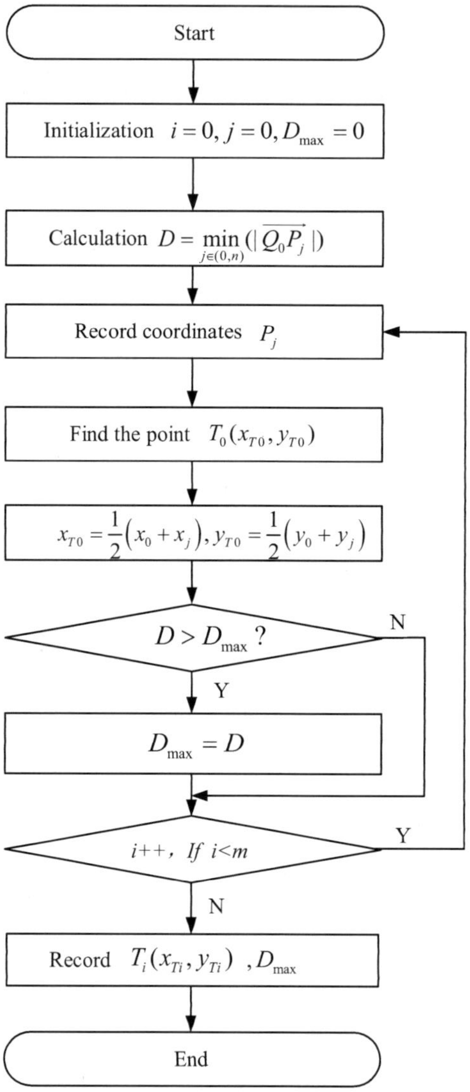

Figure 2. Flow chart of mid-arc and maximum.

2.2. Extraction of front/rear edge parameters

The laser sensor used in the experiment has a larger back diameter (0.08 mm). In the measurement of the front/rear edge, the laser beam is prone to bouncing when illuminated on a small arc, and the CCD matrix cannot accurately capture the reflected beam. This makes it difficult to measure the leading/trailing edges of the blades or the measurement data inaccurate, making it impossible to measure and solve the front/rear edges of the blades.

We set the leaf back point column of the blade profile as $B_1(x_1, y_1)$ and the blade back point column as $P_0(x_2, y_2)$, calculate the shortest distance $\overline{B_1P_1}$ and $\overline{P_0B_1}$ from the two ends to the point cloud of another curve, and then compare their distances. If $\overline{B_1P_1} \le \overline{P_0B_1}$, we will use the coordinates of Point B_1 and the slope at Point B_1 as the constraint conditions to make the inscribed circle of the blade basin and back; If $\overline{B_1P_1} > \overline{P_0B_1}$, we will use the coordinates of Point P_0 and the slope at Point P_0 as constraints to make the inscribed circle of the leaf basin and the back of the blade. Therefore, the fitted circle is the circle of the front/rear edge sought. Figure 3 shows the legend of the front/rear edges fitted from a set of blade profile measurement data.

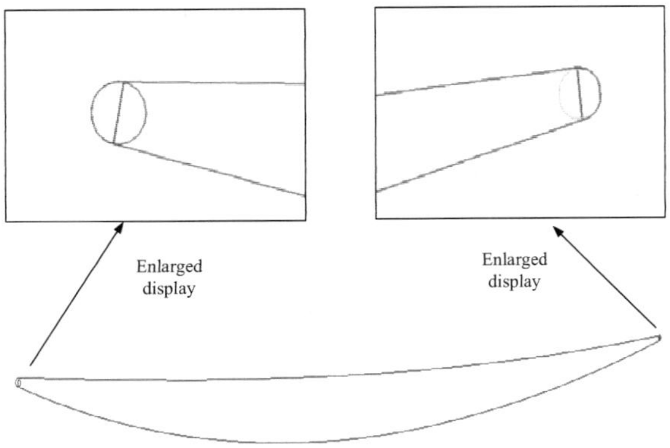

Figure 3. Arc fitting of the front/rear edges of the blade cross-section profile.

2.3. Extraction of front/rear edge thickness

It can be seen from the above discussion that the quality of the front/rear edge can be evaluated by the diameter of the inscribed circle at L from the front/rear edge point on the arc line in the blade section. Based on the geometric relationship in Figure 4, first, we obtain the tangent equation at Point O of the mid-arc edge in the blade, and then find the intersection point $C(x_1, y_1)$ with Circle O and take Point C as the center of the circle, with a radius of L=1 mm. We make the mid-arc at Point D and then make the perpendicular line of CD at Point D to intersect the back curve of the blade basin at Points E and F. Therefore, line segment EF can be considered as the thickness of the front/rear edge.

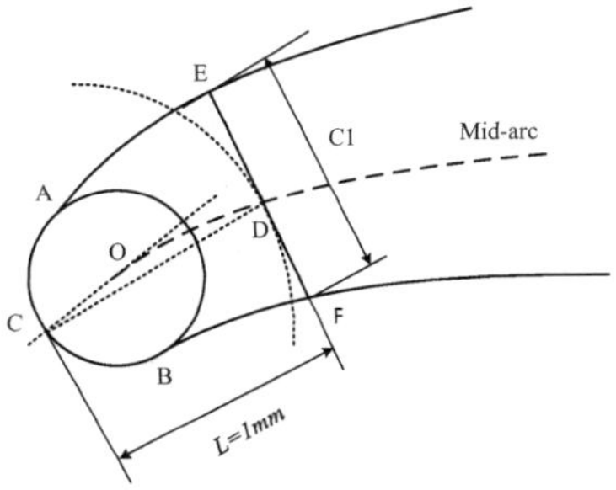

Figure 4. Calculation of front/rear edge thickness.

The simplified algorithm process is as follows:

(1) It is assumed that the endpoint of the leaf back curve is $A(x_a, y_a)$ and the endpoint of the leaf basin curve is $B(x_b, y_b)$, and then the coordinates of the center of the edge circle O are:

$$x_o = \frac{1}{2}(x_a + x_b), y_o = \frac{1}{2}(y_a + y_b) \tag{2}$$

(2) Since Point O is on the mid-arc, the middle arc can be obtained by a cubic spline interpolation curve. As the first-order partial derivative of the middle arc at Point O, the slope K can be obtained.

(3) Through the tangent of the central arc of the circle O, the intersection head arc is located at Point $C(x_1, y_1)$, where Point C is determined by the following equations:

$$\begin{cases} (x-x_o)^2+(y-y_0)^2=R^2 \\ y-y_o=\frac{1}{K}(x-x_0) \end{cases} \tag{3}$$

(4) We make a circle with a radius of L through Point C and intersect the mid-arc at Point D.

We make the normal of the mid-arc through Point D and place the posterior leaf basin of the intersecting leaves at Points E and F, and then EF is the thickness of the edge head obtained.

2.4. Extraction of blade chord length

The chord line is a straight line on the cross-section of the blade profile that is tangent to the front and rear edge arcs. The solution of the chord can be obtained by the common tangent of the leading/trailing edge arc or by obtaining the two tangent points of the chord on the front/rear edge profile. By using the algorithm in the previous section, the center coordinates and radius of the front/rear edge arc can be obtained. Based on the definition of the chord line, the common tangent can be solved through equations. We set the common tangent equation as:

$$y = kx + b \tag{4}$$

According to Figure 6, it can be seen that the chord line is tangent to the front/rear edge arc, so the distance from the centre $O_1(x_1, y_1)$ to the line is Radius R_1, and the distance from the centre $O_2(x_2, y_2)$ to the line is Radius R_2. Based on the equation for the distance from the point to the line, the following equations set can be obtained:

$$\begin{cases} R_1=\frac{|y_1-kx_1-b|}{\sqrt{1+k^2}} \\ R_2=\frac{|y_2-kx_2-b|}{\sqrt{1+k^2}} \end{cases} \tag{5}$$

By using the above equations to obtain the values of k and b, the cross-section chord equation of the blade can be obtained.

The calculation principle is shown in Figure 5. Circles O_1 and O_2 are the fitting circles of the front/rear edges of the blade section, respectively, while T_1 and T_2 are the tangent points. To determine the chord length, two arcs are made on the outside, which are perpendicular to the chord line and tangent to the arcs at Points M and N. We set the chord length as L, and based on the definition of the blade chord length, the projection length L of the blade profile on the chord length is the length of the segment T_1T_2 plus the front/rear edge radius. The specific algorithm is shown in the following equations:

$$\begin{cases} L=R_1+R_2+O_2H \\ O_2H=\sqrt{(O_1O_2)^2-(R_1-R_2)^2} \end{cases} \tag{6}$$

2.5. Extraction of blade chord inclination angle

The chord inclination angle refers to the angle between the chord line and the X-axis in the current blade profile coordinate system. As shown in Figure 5, the radius of the centre $O_1(x_1, y_1)$ of the trailing edge is R_1, and the radius of the centre $O_2(x_2, y_2)$ of the leading edge is R_2. The chord intersects the common tangent of the front/rear edge arc at two points T_1 and T_2, and O_2H intersects O_1T_1 parallel to the chord at Point H, passing through Point H to create an auxiliary line HE parallel to the X-axis, and passing through Point O_1 to create an auxiliary line O_1F parallel to the X-axis. According to the definition of chord inclination angle, the angle α_1 is the inclination angle of the blade cross-section chord. The centre coordinates and radius values of the front/rear edges can be obtained through the algorithms in the previous sections. According to the geometric relationship in the figure below, the blade chord angle can be calculated by using the following equations:

$$\begin{cases} \alpha_1 = \alpha_3 - \alpha_2 \\ \alpha_3 = \arctan\dfrac{y_2 - y_1}{x_2 - x_1} \\ \alpha_2 = \arccos\dfrac{R_1 - R_2}{O_1O_2} \end{cases} \tag{7}$$

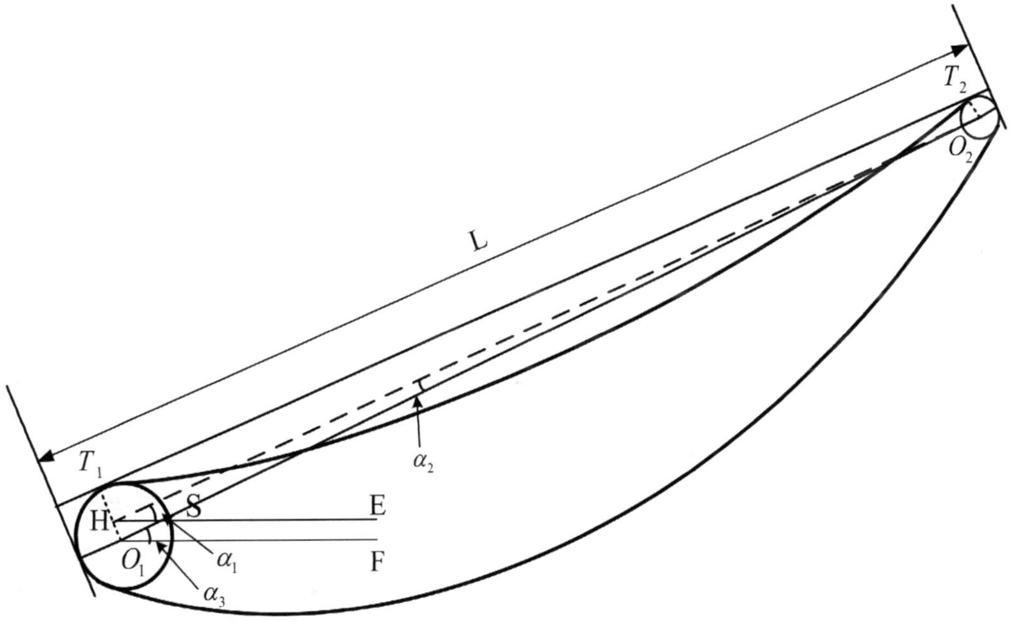

Figure 5. Schematic diagram of chord inclination calculation.

3. Measurement experiments

The experiment was conducted on a four-coordinate measuring device for aircraft engine blades developed by the research group, as shown in Figure 6. Taking a 4-stage compressor of a certain model as the object, the sensor scans multiple cross-sections of the blade profile along the blade profile in order of equal height. After filtering the collected data, the feature parameters are quickly extracted by using the above algorithm. The specific data is shown in Table 1. From the results, it can be seen that the extracted parameters of the blade cross-section are relatively accurate, and the error is controlled within the allowable range.

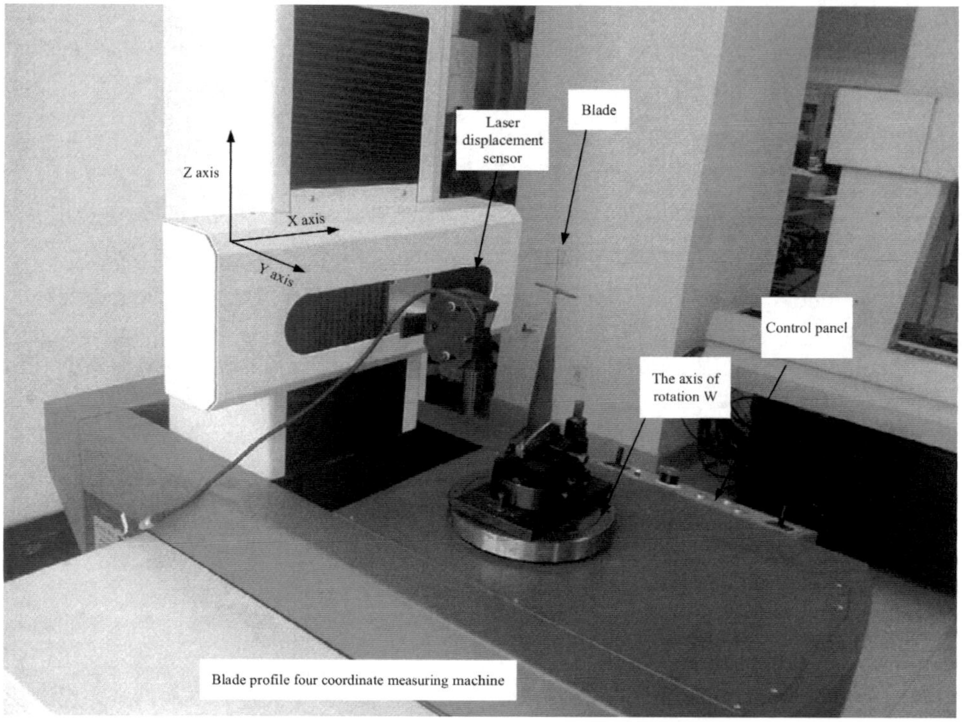

Figure 6. Aircraft blade measurement device.

Table 1. Measurement data of characteristic parameters of a certain model of compressor blade profile (mm).

No.	Section height	Maximum thickness		Front edge thickness		Rear edge		Chord length		Fitted front edge radius	Fitted rear edge radius
		Theoretical	Error	Theoretical	Error	Theoretical	Error	Theoretical	Error		
1	28	5.2902	-.0024	0.5807	0.0312	0.7803	0.0381	60.6403	0.0091	0.3511	0.3650
2	48	4.8403	0.0081	0.5604	0.0407	0.7402	0.0543	62.5102	-.0013	0.3378	0.3517
3	68	4.4005	0.0116	0.5401	0.0352	0.6906	0.0586	64.3905	0.0035	0.2683	0.3065
4	88	3.9601	0.0335	0.5305	0.0301	0.6505	0.0420	66.2707	-.0027	0.2351	0.2533
5	108	3.5202	0.0286	0.5206	0.0550	0.6002	0.0453	68.1401	-.0152	0.2023	0.2097

4. Conclusion

This manuscript studies and innovatively proposes a fast extraction algorithm for geometric parameters such as blade centreline, cross-sectional line, front/rear edge center, front/rear edge radius, blade chord length, and chord angle. By using this fast blade profile detection method, while meeting the measurement accuracy, the measurement process of each aircraft blade is controlled within 30 minutes, which is half of the traditional measurement time and improves the detection efficiency by over 40% compared to traditional coordinate measuring machines.

Acknowledgments
The project is supported by the National Natural Science Foundation of China (Grant No. 62263028) and the Natural Science Foundation of Shaanxi Provincial Department of Science and Technology (Grant Nos. 2023-JC-QN-0633 and 2023-JC-YB-592).

References

[1] Li L., Kun C., Jianmin G., et al. Research on optimizing-assembly and optimizing-adjustment technologies of aero-engine fan rotor blades *J. Advanced Engineering Informatics*, 2022, 51: 101, 506.

[2] Hw A., Fld A., Zj A., et al. Electrical insulation improvements of ceramic coating for high-temperature sensors embedded on aeroengine turbine blade *J. Ceramics International*, 2020, 46 (3): 3, 600-3, 605.

[3] Zhang Kaiyao, Meng Qingheng, Ma Rui, Yu Tao, and Wang Jiao. Process analysis of cutting force affecting precision forged blades of aircraft engines *J. Manufacturing Technology and Machine Tool*, 2020 (09): 35-39.

[4] Tw A., Lai Z. A., Qw A., et al. A high-precision prediction model of surface roughness in abrasive belt flexible grinding of aero-engine blade *J. Journal of Manufacturing Processes*, 2021, 66: 364-375.

[5] Praveenkumar K., Mylavarapu P., Sarkar A., et al. Residual stress distribution and elevated temperature fatigue behavior of laser peened Ti-6Al-4 V with a curved surface *J. International Journal of Fatigue*, 2022, 156: 106, 641.

[6] Gao Y., Lin X. J., Zhang J. Q., et al. A region-division method of measurement point sampling for thin-walled blade sections based on CMM *J. Aeronautical Manufacturing Technology*, 2020, 63 (7): 14-20.

[7] Zhang Y., Chen Z., Zhu Z., et al. A sampling method for blade measurement based on a statistical analysis of profile deviations *J. Measurement*, 2020, 163 (1): 107, 949.

[8] Sun B., Du H. B., Wang J. H., et al. Analysis of aeroengine blade surface laser point cloud 3D reconstruction error *J. Acta Photonica Sinica*, 2018, 47 (5): 1-9.

[9] Liu J., He X., and Huang X. An improved registration strategy for aligning incomplete blade measurement data to its model *J. Optik*, 2021, 243: 167, 304.

[10] Feng G., Ziyue P., Xutao Z., et al. An adaptive sampling method for accurate measurement of aeroengine blades *J. Measurement,* 2021, 173: 108, 531.

Research on the UOSM four-port Calibration algorithm and determination method of the Transmission error term

Jinghui Du[1], Xiaolong Chen[1, 2], Jianqiu Wu[1] and Yingquan Wang[1]

[1]Xi'an Key Laboratory of Intelligent Instrument and Packaging Test, School of Mechano-Electronic Engineering Xidian University, Xi'an Shaanxi 710071, China

[2]Corresponding author's e-mail: xlchen@mail.xidian.edu.cn

Abstract. As one of the key techniques for vector network analyzers to achieve high precision measurement, calibration, and error correction technology has always been the focus of continuous exploration and innovation. UOSM calibration algorithm uses unknown straight-through transmission standards that do not need to be defined and known open-circuit standards, short-circuit standards, and matching standards to calibrate microwave and millimeter wave bands. Based on the study of the mature UOSM calibration algorithm, this paper proposes a method of self-calibration and comparison of the measurement data based on the unknown cut-through standard, which no longer needs the group delay parameters of the unknown straight-through standard and reduces the dependence on the calibrator. The accuracy of the algorithm and error item determination method is verified by ADS simulation and instrument actual test, and a good calibration effect is achieved. It is proved that the algorithm has the ability of accurate calibration. Due to the reduction of parameters, the stability and applicability of the algorithm are improved.

1. Introduction

Vector network analyzer is a kind of precision instrument based on frequency sweep measurement. In recent years, because of its wide frequency measurement range and high testing accuracy, it has been widely used in various fields of microwave engineering [1]. Due to the emergence of multi-port and multi-function components and modules and the rapid growth of demand, the scattering parameters of multi-port network analyzers are required to test and analyze these devices accurately and rapidly. Therefore, the four-port vector network analyzer plays a more and more important role in modern microwave measurement technology. However, any instrument will have errors in measurement.

At this time, the use of a reasonable calibration algorithm can effectively improve the measurement accuracy and obtain reliable results. The calibration algorithm establishes the error model and signal flow diagram according to the hardware structure of the instrument and the connection relationship of the devices in the measurement process and uses the generalized node method or Mason formula to obtain the relationship between the measured data, the error term and the true value. in practice, the instrument is used to measure the relevant calibration parts, and the system error term is obtained according to the measurement results. Then these error terms are used to process the measurement results of the target device to complete data calibration, in which the system error terms have many types, such as directivity error (Ed), crosstalk error (Er), source matching error (Es) and transmission error (Et), which are caused by the non-ideality of the internal hardware of the instrument.

At present, among the existing calibration algorithms, the two-port vector network analyzer is the most abundant and mature, such as SOLT with 12 error models and UOSM and TRL with 8 error models. When the instrument is expanded to four ports, all kinds of error terms increase with the increase in the number of ports. According to certain methods, the two-port calibration algorithm can be extended to the four-port calibration algorithm. Currently, there are GSOLT [2] and Non-Ideal TRM [3] about 48 error models. In order to reduce the complexity of the calibration process and the dependence on the calibration parts, the algorithm is constantly updated. Some scholars have proposed the four-port optimization calibration technology, downgraded the calibration step, and do not need to measure the reflectometer at each test port. A multi-port RF microwave calibration method based on a self-calibration algorithm is proposed in this paper [4]. The calibration of multi-port RF microwave devices is completed based on an error correction matrix and vector network analyzer. The multi-port RF microwave calibration method effectively solves the problem of insufficient accuracy of multi-port RF microwave calibration and has the advantages of convenient operation and strong applicability. In this paper [5], a hybrid calibration method for a multi-port vector network analyzer is proposed. firstly, the calibration parts used by the port are set up, and then the straight-through method and calibration method are extracted automatically for different calibration parts. and create the connection, measure the message, and then through the measurement standard, calculate the system error, finally measure the target device and correct the measured value, and complete the hybrid calibration. The complex connection and measurement in the process of four-port calibration seriously affect the calibration speed. In order to understand the speed problem of multi-port calibration, under the premise of four-port calibration, two efficient and fast new calibration methods are introduced in this paper [6], namely, the fast calibration method and redundant S-parameter algorithm. These two new techniques simplify the calibration process and the measurement of S-parameters to the maximum extent and improve the calibration speed. However, there is an essential difference between the two. The former only simplifies the reflection measurement, but not the transmission measurement, while the latter simplifies the transmission measurement rather than the reflection measurement.

This paper first introduces the calibration principle of a vector network analyzer, the types of error terms, and the research status of the four-port calibration algorithm. Then it analyzes the error model of the two-port UOSM algorithm and the characteristics of calibration parts used, introduces in detail the error term calculation and data processing process of the algorithm in the four-port application, in which the symbol of the error term needs to be judged in the calculation of transmission error term. If the inaccurate judgment of symbols will lead to the phase oscillation of the transmission parameters of the parts to be tested, the simplest known method still requires the group delay of unknown straight-through calibrators as a known condition. Due to the non-ideality of the actual calibrator and the loss caused by the use of the device, the group delay data will deviate, which will affect the calibration results to a certain extent. In this paper, a new method to judge the phase of the transmission error term is proposed, which does not need additional data and improves the accuracy of the algorithm. Finally, the feasibility and reliability of the method are verified by experiments.

2. Formatting research and improvement of UOSM calibration technology

UOSM algorithm plays a very good role in calibrating the measurement of non-pluggable devices. It was first applied to the two-port vector network analyzer and then extended to the four-port vector network analyzer. For error calibration, firstly, the signal flow diagram of the system error term is established according to the internal hardware structure of the vector network analyzer or at the measurement port containing the measurement transfer device, and then based on the signal flow diagram, the corresponding formula is deduced by using Mason formula or generalized node method, and the relationship between the system error term and input and output signal is simplified. Finally, these error terms are brought into the error correction formula to modify the result data [7], and the actual value of the tested part is obtained.

2.1. Error model

The error model of the two-port vector network analyzer when measuring the device is shown in Figure 1. There are three groups of error networks between the two ports. e^{00}, e^{01}, e^{10}, and e^{11} constitute a one-port error term network, S11, S21, S12, and S22 constitute the scattering parameter network of the devices to be tested, and e^{22}, e^{23}, e^{32}, and e^{33} constitute a two-port error term network. The directional error, source matching error, and reflection tracking error of the two ports can be known through the single-port OSM calibration algorithm, and then the transmission error term is calculated by the cascade of the transmission network. The two-port UOSM calibration algorithm involves a total of 8 error items, also known as 8 error models.

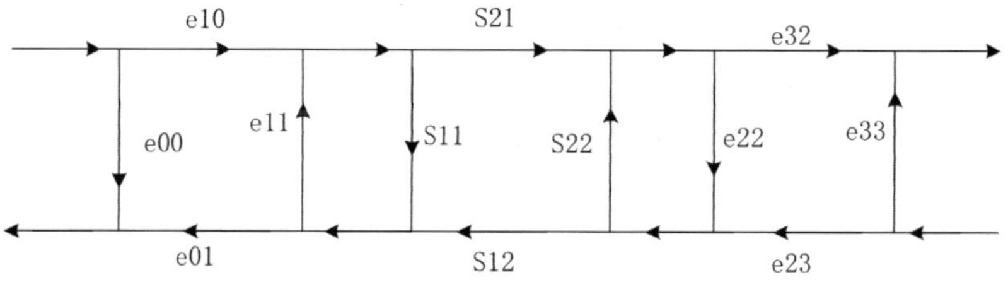

Figure 1. Error model of two-port vector network analyzer.

The error model of the four-port vector network analyzer in measuring the device is shown in Figure 2, in which 1, 2, 3, and 4 are the measuring ports, a_{im} and b_{im} are the parameters measured by the instrument port, the actual scattering parameters of a_i and b_i are the actual scattering parameters of the device to be tested, e_i^{00}, e_i^{01}, e_i^{01}, and e_i^{11} are the systematic errors between the original parameters and the actual parameters, where e_i^{00} is the directional error and e_i^{11} is the source matching error. e_i^{01} e_i^{01} is a crosstalk error, and e_i^{01} e_j^{10} is the transmission error between two ports. There are 16 errors for the four-port vector network analyzer, so it is also called the 4N error model.

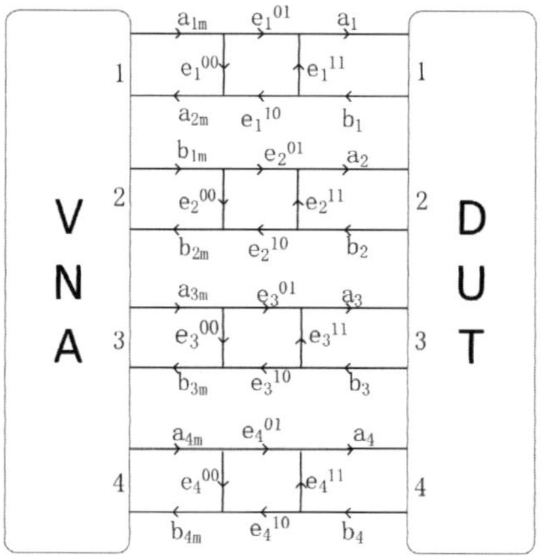

Figure 2. Error model of four-port vector network analyzer.

2.2. Calibration standard

Calibration standards, also known as calibration parts, mainly include mechanical calibration, electronic calibration, and on-wafer calibration. In cases where on-wafer calibration is mainly used for wafer measurement [8], mechanical calibration is a single-port or two-port device with known characteristics. For each calibrator, it is not necessary to have an accurate and known definition to satisfy some characteristic relations. The calibration standard used by the UOSM calibration algorithm is to define known circuit breakers (Short), circuit breakers (Open), matchers (Match), and unknown straight-through connectors (Unknown thru) that are unknown but need to meet the S21=S12 condition.

2.3. Error term calculation

The parameter signal flow diagram established according to the error model is shown in Figure 3. The relationship between each error term and the measured scattering parameters and actual scattering parameters of the calibration part is deduced by using the Mason formula. By connecting the calibration part between a single port and two ports, the value of each error term is calculated, and the connection calibration part and the calculation process are described in detail as follows.

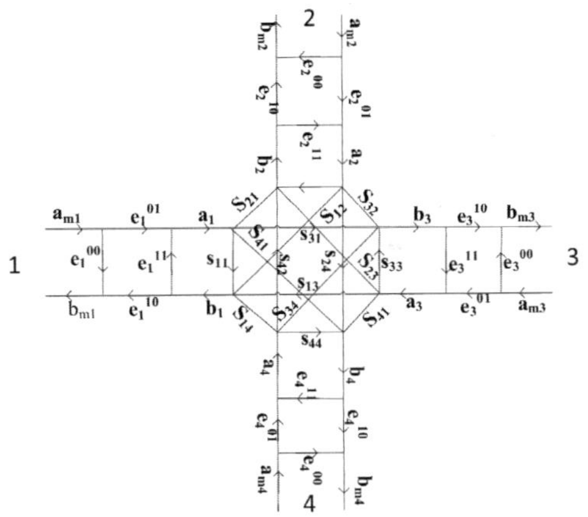

Figure 3. Parametric signal flow graph.

2.3.1. Calculation of single port error term. Using OSM [9] single port calibration method one by one, O, S, and M calibrators are connected at each port respectively. Four groups of error terms e_i^{00}, e_i^{01}, e_i^{01}, and e_i^{11} are obtained by calibration algorithm, of which 12 error terms are known at this time, and the calculation formula of error term for each port is as follows. Smo, Sms, and Smm are the measured values of the instrument connecting O, S, and M calibration parts respectively, and So, Ss, and Sm are the nominal values of the three calibration parts.

$$e_i^{11} = \frac{(Sm \cdot (Smo - Sms) + So \cdot (Sms - Smm) + Ss \cdot (Smm - Smo))}{(Ss \cdot Sm \cdot (Smm - Sms) - So \cdot Sm \cdot (Sm - So) - Ss \cdot So \cdot (Smo - Sms))} \tag{1}$$

$$e_i^{01} e_i^{10} = \frac{(Sm - Ss) \cdot (1 - Es \cdot So) \cdot (1 - Es \cdot Ss)}{(So - Ss)} \tag{2}$$

$$e_i^{00} = \frac{(Smo - Er \cdot So)}{(1 - Es \cdot So)} \tag{3}$$

2.3.2. Calculation of transmission error term. U calibrator is connected between any two ports. The relationship between the error term matrix of the two ports and the measurement data matrix of the calibrator can be obtained by cascading the transmission matrix between the ports. Under the condition that S21=S12 is known, the transmission error term ei01 ej 10 between ports can be calculated, where i≠j, the calculation methods of the error network matrix and transmission error term are as follows.

$$TX = \frac{1}{e_i^{10}} \cdot \begin{bmatrix} e_i^{01} \cdot e_i^{10} - e_i^{00} \cdot e_i^{11} & e_i^{00} \\ e_i^{11} & 1 \end{bmatrix} \tag{4}$$

$$TY = \frac{1}{e_j^{10}} \cdot \begin{bmatrix} e_j^{01} \cdot e_j^{10} - e_j^{00} \cdot e_j^{11} & e_j^{00} \\ e_j^{11} & 1 \end{bmatrix} \tag{5}$$

$$TDUT = TX \cdot TR \cdot TY \tag{6}$$

2.3.3. The determination of the symbol bit of the transmission error term. Open root sign processing is needed in the calculation of the transmission error term, which involves the judgment of the positive and negative signs of the transmission error term. If the judgment is wrong, it will cause the oscillation of the phase result [10], of course, if the phase situation is not considered, if only the amplitude of the scattering parameter is considered, there is no need to deal with it. The new judgment method proposed in this paper is based on the principle of phase continuity change of unknown straight-through calibration parts, and the phase of the transmission error term is judged by calibrating and comparing the measurement data of unknown straight-through calibration parts. firstly, it is assumed that the symbol of the transmission error term of the frequency point is positive or negative, and the calibration data of this frequency point is compared with the data results of the previous frequency point. When the change of phase continuity is satisfied, it is the correct error term symbol.

2.4. Calibration processing

Similar to the two-port measurement, there are four error term matrices between the four-port vector network analyzer and the equipment being tested. according to the signal flow diagram, the relationship between the measured values and the real values of the scattering parameters of the four-port devices is.

$$\begin{bmatrix} b_{m1} \\ b_{m2} \\ b_{m3} \\ b_{m4} \end{bmatrix} = \begin{bmatrix} e_1^{00} & 0 & 0 & 0 \\ 0 & e_2^{00} & 0 & 0 \\ 0 & 0 & e_3^{00} & 0 \\ 0 & 0 & 0 & e_4^{00} \end{bmatrix} \cdot \begin{bmatrix} a_{m1} \\ a_{m2} \\ a_{m3} \\ a_{m4} \end{bmatrix} + \begin{bmatrix} e_1^{01} & 0 & 0 & 0 \\ 0 & e_2^{01} & 0 & 0 \\ 0 & 0 & e_3^{01} & 0 \\ 0 & 0 & 0 & e_4^{01} \end{bmatrix} \cdot \begin{bmatrix} a_1 \\ a_2 \\ a_3 \\ a_4 \end{bmatrix} \tag{7}$$

$$\begin{bmatrix} b_1 \\ b_2 \\ b_3 \\ b_4 \end{bmatrix} = \begin{bmatrix} e_1^{10} & 0 & 0 & 0 \\ 0 & e_2^{10} & 0 & 0 \\ 0 & 0 & e_3^{10} & 0 \\ 0 & 0 & 0 & e_4^{10} \end{bmatrix} \cdot \begin{bmatrix} a_{m1} \\ a_{m2} \\ a_{m3} \\ a_{m4} \end{bmatrix} + \begin{bmatrix} e_1^{11} & 0 & 0 & 0 \\ 0 & e_2^{11} & 0 & 0 \\ 0 & 0 & e_3^{11} & 0 \\ 0 & 0 & 0 & e_4^{11} \end{bmatrix} \cdot \begin{bmatrix} a_1 \\ a_2 \\ a_3 \\ a_4 \end{bmatrix} \tag{8}$$

The e parameter matrix is the value of the error term of each port, an and b are the real data matrix of each port of the part to be tested and am and bm are the measured values of each port of the instrument. after proper matrix operation, the scattering parameters of the four-port device are as follows.

$$[S]=A \cdot (I+ED_{11} \cdot A)^{-1} \tag{9}$$

$$A=\begin{bmatrix} \frac{S_m^{11}-e_1^{00}}{e_1^{01}e_1^{10}} & \frac{S_m^{12}}{e_1^{01}e_2^{10}} & \frac{S_m^{13}}{e_1^{01}e_3^{10}} & \frac{S_m^{14}}{e_1^{01}e_4^{10}} \\ \frac{S_m^{21}}{e_2^{01}e_1^{10}} & \frac{S_m^{12}-e_2^{00}}{e_2^{01}e_2^{10}} & \frac{S_m^{23}}{e_2^{01}e_3^{10}} & \frac{S_m^{24}}{e_2^{01}e_4^{10}} \\ \frac{S_m^{31}}{e_3^{01}e_1^{10}} & \frac{S_m^{32}}{e_3^{01}e_2^{10}} & \frac{S_m^{33}-e_3^{00}}{e_3^{01}e_3^{10}} & \frac{S_m^{34}}{e_3^{01}e_4^{10}} \\ \frac{S_m^{41}}{e_4^{01}e_1^{10}} & \frac{S_m^{42}}{e_4^{01}e_2^{10}} & \frac{S_m^{43}}{e_4^{01}e_3^{10}} & \frac{S_m^{41}-e_4^{00}}{e_4^{01}e_4^{10}} \end{bmatrix} \tag{10}$$

The real parameter values of the device are known through the operation, and the focus of the calibration algorithm lies in the calculation of the error term, especially the determination of the symbol of the transmission error term. At the same time, the non-ideality of the calibration part will also bring some errors in the calibration results.

3. Experimental analysis and verification

In order to verify the accuracy of the UOSM four-port calibration algorithm and the effectiveness of the transmission error term determination method, two test schemes are designed. In the experiment, the MATLAB software editing algorithm is used to import all kinds of measurement data into the algorithm to calculate and calibrate the error term.

3.1. Simulation S-parameter experiment with ADS

In experiment 1, the accuracy of the UOSM four-port calibration algorithm is verified by ADS2020 software simulation. The scattering parameters of the calibrator and the device to be tested are obtained by software simulation, and then the measurement data are processed by the algorithm proposed in this paper and the existing TOSM four-port calibration algorithm. Finally, the calibration results of the algorithm proposed in this paper are compared with the real values of the devices to be tested and the results of the TOSM four-port algorithm to verify the accuracy of the algorithm. The experimental flow is shown in Figure 4.

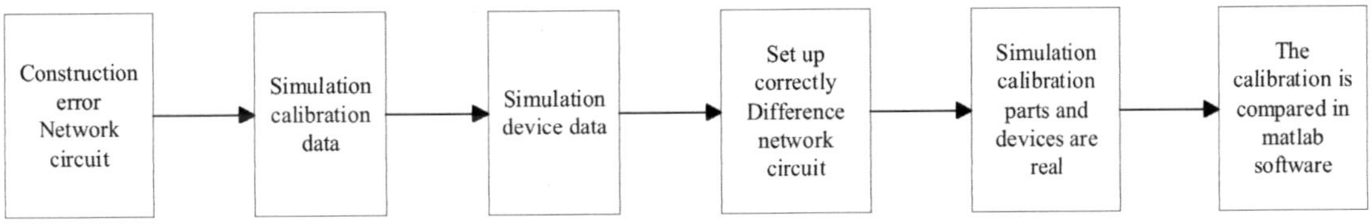

Figure 4. Experiment 1 flow chart.

Figure 5 below is a simulation schematic diagram built in ADS2020. Each port is connected to a RLC circuit to build an error network. The device to be tested is a four-port lange directional coupler. The unknown straight-through calibration part adopts a transmission line with a certain delay, the open circuit calibration part is in the form of disconnection, the short circuit adopts the calibration part grounding, and the matching calibration part adopts 50 Ω load impedance. The actual test situation is simulated by connecting each calibrator and the tested part between different ports. The measuring frequency band is 20 GHz to 50 GHz, the frequency step is 55.2 MHz, and the number of measuring frequency points is 544.

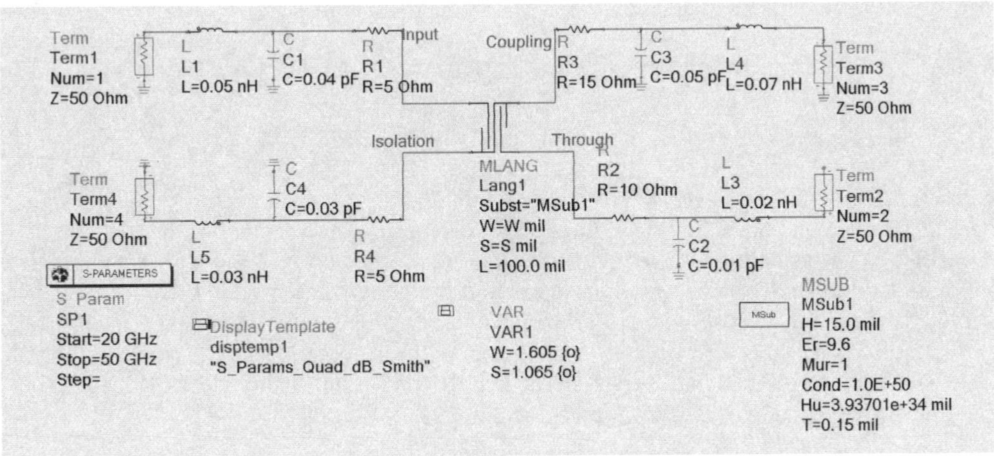

Figure 5. Simulation schematic diagram of S parameters of the attenuator.

The simulated scattering parameters are compared with the other two kinds of data, as shown in Figures 6 and 7, which are the amplitude and phase of the scattering parameters, respectively. From the result diagram, we can see that the new method of judging the transmission error term sets up the error network circuit in the ADS software, replaces the calibration part, obtains the simulated data, designs the test piece, and obtains the original measurement data. In the ADS software, the error-free network circuit is built, the parameters of the calibration part and the part to be tested are used for calibration and comparison, and all the data are imported into matlab. Compared with the UOSM calibration algorithm applied to the four-port vector network analyzer, good calibration results can be achieved.

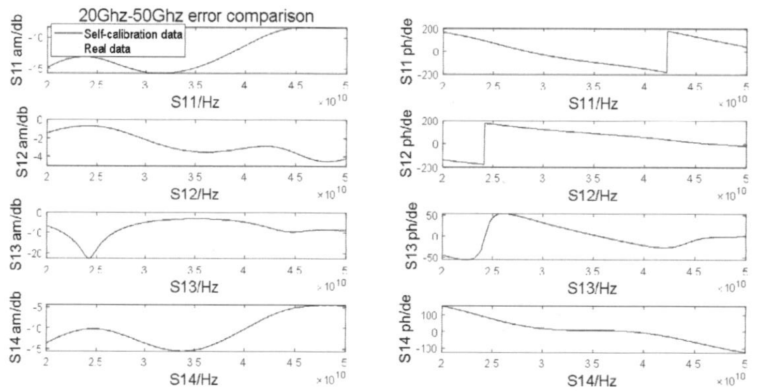

Figure 6. Comparison between the calibration result and the real value of DUT.

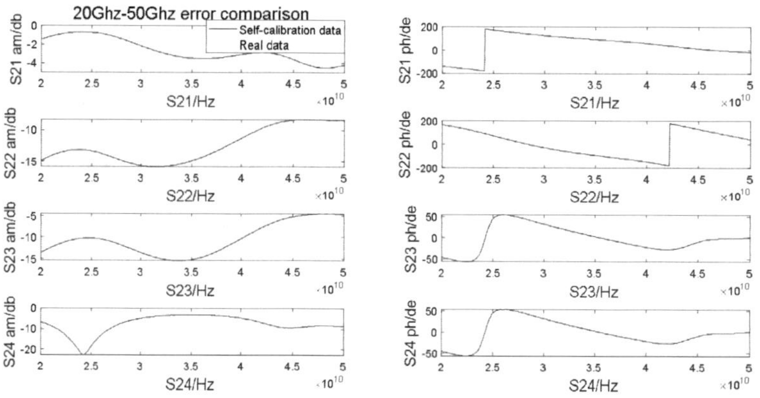

Figure 7. Comparison of calibration results with TOSM algorithm.

3.2. Measured S-parameter experiment

In experiment 2, the validity of the square method for determining the symbol of the transmission error term is verified by the actual measurement of the vector network analyzer. The scattering parameters of the calibrator and the device to be tested are measured by the instrument, and then the measured data are calibrated. The method proposed in this paper is used to determine the transmission error term. Finally, the calibration results are compared with the actual measurement of the instrument. In particular, the phase of the transmission parameters is compared to verify the effectiveness of the determination method. The experimental flow is shown in Figure 8 below.

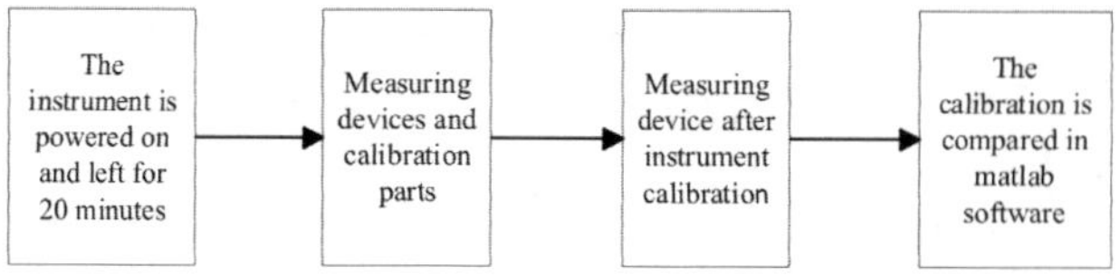

Figure 8. The second flow chart of the experiment.

The S parameters are obtained by testing the 3dB attenuator with the vector network analyzer RS ZNB 20 of Roeder and Schwartz. The code is edited according to the existing algorithm, and the new method proposed in this paper is selected to determine the transmission error term. The algorithm calibration results are compared with the instrument measurement results as shown in Figure 9 below.

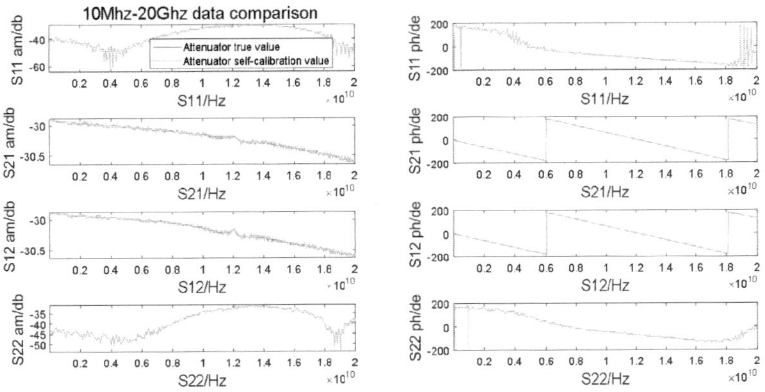

Figure 9. Calibration results of two-port UOSM.

From the result diagram, it can be seen that the symbol bit of the transmission error term is still accurate when the algorithm is calibrated by the measured S parameters, which proves that the method has the ability to accurately judge the symbol bit of the transmission error term, and verifies the stability and feasibility of the new method for judging the transmission error term from the side.

4. Conclusion

The calibration algorithm, as one of the core technologies of a vector network analyzer, has been constantly studied by scholars, from the initial single-port calibration to the current multi-port calibration. How to achieve a convenient and accurate calibration algorithm is the purpose of the research. In order to solve this problem, based on the study of the UOSM calibration algorithm, this paper focuses on the calculation process of the four-port UOSM calibration algorithm and puts forward a new method to judge the symbol of the error term, which reduces the dependence on calibration parts and improves the stability and applicability of the algorithm. In the experiment, simulation and actual testing verify the accuracy of the UOSM four-port algorithm and the effectiveness of the transmission error determination method. Because the method in this paper has some requirements for OSM calibration parts, and the real parameters of the three devices need to be known on time, the calibration accuracy will be affected by the inaccurate real parameters of calibration parts..

Acknowledgments

The research was supported by the National Key R&D Program of China(2021YFF0701500).

References

[1] X Li, Z Liu, L Lin, H Fan, X Liang, J Xu. (2021) A Novel Method for the Accurate Measurement of Soil Infiltration Line by Portable Vector Network Analyzer. Sensors 2021, 21(21): 7201.https://doi.org/10.3390/s21217201.

[2] W Zhao, X Yang, J Xiao, H Qin. (2012) Uncertainties of Multiport VNA S-Parameter Measurements Applying the GSOLT Calibration Method. IEEE Transactions on Instrumentation and Measurement,2012,12(61): 3251 - 3258. 10.1109/TIM.2012.2210332.

[3] M Wang, Y Zhao, Y Zhou. (2017) The Residual Errors of Multiport VNA Measurements Due to Non-Ideal TRM Calibration. In: 2017 International Workshop on Electromagnetics:

Applications and Student Innovation Competition. London. pp.154-156. 10.1109/iWEM. 2017.7968815.

[4] X Ding, L Wang. (2020) A Multi-port RF Microwave Calibration method based on self-Calibration algorithm. Chinese Patent: CN 112564823 A. 2020.12.03.

[5] Z Zhuang, Y Guo, S Li. (2020) A Hybrid Calibration method for Multi-Port Vector Network Analyzer. Chinese Patent: CN 109254257 A. 2020.08.04.

[6] Y Guo, S Li, Z Zhuang. (2017) Research on the latest calibration technology of vector Network Analyzer. In: 2022 37th Youth Academic Annual Conference of Chinese Association of Automation (YAC). Beijing.pp.852-856. 10.1109/YAC57282.2022.10023671.

[7] G Liu, X Zou, Q Liang.(2019) Measurement error Analysis and Calibration method of Vector Network Analyzer. Metering and testing technology,2019, 46(08):113-118. 10.15988/ j.cnki.1004-6941.2019.8.038.

[8] H Votsi, J Urbonas, S Iezekiel. (2021) Electromagnetic Field Measurements Above On-Wafer Calibration Standards.In: 96th ARFTG Microwave Measurement Conference (ARFTG). San Diego.pp.1-5. 10.1109/ARFTG49670.2021.9425326.

[9] J Sun, C Liu, F Liang. (2021) Extraction of S-parameters of two-port network using single-portcalibration error term of vector network analyzer. Journal of Metrology 2021, 42(11):1499-1503.https://doi.org/10.3390/s21217201. https://kns.cnki.net/.

[10] N Zhang, G Zhang, T Chen, J Liu. (2014) Study on the Unknown Thru Calibration Technique.In: 2014 XXXIth URSI General Assembly and Scientific Symposium (URSI GASS). Beijing.pp.1-4. 10.1109/URSIGASS.2014.6929005.

Research on FPGA-based dual-channel time synchronization system

Ang Gao[1,2], Junxia Gao[1], Yiming Zhang[1] and Xuhong Wang[1]

[1]Faculty of Information Technology, Beijing University of Technology, Beijing, China

[2]Corresponding author's e-mail: ang_gao_bjut@163.com

Abstract. As geological exploration progresses in depth and scope, the distance between transmitting and receiving systems also increases. Consequently, the synchronization of time between the two becomes an increasingly critical issue. In response, this paper presents a proposed design for a time synchronization system that is suitable for use between transmitting and receiving equipment. The system uses GPS as the main clock source. The information within the GPS system will be encoded and decoded using the FPGA. This allows us to obtain both UTC time and the time of the IRIG-B code format, enabling compatibility with the various system interfaces. Our experiments have confirmed that the system has a time error at the microsecond level, and the precision satisfies the design requirements. This verifies the validity and reliability of the system.

1. Introduction

In contrast to small-scale and shallow geological surveys, large-scale surveys require synchronized data reception from multiple devices distributed throughout the survey area. Due to the large geographical distances between them, the previous practice of achieving time synchronization through wired communication is no longer applicable. Based on the above issues, mainstream time communication protocols are used, and therefore time synchronization requirements must be strictly adhered to ensure the reliability of data acquisition.

The GPS synchronization method is to receive the standard time signal transmitted by the satellite in orbit through the GPS receiver. Then, the time information is decoded and encoded according to specific requirements and sent to each device to realize the time synchronization function. At home and abroad, IRIG-B has been widely used in power systems and is a mature mainstream time synchronization[1]. And with the development of chip technology, there are a variety of chips as the choice of processing unit[2], can be selected by the designer according to power consumption, cost, use of scenarios, and other practical needs of different programs, such as ZYNQ, FPGA + DSP, FPGA + ARM, MCU and other different programs[3]. The FPGA master chip program with a short development cycle, fast parallel processing speed and reasonable cost is used in this design. According to the actual demand, the second pulse signal is used as the reference flag, and the IRIG-B code is the final output signal.

2. Introduction to IRIG-B Code

In this paper, we have designed a time synchronization system based on the FPGA-based IRIG-B code. The reliability of the system has been verified through field exploration operations and proved to be well-suited for complex environmental conditions in the field. It is worth noting that the IRIG time code has six formats, each with different code periods, code rates, and time information. Among these formats, the most widely used and accurate one is the IRIG-B code.

Figure 1 illustrates the three code element formats. IRIG-B code period lasts for 10 milliseconds. During this time, high and low levels are controlled to indicate “1 code element”, “0 code element”, or “P code element”. It is crucial to maintain the output level within this 10 ms window. Consequently, it is necessary to control the output high and low levels within 10 ms, define the output array as 1000 bits, and transmit it at a rate of 1000 bits per second[4]. When outputting a “P” code element, it is necessary to output a high level in the first 8 milliseconds and a low level in the last 2 seconds of the code element cycle[5]. When outputting a “1” code element, it is necessary to output a high level in the first 5 milliseconds and a low level in the last 5 seconds of the code element cycle[6]. When outputting a “P” code element, it is necessary to output a high level in the first 2 milliseconds and a low level in the last 8 seconds in one code element cycle.

After obtaining the year, hour, minute, second, and time data, generate the corresponding array based on each value, and output it in accordance with the format of the IRIG-B code[7].

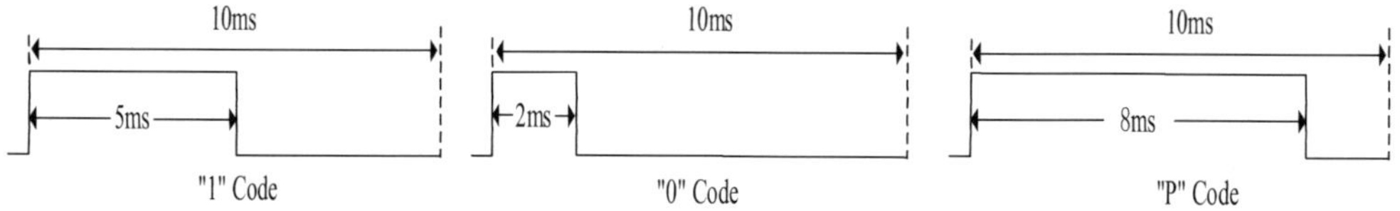

Figure 1. IRIG-B basic code elements.

3. Hardware Structure

The hardware portion of the system is shown in Figure 2. The system’s hardware circuit comprises four components. The GPS antenna receives GPS signals and sends them to the time receiving circuit for processing[8]. Once the circuit receives the satellite signals, it transmits time information to the FPGA via a serial port. The FPGA extracts, decodes, and encodes the information before outputting it to the equipment requiring time synchronization. Then, the FPGA extracts, decodes, and encodes the initial time data to obtain the IRIG-B code, which is ultimately transmitted to the equipment requiring time synchronization.

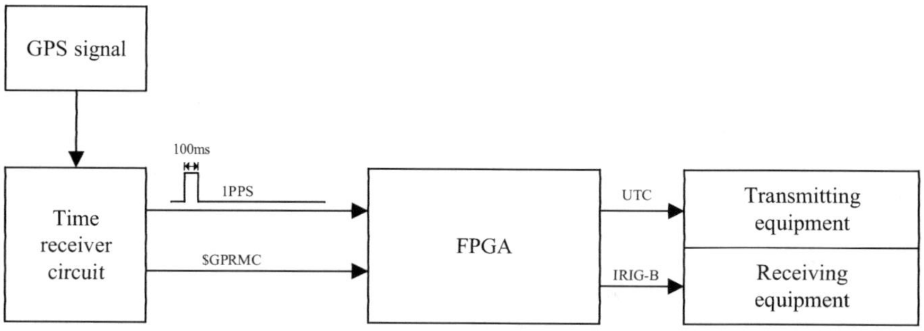

Figure 2. Hardware flowchart.

3.1. Time Receiver Circuit

The schematic diagram of the time receive circuit is shown in Figure 3. The time receiving circuit employs a custom GPS timing chip, the LEA-6T, manufactured by U-Blox in Switzerland. The circuit produces a second pulse 1PPS signal with a GPS time error of less than 15 ns. Its functions include acquiring current time information and accessing the second pulse signal to calibrate time with high accuracy.

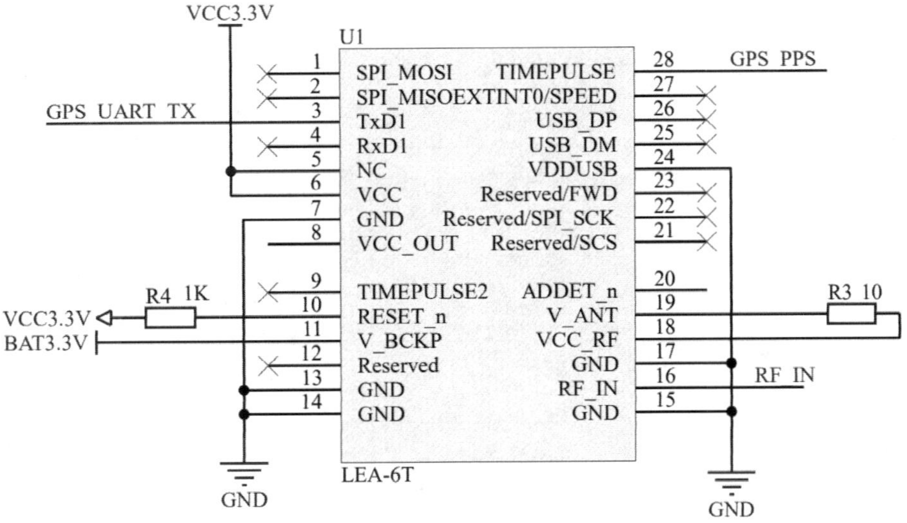

Figure 3. Time receiver circuit schematic.

3.2. FPGA Selection

This design utilizes the EP4CE15E22C8N chip as the primary control chip. The chip is capable of achieving a maximum processing speed of 400 MHz and features four internal PLL clocks, each of which has five clock outputs for frequency reduction and frequency doubling to maintain clock synchronization as per the project's requirements. Additionally, the chip contains an embedded multiplication unit that, when coupled with existing IP cores, significantly shortens the R&D cycle for the project[9]. Furthermore, the external IO ports are diverse, enabling communication with various devices for data processing[10]. The chip features an included multiplier unit that significantly shortens the project development time, along with the current IP cores.

The FPGA acquires GPS message data in GPRMC format via the serial port. It verifies the accuracy and completeness of the frame header $GPRMC and the time valid flag bits, completes 30 detection cycles, and confirms the validity of GPS time. It then extracts time information from the message format's various positions, providing the precise UTC time.

4. System software design

The system flow is shown in Figure 4. After the system is turned on, the time receiving circuit sends GPS data to the FPGA via the serial port. However, the initial GPS signal is unstable due to the insufficient number of locked satellites, so the FPGA continuously detects the received data until the valid bits are identified as such. The system's counter will tally the data, and once the counting time reaches three minutes, it can be confirmed that the system has achieved stable operation. From this point, the FPGA will commence the conversion of the time data format. The ASCII information received is first converted to 8421BCD code format before combining with the TOD value and adhering to the three basic code elements format of the IRIG-B code.

It is known that a basic IRIG-B code element period is 10 milliseconds, and it is necessary to limit the high and low level time in it to achieve the purpose of outputting different code elements, so a code element period is divided into 10 parts, the length of time of each part is 1 millisecond, and then the high and low level of each 1 millisecond is controlled according to the type of code element to be output.

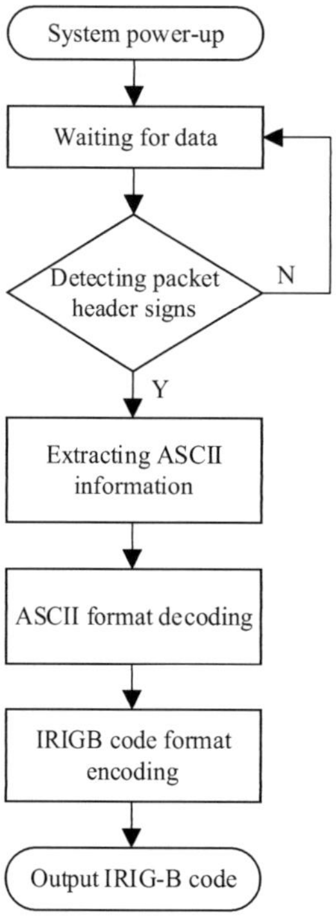

Figure 4. Software flow chart.

5. System testing

Figure 5 displays the actual IRIG-B code waveform outputted by synchronously outputting the code using the DL850 Yokogawa oscilloscope recorder while observing the waveform at the rising edge of the 1PPS signal. The first code element is the "P" code element, used to locate the flag bit. Technical term abbreviations will be explained when introduced. The 8-bit code element displays time information that corresponds with the "1" code element's output. Each of the next ten bits of the code element appears once in the middle of the two "P" code elements. The "0" code element directly follows the second information expressed in the 8-bit code element. The minutes, hours, years, and other information follow in turn.

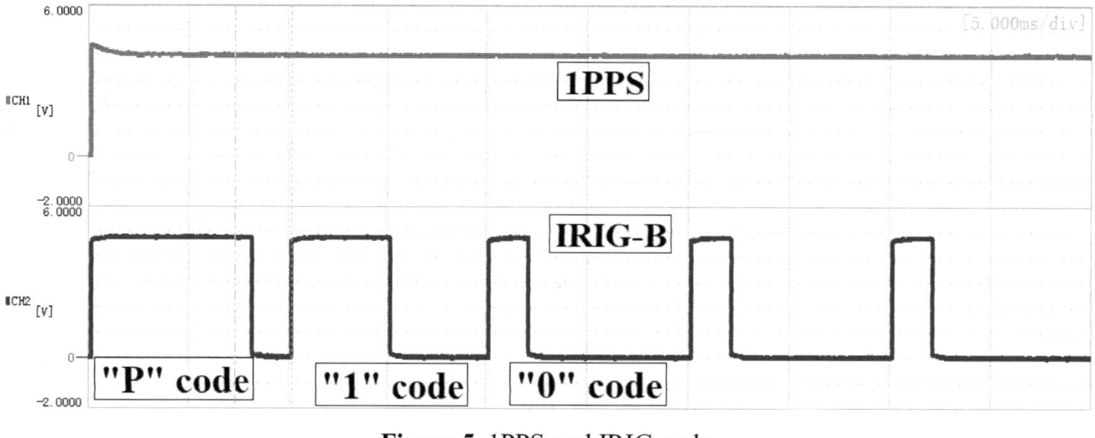

Figure 5. 1PPS and IRIG code.

Zooming in on Figure 5 reveals Figure 6, which shows the rise time of the 1PPS and IRIG-B signals from a low level to a high level, as well as the times at which the levels of the two signals start and stop changing. Analysis of the details of the waveforms shows that the rise time for both is 5 microseconds and the start of the voltage change is strictly synchronized, ensuring the accuracy of the time information fed to the equipment.

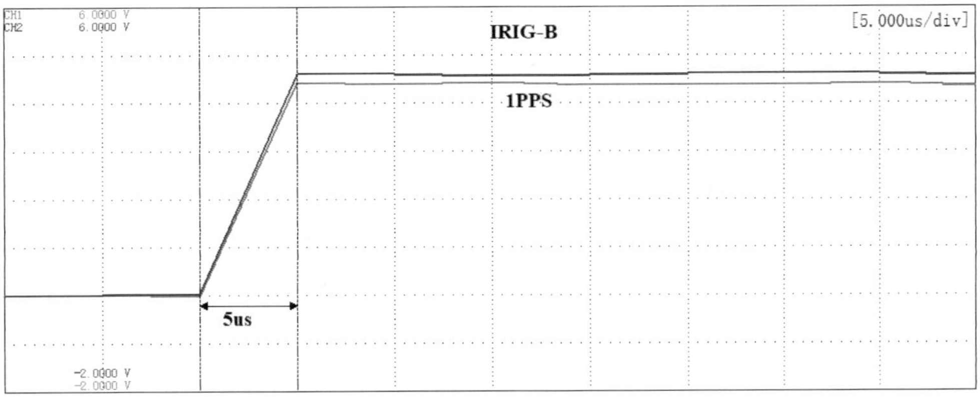

Figure 6. Rise time of 1PPS and IRIG-B.

The output of the IRIG-B code from the system is connected to the external time interface of the DL850 device via a coaxial cable to reduce external interference. Once connected, the equipment status is monitored. Figure 7 demonstrates the equipment's time display when the system is not operating, indicating that it is not locked while Figure 8 shows the equipment's display once the system is operating regularly, demonstrating that it is locked. This concludes that the IRIG-B code output from the system is valid.

Figure 7. Equipment unlocked.

Figure 8. Equipment locked.

6. Conclusion

The time synchronization system presented in this paper utilizes GPS as the signal source and FPGA as the primary control chip. It has the capability to produce UTC time, IRIG-B code, and pulse per second signal. By measuring the output waveform of the system and verifying equipment accessibility, the time synchronization function can be achieved, allowing for stable, long-term operation. Further, measurement of the rising edge of the 1PPS and IRIG-B time waveforms demonstrates that the system error aligns with design requirements, making it suitable for a broad range of industrial and scientific exploration applications.

Acknowledgments
This study is supported by the National Key R&D Program of China (Grant No. 2022YFF0706202-1).

References

[1] Zhao, W. G., &Ren, X. Y. (2010) Time Synchronization with IRIG-B Code in Smart Electronic Devices. J. Automation of Electric Power Systems, 34, 113-115. https://doi.org /10.1109/ CCECE.2010.5575154.

[2] Sangeetha, R. S., Kamal, C., & Jayashree, M. M. (2021) Implementation of IRIG-B Time codes on ZYNQ SoC Platform. In: 2021 IEEE Mysore Sub Section International Conference. Hassan. pp. 634-637. https://doi.org/10.1109/MysuruCon52639.2021. 9 641.
[3] Wang, H., Zhu, G., Hou, M., & Wang, S. (2016). Time synchronization based on multiplexing RPR channel and IRIG-B time code. In: International Conference on Electric Utility Deregulation & Restructuring & Power Technologies. Changsha. pp. 869-872. https://doi.org/ 10.1109/DRPT.2015.7432396.
[4] Song, Y., Li, X., & Cao, S. (2021) Design of IEEE 802.1AS time synchronization system based on FPGA. In: Society of Photo-Optical Instrumentation Engineers Conference Series. Qingdao. pp. 412-418. https://doi.org/10.1117/12.2614126.
[5] Qiao, Z. T., & Xu, C. (2020) Research on the Technology of Wireless Time Synchronization System. In: Cross Strait Radio Science & Wireless Technology Conference. Fujian. pp. 1-2. https://doi.org/10.1109/CSRSWTC50769.2020.9372593.
[6] Zhao, Z., Liu, H., Bi, T., & Yang, F. (2020) Data analysis and application platform for generators-grids-loads full-view synchronized measurement system. In: International Conference on Power and Energy Engineering. Xiamen. pp. 64-69. https://doi.org/10.1109/ICPEE51316.2 020. 9311086
[7] Yan, S., Yang, J., & Li, J. (2019) Design and Its Application of Power Time Synchronization Management System Based on TMU and IEC61850 Protocol. In: International Conference on Environmental Science and Material Application. Xi'an. pp. 3-5. https://doi.org/10.1088/ 1755-1315/252/3/032201.
[8] Zhang, C. Y. (2019) Research and implementation of substation clock unsynchrony alarm system based on GPS pulse. In: International Conference on Environment and Renewable Energy. Ho Chi Minh City. pp. 1-6. https://doi.org/10.1088/1755-1315/ 307/1/012003.
[9] Yang, Y., Yang, Y. F., Zheng, K., & Jia, Y. (2016) Research of Satellite and Ground Time Synchronization Based on a New Navigation System. In: China Satellite Navigation Conference. Changsha. pp. 15-25. https://doi.org/10.1007/978-981-10-0940-2_2.
[10] Wang, S., Hou, Y., Gao, F., & Ma, S. (2016) A novel clock synchronization architecture for IoT access system. In: International Conference on Computer and Communications. Chengdu. pp. 1456-1459. https://doi.org/ 10.1109/Comp Comm.2016.7924944.

Journal of Physics: Conference Series **2724** (2024) 012006 doi:10.1088/1742-6596/2724/1/012006

Design of control system for six-degree-of-freedom robotic arm

Chaoda Chen[1], Zhisheng Lv[1,2], Siyang Wu[1], Kangxing Li[1], Ziyang Chen[1] and Zhixing Huang[1]

[1]School of Naval Architecture and Ocean Engineering, Guangzhou Maritime University, Guangzhou, 510725, China

[2]Corresponding author's e-mail: cdchen@gzmtu.edu.cn

Abstracts: To study the robotic arm trajectory planning and motion, the six-degree-of-freedom robotic arm control system design is proposed. In this paper, we choose a multi-joint type of robotic arm. We analyze the design of the robotic arm at home and abroad and determine the specific servomotor model and the specific robotic arm mechanism. We analyze the torque relationship of each joint and the size of the servomotor. Using SolidWorks and Robot Studios, we carry out the three-dimensional design modeling and simulation of the robotic arm, design the whole set of robotic arms, and use the Arduino Uno development board for precise control of the servo motor. The control program was written using C language, and various control tests were carried out on the robotic arm to make sure that the system could satisfy the debugging and research of trajectory planning type.

1. Introduction

As one of the most widely used automated mechanical devices in the field of robotics, robotic arms are still in the "neck" stage due to the late start and weak foundation of this field in China [1-5]. In this project, a six-degree-of-freedom robotic arm control system is designed, which is simple in operation, low in manufacturing cost, and suitable for debugging and research of trajectory planning type. Therefore, the design of a six-degree-of-freedom robotic arm control system is of great significance in solving the current situation of robots in China.

Yao et al. [6] designed a new six-degree-of-freedom service robot arm using a partially decoupled structure, which can bring the advantage of a fast inverse kinematics solution without a significant increase in volume. The D-H method is used to obtain the positive kinematics model of the robotic arm for positive kinematics solution, and a new geometric method is proposed to solve the inverse kinematics problem according to its structural characteristics. Liu et al. [7] used the Bayesian optimization algorithm based on the Random Forest Probabilistic Agent Model with the ABBIRB1410 robotic arm as the experimental object. They concluded that the Bayesian optimization algorithm improves the comprehensive effect of the robotic arm kinematic parameter optimization problem by 15% compared with the traditional algorithm. Zhang et al. [8] analyzed the fuzzy control method for the end position accuracy of an articulated hinge gap robotic arm using a spatial 3R hinge gap-containing robotic arm as a research object. Chao et al. [9] used software to optimize the topology of the arm and found that the optimized arm achieved lightweight while improving strength, stiffness, and minimum safety factor. Xie and lv [10] created a linkage coordinate system by improving the D-H method and derived the kinematic equations, which verified the feasibility and validity of the modified design and motion planning of the robotic arm.

In this paper, a six-degree-of-freedom robotic arm is designed, and the base, end-effector, large arm, and small arm of the arm are modeled using SolidWorks. The appropriate motor and servo are selected based on the torque and rotational accuracy required for the work. The control system is programmed using Arduino IDE. 3D printing is used to create the arm parts and assemble them. Finally, the arm is used for experimentation and debugging to verify the feasibility of this design.

2. Hardware system design of six-degree-of-freedom robotic arm

The lower end of the arm is compatible with the MG996R servo. One side of the circular stretching excision can be fixed with the servo output axis of the rudder disc output to drive the movement of the arm; the other side of the excision out of a circular hole, with a removable protrusion and M3 * 8 mm self-tapping countersunk head screws fixed to play a role in supporting the rotation of the arm and facilitate the role of the installation. Because the servo is used for the single output servo in the middle of the arm plus two beams, one is convenient for the installation and support of the servo. The second can connect the active side of the servo output shaft to drive the auxiliary side of the movement. The third can reinforce the arm to avoid the problem of too much tension and low transmission efficiency. Because the design principle of the large and small arms is the same, only the design of the large arm is elaborated here. The uppermost side of the arm connected to the servo output shaft has two semi-circular stretching excision. The excision can just be stuck in the MG90S servo output shaft protruding part, the size of the servo output shaft protruding part of the half, to avoid the servo sliding due to only one screw hole fixing and other fixing unstable problems.

There are three types of end-effectors used in robotic arms: gripper, pneumatic, and magnetic end-effector. Considering the convenience and practicability, the gear-driven gripper end-effector is used, and its fittings include a "π"-shaped one-piece connecting piece and an "L" shaped gripper fixing piece. The fittings include a "π"-shaped one-piece connector and an "L"-shaped clamping jaw fixing.

The end of the clamping jaw is driven by a servo as the active part of the clamping jaw on one side, and the other side of the follower of the clamping jaw is driven by gear transmission to realize the function of simultaneous opening or clamping. The outermost part of the jaws has a rounded notch with a saw-like non-slip pattern near it, which is mainly designed to grip small and smooth objects such as a marker pen more firmly and to avoid slipping off when handling the objects. The assembly diagram is shown in Figure 1.

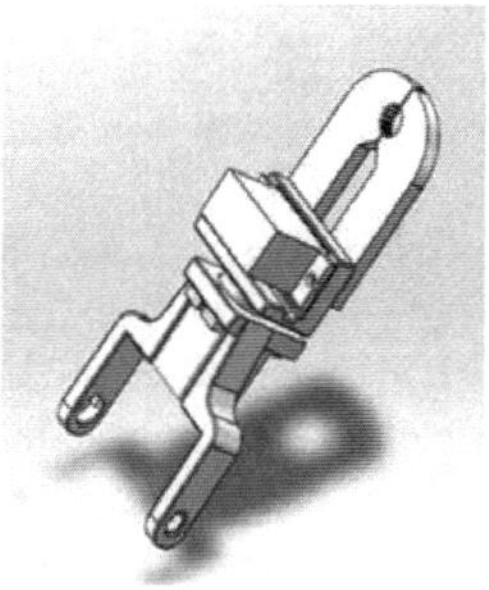

A. Assembly diagram of end-effector part

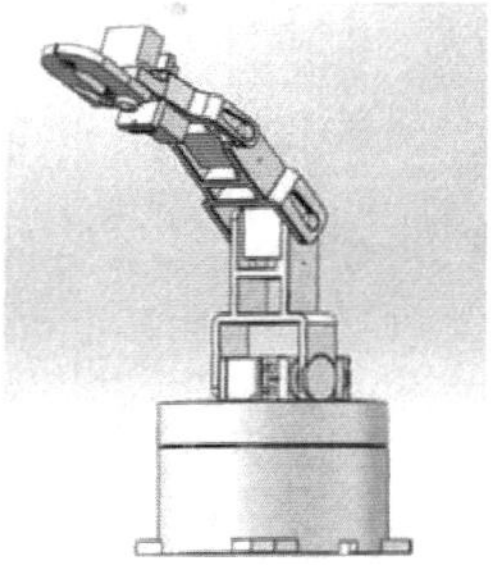

B. 3D design assembly drawing

Figure 1. Assembly drawing.

3. Robotic arm path planning

The MG90S servo and MG996R servo have a rotatable range of 0~180 degrees. However, the large arm, small arm, and other components can have a rotation angle much larger than 180 degrees. Thus, to make better use of the limited 180 degrees of the servo, it is necessary to carry out the assembly of software and hardware cooperation. To debug the six-degree-of-freedom robotic arm so that it can operate normally, reference to the movement of the handling robotic arm, the use of Robot Studios to make the following a complete motion path planning simulation, path planning shown in Figure 2.

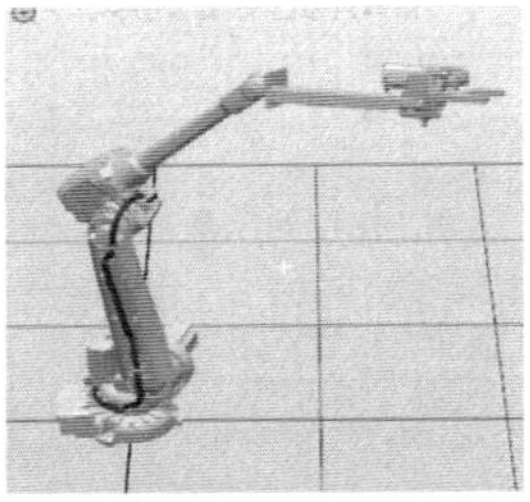

A. Initial position

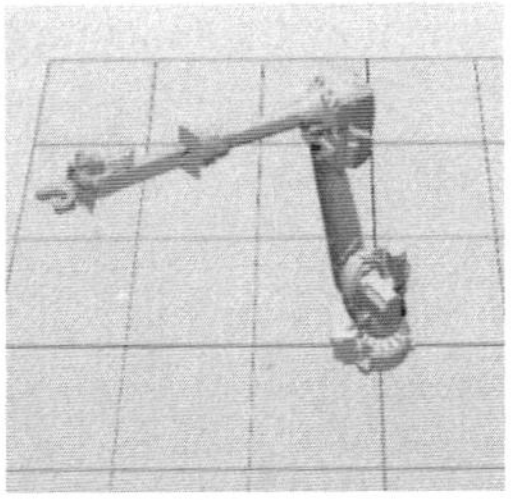

B. Above grab position

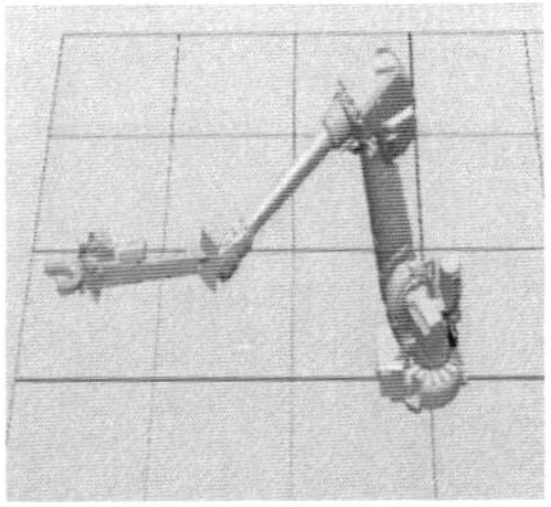

C. Gripping position

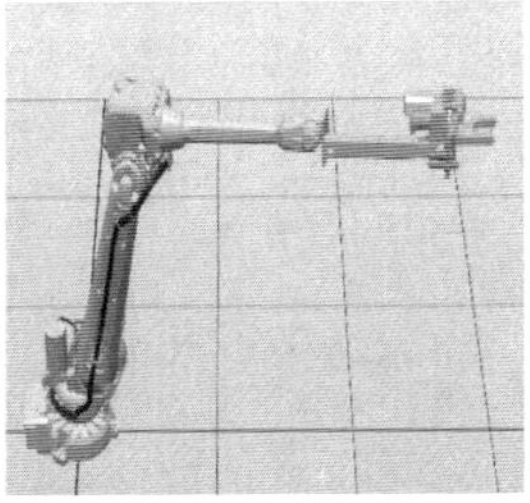

D. Above release position

Figure 2. Motion path simulation diagram.

(1) The vertical position is designed according to the robot arm when the robot arm is energized and can receive signals for a moment, all servos rotate to the first signal position output in the control program code at a very fast speed. If the first position signal is the initial position, the initial attitude of the robot arm is difficult to artificially adjust to the position close to the initial position, so it is designed to have a vertical position. The attitude at this time is the overall vertical of the robot arm Upwards, easy to adjust, to solve the above problem.

(2) The robotic arm is in the vertical upward state when it is not controlled. The first action is to move from the vertical upward state to the initial position at a uniform speed. The initial position is set concerning the Phome position of the robotic arm in the laboratory, i.e., the position where the robotic arm is located before the movement and after completing a set of movement processes.

(3) The robot arm rotates each servo to move the clamping jaws from the initial position to the top of the clamping position, avoiding collision with the clamped objects during the process of the clamping jaws going directly to the clamping position, which may lead to the problem of the objects' position being shifted or damaged.

(4) The clamping jaws are kept horizontal and open to drop vertically from above the clamping position to the clamping object position. The clamping function of the clamping jaws is carried out and maintained. Then, the robotic arm is allowed to clamp the object while rotating the servos so that the three servos return to the angle of rotation at the initial position.

(5) After that, the servo is rotated to make the robot arm gripper part move to the top of the object release position. The object is dropped vertically to release, and then vertically back to the top of the position, and finally back to the initial position.

4. Robotic arm experiment and debugging

Using 3D printing technology to print the robotic arm parts, six degrees of freedom robotic arm is assembled using for loop statements for experimentation and debugging to realize the above path planning so that the robotic arm can move according to the planned path to achieve the purpose.

4.1 Recording of fixed points and angles under trajectory planning

After the attitude adjustment, the servo angles of vertical position, initial position, off (grasping position), grasping position, off (releasing position), and releasing position are obtained for each servo in 6 positions. The servo output angle of each position is shown in Table 1.

Table 1. Servo output angle for each position.

	Servo 1	Servo 2	Servo 3	Rudder 4	Servo 5	Servo 6
Vertical position	85 degrees	95 degrees	50 degrees	27 degrees	90 degrees	90 degrees
initial position	85 degrees	95 degrees	110 degrees	90 degrees	65 degrees	90 degrees
Off (grabbing position)	110 degrees	95 degrees	53 degrees	100 degrees	110 degrees	45 degrees
Grab Location	110 degrees	95 degrees	13 degrees	130 degrees	120 degrees	45 degrees
Off (release position)	88 degrees	95 degrees	53 degrees	100 degrees	110 degrees	135 degrees
Release position	88 degrees	95 degrees	13 degrees	130 degrees	120 degrees	135 degrees

To calculate the servo output angle from the initial position to reach off (gripping position) as an example, so that the robotic arm by the trajectory planning from the initial position to reach off (gripping position): servo 1 output angle from 85 degrees to 110 degrees, servo 3 output angle from 110 degrees to 53 degrees, servo 4 output angle from 90 degrees to 100 degrees, servo 5 output angle from 65 degrees to 110 degrees The servo 6 output angle is from 90 degrees to 45 degrees, and the servo waits 15 milliseconds for every one degree of rotation to control the servo rotation speed.

4.2 Experimental debugging

Experimentation and debugging are done using an assembled six-degree-of-freedom robotic arm and a for-loop statement to implement the above path planning so that the robotic arm can move according to the planned path and thus achieve the objective. The parameter of the original initial position is that each servo is in the 90-degree position. However, in the debugging, it was found that when servo 6 was allowed to rotate in the original initial position, the whole six-degree-of-freedom robotic arm appeared to have a large degree and was visible to the naked eye of the violent jitter. It would even affect the rotation of the robotic arm. After analysis and research, it was found that it was because the position of the original initial position was biased towards the front of the arm, which generated a certain amount of radial force, thus causing the rotating platform connected to the fixed servo 5 in the base to rub against the bearings. The larger friction coefficient of the contact surface of the rotating platform resulted in more violent shaking. Therefore, today's initial position is adopted to solve the problem of jitter caused by excessive radial force. The six-degree-of-freedom robotic arm carries out the movement schematically according to the trajectory planning, as shown in Figure 3.

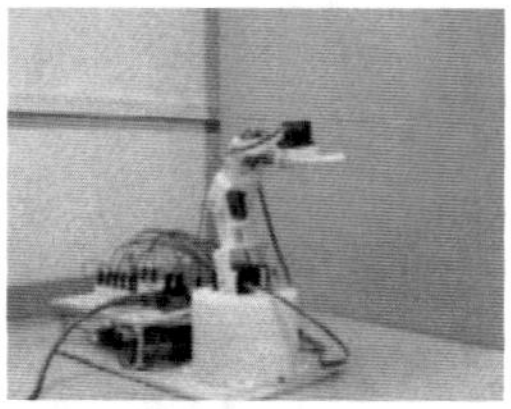

A. Initial position

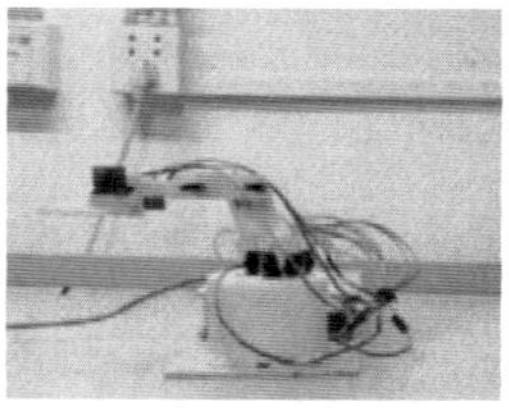

B. Above the gripping position

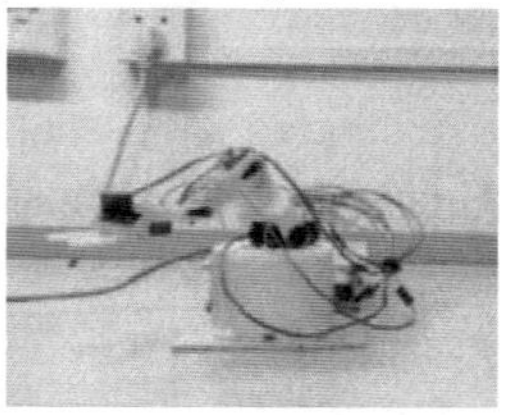

C. Gripping position

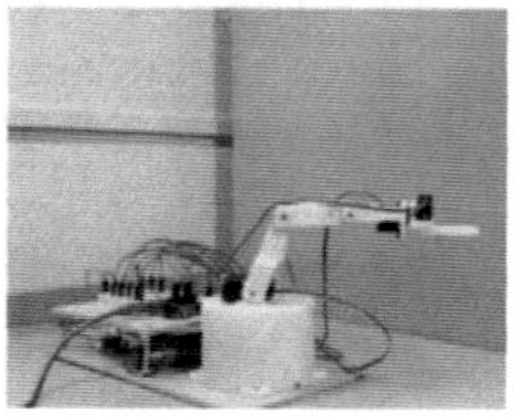

D. Above the release position

Figure 3. Schematic diagram of the trajectory of the six-degree-of-freedom robotic arm.

5. Conclusion

The main work accomplished in this paper is as follows:

This paper is based on a six-degree-of-freedom mechanical arm to adjust the design through SolidWorks three-dimensional modeling software for the development of six-degree-of-freedom mechanical arm modeling and simulation. The use of 3D printing technology will be designed to print the mechanical arm parts. The software of the robotic arm is designed to calculate the relative coordinates of the target position, and the software is used to program and control the servo through the control development board. The simulation is carried out through the simulation software Robot Studio. The six-degree-of-freedom robotic arm is realized in the software for the movement clamping and placing functions. Through the development board to the servo position to achieve the rotation position initialization, each servo output is connected to the axis of different angles so that the six degrees of freedom robotic arm can achieve the function of arriving at any position in space.

Acknowledgments

This work was supported by Foundation of No.2023CJRHJD002 and No.2023ZDZX2051.

References

[1] Department of Equipment Industry, Ministry of Industry and Information Technology. Promoting the development of robots in the series of interpretations of the planning of Made in China 2025[J]. *Science and Technology Herald*, 2015, 33(21).
[2] WANG Tianmiao, TAO Yong. Status quo of China's industrial robot technology and industrialisation development strategy[J]. *Journal of Mechanical Engineering*, 2014, 50(9).
[3] WANG Yuqi, ZHOU Lei, CAO Jinjiang and WANG Lei. Introduction to robotics course - practical teaching ability and teaching quality improvement[J]. *China Technology and Equipment*, 2021(2): 129-131.
[4] HOU Zhi, ZHANG Ai. Analysis of the development status quo and talent demand of the industrial robotics industry[J]. *Robotics industry*, 2023(4): 95-100.
[5] HU Yuhang, ZHANG Shangying and ZHANG Xiaoyan. Simulation study on the kinematics of 6R robotic arm based with MATLAB[J], *Manufacturing Technology and Machine Tools*, 2022(11): 27-32.
[6] YAO Xingtian, XU Dan, ZHANG Maku and SHUAI Liguo. Structural design and kinematic analysis of a new six-degree-of-freedom service robot[J], *Modern Manufacturing Engineering*, 2015(4): 27-32.
[7] LIU Chenhui, ZHANG Weimin and XUE Feng. Optimisation of robotic arm motion parameters based on random forest Bayesian optimisation[J], *Manufacturing Technology and Machine Tools*, 2023(1): 83-90.
[8] ZHANG Fajun, ZHANG Sujun, YU Chengjun and HE Kongde. Considering fuzzy control of end position accuracy of joint hinge gap robotic arm[J], *Manufacturing Technology and Machine Tools*, 2020(6): 148-154.
[9] CHEN Chao, CHEN Hehe, WU Shuting and WANG Zhen. Structural optimization of robotic arms for industrial robots[J], *Machine Tools and Hydraulics*, 2021, 49(3): 25-29.
[10] Xie Xiaozheng, Lv Weizian. Motion performance of 6R welding robotic arm and its workspace simulation[J], *Mechanical Design and Manufacturing Engineering*, 2023, 52(8).

Research on high precision flow control technology of low speed wind tunnel air supply system

Ling Tang [1], Jian Ye [1] and Hongming Kang [1,2]

[1]China Aerodynamics Research and Development Center Mianyang, China

[2] Corresponding author's e-mail: 657461645@qq.com

Abstract: Aiming at the shortcomings of traditional analog control valves and digital valves, a high-pressure air supply system with accurate flow control is designed. Firstly, the ER5000 controller and pilot proportional valve drive several diaphragm reducing valves to achieve the preliminary adjustment of air pressure. Secondly, a needle valve based on the PCM digital valve and equal percentage flow characteristics is used to achieve accurate control of gas flow. Then, the Venturi flow meter with a honeycomb and damping network is used to accurately measure the airflow. Finally, a PID flow control algorithm based on nonlinear self-tuning is proposed. The results show that the flow regulation range of the system reaches 0.1~8 kg/s, the absolute control accuracy is better than ±3 g/s, and the relative control accuracy is better than ±0.04%. Moreover, the system has the characteristics of short regulation time, small overshoot, and good dynamic performance, which can achieve strong support for wind tunnel air supply tests.

1. Introduction

Wind tunnel tests on items such as inlet, flow control, dynamic simulation, and thrust vector provide important performance parameters for the design and improvement of the aerodynamic force of engine inlets in China and the optimization of the aerodynamic layout components of aircraft [1]. With the deepening of experimental technology research, the refinement level of the high-pressure air supply system, which is the most basic and important part of wind tunnel tests, directly affects the accuracy of test data [2].

At present, the comprehensive error of the traditional analog control valve is 0.5%, and the actual adjustable ratio is usually lower than the design value (usually 30~60) [3-4]. In addition, the control valve has the problems of dead zone control, regulation lag, and easy wear and tear of seals, so it can hardly meet the test requirements of newly developed flow control valves—high-precision, wide-range, and high-reliability PWM, PCM, and other digital valves, which have been widely used in various industries and played an important role [5-8]. However, they still have some shortcomings. The solenoid valve used by PWM digital valves has always been in a high-frequency switching state, so the parts are easy to wear, and the flow rate is relatively small, which leads to loud high-frequency noises, and the need for large flow rates cannot be met. The PCM digital valve is a kind of step control, and the output flow is discrete. The machining error of the nozzle will lead to oscillation when the step is advanced and backward. In addition, the response time of solenoid valves with different diameters is also different, which will reduce the synchronization of the switch and reduce the control accuracy.

To solve the above problems, a high-pressure air supply system with accurate flow control is designed, which can effectively compensate for the shortcomings of analog control valves and digital valves, improve the control accuracy and adjustment range of airflow, and provide strong support for high-pressure air supply tests in wind tunnels.

2. Working principle

The high-pressure air supply system adopts a two-stage adjustment scheme, and its working principle is shown in Figure 1. In the first stage, the ER5000 controller and pilot proportional valve are used to driving two thin film pressure relief valves to reduce the 13 ~ 22 MPa high-pressure gas source to 6 ~ 10 MPa. The adjustment pressure is set according to the demand so that the pressure can remain constant while sufficient flow capacity is ensured. In the second stage, the flow regulating device composed of a PCM digital valve and a needle valve with equal percentage flow characteristics is used to achieve accurate flow control.

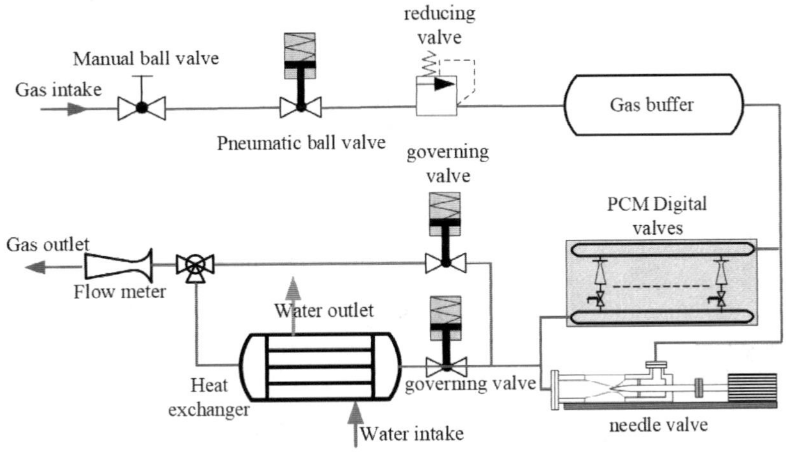

Figure 1. Diagram of the working principle of the high-pressure air supply system.

At the second stage of the flow regulation process, the PCM digital valve provides reference flow, which is the basis for needle valve fine-tuning. Therefore, it is not necessary to design many nozzles with small valve positions and throat areas for the digital valve. During a certain flow regulation interval, the switching state of all solenoid valves remains unchanged, which can avoid the control oscillation caused by frequent switching. Due to the poor adjustment performance of the needle valve in the two transitional stages of the opening moment and full opening, the adjustment range is compressed to 30% ~ 80% during the design, which is the best adjustment range of the control valve, yielding the best performance [9]. Both the digital valve and the needle valve adopt a critical flow Venturi nozzle to achieve accurate flow regulation by controlling the total throat area.

In order to prevent large fluctuation of the pressure behind the valve of the high-pressure relief valve caused by the wide range adjustment of flow in the test process, an air buffer with a large capacity is added between the relief valve and the flow adjustment device, which can effectively filter and improve the stability of flow control. In order to adjust the outlet temperature of the airflow, two pneumatic control valves are added at the back end of the flow regulating device. The airflow is divided into two branches, one of which has a heat exchanger to heat the airflow, and the airflow temperature after mixing is quickly adjusted by controlling the airflow ratio of the two branches.

The flowmeter is mainly composed of a honeycomb and damping network, a throat, a total pressure probe, a static pressure probe, and so on. Its structure is shown in Figure 2. Among the components, the rectification device composed of honeycomb and damping networks can achieve the purpose of airflow guidance and vortex breaking to improve the flow field quality and the flow measurement accuracy. Due to the need to take into account the accuracy requirements of different measurement ranges and to meet the conditions of the throat to achieve sound speed under different test conditions, a number of flowmeters with different throat areas are designed to switch and match, which can ensure better measurement accuracy in different intervals.

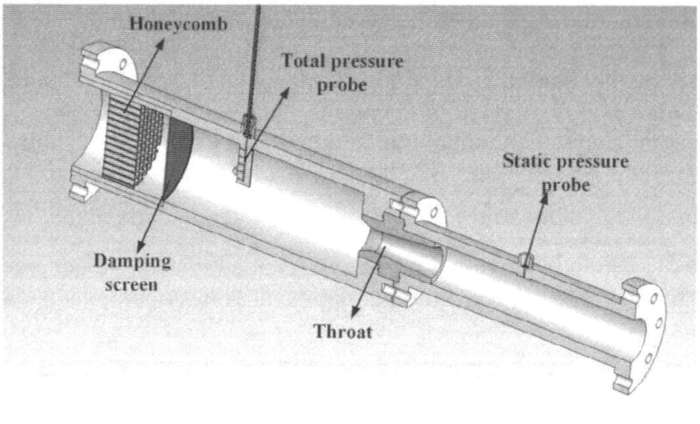

Figure 2. Diagram of the Venturi flow meter structure.

3. Key equipment

3.1 High-Pressure relief valve

The working principle of the high-pressure relief valve is shown in Figure 3, the remote controller sends the set value of the pressure signal to the electronic pressure controller ER5269, which convert the electrical signal into the pressure signal, and then amplify the pressure to 6 ~ 10 MPa through the pilot proportional valve 26-2065, and then output it to the control chamber of the two thin film pressure relief valves 26-1261, respectively. The gap between the pressure regulating valve core and the valve seat increases or decreases the pressure behind the valve. At the same time, the feedback chamber of the regulating valve applies an effect opposite to that of the control chamber. When the two functions are balanced, the pressure behind the valve is stable.

Because of the large-capacity air buffer at the back end of the pressure relief valve 26-1261, the pressure regulation will be delayed. Therefore, the pressure relief valve adopts the way of external pressure. The pressure at the top of the air buffer is measured and used as the feedback pressure of the regulating valve. At the same time, the pressure signal is collected in real time by the pressure sensor, and the PID control is carried out by the remote controller. The outlet pressure of the valve can be accurately adjusted. In order to ensure the safety of the system and equipment, the high-pressure pneumatic ball valve is equipped before the pressure relief valve to quickly cut off the upstream pressure.

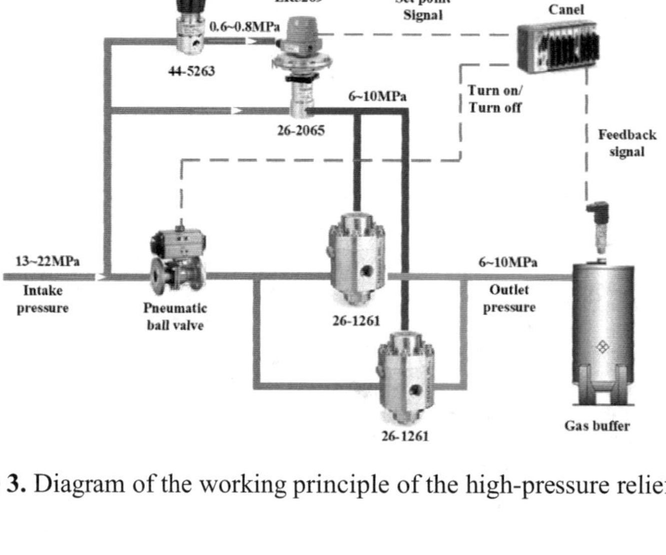

Figure 3. Diagram of the working principle of the high-pressure relief valve.

3.2 PCM digital valve

The structure of each branch of the PCM digital valve is shown in Figure 4. The airflow flows through the Venturi nozzle, high-pressure hose, and solenoid valve in turn. The control system controls the opening and closing of each branch by changing the switching state of the solenoid valve to combine different flows. When all solenoid valves are fully opened, the flow through all nozzles is the maximum flow, and the flow through the lowest nozzle is the minimum flow.

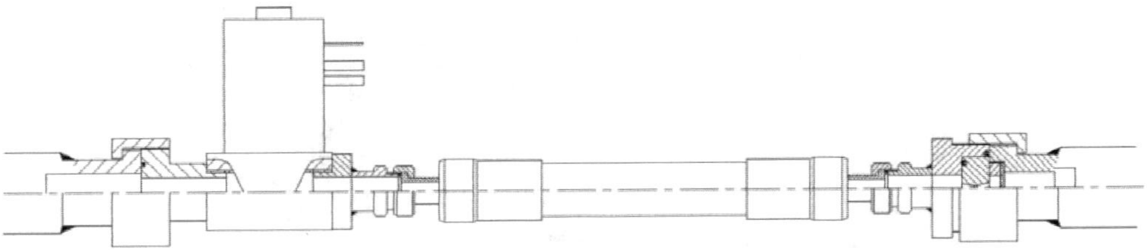

Figure 4. Diagram of the digital valve branch structure.

The adopted nozzle of the digital valve is a critical flow Venturi nozzle with a diffusion tube. Compared with the standard nozzle, the critical outlet pressure can be significantly restored to avoid excessive pressure loss. According to Wu and Liu [10], the ratio of diffusion section area to throat area in the selected nozzle is 4, and the critical pressure ratio can be increased from 0.528 to 0.89, that is when the outlet pressure of the digital valve is less than 0.89 times the inlet pressure, the critical state can be achieved. The flow rate of the nozzle throat is the sound velocity. In the ideal state, the flow rate is only determined by the inlet pressure, inlet temperature, and throat area. The calculation method is shown in Formula (1).

$$\dot{m} = \frac{0.04042 \times A^{*} \times P_0}{\sqrt{T_0}} \qquad (1)$$

where P_0 is the nozzle inlet pressure (Pa), T_0 is the nozzle inlet temperature (K), and A^* is the throat area (m^2).

The PCM digital valve consists of four branches, and its throat area is arranged in a binary sequence. Under the design conditions of inlet pressure 10 MPa, inlet temperature 20°C, and total flow 7.5 kg/s, the throat area of each nozzle can be obtained by four-bit binary calculation, as shown in Table 1.

Table 1. Throat area distribution of the digital valve.

Valve position (bit)	design flow rate (kg/s)	throat area (mm^2)
1	0.5	21.17
2	1	42.35
3	2	84.7
4	4	169.4
Total	7.5	317.62

3.3 Needle valve

The needle valve structure, as shown in Figure 5, mainly includes a cavity, an actuator, an orifice plate, a sleeve, a throat, a valve core, and so on. The flow control of the needle valve is realized by changing the throttling area of the throat, which is completed by the linear motion of the valve core. Therefore, the surface design and effective stroke of the valve core are directly related to the resolution and control accuracy of the flow. In addition, in order to improve the stability of the valve core operation, a cylinder orifice plate is used to rectify the upstream of the valve core, which can effectively reduce the unevenness of the airflow in the cavity. In order to avoid the installation of the actuator and other components in the pipeline, which will increase the difficulty of installation and maintenance, the inlet and outlet of the needle valve are misaligned. The actuator is installed outside the pipeline, connected to the tail of the valve core through the sleeve, and sealed with a double-layer O-ring.

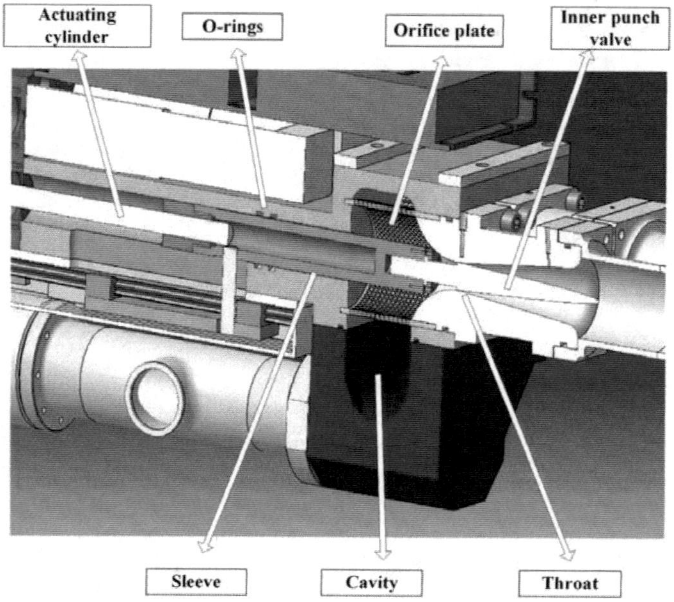

Figure 5. Diagram of the needle valve structure.

The flow characteristics of the control valve usually refer to the relationship between the relative flow of the valve and the relative displacement of the valve core, as shown in Formula (2). Compared with other flow characteristics such as parabola and straight line, the regulation performance of equal percentage characteristics is stable and the effect is better [11]. Its main feature is that the flow change caused at each point is proportional to the flow at that point, and the percentage of flow change is equal. When the flow is small, the flow change is small, and when the flow is large, the flow change is large. Therefore, the flow characteristics with an equal percentage have the same adjustment accuracy at different openings, which suits the design requirements of needle valves well

$$\frac{q_V}{q_{V\max}} = R^{\left(\frac{l}{L}-1\right)} \tag{2}$$

where $q_V/q_{V\max}$ is the ratio of the flow rate of the control valve in a certain opening state to that in a full opening state, l/L is the ratio of the core displacement of the control valve in a certain opening state to that in a full opening state, and R is the ideal adjustable ratio of the valve.

Since the throttling area of the needle valve is the difference between the throat area and the cross-sectional area of the valve core at the throat, the surface curve of the valve core can be designed as:

$$r = \frac{D}{2} \times \sqrt{1 - R^{\left(\frac{l}{L}-1\right)}} \tag{3}$$

where D is the throat diameter (mm), and r is the radius (mm) of the valve core at a certain opening.

The maximum flow rate of the needle valve is 1 kg/s, and the throat diameter is calculated at 7.34 mm by Formula (1). The design effective stroke of the valve core should not be too small, otherwise, the resolution will be reduced. Also, it should not be too large, as a stroke too large will increase the overall size of the valve. After comprehensive consideration, the length of 10 times the throat diameter is selected as the effective stroke of the valve core in practical application.

4. Control strategy

4.1 Flow control principle

The control principle of the high-pressure air supply system is shown in Figure 6. The remote controller gives the regulating pressure signal of the high-pressure relief valve and the flow signal required by the test. The pressure regulator realizes the closed-loop control of the regulating pressure of the relief valve. The flow regulator realizes the opening and closing of each valve position of the PCM digital valve and the precise control of the opening of the needle valve. When the test requires the system to provide a stable outlet pressure, the pressure can be converted to a flow, so that the pressure can be controlled by controlling the flow.

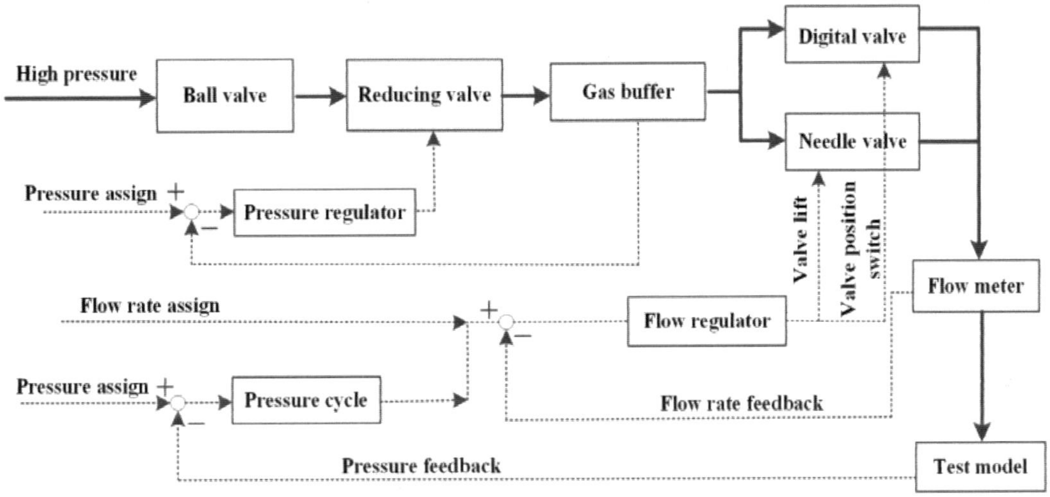

Figure 6. Schematic diagram of the control principle of the air supply system.

In the flow control, firstly, the flow value is directly given or converted from the pressure value; secondly, the difference between the actual flow measured by the flowmeter and the flow value is calculated; thirdly, the control value is output by the flow regulator; finally, according to the specified needle valve regulation interval of 30% ~ 80%, the flow of the PCM digital valve and needle valve is determined. Under the design conditions, when the flow control value is less than 0.8 kg/s, the digital valve should be fully closed, and only the needle valve should be adjusted. When the flow control value is higher than 0.8 kg/s, the digital valve controls each valve position according to different flow ranges without advanced or backward operation and then adjusts the opening of the needle valve on this basis. If the flow rate required by the test is less than 0.3 kg/s, the output pressure of the high-pressure relief valve can be reduced to ensure that the needle valve can also be in the optimal adjustment range. Due to the needle valve in the same position, the front pressure is reduced, and the flow will be proportionally reduced, so the accuracy requirements of small flows can be met.

4.2 Flow Control Algorithm

By analyzing the principle and structure of the high-pressure air supply system and the flow response curve, the transfer function of the flow control object can be simplified as a first-order inertial plus pure lag link, and its time constant is related to the length and volume of the pipeline and the heat exchanger, filter, and flowmeter at the back end. Due to the wide range of flow regulation, pressure, and flow rate have a great influence on the parameters of the transfer function, and the change of load during the test will also bring great disturbance, so it is difficult to accurately establish the control model.

In order to meet the test requirements for a fast, accurate flow with a wide range, a nonlinear self-tuning PID control algorithm is designed for the system, which can realize the self-tuning of the three parameters of proportional, integral, and differential in a certain range. The basic principle of the control algorithm is as follows:

(1) Proportional gain parameter k_p: when the deviation is large, k_p is also large, which can improve the response speed; when the deviation is small, k_p is also small, which can reduce overshoot. Therefore, according to the above variation law, the nonlinear function of k_p is constructed as follows:

$$k_p = k_{p1} + k_{p2}\left[\exp(k_{p3}e(t) + \exp(-k_{p3}e(t)\right] \quad (4)$$

where k_{p1}, k_{p2}, and k_{p3} are positive real constants, the maximum value of k_p is $k_{p1} + k_{p2}$, and the minimum value is k_{p1}. k_{p3} determines the change rate of k_p.

(2) Integral gain parameter k_i: when the deviation is large, k_i should not be too large so that integral saturation can be prevented and the impact on dynamic performance and stability can be reduced; when the deviation is small, k_i increases appropriately, which can effectively eliminate the steady-state error. Therefore, according to the above variation law, the nonlinear function of k_i is constructed as follows:

$$k_i = k_{i1} \times \frac{2}{\exp(k_{i2}e(t)) + \exp(-k_{i2}e(t))} \quad (5)$$

where k_{i1} and k_{i2} are positive real constants, and the maximum value of k_i is k_{i1}. k_{i2} determines the change rate of k_i.

(3) Differential gain parameter k_d: when the negative deviation is large, k_d should gradually increase without affecting the response speed, which can inhibit the generation of overshoot; when the positive deviation is large, and k_d should be slowly reduced and limited. Therefore, according to the above variation law, the nonlinear function of k_d is constructed as follows:

$$k_d = k_{d1} + \frac{k_{d2}}{1 + k_{d3}\exp\left(k_{d4}e(t)\right)} \quad (6)$$

where k_{d1}, k_{d2}, k_{d3}, and k_{d4} are positive real constants, the maximum value of k_d is $k_{d1} + k_{d2}$, and the minimum value is k_{d1}. k_{d4} determines the change rate of k_d.

The nonlinear self-tuning control algorithm involves many parameters and needs some experience and debugging ability. If the parameters are selected properly, the control system can have a faster response speed and stronger anti-interference ability to effectively improve the flow control accuracy and dynamic performance.

5. Performance Testing

After the installation and commissioning of the high-pressure air supply system in the FL-12 wind tunnel, a comprehensive performance test was carried out. Figures 7 and 8 show the characteristic curves of dynamic and steady flow regulation, respectively.

It can be seen from Figure 7 that the flow regulation range of the air supply system is 0.1 ~ 8 kg/s, the regulation time is within 15 s, the overshoot is small, and there is no control oscillation phenomenon. It can be seen from Figure 8 that the flow steady-state performance is good. Under the given flow rate of 7 kg/s, the absolute control accuracy is better than ±3 g/s, and the relative control accuracy is better than 0.04%.

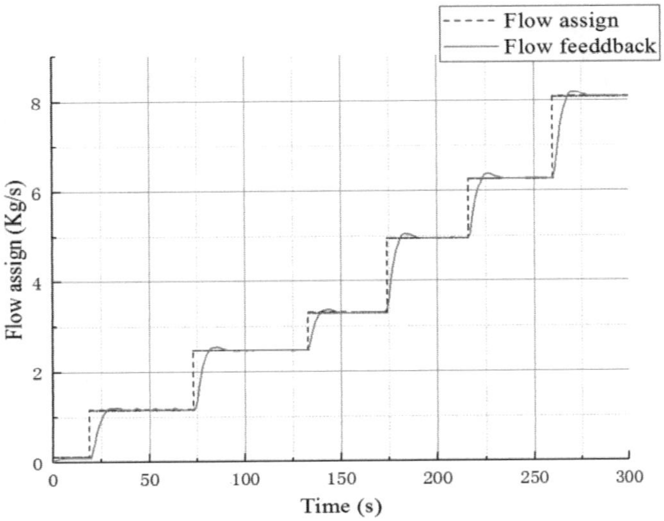

Figure 7. Dynamic flow regulation curve.

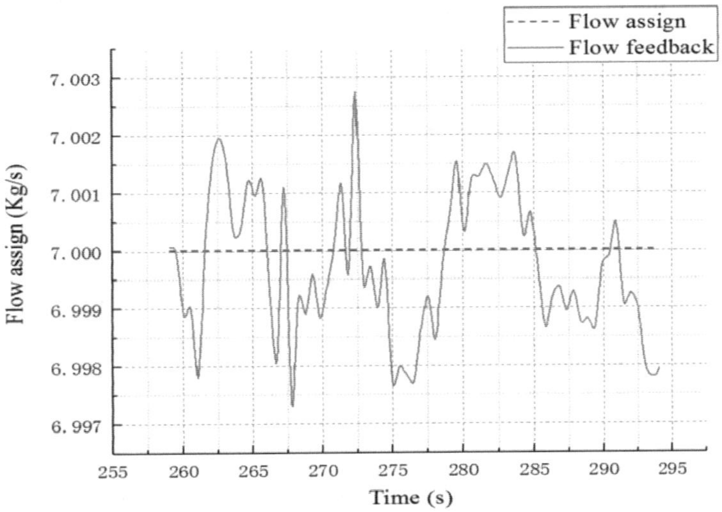

Figure 8. Steady flow regulation curve.

6. Conclusions

(1) The design of the high-pressure air supply system is successful. It has the characteristics of a wide flow regulation range, high control accuracy, and good dynamic characteristics. At present, it has been successfully applied to many wind tunnel tests, such as tests on inlets and dynamic simulation, which provide reliable experimental data and strong support for the design and optimization of inlets and aerodynamic layouts in China.

(2) The two-stage control scheme is used to comprehensively utilize the accurate pressure regulating the ability of the high-pressure relief valve, the flow expansion ability of the PCM digital valve, and the flow control ability of the needle valve in the middle area, which can effectively compensate for the shortcomings of the single traditional control valve, PWM digital valve, and PCM digital valve.

(3) The regulating interval of the needle valve is limited to 30% ~ 80%, which belongs to the optimal regulating interval of regulating valves and can avoid the adverse effects of the valve in the two

transitional stages of the opening moment and full opening. In addition, the equal percentage flow characteristics can ensure high flow control accuracy at different valve core positions.

(4) It is not necessary to consider the influence of the response time difference of different types of solenoid valves and throat machining accuracy. By continuing to increase the digital valve position, the flow adjustment range can be further widened, which has great application prospects.

References

[1] Li Cong, Xu Tiejun. Calibration and test for ejector and TPS [J]. Journal of Experiments in Fluid Mechanics, 2006, 20(1): 17-22.
[2] Wang Xunnian, Wu Chaojun. Investigation on test technique of fighters with propulsion simulation in a low speed wind tunnel [J]. Journal of Experiments in Fluid Mechanics, 2011, 06(3): 46-49.
[3] Xu Ou, Influence of control valve flow characteristics on the system control quality [J]. Journal of Changchun Institute of Technology, 2009, 10(2): 27-30.
[4] Yang Jiwei. Analysis of characteristics of adjustable ratio of regulator [J]. Valve, 2000(5): 11-14.
[5] Xu Yangzeng, Chen Weining, et al. Hydraulic High-speed On-off Valve and Its Application on Industry [J]. Hydraulics Pneumatics and Seals, 2011: 210-221.
[6] Wang Xuanyin, Lin Jianya. A study of pneumatic position servo system with pulse code modulation [J]. Journal of Zhejiang University, 1997, 31(4): 525-532.
[7] LIN Yizhong, ZENG Dele, FENG Zhe. Simulation Analysis of Dynamic Characteristics for High-speed Switch Electromagnetic Ball Valve Based on PWM [J]. Machine Tool & Hydraulics, 2014, 42(3): 152-154.
[8] Yu Juntao, Jiao Zongxia, Wu Shuai. Design, Simulation and Test of High-flow High-speed On/Off Valve Driven by Piezoelectric [J]. Journal of Mechanical Engineering, 2020, 56(18): 226-234.
[9] Cao Wenbin. Research of the flow performance and dynamic characteristics in control valve [D]. Lanzhou University of Technology, 2013: 28-29.
[10] Wu Mingqing, Liu Qinghua. Maximum allowable back pressure ratio of the critical flow Venturi-type nozzle [J]. Modern Measuring Instruments and Techniques, 2012, 10(6): 76-78.
[11] KANG Hongming, HUO Guo, CHENG Lei. Flow regulating device based on PCM digital valve and needle valve [J]. Chinese Hydraulics & Pneumatics, 2022, 46(4): 139-146.

Simulation and experimental study of die-casting pickup robot

Chaoda Chen[1], Zhisheng Lv[1,2], Zhuoyuan Yu[1], Biaoqing Xu[1], Junxian Xie[1] and Yayuan Luo[1]

[1]School of Naval Architecture and Ocean Engineering, Guangzhou Maritime University, Guangzhou, 510725, China

[2]Corresponding author's e-mail: 28477850@qq.com

Abstract: To improve the efficiency of die-casting machine pickup, the combination of SolidWorks and RobotStudio is proposed to establish a simulation workstation. In this paper, the key mechanism analysis and design of the die-casting machine pickup robot are used to determine the robot body and end-effector. SolidWorks three-dimensional software is used for die-casting pickup robot's overall mechanical structure and workstation peripheral equipment modeling. The RobotStudio software is used for the corresponding model planning layout and Smart components to write to achieve die-casting robot workstation efficient pickup work program writing and simulation debugging. Experimental results show that the die-casting machine pickup robot completed the pickup behavior. The system has the advantages of fast response speed, high positioning accuracy, and reliable operation, effectively improving the efficiency of the production line.

1. Introduction

China's small and medium-sized die-casting production relies on domestic equipment, and large-scale die-casting production relies on imported equipment. Technology content is low, the degree of mechanization and automation is low, and die-casting efficiency is low. The use of a die-casting pickup workstation model can effectively solve the above problems. Therefore, the establishment of a die-casting pickup workstation model for improving the degree of automation and mechanization of die-casting plants is important significance [1-5].

Taking the welding robot as the research object, in Sun et al.'s study [4], the welding robot workstation was built by using the SolidWorks modeling function and RobotStudio simulation function. Experiments show that the simulation design of the welding workstation is helps improve labor conditions and productivity. In Li's work [7], to ensure the stability of the automated assembly process and reduce the robot station offset, an automated assembly robot machining station tracking control method is designed. Test results show that the method can determine the best robot machining station and has good machining station optimization tracking ability and control effect. Zhang et al. [8] conducted research on multi-robot physical production line research investment, which is not easy to plan and has other issues. They proposed the combination of physical and simulation software to build a multi-robot virtual and real integration of programming methods; the results show that the multi-robot virtual and real integration of coordinated control and beat optimization method can solve the problem of existing hardware resource bottlenecks. It can further guide the design of the actual production line and the adjustment of the production plan. In Feng et al.'s work [8], to improve the design efficiency, functional analysis, and module division of industrial robots, the appropriate product series and module types need to be chosen according to a certain way of module assembly to complete the robot design, effectively shorten the robot design cycle and design difficulty. Xie and Lv [8] carried out the experimental simulation of the robotic arm for planning welding tasks to study the motion performance of the IRB4600 robotic arm modified into a spot welding robotic arm and its working domain, which verifies the feasibility and validity of the design of robotic arm modification and motion planning.

In this paper, the end of the packaging production line die-casting pickup work is researched. SolidWorks is used to establish based on the IRB4600 robot die-casting pickup robot body, end-effector, and workstations. RobotStudio modeling and die-casting robot workstations are used for programming and simulation simulation. The laboratory die-casting robot platform is used for die-casting robot pickup experiments. The data is collected and analyzed to provide a theoretical basis for the actual die-casting workstation in the enterprise.

2. Key mechanism design of die-casting pickup robot

In the die-casting pickup robot hardware system, the choice of the robot body and the reasonable design of the end-effector can improve the working efficiency as well as reliability and stability. It can reduce the breakage rate in the transportation process, which is a key technical factor in logistics and transportation.

2.1. Die-casting pickup robot body selection

In this paper, the IRB4600 robot is selected, which is a 6-axis + 3 external axes to satisfy the need for flexibility, followed by a working range of up to 2.55 m, which allows access to the inside of the die-casting machine for picking parts. The characteristics of the IRB4600 robot are shown in Table 1.

Table 1. Characteristics of the IRB 4600 robot.

Specificities	Clarification
Short cycle	Adopting an innovative optimized design, the body is compact and lightweight, and the speed reaches the highest of its kind, ultra-fast running speed, cycle time can be shortened by 25% compared with the peer.
Large-scale	The large operating range of the IRB 4600 allows for comprehensive optimization of reach, cycle time, auxiliary equipment, and much more.
Well-protected	The Casting Expert system is IP 67 rated and includes a corrosion-resistant coating and rust-resistant mounting flanges.
Adapt according to needs	The excellent range of IRBP translators, IRBT trajectory motion systems, and motors enhance the adaptability of the IRB 4600 to the target application in every way.

2.2. End-effector design for die-casting pickup robots

An end-effector is a device specially designed for completing special operations. Nowadays, end-effectors are widely used in painting, welding, handling, measuring, cutting, and so on. The end-effector of a die-casting pickup robot is a mechanism mounted on the robot arm or mobile equipment to enable it to perform the required function. The end-effector of this project adopts the clamp type reclaimer based on the E22 CJES-7 reclaimer. The design has the following advantages: 1) simple structure, easy to install and maintain, low cost. 2) high adaptability to the items: it can satisfy the gripping of our cylindrical parts, and the process of realizing it is also easier. It is a kind of mechanical material most suitable for gripping the die-castings. 3) strong anti-slip: the clamping jaws of the E22 CJES-7 reclaimer have four similar rectangular jaw teeth, which are made of rubber material with certain protective clamping functions. The claw has four similar rectangular claw teeth.

The claw teeth are made of rubber material, with a certain degree of protection of the clamping function, and the clamping claw as a whole is a linkage structure through the movement of the intermediate linkage to control the clamping claw opening and closing. 4) small angle to produce a large angle: based on the E22 CJES-7, reclaiming the hand of the robotic claw can be a very small angle to produce a large angle. The picking work can be accomplished with a small amount of opening. A copycat gripper-type picker based on the E22 CJES-7 picker is shown in Figure 1.

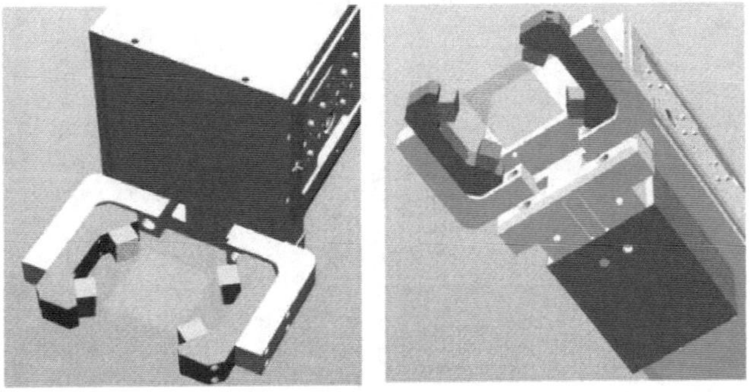

Figure 1. Model.

3. RobotStudio-based modeling simulation analysis

3.1. Die-Casting pickup robot workstation construction

RobotStudio software is used to build a die-casting pickup robot workstation modeled after IZM DM180, which has the following basic components: robot body, end-effector, conveyor belt, frame holder, safety guardrail, control cabinet, etc., as shown in Figure 2.

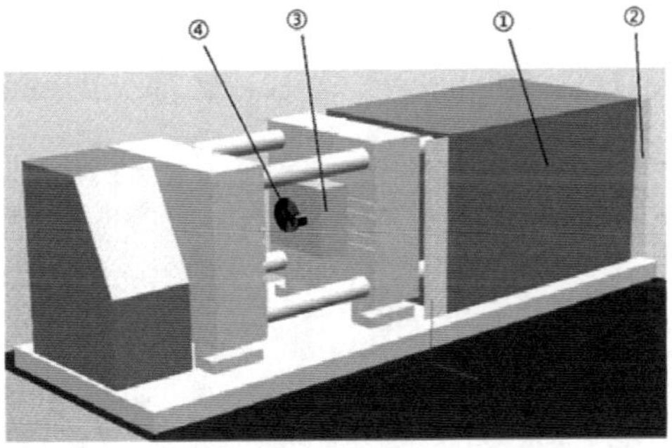

①Body (casing) ②Shield ③Dynamic mold ④Die-castings

Figure 2. Die-casting machine model and components.

The various work equipment of the die-casting pickup robot includes the pickup robot, the cooling table, the conveyor, the die-casting machine, the scrap box, the safety fence, and the control cabinet. About the top view of the workstation and each composition of the workstation.

The workstation covers a small area, saving space resources, and its control cabinet is located outside the workstation fence. This arrangement is designed to facilitate the staff to operate the control cabinet more conveniently, without having to enter the workstation inside, with security.

The face sensor PlaneSensor is inserted in the smart component in the RobotStudio software in the center of the robotic hand claw to detect whether it is in the position to be grasping the object block. When the face sensor PlaneSensor detects the object block, it sends out a grasping signal to the robotic hand claw. The robotic hand claw is closed to clamp the cylindrical residue of the casting, and then the object block will be removed from the static mold of the die-casting machine, completing the fetch work.

3.2. Workstation program code and debugging
The programmed points are shown in Figure 3 below.

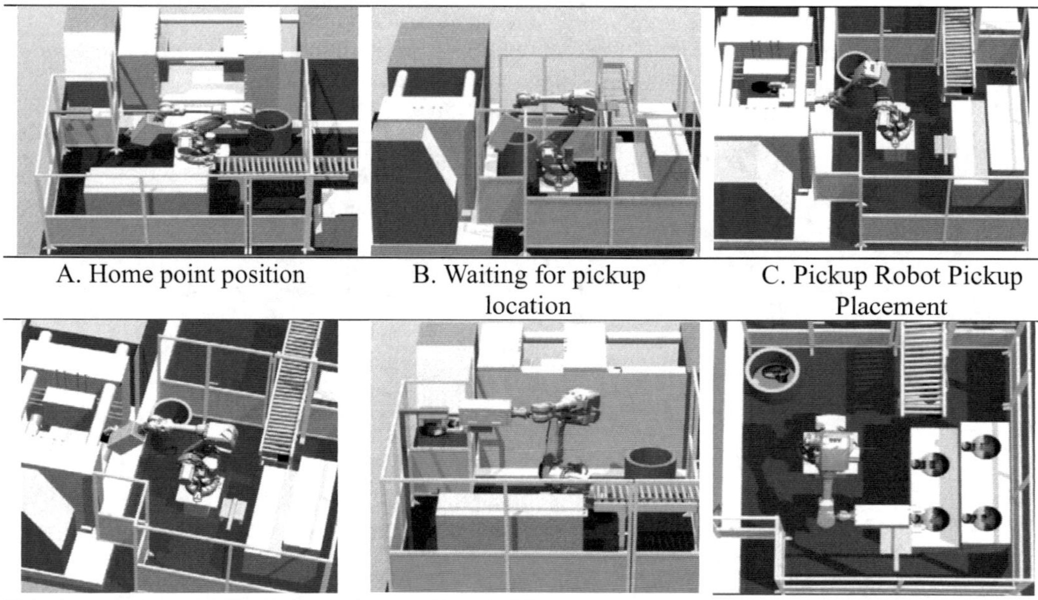

A. Home point position | B. Waiting for pickup location | C. Pickup Robot Pickup Placement

D. Exit die-casting machine position | E. Quality testing locations | F. Cooling table placement

Figure 3. Program setting points.

After the program is started, whatever state the robot is in, it will preferentially return to the Home point position. After the robot returns to the Home point, the pickup routine will be executed, and the pickup robot will move to the waiting pickup position. When the die-casting machine completes the die-casting work, and the safety gate opens, a safety gate open signal is sent, after which the pickup robot begins to enter the appropriate position between the two molds to pick up the parts. When the pickup robot is in the pickup position, the surface sensor on the end gripper will recognize whether there is an object or not; if there is an object, the robotic gripper will pick up the casting, and then the pickup robot exits the die-casting machine, along a certain path movement to the appropriate position near the quality inspection equipment. The die-casting quality inspection is completed. The workstation will issue a qualified or not signal; if the quality is qualified, the pickup robot will be placed in the die-casting cooling table; if the quality is not qualified, the pickup robot will be placed in the scrap box blocks. After the four positions of the cooling table are full, the workstation sends out a signal that the cooling table is full, and the pickup robot will clip and place the earliest die-castings placed on the cooling table to the conveyor chain. Then, the conveyor chain will transport the die-castings to the specified position, and so on. Each time the pickup robot places new die-castings in the cooling table, the earliest die-castings will be clipped and placed on the conveyor chain from the cooling table.

4. Analysis of experiments and studies
The end-effector used in the die-casting robot experiment platform is a gripper. There is a conveyor belt and a platform on each left and right side to simulate a die-casting conveyor belt in a real working environment. Due to the limitation of using the gripper as the end-effector, the palletization pattern of the blocks palletized on the platform is somewhat different from reality. Experiments were conducted using 40 mm x 40 mm x 20 mm cubes instead of die-castings.

The main robot positions in this experiment are pHome position, pickup position, quality inspection position, cooling table position, conveyor position, cooling table block placement position, conveyor conveying blocks, and the rotary position added due to the shape characteristics of the blocks are different from the die-casting designed in this paper. There are 6 key points in this experiment, and each point is shown in Figure 4.

A. PHome points

B. Pickup location

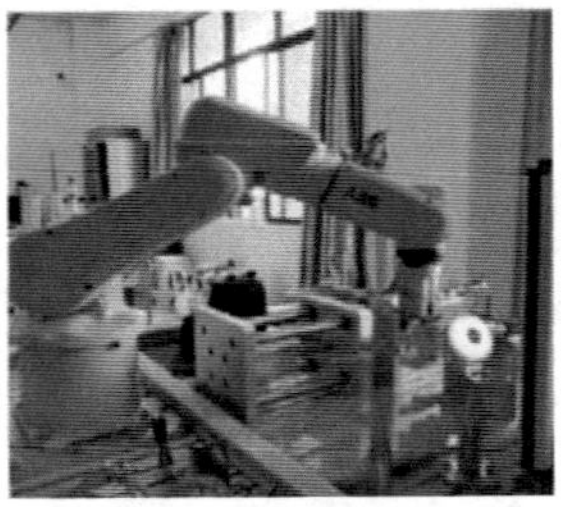

C. Quality testing locations

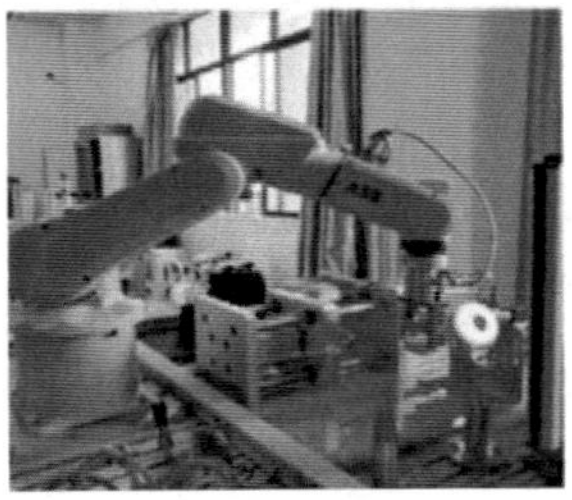

D. Rotational position

E. Cooling table position

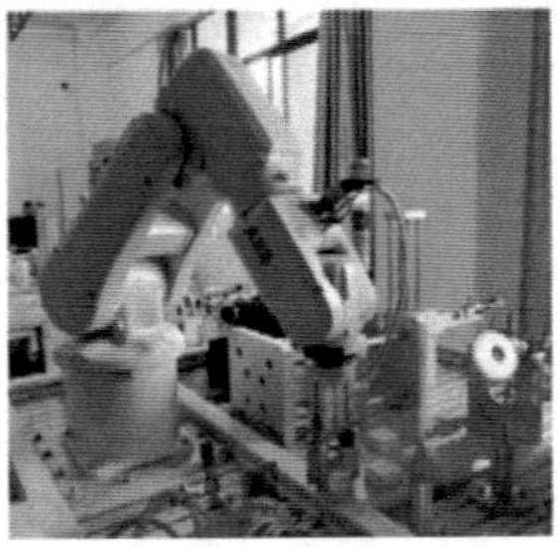

F. Conveyor position

Figure 4. Experimental site location.

Since the workstation is not equipped with a conveyor module, a suitable location is used to simulate the conveyor location, and after confirming the above points, the experiment is started. After first recovering to the PHome point position, the interval position is reached. After completing the pickup, the object is moved to the quality inspection position, after which the object is placed on the cooling table. The process of placing the object on the cooling table utilizes the offs offset function to offset the cooling table position in the x and y directions. Only one point position of the cooling table position needs to be determined to complete the positional localization of the cooling of the four die-castings. After the conveyor palletizes the four blocks, the pickup robot returns to the mechanical origin, and the experiment is complete.

5. Conclusion

In this paper, a simulation and experimental study of a RobotStudio-based die-casting pickup robot is carried out, and the following conclusions are drawn:

(1) For the key mechanism of die-casting robot design, the IRB4600 robot is selected as the die-casting robot body, and the clamp-type pickup hand is designed as the die-casting pickup robot end-effector. The experimental results show that the die-casting robot response speed is fast, and work stability is high.

(2) The 3D model is imported into RobotStudio's workstation, and the code is written and debugged using the Smart component for the IO connection, making the modeling workstation convenient and accurate.

(3) Experimental testing using the die-casting robot R&D platform is able to complete a 4x1 stacking pattern, i.e., the molds are placed as 1 layer with 4 items per layer. The equipment runs stably, verifying the feasibility of the program.

(4) The design of this die-casting workstation provides a theoretical basis and test platform for the actual die-casting workstation of the enterprise and can also plan the optimal die-casting path, reduce the cost of production line design and debugging, and provide a safe production environment for workers. Through the die-casting pickup robot workstation simulation design, it is of great significance in improving the quality and efficiency of die-casting.

Acknowledgments
This work was supported by Foundation of No.2023CJRHJD002 and No.2023ZDZX2051.

References

[1] LI Benwang, WU Zengping, WANG Maolin, et al. A review of the research on performance evaluation system of industrial robots[J]. Modern Manufacturing Engineering,2020 (09):149-155.
[2] Hou Yuanshao. Model construction and evolution of die-casting robot based on digital twin[J]. Journal of Anyang Institute of Technology,2023,22(04):68-72.
[3] HUANG Bo, ZHAO Fei, WANG Jia, et al. Research on automated assembly systems based on machine vision[J]. Manufacturing Technology and Machine Tools,2022(09):117-122.
[4] DONG Boatship, ZHANG Ji, WANG Danfeng, et al. Research on the design of elastic finger joint structure of high simulation anthropomorphic manipulator[J]. Mechanical Design and Manufacturing,2023(09):253-257.
[5] Zheng Jinsong. Design and application of control system of warehouse logistics handling robot[J]. Packaging and Food Machinery,2023,41(3):59-64.
[6] SUN Zengguang, WANG Shijun, MENG Lingjun et al. Simulation design of welding robot workstation based on RobotStudio[J]. Machine Tools and Hydraulics,2020,48(05):29-33.
[7] Li Xiaodong. Tracking control method for machining stations of automated assembly robots[J]. Manufacturing Automation,2021,43(12):74-77.
[8] ZHANG Xiaofang, CHENG Ruilong, ZHANG Ziyu. Design and realization of multi-robot virtual-real integration collaborative workstation[J]. Modern Manufacturing Engineering, 2022(10):27-33.
[9] FENG Huijuan, MIAO Qing, FAN Shengqiu et al. Modular parametric design of industrial robot mechanical structure[J]. Machine Tools and Hydraulics,2019,47(21):60-63.
[10] Xie Xiaozheng, Lv Weizian. Motion performance of 6R welding robot arm and its workspace simulation[J]. Mechanical Design and Manufacturing Engineering,2023,52(08):1-6.

Design of fault degree diagnosis algorithm for circuit breaker spring based on fuzzy clustering

Xibo Zhang[1,3], Jianguo Ying[1], Junhua Wang[2], Rongwei Zhang[2], Zhou Hong[2] and Haibo Bi[2]

[1]Ningbo Xinsheng Medium Voltage Electrical Appliance Co., Ningbo 315000, China
[2]Ningbo Yongyao Power Investment Group Co., Ningbo 315000, China

[3]Corresponding author's e-mail: nbnve51109770@126.com

Abstract. This paper proposes a design of a fault diagnosis algorithm for circuit breaker springs based on fuzzy clustering. The features of the fault state signal are extracted by combining the methods of Intrinsic Time-Scale Decomposition and Singular Spectrum Analysis. Using the fuzzy clustering method, this study classifies circuit breaker spring faults, extracts fault features, and achieves fault degree diagnosis. The experimental results show that the algorithm has high accuracy in fault diagnosis.

1. Introduction

A circuit breaker is an automatic switching device used to protect power systems and electrical equipment [1]. It can detect and disconnect situations where the current exceeds the rated value or malfunctions in the circuit to prevent damage to the circuit and equipment. When the current exceeds the rated value or a fault occurs, the protection unit will detect an abnormal situation and send a signal to the triggering mechanism [2]. The triggering mechanism transmits the triggering signal to the main body of the circuit breaker, enabling it to quickly disconnect the circuit to protect electrical equipment and systems from overload, short circuits, and other faults [3]. At the same time, the arc extinguishing system will eliminate the arc generated when the contact is disconnected. Circuit Breaker Spring is an important component commonly used for mechanical operation control in circuit breakers. It is mainly used to provide the force and rebound force for the opening and closing operation of the circuit breaker, ensuring the normal operation of the circuit breaker. Reasonable maintenance of circuit breaker springs helps to ensure normal operation and safe and reliable electrical protection of the circuit breaker. When the circuit breaker spring fails, it is difficult to control the opening and closing of the circuit breaker, leading to a threat to the safety of electrical equipment. Therefore, it is necessary to diagnose the degree of circuit breaker spring failure in order to determine the necessity of circuit breaker maintenance.

Tahvilzadeh et al. [4] put forward the simulation of a spring-driven operating mechanism, and used mechanical analysis software to obtain the stroke curve as the main signal for diagnosing mechanical faults. They established the model and verified it, solved the problem of data acquisition, and improved the accuracy and efficiency of fault detection. Shi et al. [5] put forward a hybrid method of spring energy storage state identification, which uses the Glaman angle field to represent the evolution process of dynamic characteristics. They combined the convolution block attention module and residual network to identify the state, thus realizing rapid monitoring. However, there are still some challenges and shortcomings in the current research. For example, the failure modes of circuit breaker springs are diverse and influenced by many factors, so these complexities need to be considered in the diagnosis process. In addition, the accuracy of data acquisition and processing, as well as the stability and reliability of the algorithm are also problems to be solved in current research.

Fuzzy clustering, as an unsupervised learning method, is applied in the diagnosis of circuit breaker spring fault severity [6]. It can automatically learn and adapt to different fault modes and features based on the characteristics of the data. This gives it flexibility when dealing with complex circuit breaker spring faults. The fuzzy clustering algorithm can handle incomplete or noisy data, reducing the requirements for data quality. In the diagnosis of circuit breaker spring faults, due to various interferences in the data collection process, the tolerance of fuzzy clustering can make the diagnostic results more stable and reliable.

2. Feature extraction of 1 circuit breaker spring fault state

With the help of Internet of Things technology, multiple sensors are installed on circuit breakers [7] to monitor the fault status of circuit breaker springs and collect fault status signals. This signal has the characteristics of nonlinearity, low signal energy, and being easily submerged in noise, which is not conducive to monitoring the degree of circuit breaker spring failure. Therefore, this paper proposes a method that combines Intrinsic Time-Scale Decomposition (ITD) and Singular Spectrum Analysis (SSA) to extract the features of fault state signals and reconstruct them, thereby obtaining effective signal features of the fault degree of circuit breaker springs. The specific process is:

(1) The circuit breaker spring fault status signal is set as $c(y)$, which is a time series signal with a length of M, and ITD decomposition is performed on it. The calculation formula is as follows:

$$c(y)=\sum_{i=1}^{q} J_i(y)+Z(y) \tag{1}$$

In the formula, q represents the nesting dimension, $q \le \frac{M}{2}$; $J_i(y)$ represents the i-order inherent rotation component in ITD decomposition; $Z(y)$ represents the trend component of the initial circuit breaker spring fault status signal.

By using the above formula to perform frequency domain analysis on the inherent rotation component $J_i(y)$, components consistent with the characteristic frequency information are obtained. Furthermore, components of the same frequency component are collected together to obtain the signal $u_t(y)$ to be analyzed and processed.

(2)Using the SSA algorithm, the signal $u_t(y)$ to be analyzed and processed is reconstructed into a $q \times m$ dimensional phase space to obtain a trajectory matrix F_c composed of m coordinate points in the phase space:

$$F_c=\begin{bmatrix} c_1 & \cdots & c_m \\ \vdots & \cdots & \vdots \\ c_{1+(q-1)_y} & \cdots & c_{m+(m-1)_y} \end{bmatrix} \tag{2}$$

The covariance matrix Y_c of trajectory matrix F_c is further calculated:

$$R=I\Lambda B^{A} \tag{3}$$

Where I and B respectively represent two orthogonal left and right singular matrices; Λ represents the singular value of the matrix, and $\Lambda = diag\left(r_1, r_2, \ldots, r_q\right)$ represents the eigenvector $A = \left(A_1, A_2, \ldots, A_q\right)$ corresponding to each singular value. Therefore, the original time series C composed of q th order components is:

$$C = R_1 + R_2 + \cdots + R_q \tag{4}$$

(3) Frequency domain analysis is performed on all components R_i to obtain the component R_j where the characteristic frequency of the circuit breaker spring fault signal is located. According to the principle of inverse reconstruction in singular spectrum analysis, the signal is superimposed and reconstructed to obtain a new matrix U. The time series after feature extraction is obtained, which is the feature of the circuit breaker spring fault signal:

$$U = \sum R_j, 1 < j < q \tag{5}$$

3. Fault degree diagnosis algorithm based on fuzzy clustering algorithm

The fuzzy clustering algorithm used in this paper is the short-term fuzzy C-means clustering algorithm, which is further optimized on the basis of the fuzzy C-means A algorithm (FCMA) [8]. The algorithm first performs segmented processing on the circuit breaker spring fault signal by frame [9]. Then, it selects the characteristics of the circuit breaker spring fault signal in each segment, selects appropriate parameters, and implements spatial mapping [10]. Finally, the fuzzy C-means clustering algorithm is used to implement pattern classification and recognition of the fault signal features of the circuit breaker spring, achieving the diagnosis of fault severity.

On the basis of obtaining the fault signal characteristics of the circuit breaker spring, this method uses this algorithm to classify and diagnose the fault degree of the signal features. The specific steps are as follows:

(1) Feature parameter space mapping of circuit breaker spring fault signal

The original time series are segmented into frames. The average and standard deviation of the characteristic parameters of the circuit breaker spring fault signal for each segment are calculated. The spatial mapping of the characteristic parameters of the circuit breaker spring fault signal is completed, and feature vectors are generated described by Q indicators:

$$C_i = \left(C_{i1}, C_{i2}, \ldots, C_{iQ}\right) \tag{6}$$

To avoid the phenomenon of indicator data being too small and submerged, each indicator is regularized:

$$C'_{ij} = \left(C_{iQ} - \min_i C_{iQ}\right) / \left(\max_i C_{iQ} - \min_i C_{iQ}\right) \tag{7}$$

From this, we obtain the normalization matrix $C = \left(C'_{ij}\right)_{M \times Q}$, where each row is used to describe the fuzzy set of the classification object on the indicator set. The expression is as follows:

$$C_i = \left(C'_{i1}, C'_{i2}, \ldots C'_{iQ}\right), i = 1, 2, \ldots, w \tag{8}$$

(2) Sample set construction and parameter initialization

A sample set of $C=\{C_1,C_2,...,C_w\}$ circuit breaker spring fault test signals is established, where w represents the number of signal samples, $V\left(2\le V\le w\right)$ describes the clustering number, and the number of iterations is $l=1$. On this basis, the signal sample set C is divided into V categories, described by $C_1,C_2,...,C_V$; V initial clustering centers are selected and described with $\delta_1(l),\delta_2(l),...,\delta_V(l)$.

(3) Defining modal bundles

The distance between all circuit breaker spring fault signal samples and each cluster center is calculated, and the objective function is determined as:

$$K(I,B)=\sum_{k=1}^{M}\sum_{i=1}^{V}(o_{ji})^{\gamma}(f_{ji})^{2} \tag{9}$$

In the formula, $O=\left[o_{ji}\right]$ represents the fuzzy classification matrix; $o_{ji}\in[0,1]$ represents the membership degree of circuit breaker spring fault signal sample C_j to the i th type signal sample set; $\gamma\in[0,\infty)$ represents the weighted index; $f_{ji}=\left\|C_j-q_i(l)\right\|$ represents the distance between signal sample C_j and the center of the i th type signal sample; $K(O,B)$ represents the sum of squared weighted distances between each signal sample and the clustering center, and C_j represents the weight of the signal sample.

The main criterion for clustering is to achieve the minimum value of the objective function $K(O,B)$. Firstly, it is necessary to construct a Lagrange function to calculate the membership function o_{ji}. The calculation method is as follows:

$$Z(\mu,o_{ji})=\sum_{i=1}^{V}(o_{ji})^{\gamma}(f_{ji})^{2}+\mu\left(\sum_{i=1}^{V}o_{ji}-1\right) \tag{10}$$

$$o_{ji}=1/\sum_{i=1}^{V}(f_{ji}/f_{jl})^{\frac{2}{\gamma-1}} \tag{11}$$

(4) The modal bundle is updated, the cluster center $\delta_i(l+1)$ is recalculated, and the following equation is derived:

$$\delta_i(l+1)=\sum_{j=1}^{M}\left[(o_{ji})^{\gamma}C_j\right]/\sum_{j=1}^{M}(o_{ji})^{\gamma} \tag{12}$$

If $i\in\{1,2,...,V\}$ and $\delta_i(l+1)\ne\delta_i(l)$, then $l=l+1$, Step (3) is returned to repeat clustering. On the contrary, clustering ends until the optimal clustering result is achieved, and the recognition of the fault signal characteristics of the circuit breaker spring is completed, achieving the diagnosis of the degree of the circuit breaker spring fault. The overall algorithm flow is shown in Figure 1.

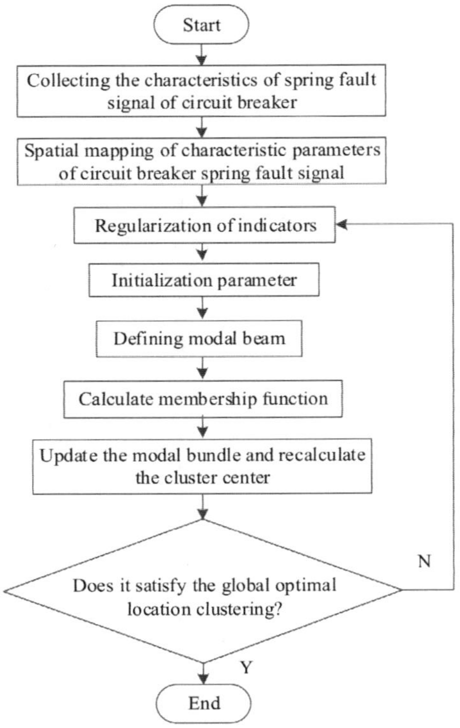

Figure 1. Overall flow chart of the algorithm.

4. Experimental results and analysis

MATLAB simulation software is used to simulate the working process and parameter changes of circuit breakers and a simulation environment is built to simulate circuit breaker spring faults. An AC power supply with an input voltage of 220 V is the power supply. A load simulator or current source is used to simulate the load and current of the circuit breaker. Current sensors, voltage sensors, temperature sensors, etc., are used to obtain relevant parameters. The experimental simulated circuit breaker spring is shown in Figure 2.

Figure 2. Actual diagram of circuit breaker spring.

Corresponding data collection and recording systems are developed using the LabVIEW programming language. The NumPy library in Python is used for data manipulation and visual analysis. Among them, the current sensor has an accuracy of 0.1 A, the voltage sensor has an accuracy of 0.1 V, the temperature sensor has an accuracy of 0.1°C, the simulation software has a simulation time step of 0.001 seconds, and the data collection frequency is 100 Hz. In the fuzzy clustering algorithm, the number of clusters is set to 3, the fuzzy index is set to 2, and the maximum number of iterations is 1000. A threshold is set to determine whether the algorithm is convergent, which is 0.001.

According to the above experimental environment, there are five types of faults: spring deformation or fracture, spring relaxation or loss of tension, uneven spring compression force, insufficient spring rebound force, and spring misalignment or jamming. Each type of fault occurs randomly. The operating data of the circuit breaker is collected within 24 hours of operation, which is used as the dataset. It is randomly divided into 60% as the training set to train the method in this paper, and the remaining 40% as the testing group. The testing group data is randomly divided into 5 groups for performance verification.

The accuracy of diagnosing the degree of circuit breaker spring failure under different methods is shown in Table 1:

Table 1. Diagnosis accuracy of circuit breaker spring fault severity using different methods/%.

Group	The method of this paper	Methods in [4]	Methods in [5]
1	97.7	92.3	90.7
2	96.8	93.6	88.6
3	98.1	91.1	89.7
4	96.3	90.8	86.4
5	97.2	92.7	87.5

Analysis of Table 1 shows that different methods can diagnose the degree of circuit breaker spring fault, but the accuracy of traditional methods is relatively low, with an average of 92.1% and 88.58%, respectively. However, the accuracy of the circuit breaker spring fault degree diagnosis proposed in this paper remains stable at a high level, with an average of 97.22%. This indicates that the method proposed in this paper has certain reliability in identifying the degree of circuit breaker spring failure.

To further verify the accuracy of the fault diagnosis method proposed in this paper, the deviation during the spring opening and closing process was used as the validation indicator to compare the error of the deviation diagnosis values under different methods, as shown in Table 2:

Table 2. Deviation diagnostic values for the degree of circuit breaker spring failure using different methods/mm.

Group	The deviation is the actual value	Deviation diagnosis values of the method used in this paper	Deviation diagnostic values of methods in [4]	Deviation diagnostic values of methods in [5]
1	2.5	2.7	3.5	3.3
2	3.2	2.6	3.3	3.4
3	3.4	2.9	3.6	3.5
4	2.8	3.1	3.1	2.9
5	3.4	3.3	2.7	2.8

By analyzing Table 2, it can be seen that the diagnostic range of the deviation of the method in this paper for the degree of circuit breaker spring failure is within 0.6 mm, which is much smaller than other methods. This indicates that the method proposed in this paper can effectively diagnose the degree of circuit breaker spring fault and ensure the operation detection of circuit breaker springs.

Based on the above experimental results, it can be concluded that the method proposed in this paper exhibits high accuracy and precision in diagnosing the degree of circuit breaker spring failure. This plays an important role in ensuring the operation and detection of circuit breaker springs, providing engineers and maintenance personnel with a reliable means of assessing the degree of failure.

Journal of Physics: Conference Series **2724** (2024) 012009 doi:10.1088/1742-6596/2724/1/012009

5. Conclusion

This paper successfully designs a fault degree diagnosis algorithm for circuit breaker springs by using the fuzzy clustering method, combined with inherent time scale decomposition and singular spectrum analysis. This algorithm not only has high accuracy but also can effectively extract and identify the characteristics of circuit breaker spring faults, providing a more reliable guarantee for the stable operation of the power system. In the future, we will continue to optimize the algorithm, improve its diagnostic accuracy and efficiency, and explore more application scenarios in order to make greater contributions to the development of related fields.

References

[1] Zhang G, Gao F, Xian Y 2021 Topology and Simulation Analysis of Novel DC Microgrid Hybrid Circuit Breakers. Computer Simulation, 38(3): 206-210+315.

[2] Hsu C C, Frusque G, Muratovic M 2022 Vacuum Circuit Breaker Closing Time Key Moments Detection via Vibration Monitoring: A Run-to-Failure Study. 2022 IEEE International Conference on Systems, Man, and Cybernetics (SMC). IEEE, 2022: 254-260.

[3] Terada M, Urai H, Ishikawa M, et al 2022 Method of Evaluating Dielectric Interruption Capability of High-voltage Circuit Breaker. IEEE Transactions on Power Delivery, 47(1): 3885-3892.

[4] Tahvilzadeh M, Aliyari-Shoorehdeli M, Razi-Kazemi A A 2023 Model-Aided Approach for Intelligent Fault Detection System for SF 6 High-Voltage Circuit Breaker with Spring Operating Mechanism. IEEE Transactions on Power Delivery, 38(5):3356-3365.

[5] Shi Y, Zhou Y, Ren Y, et al 2023 A Hybrid Method for Identifying the Spring Energy Storage State of Operating Mechanism in Circuit Breakers. IEEE Transactions on Instrumentation and Measurement, 72(1): 1-9.

[6] Farsoni S, Simani S, Castaldi P 2021 Fuzzy and neural network approaches to wind turbine fault diagnosis. Applied Sciences, 11(11): 5035.

[7] Nath A G, Udmale S S, Raghuwanshi D 2021 Improved structural rotor fault diagnosis using multi-sensor fuzzy recurrence plots and classifier fusion. IEEE Sensors Journal, 21(19): 21705-21717.

[8] Elshenawy L M, Chakour C, Mahmoud T A 2022 Fault detection and diagnosis strategy based on k-nearest neighbors and fuzzy C-means clustering algorithm for industrial processes. Journal of the Franklin Institute, 359(13): 7115-7139.

[9] Rajan M S, Dilip G, Kannan N 2023 Diagnosis of fault node in wireless sensor networks using adaptive neuro-fuzzy inference system. Applied Nanoscience, 13(2): 1007-1015.

[10] Xiong Y, Cheng J, Feng Z 2022 Transformer fault diagnosis based on an improved sine cosine algorithm and BP neural network. Recent Advances in Electrical & Electronic Engineering, 15(6): 502-510.

Design of ammonia leakage online monitoring system based on tunable diode laser absorption spectroscopy

Ribo Jia [1,2], Junbo Gao [1], Shenhao Yang [1] and Junna Cheng [1]

[1]Purification Equipment Research Institute of CSSC, Handan, China

[2] Corresponding author's e-mail: heartlawyer@163.com

Abstract: A real-time online ammonia volume fraction monitoring system based on tunable diode laser absorption spectroscopy (TDLAS) is designed to meet the demand for ammonia gas leakage monitoring in ammonia-fueled ships. The system uses a 1, 512 nm distributed-feedback (DFB) laser as the light source and a 5 m high-temperature Boss-Gas-Cell as the absorption cell. By applying a 4 Hz scan signal and a 31.4 kHz modulation signal to the laser current, the effective absorption signal of the gas is transferred to the high-frequency band, then the second harmonic (2f) signal about gas volume fraction is extracted by a digital lock-in amplifier from the gas absorption signal, and ammonia volume fraction can be calculated from the amplitude of 2f signal. The test results show that the ammonia detection accuracy in the range of 0-500 ppm can reach ±1% FS, the response time is less than 30 s, and the system measurement repeatability is less than 1%, which meets the needs of ammonia leakage monitoring in ammonia-fueled vessels.

1. Introduction

As a zero-carbon energy carrier, ammonia, with the advantages of high energy density, and convenience in manufacturing, storage, and transportation, can be used in fuel cells with the development of existing technologies. To reduce the climate impact of shipping, ammonia fuel has recently received a lot of attention [1]. However, ammonia is a kind of flammable, explosive, and toxic gas, and ammonia leakage will bring about safety problems such as poisoning and explosion. Therefore, designing a fast, safe, and effective ammonia leakage monitoring system is of great significance for the promotion and application of ammonia fuel.

Existing ammonia gas detection methods mainly include electrochemistry, non-dispersive infrared (NDIR), and tunable laser diode absorption spectroscopy (TDLAS), etc. Compared with the optical class of detection methods, the traditional electrochemical sensors of ammonia have poor immunity to interference, poor stability, and a short life span [2]. As one of the optical detection methods, the NDIR sensor has a simple structure and low cost, but the detection accuracy and detection limit are relatively poor, due to its broadband light source, and it only has good detection performance for specific gases, such as CO_2 and CH_4. TDLAS utilizes the narrow-band absorption characteristics of gases and offers good specificity and sensitivity necessary to investigate the concentration of gas, with a single-wavelength laser as the light source. Therefore, this paper designs an ammonia leakage monitoring system based on the TDLAS principle according to the requirements of China Classification Society's "Guidelines for the Application of Ammonia Fuel in Ships", which is characterized by high accuracy, fast response, and good repeatability and meets the ammonia leakage monitoring needs of ammonia-fueled vessels.

2. Theory

2.1. Principle

The basic principle of tunable laser absorption spectroscopy is Lambert-Beer law, which states that gas molecules have absorption properties for light of a specific wavelength. When a laser light with an intensity of I_0 and wavelength of λ is absorbed by the gas, its intensity becomes I. The mathematical relationship between them is as follows:

$$\frac{I(\lambda)}{I_0(\lambda)} = e^{-\alpha(\lambda)CL} \tag{1}$$

where C is the gas volume fraction, $\alpha(\lambda)$ is the gas absorption coefficient, and L is the optical path length. The gas absorption coefficient can be expressed as:

$$\alpha(\lambda) = S * \varphi(\lambda) \tag{2}$$

where S is the line intensity of the gas absorption spectrum and $\varphi(\lambda)$ is the normalized line shape function. For a given absorption line of a gas molecule, the absorption line strength S is only related to the gas temperature. The normalized line shape function $\varphi(\lambda)$ describes the broadening form of absorption spectral lines, which can be divided into natural broadening, collision broadening, and Doppler broadening, according to the reason of the formation of broadening. The collision broadening of molecules dominates at room temperature and normal pressure, and the line shapes are determined by the Lorentzian line shapes [3-6]. Therefore, after selecting the absorption spectral lines of the gas molecules, the gas volume fraction can be inverted according to Formula (1), by measuring the light intensity change before and after the absorption by gas and the optical length of the gas absorption.

2.2. Harmonic detection

The methods of detecting gas volume fraction using laser absorption spectroscopy can be categorized into direct absorption spectroscopy (DAS) method and harmonic detection method. The DAS method measures the attenuation of the laser light intensity before and after absorption by the gas by superimposing a low-frequency scanning signal on the laser driver current and obtaining the absorption peak signal of the gas. This method can directly obtain the gas concentration through the calculation of known parameters, without calibration, but it is very sensitive to system noise and has low detection sensitivity.

The harmonic detection method modulates the wavelength of the laser by superimposing a high-frequency modulation signal and a low-frequency scan signal on the laser driver current. This method shifts the gas absorption signal to the high-frequency band for processing, extracts the harmonic signal which contains the gas concentration information through phase-locked amplification technology, and removes system low-frequency noise to improve the detection sensitivity of the system [7-12].

3. System design

The ammonia leakage online monitoring system is mainly composed of the light source, absorption cell, sampling pump, control board, and serial port screen, and the system structure block diagram is shown in Figure 1. The control board generates the laser drive signal and modulates the laser wavelength, and the modulated laser is reflected several times in the absorption cell and absorbed by the target gas. The outgoing light is converted into the photo-current signal by the photoelectric detector and sent to the control board for processing, which extracts the harmonic signal, inverts the gas volume fraction, and is displayed on the serial port screen. The pump is used to extract the target gas to ensure the flow of gas in the absorption cell.

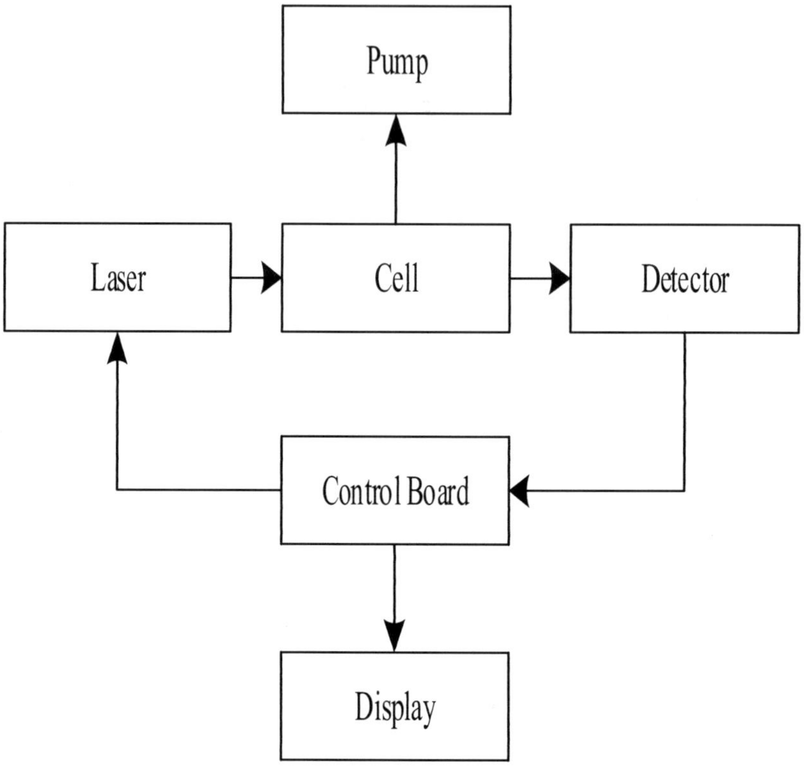

Figure 1. Block diagram of system structure.

3.1. Light source selection

The absorption peaks of ammonia in the entire infrared band can be found in the HITRAN database [13]. Among them, the absorption spectral line intensity of ammonia in the mid-infrared band is the largest, but the lasers in the mid-infrared band are expensive and bulky. The lasers in the near-infrared band have been commercialized on a large scale, which is cheap and easy to use. Although the absorption intensity is reduced, high-precision gas detection can also be achieved by increasing the optical path. Therefore, a 1, 512 nm DFB laser is selected as the light source for the ammonia leakage monitoring system.

3.2. Absorption cell selection

According to the system's detection range of 0-500 ppm and accuracy of ± 2% FS, as well as the absorption line strength of ammonia at 1, 512.2 nm, the minimum optical path of the gas absorption is approximately 1 meter. In addition, considering the strong adsorption of ammonia, to ensure the response time of the system, the absorption cell needs to be heated, so the 5-meter high-temperature Boss-Gas-Cell from Xuhai optoelectronics is selected as the absorption cell of the system. The optical path of this absorption cell is 5 meters. It is small in size, integrates the laser interface and photoelectric detector, and does not require light, which is also easy to use.

3.3. Design of the control board

The control board of the ammonia leakage online monitoring system includes generating a laser driver signal, controlling the temperature stability of the laser, collecting the photo-current signal from the detector, extracting the harmonic signal, inverting the gas concentration, and driving the serial screen.

The TDLAS control board from Weirui Technology is used to output the laser driver signal, collect the detector signal, and extract the second harmonic signal. The laser modulation parameters are debugged by its supporting software. The frequency of the saw-tooth scanning signal is set to 4 Hz, and the frequency of the sinusoidal modulation signal is set to 31.4 kHz. The drive circuit of the laser TEC module is designed based on the microcontroller of GD32F103xx and the MAX8521 chip to realize the setting and stability of the laser temperature. The second harmonic signal output by the TDLAS control board is collected by the ADC module inside the microcontroller. To ensure the stability of the sampling, a 100 Hz analog low-pass filter is added in front of the ADC input port, which effectively reduces the noise in the harmonic signal and improves the sampling accuracy. The circuit block diagram of the control board is shown in Figure 2.

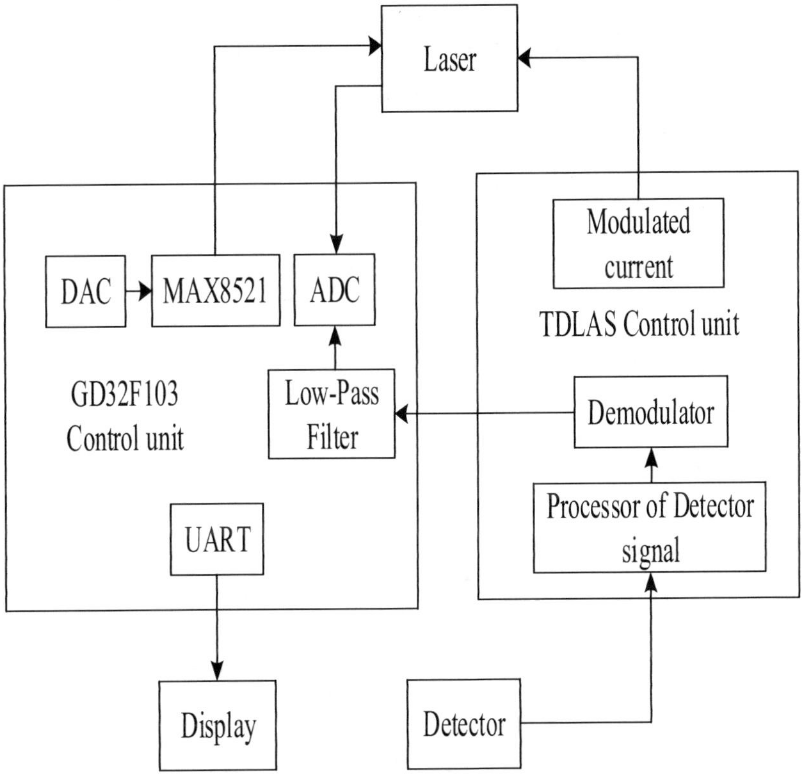

Figure 2. Block diagram of control board circuit.

3.4. Design of software

The flow chart of the control board driver designed based on GD32F103 is shown in Figure 3. After power-on, the peripheral modules of the microcontroller are initialized, and then the ADC is started to collect the voltage on the NTC resistor in the laser, calculate the current temperature of the laser, and compare it with the set value of temperature. The DAC output voltage is calculated by the PID algorithm to control the MAX8521 to adjust the working current of the TEC of the laser to stabilize its temperature at the set value. Timer 3 is used to update the serial screen data every second. If the serial port receives the command data from the serial screen, the instruction response program will be launched. The second harmonic signal is collected by ADC + Timer trigger + DMA. Channel 1 of Timer 1 outputs a 1 kHz PWM as the trigger source of ADC sampling, controlling ADC to collect 1, 000 samples of harmonic signal voltages every second and directly store them in the DMA buffer. When the DMA transmission is completed and the interrupt is triggered, the sampling is completed. The maximum and minimum values of the sampling data are found and the harmonic peak value is calculated. According to the calibration coefficient, the gas volume fraction is calculated. The above program is executed recursively.

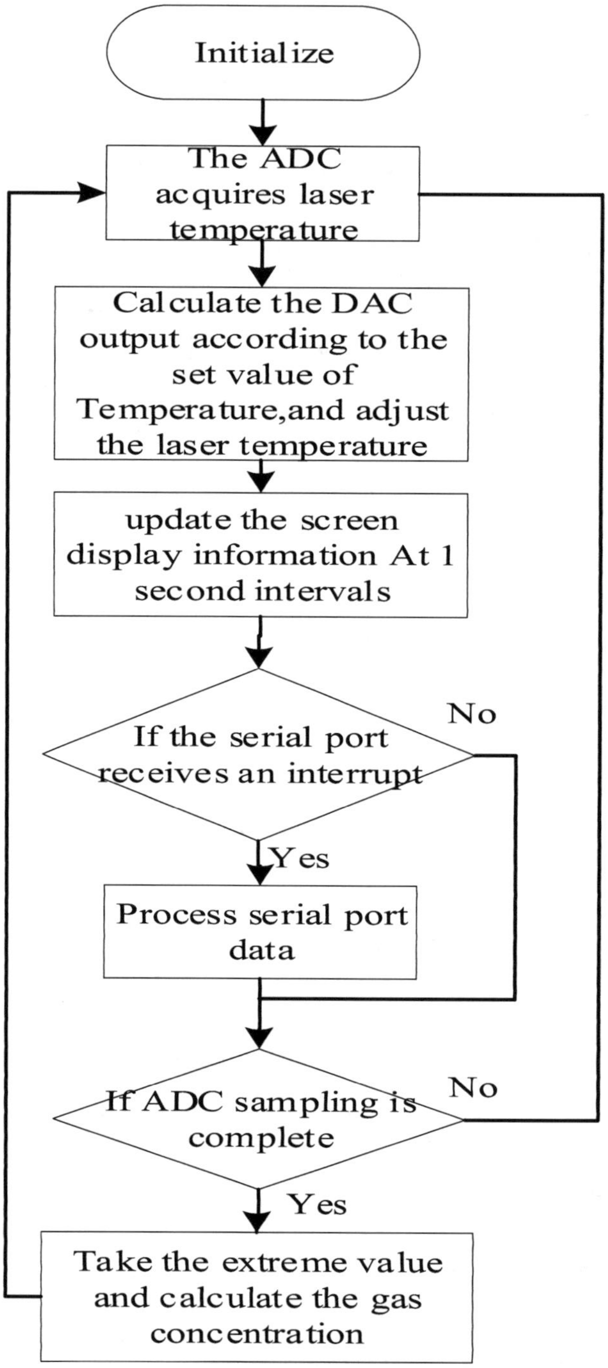

Figure 3. Program flow chart of GD32 control unit system tests.

3.5. Test methods

A dynamic gas mixing device was built by using the flow meter from Sevenstar, which is used to prepare test gas for the ammonia online monitoring system. The test arrangement is shown in Figure 4.

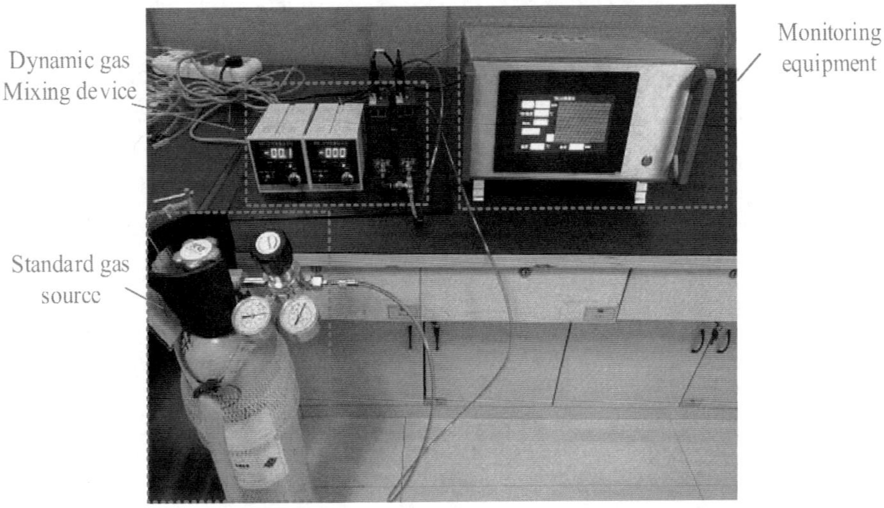

Figure 4. Experimental layout.

Dalian special gases produced 500 ppm ammonia standard gas and high-purity nitrogen as the gas source. Ammonia standard gas with 0 ppm, 100 ppm, 200 ppm, 300 ppm, 400 ppm, and 500 ppm were respectively prepared to calibrate the output of the ammonia online monitor. A triple-fitting curve was used to obtain the calculation relationship between the harmonic peak value and the gas concentration, as shown in Figure 5. Then, the calibration coefficient was set through the serial screen, and the calibration was completed.

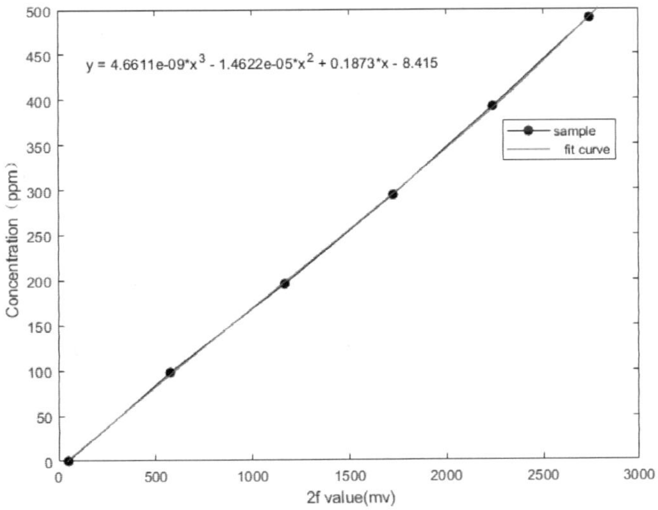

Figure 5. Fitting curve of harmonic peak value and gas volume fraction.

3.6. Test results

After calibration, the ammonia sample gas used in calibration was prepared by the dynamic gas mixing device to test the detection indicators of the ammonia online monitoring system. The measured results of ammonia volume fraction within the range of 0-500 ppm are shown in Table 1. Among all the test concentration points, the maximum measurement error is 2.5 ppm, which is less than ± 1% FS (5 ppm) and meets the design requirements.

Table 1. Experimental results.

Sample gas volume fraction (ppm)	Measurement result (ppm)	Measurement error (ppm)
0	1	+1
100	98.3	-1.7
200	202.5	+2.5
300	300.6	+0.6
400	399.2	-0.8
500	501	+1

The response time of the ammonia online monitoring system was tested by using 500 ppm standard gas, and the test results are shown in Figure 6. The system response time (T90) is 22 s.

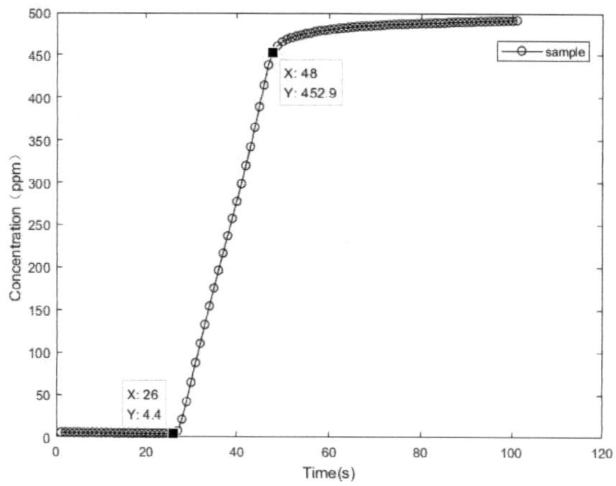

Figure 6. Response time test.

In a short period, 200 ppm standard gas was repeatedly measured 10 times, the average measured value was 201.6 ppm, and the measurement standard deviation was 0.56 ppm. The system measurement repeatability was calculated to be 0.28%.

3.7. Experimental conclusion

The above test results show that the measurement accuracy, response time, and measurement repeatability of the ammonia online monitoring system meet the design requirements and can realize real-time online monitoring of ammonia leakage on ammonia-fueled ships.

4. Conclusions

This paper aims to monitor demand for ammonia gas leakage in ammonia-fueled ships. Based on TDLAS technology, we designed an ammonia online monitoring system. It described the system composition and design ideas in detail, and proved through standard gas testing that the detection indicators of the system for ammonia gas meet the design requirements. However, in the test, the system calibration and testing only used 6 concentration points. If the system response is linear, the method will be correct. However, the response of the infrared absorption signal is nonlinear, especially in the low concentration range (below 10 ppm), limited by the detection sensitivity of the system, the measurement results have a large deviation, which can be improved by increasing the calibration data points.

References

[1]Gao H., Liu F., and Li H. (2022). Opportunities, challenges, and application prospects of ammonia fuel under carbon neutralization target. *Power Generation Technology*, 43 (3): 462-467.
[2]Dong M. (2019). Research on key issues of infrared laser gas analyzer. Jilin University, Jilin.
[3]Zhang R. (2014). Research on high-precision gas detection technology in tunable diode laser absorption spectroscopy. Tianjin University, Tianjin.
[4]Reid J. and Labrie D. (1981). Second-harmonic detection with tunable diode lasers-comparison of experiment and theory. *Applied Physics B*, 26 (3): 203-210.
[5]Roller C., Fried A., Walega J. G., et al. (2006). Advances in hardware, system diagnostics software, and acquisition procedure for high-performance airborne tunable diode laser measurements of formaldehyde. *Applied Physics B*, 82 (2): 247-264.
[6]Pan W. D. (2013). Research on trace ethylene gas detection technology based on TDLAS. Harbin Institute of Technology, Harbin.
[7]Liu Y. F. (2019). Based on TDLAS-WMS gas detection and its application in oil-immersed transformer fault detection. Northeast Petroleum University, Daqing.
[8]Wang Z. (2015). Research on ammonia volume fraction detection system based on tunable diode laser absorption spectroscopy. Tianjin University, Tianjin.
[9]Zhang H. D, Wang F., and Xing D. W. (2013). In situ measurements of CO in combustion gases based on TDLAS and analysis of relevant influencing factors. *Journal of Chinese Society of Power Engineering*, 06: 448-454.
[10]Yao S. C, Guo S. J., Yang Y., et al. (2023). Research and application of flue gas ammonia slip detection based on tunable diode laser absorption spectroscopy (Invited). *Acta Photonica Sinica*, 52 (3): 0352101.
[11]Qin C. K. (2022). Design of multi-point monitoring system for CH_4 volume fraction based on TDLAS. Xi'an Technological University, Xi'an, Shaanxi, China.
[12]Qian J. R., Zhang F. C., Ding Y. J., et al. (2022). Research on space dynamic distribution monitoring technology and application of natural gas leak based on TDLAS. *Laser & Infrared*, 52 (3): 360-367.
[13]Rothman L. S., Gordon I. E., Barbe A., et al. (2009). The HITRAN 2008 molecular spectroscopic database. *Journal of Quantitative Spectroscopy & Radiative Transfer*, 110: 533-572.

Prediction of milling surface roughness based on DGLCM and neural network

Chenghui Zhang[1,2], Zaojun Fang[2,3], Hongyuan Lian[2], Qiping Zhang[2], Chi Zhang[2] and Guilin Yang[2]

[1] Faculty of Mechanical Engineering & Mechanics, Ningbo University, Ningbo, China
[2]Zhejiang Key Laboratory of Robotics and Intelligent Manufacturing Equipment Technology, Ningbo Institute of Materials Technology, CAS, Ningbo, China

[3]Corresponding author's e-mail: fangzaojun@nimte.ac.cn

Abstract. With the manufacturing industry's increasingly automated and intelligent development, traditional workpiece roughness measurement methods can no longer meet the requirements for short time, high efficiency, and online measurement. Therefore, this article proposes a roughness evaluation model for vertical milling workpiece roughness measurement and improves upon the traditional Gray Level Co-occurrence Matrix (GLCM) method to achieve high accuracy and robustness in measuring milling surface roughness. First, the milled surface is illuminated by a coaxial light source, and an image is captured using an industrial camera and a telecentric lens. Then, four features of the image are extracted using the Direction Measure-based Gray Level Co-occurrence Matrix (DGLCM) to construct a feature descriptor. Finally, Finally, create an RBF network to predict surface roughness. Through experimental comparison with traditional GLCM feature extraction methods, our proposed feature extraction method has a prediction error of only 2.01%, which is superior to traditional methods.

1. Introduction

The roughness of a workpiece's surface is a crucial factor in assessing the surface quality of components in modern mechanical manufacturing. The magnitude of surface roughness significantly impacts the workpiece's performance and service life, especially in terms of wear resistance, fatigue resistance, and assembly performance. Therefore, accurate measurement of the size of the workpiece roughness is of great significance [1].

Currently, there are two main surface roughness measurement methods: contact and non-contact. Contact measurement primarily involves using a stylus-type roughness measuring instrument, which can measure the rough surface contour and record the results multiple times. However, this method has low measurement efficiency, and the diamond stylus can easily scratch the smooth and soft surface of the workpiece, leading to inaccurate results. Non-contact measurement methods typically utilize optical or electronic measuring instruments, offering high accuracy. However, these methods have lower measurement efficiency and require expensive equipment that occupies a large area.

With the development of technology and improved industrial production, non-contact roughness measurement methods based on optics and machine vision have been widely researched and applied [2–5]. Kamguem et al. [6] used optical microscopy and machine vision to extract the image texture gradient factor that characterizes roughness from the collected images. However, their features were still established using traditional regression methods, and the accuracy needs to be improved. Nathan et al. [7] studied the surface texture characteristics of AA6061, and it was further demonstrated that there is a certain statistical relationship between them and the surface roughness. However, no further research was conducted on roughness prediction models. Pontes et al. [8] studied the structure of the MLP in depth and proposed a new MLP model for predicting average surface roughness, which provides an effective choice for designing ANN-based system models. D.V.N. Prabhakar et al. [9] extracted features that characterize roughness by combining multiple frequencies and predicted roughness using artificial neural networks without considering the influence of image texture features on roughness. Wei et al. [10] used a BP network to predict milling surface roughness, but its error needs to be reduced.

This article delves into the GLCM texture feature extraction method and improves the traditional GLCM method. A gray-level co-occurrence matrix method based on directional measurement (DGLCM) is proposed, and an RBF network model is established to compare the prediction results of the two feature extraction methods. Experiments have shown that our proposed method can reduce measurement errors.

2. Methodology

2.1 Measurement principle

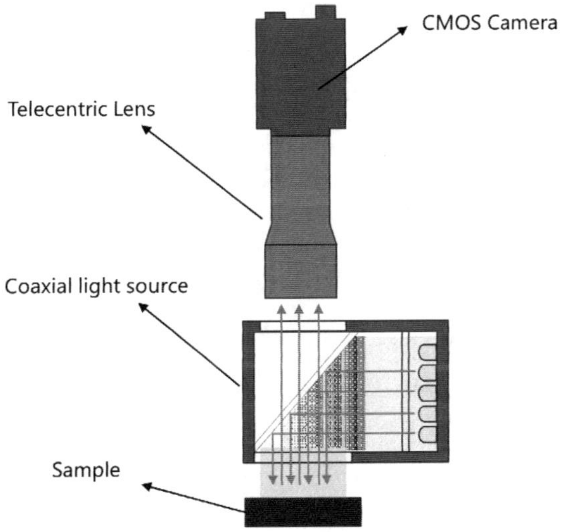

Figure 1. Schematic diagram of the experimental device.

The schematic diagram of the experimental device is shown in Figure 1. This acquisition system consists of a CMOS camera, a telecentric lens, and a coaxial light source. Telecentric lenses have the advantages of high accuracy, large depth, and great advantages in capturing rough surface details. The structure of the coaxial light source is shown in the figure, consisting of a light source and a semi reflective semi lens. This special structure allows light to not only shine on the surface to be tested, but also screen the reflected light, ensuring that the light entering the lens is coaxial, reducing the interference of ambient light, and achieving high-precision and robust image acquisition.

2.2 Samples preparation

Small uniform pieces of the cold rolled strip were collected from an industrial site and machined into sample blocks with varying degrees of roughness using vertical milling. The roughness of the sample blocks was measured using a stylus roughness measuring instrument, and the measured roughness values were 0.4516 μm, 0.5146 μm, 0.5412 μm, 0.5656 μm, 0.6174 μm, 0.6782 μm, 0.7106 μm, 0.7560 μm, 0.8602 μm, and 1.0658 μm. The images of surfaces with different roughness obtained by industrial cameras are shown in Figure 2.

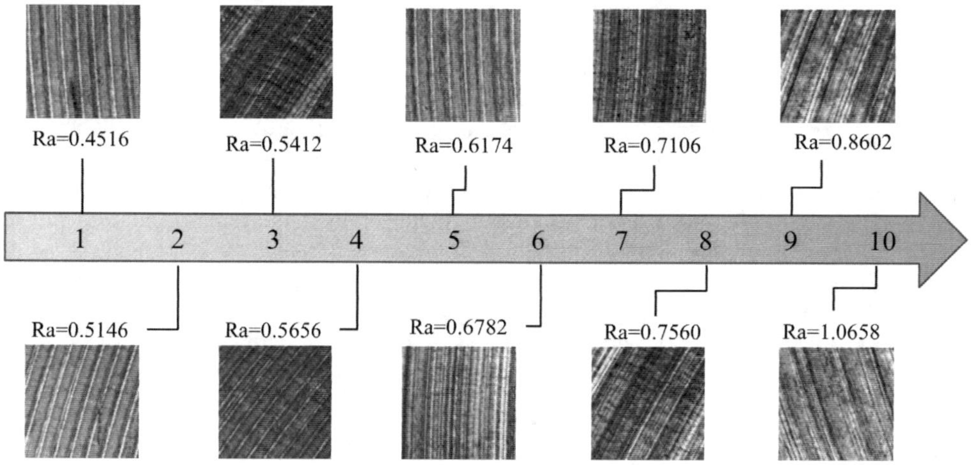

Figure 2. Images with different roughness levels.

2.3 Feature extraction method

Before extracting image features, a series of preprocessing operations need to be performed on the image to minimize the impact of interference factors during shooting. Therefore, the original image should be converted to a grayscale image. The Hikvision industrial camera used in this paper is grayscale, so additional grayscale processing is unnecessary. Image texture features contain a large amount of information, and the commonly used method for texture feature extraction is the GLCM method [11]. The GLCM represents the probability of a grayscale value *j* at points separated by a distance d and direction θ from a pixel point with a grayscale value *i*. The matrix is composed of the frequency of occurrence of all gray value pairs, which is defined by the following formula:

$$P(i,j,d,\theta)=\{(x,y)|f(x,y)=i,f(x+dx,y+dy)=j;x,y=0,1,2...N-1\} \quad (1)$$

In the formula, (x, y) represents the position of a point in the image with a gray value of *i*, *d* is the relative distance expressed in pixel pairs, *i*, *j*=0,1,2..., *L*-1, *L* is the quantized gray level of the pixel's gray value in the image, θ is the four directions in the image, which are 0°, 45°, 90°, 135°.

The GLCM extracts fourteen different texture features from the image. After research, it was found that only four of the features are non-linearly correlated, and the other ten features can be represented by these four features. These four features are the most capable of representing the texture in the image. Therefore, this paper extracted these four features, and their formulas are as follows:

$$Contrast=\sum_{i=0}^{L-1}\sum_{j=0}^{L-1}(i-j)^2P(i,j,d,\theta) \quad (2)$$

$$Correlation=\sum_{i=0}^{L-1}\sum_{j=0}^{L-1}\frac{(1-\mu_x i)(1-\mu_y j)P(i,j,d,\theta)}{\sigma_i\sigma_j} \quad (3)$$

$$Energy=\sum_{i=0}^{L-1}\sum_{j=0}^{L-1}p(i,j,d,\theta)^2 \quad (4)$$

$$Homogeneity=\sum_{i=0}^{L-1}\sum_{j=0}^{L-1}\frac{p(i,j,d,\theta)}{1+|i-j|} \quad (5)$$

Where $\mu_x = \sum_{i=0}^{L-1}\sum_{j=0}^{L-1} i * P(i,j,d,\theta)$, $\mu_y = \sum_{i=0}^{L-1}\sum_{j=0}^{L-1} j * P(i,j,d,\theta)$

$$\sigma_x{}^2 = \sum_{i=0}^{L-1}\sum_{j=0}^{L-1}(i - \mu_x{}^2) * P(i,j,d,\theta),\ \sigma_y{}^2 = \sum_{i=0}^{L-1}\sum_{j=0}^{L-1}(i - \mu_y{}^2) * P(i,j,d,\theta)$$

2.4 GLCM based on the directional measure

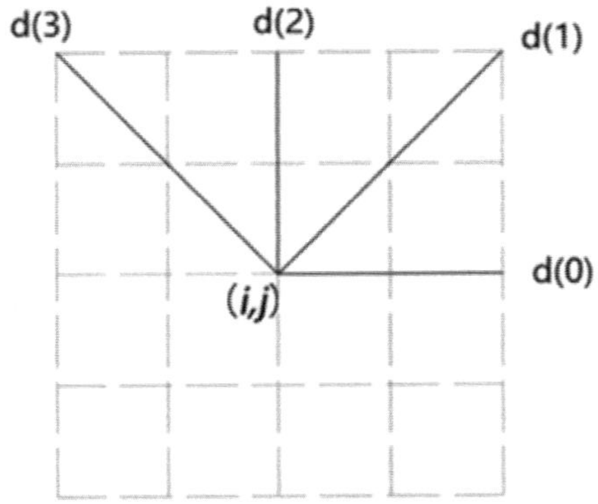

Figure 3. 5 × 5 grayscale co-occurrence matrix direction measurement window.

The grayscale co-occurrence matrix solely accounts for regional texture characteristics at a macro level, neglecting the evaluation of texture significance and alteration in each direction. To capture the nuanced variations among diverse directions of the texture, this study employs directional measures to compute the pixel alteration rates and pixel alterations in the image across different directions (0°, 45°, 90°, 135°) in the grayscale co-occurrence matrix, thereby delineating the micro directional features of the texture.

The mixed higher-order directional measure window of size 5×5, centered on any point (i, j) in an N × M image, is illustrated in Figure 3. The directional measures of the image, starting from the center point (i, j) in the 0°, 45°, 90°, and 135° directions, are denoted by $\{d(k) \mid k=0, 1, 2, 3\}$.

The expression for the mixed high-order directional measure obtained by the fourth-order difference and the second-order difference on the 5×5 directional window in Figure 3 is:

$$d(0) = \sum_{i=1}^{N}\sum_{j=1}^{M} |f(i,j+2) - 4f(i,j+1) + 6f(i,j) - 4(i,j-1) + f(i,j-2)| \quad (6)$$

$$d(1) = \sum_{i=1}^{N}\sum_{j=1}^{M} |f(i-2,j+2) - 4f(i-1,j+1) + 6f(i,j) - 4(i+1,j-1) + f(i+2,j-2)| \quad (7)$$

$$d(2) = \sum_{i=1}^{N}\sum_{j=1}^{M} |f(i-2,j) - 4f(i-1,j) + 6f(i,j) - 4(i+1,j) + f(i+2,j)| \quad (8)$$

$$d(3) = \sum_{i=1}^{N}\sum_{j=1}^{M} |f(i-2,j-2) - 4f(i-1,j-1) + 6f(i,j) - 4(i+1,j+1) + f(i+2,j+2)| \quad (9)$$

The calculation formula for feature T after the fusion of direction measures:

$$T = \sum_{t} \omega_t T_\theta \quad (10)$$

$$\theta = 45^{\circ} t \tag{11}$$

$$w(t) = \frac{1}{\sum_{t=0}^{3} \frac{1}{d(t)^2}} \frac{1}{d(t)^2} \tag{12}$$

$$\sum_t \omega_t = 1 \tag{13}$$

In Formula 5, T = (*Contrast*, *Correlation*, *Energy*, *Homogeneity*).

2.5 Roughness prediction model

The RBF network is a classic artificial neural network with good generalization capabilities. It has been widely used in various fields, such as nonlinear function approximation and image processing. RBF networks are relatively widely used, and their basic concepts will not be described in detail here. Compared to a Backpropagation (BP) neural network, the calling format of an RBF network is relatively simple. To achieve the desired prediction accuracy, adjusting the spread speed of the radial basis function propagation is necessary. BP neural networks are global approximation networks with slow learning speeds that cannot meet real-time requirements. In contrast, RBF networks are local approximation networks with faster learning speeds that can meet real-time requirements. Therefore, this article focuses on the use of RBF networks.

As shown in Figure 4, the structure of an RBF network designed in this study adopts the classic three-layer structure and uses the Gaussian function as the kernel function. The input layer has four nodes representing the features T(*Contrast*, *Correlation*, *Energy*, *Homogeneity*) extracted by DGLCM. The number of neurons in the hidden layer equals the number of input samples. The output of the *jth* node of sample X in the hidden layer is shown in the following formula.

$$\Phi_i(x) = \exp\left[- \frac{\|X - C_i\|}{2s^2} \right] (j = 1, 2, 3 \ldots, p) \tag{14}$$

where training sample $X = [x_1, x_2, x_3, x_4]^T$ is the input feature vector, i.e., the four features *Contrast*, *Correlation*, *Energy*, *Homogeneity*, C_i is the centroid of the basic function of the *ith* neuron in the hidden layer, and s is the standard deviation of the basis function.

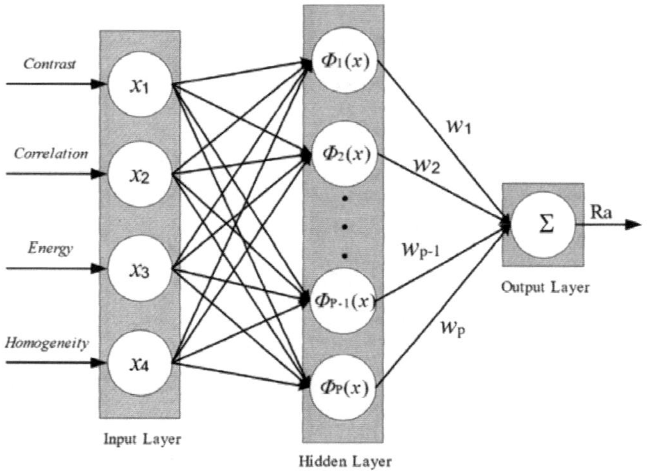

Figure 4. Structure diagram of RBF network.

The output layer completes the linear mapping from $\Phi_i(x)$ to Ra, and the output is shown in Formula (15):

$$\mathrm{Ra} = \sum_{i=1}^{p} w_i \Phi_i(x) \tag{15}$$

where Ra is the output sample, i.e., the roughness value, $\Phi_i(x)$ is the output of the *ith* neuron, and w_i represents the connection weight of the *ith* neuron in the hidden layer to the output layer.

3. Experimental results and analysis

The image acquisition platform built in this study is shown in Figure 5, the telecentric lens used is XF-MT08X110D, and the industrial camera used is the Hikvision HIKROBOT MV-CE120-10UM industrial array camera with a resolution of 4000 × 3036. The Milling samples are shown in Figure 6.

The experiment collected images of 10 different roughness milling surfaces, as shown in Figure 2. Image acquisition is conducted to expand the data at different positions and angles on the same roughness surface. Each roughness image is expanded to 90 images, and the dataset consists of 900 images. Perform feature extraction on the dataset, totaling 900 sets of feature values.

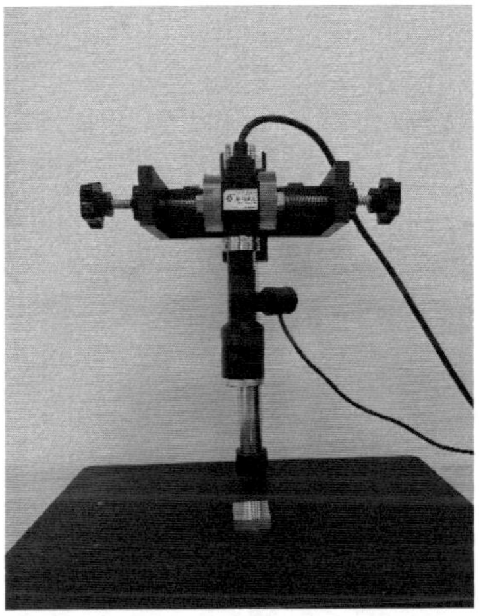

Figure 5. Experimental device.

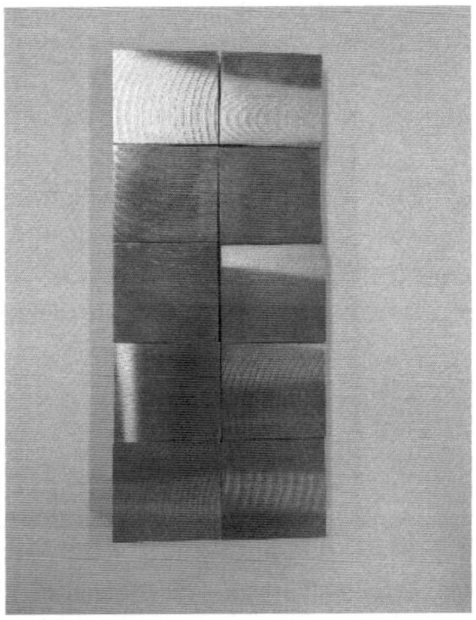

Figure 6. Milling samples.

The expanded dataset of 900 images is divided into a training set and a test set in an 8:2 ratio. The Radial basis function spread value is set to 4. The comparison between the average and true values of the prediction results of DGLCM and GLCM on each data set is shown in Table 1.

Table 1. DGLCM, GLCM results, and real values.

Data Set	DGLCM (μm)	GLCM (μm)	Real value (μm)
1	0.4735	0.5027	0.4516
2	0.5258	0.4983	0.5146
3	0.5291	0.5359	0.5412
4	0.5814	0.5391	0.5656
5	0.6046	0.5716	0.6174
6	0.6718	0.7352	0.6782
7	0.7207	0.7122	0.7106
8	0.7462	0.7614	0.7560
9	0.8735	0.8238	0.8602
10	1.0574	1.0376	1.0658

The average error on each data set is shown in Figure 7. According to the analysis of the experimental results in Table 1 and Figure 7, the roughness prediction values obtained by the RBF network after extracting features using both the DGLCM and GLCM methods are not much different from the true values. However, the average prediction error of the DGLCM method is better than that of the GLCM method on most datasets.

To verify the superiority and robustness of the DGLCM method proposed in this study in terms of roughness regression prediction indicators, ten-fold cross-validation was used. 900 milling surface images are evenly divided into ten parts, each using one part as the test set and the others as the training set, repeating training and testing ten times. The error results were taken as the average of ten times and compared with the neural network prediction results of the features extracted by traditional GLCM.

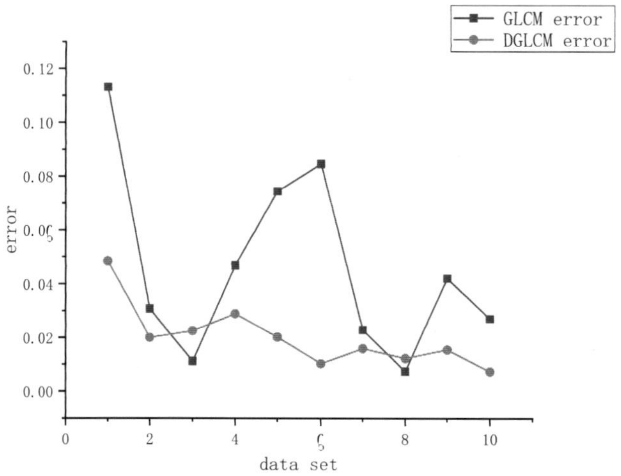

Figure 7. Average error per test dataset.

Table 2 shows the average error rate, MSE, and R^2 values of the predicted results using the GLCM method and the DGLCM method after ten cross-validations are presented. The error rate of the DGLCM method proposed in this article for roughness prediction can reach 2.01%, while the error rate of the traditional GLCM method is 4.27%. The MSE of the DGLCM method is 2.81×10^{-4}, while that of the GLCM method is 2.2×10^{-3}. The R^2 value of the DGLCM method is 0.9911, while that of the GLCM method is only 0.9287.

Table 2. Comparison of RBF and BP results.

Method	Error	MSE	R^2
DGLCM	2.01%	2.81×10^{-4}	0.9911
GLCM	4.27%	2.2×10^{-3}	0.9287

4. Conclusion

This article proposes a DGLCM feature extraction method for measuring the surface roughness of milled workpieces and constructs a milling surface roughness measurement model based on an RBF network. Compared with the GLCM feature extraction method, the experimental verification shows that the average prediction error of this feature extraction method is smaller than that of the GLCM method on most datasets. After ten-fold cross-validation, its regression indicators MSE and R2 are both smaller than those of GLCM, verifying the superiority and robustness of this method.

Acknowledgments
This work was supported in part by the National Natural Science Foundation of China under Grant (U1909215, 92048201, 52127803, U21A20121), and in part by the Zhejiang Provincial Key Research and Development Plan under Grant (2023C01176, LD22E050007), and in part by the Ningbo Major Research Plan under Grant (2021Z020).

References

[1] Cai Y, Liu Z, Shi Z, et al. Influence of machined surface roughness on thrust performance of micro-nozzle manufactured by micro-milling[J]. Experimental Thermal and Fluid Science, 2016, 77: 295-305.
[2] Lee C. A first review of optical edge-diffraction technology for precision dimensional metrology[J]. International Journal of Advanced Manufacturing Technology, 2019, 102(5-8): 2465-2480.
[3] Yildirim M, Okutucu-Ozyurt T, Dursunkaya Z. A numerical algorithm to determine straightness error, surface roughness, and waviness measured using a fiber optic interferometer[J]. Optics and Laser Technology, 2016, 85: 19-29.
[4] Yang B, Yan C, Zhang J, et al. Refractive index and surface roughness estimation using passive multispectral and multiangular polarimetric measurements[J]. Optics Communications, 2016, 381: 336-345.
[5] Hatefi S, Abou-El-Hossein K. Review of single-point diamond turning process in terms of ultra-precision optical surface roughness[J]. International Journal of Advanced Manufacturing Technology, 2020, 106(5-6): 2167-2187.
[6] Kamguem R, Tahan S A, Songmene V. Evaluation of machined part surface roughness using image texture gradient factor[J]. International Journal of Precision Engineering and Manufacturing, 2013, 14(2): 183-190.
[7] Nathan D, Thanigaiyarasu G, Vani K. Study On the Relationship between Surface Roughness of AA6061 Alloy End Milling and Image Texture Features of Milled Surface[C]//XAVIOR M A, YARLAGADDA P. 12th Global Congress on Manufacturing and Management (game - 2014): Vol. 97. Amsterdam: Elsevier Science Bv, 2014: 150-157.
[8] Pontes F J, Amorim G F, Balestrassi P P, et al. Design of experiments and focused grid search for neural network parameter optimization[J]. Neurocomputing, 2016, 186: 22-34.
[9] Prabhakar D V N, Sreenivasa Kumar M, Gopala Krishna A. A Novel Hybrid Transform approach with the integration of Fast Fourier, Discrete Wavelet, and Discrete Shearlet Transforms for prediction of surface roughness on machined surfaces[J]. Measurement, 2020, 164: 108011.
[10] Pradana A B, Prajitno P. A Portable Surface Roughness Measurement System Using Laser Speckle Imaging Based on GLCM[C]//2019 6th International Conference on Instrumentation, Control, and Automation (ICA). Bandung, Indonesia: IEEE, 2019: 100-105.
[11] Haralick R, Shanmugam K, Dinstein I. Textural Features for Image Classification[J]. Ieee Transactions on Systems Man and Cybernetics, 1973, SMC3(6): 610-621.

Research on extremely weak remanence measurement system

Chaoqun Xu [1, 3], Tian Sun [2], Kui Huang [1] and Qi Xiao [1]

[1]Laboratory of space magnetism and applications Beijing Institute of Spacecraft Environment Engineering Beijing, China
[2]Department of science and technology development Systems Engineering Research Institute Beijing, China

[3]Corresponding author's e-mail: xucq111@163.com

Abstract: To accurately obtain the extremely weak magnetic field information of the test piece, this article first introduces the basic principles of the extremely weak residual magnetic field measurement technology. Then, an ultra-high precision magnetic shielding system has been developed, which is made of high permeability materials and equipped with a demagnetization coil system, and it can regularly eliminate the inductance of the entire shielding system. The effective shielding inner diameter of the magnetic shielding system is 0.8 m, the length is 1.6 m, and the magnetic field resolution is 0.0015 nT. This system can provide strong support for the measurement of extremely weak magnetic fields in materials and biomedical magnetic research.

1. Introduction

The Earth itself is magnetic, so there is a magnetic field in the Earth and near Earth's space. This magnetic field is called Earth's magnetic field [1], and the value of Earth's magnetic field is about 0.5×10^{-4} T, its strength and direction also vary with location. The Earth's magnetic field can be approximated as the shape of a magnetic field generated by placing a magnetic rod at the center of the Earth, with its N poles roughly facing the South Pole.

The Earth's magnetic field consists of a basic magnetic field and a variable magnetic field, which is different in origin. The basic magnetic field is the main part of the Earth's magnetic field, which originates from the Earth's interior, being relatively stable and changing very slowly. The changing magnetic field includes various short-term changes in Earth's magnetic field, which mainly originates from the outside of the Earth and is very weak. The basic magnetic field of the Earth can be divided into dipole magnetic field, nondipole magnetic field, and geomagnetic anomaly. The dipole magnetic field is the basic component of the Earth's magnetic field. Its strength occupies about 90% of the total magnetic field strength. It is generated by the electromagnetic fluid mechanics process in the outer core of the Earth's liquid, that is, the self-excitation motor effect. The nondipole magnetic field is mainly distributed in eastern Asia, western Africa, the South Atlantic Ocean, and the South Indian Ocean, with an average intensity of about 10% of the magnetic field. The geomagnetic anomaly can be divided into regional anomaly and local anomaly, which are related to the distribution of rocks and ore bodies. The Earth's changing magnetic field can be divided into two types: calm changes and interference changes. The quiet change is mainly a quiet day change with solar time as the cycle, and its field sources are distributed in the ionosphere. Disturbance changes include magnetic storms, geomagnetic substorms, solar disturbances, and geomagnetic pulsations. The Earth's magnetic field plays an important role in aviation, aerospace, geological exploration, material science, biomedicine, and other fields [2-4], and the changing Earth's magnetic field will have an important impact on many studies, so how to shield and use the Earth's magnetic field has become the research focus and hotspot of many scholars at home and abroad in recent years.

At present, there are two ways to shield and use the Earth's magnetic field. One is to use Helmholtz coils to simulate and shield the Earth's magnetic field. The disadvantage is that it is impossible to avoid Earth's magnetic field fluctuations. The other is to use materials with high permeability to make shielding systems to shield the geomagnetic field, to obtain a highly stable magnetic field environment. This article uses high-permeability materials to make a shielding system for measuring extremely weak residual magnetism.

2. Basic principles

Magnetic shielding refers to putting two mediums with different permeability into the magnetic field, and the magnetic field will change suddenly at their interface. At this time, the size and direction of Magnetic flux density B will change, that is, it will cause the refraction of the magnetic induction line. When the magnetic induction line enters iron from the air, the deviation of the magnetic induction line from the normal is significant, resulting in a strong convergence effect and the formation of a magnetic shield.

The basic principle of magnetic field shielding [5-8] is to use a cavity wall made of high-permeability materials and its surrounding air as a parallel magnetic circuit. The magnetic field lines are concentrated inside the magnetic shielding body with low magnetic resistance, as shown in Figure 1. Due to the relative permeability of the air medium approaching 1, and the relative permeability of the shielding shell being at least a few thousand, $R_m=l/uS$, where R_m is the magnetic resistance, l is the length of the magnetic circuit, u is the magnetic permeability, and S is the cross-sectional area of the magnetic circuit. So the magnetic resistance R_o of the air medium is much greater than the magnetic resistance R_m of the shielding shell. At this time, the magnetic flux mainly forms a circuit along the path of low magnetic resistance. Due to the small magnetic resistance of the shielding shell, when the shielding shell is placed in a magnetic field, the magnetic flux will mainly pass through the shielding shell, and the magnetic flux passing through the air will be greatly reduced, thus playing a role in magnetic field shielding.

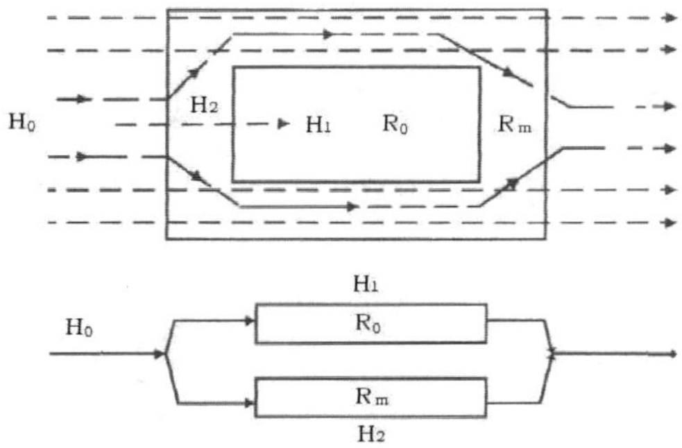

Figure 1. Magnetic shielding principle.

In practice, achieving complete shielding is extremely difficult. There is always some magnetic field that needs to leak into or run out of the shield. To achieve a better shielding effect, it is necessary to choose materials with high magnetic conductivity, such as permalloy, silicon steel sheet, etc., and not too thin. The structural design of the shielding cover should have as few joints as possible, and the joints should be tight during production to minimize air gaps. In short, the smaller the magnetic resistance of the shield is, the better the shielding effect is. If shielding is required in low-frequency alternating magnetic fields, such as in power transformers, the above principles of magnetic shielding are followed. When the shielding requirements are high, multi-layer shielding can also be used.

3. System development

Because magnetic shielding cannot completely seal the magnetic flux inside the shielding body, multi-layer shielding can be used to solve the problem of magnetic leakage, which can greatly improve the shielding effect. At the same time, to reduce the impact of mechanical stress on the permalloy layer, it is necessary to add a layer of hard aluminum to the inner and outer layers of the magnetic shielding cylinder, which can effectively reduce the mechanical stress borne by the permalloy layer on the magnetic shielding cylinder itself, as well as the shielding coil and the tested instrument. The size design of the shielding cylinder in this article is inner cavity φ 800 mm×1600 mm. At the same time, a layer of hard aluminum is added to the inner and outer layers of the magnetic shielding cylinder to effectively reduce the mechanical stress borne by the permalloy layer on the magnetic shielding cylinder itself, as well as the shielding coil and the tested instrument. The main structure is designed as a horizontal structure, with one end sealed at the bottom and the other end as a movable cover. The movable cover is a full buckle type, with measurement holes and wiring holes left at the movable cover end. Both the magnetic shielding cylinder and the movable cover are equipped with support and movement mechanisms, which bear a weight of about 1 T and can achieve the support and horizontal movement of the magnetic shielding cylinder, ensuring that the center height of the shielding cylinder is about 1 m.

For shielding materials, the type of material chosen has a significant impact on their performance and cost. When designing shielding materials, it is important to have a deep understanding of the characteristics of different shielding alloys that are commonly used. Understanding these different properties can enable you to choose suitable materials to meet the target requirements. Magnetic shielding materials should be selected based on their respective characteristics, especially magnetic permeability and magnetic saturation performance. Due to its effectiveness in changing the direction of low-frequency magnetic fields, high-permeability materials (such as Mumetal, a high-permeability iron-nickel alloy containing 80% nickel) are commonly used as shielding materials.

The development of formulas and models for most shielding materials is based on the geometric shape of circular or infinitely long cylinders. In practical applications, the practical shape of the given shielding body is determined by the device structure and the available space of the shielding body itself. When designing a shielding body, it is important to understand that it is difficult to rotate the magnetic field lines by 90°. However, it is easier to change the direction of the magnetic field lines of a circular shield, such as a cylinder or a box with circular angles, than a shield with square angles. Similarly, for magnetic field lines that have entered the shielding material and changed their direction, rounded corners are better than sharp corners. It is important to maintain a simple shape of the shield that can provide a low magnetic resistance path or a "minimum magnetic resistance path" for magnetic field movement. The shielding body should be close to all walls to avoid field leakage. This structure (even rectangular) is the closest to a circle, and it can establish a semi-closed magnetic circuit. In addition, all boxes can obtain shielding characteristics on all axes, which can ensure the best shielding performance. When special performance and import/export requirements arise, movable covers, covers, and doors can be combined into the shielding body design. The magnetic shielding cylinder is shown in Figure 2.

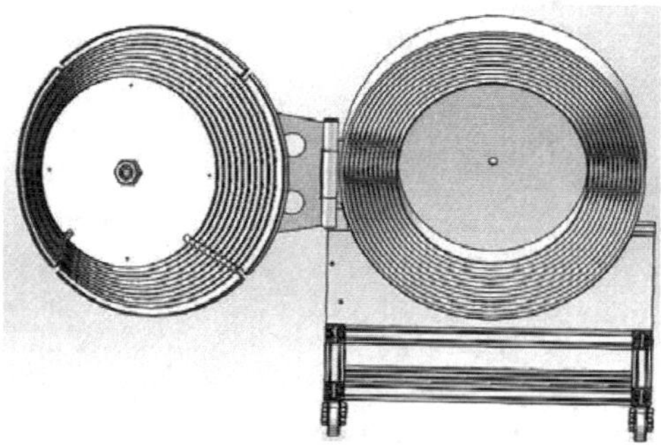

Figure 2. Measurement system.

Journal of Physics: Conference Series **2724** (2024) 012012 doi:10.1088/1742-6596/2724/1/012012

4. Simulation calculation

After modeling, we apply 0.338669 V·s/m^2 Earth's magnetic field to the model, conduct simulation analysis and calculation, and observe the shielding effectiveness and internal magnetic field uniformity of the shielding cylinder. The results are shown in Figure 3.

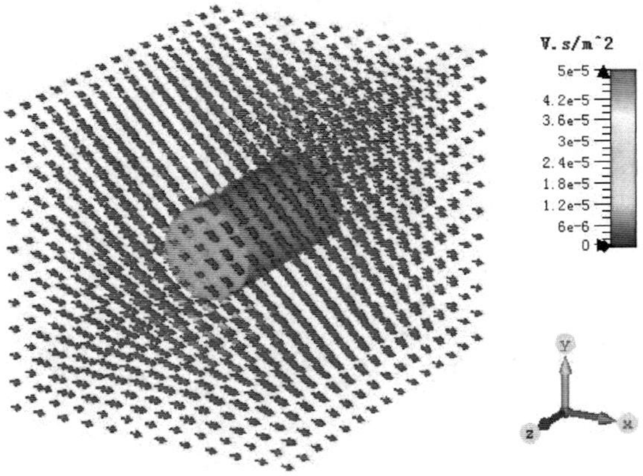

(a) Applying an external magnetic field

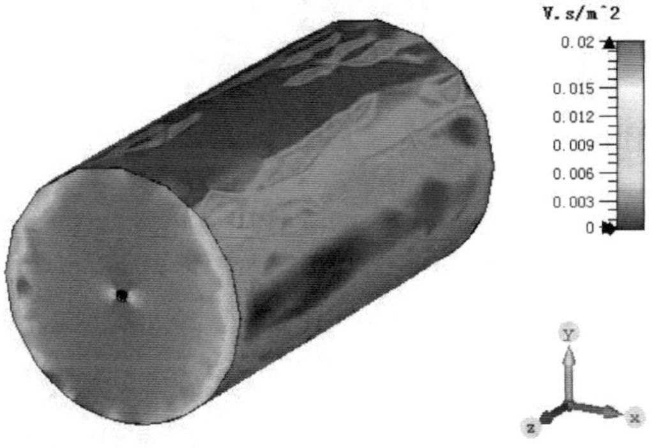

(b) Magnetic field distribution of shielding cylinder

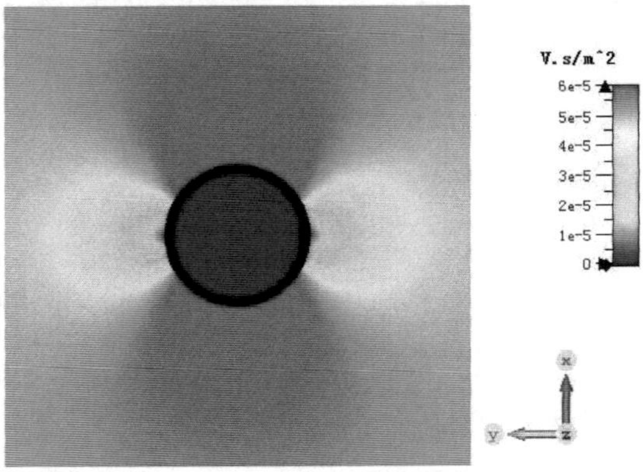

(c) *Z*-direction magnetic field distribution

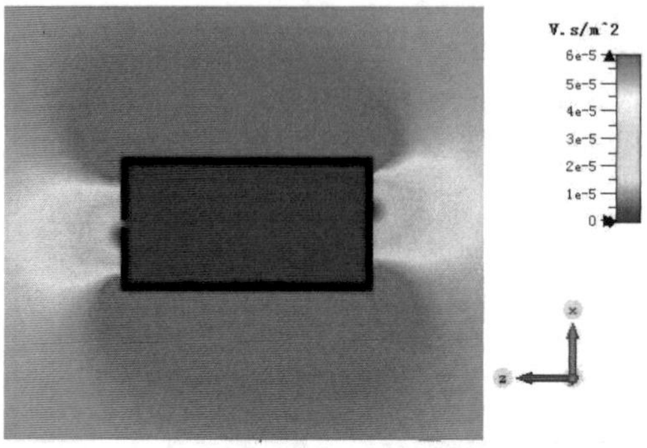

(d) *Y*-direction magnetic field distribution

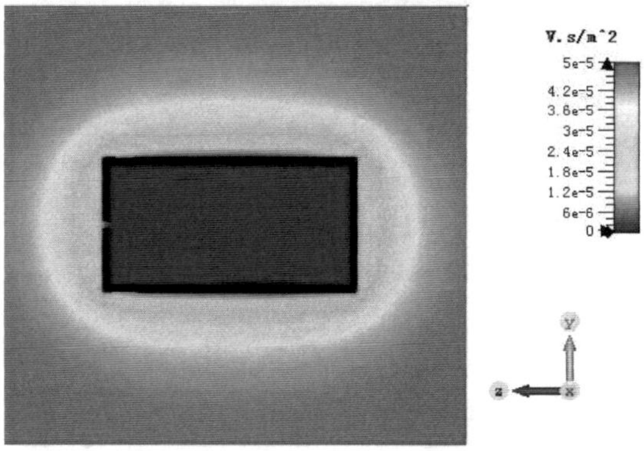

(e) *X*-direction magnetic field distribution

Figure 3. Simulation calculation results.

From the simulation results, it can be seen that when a magnetic field of 0.338669 V·s/m^2 is applied externally, the multi-layer shielding cylinder has a good shielding effect. The magnetic field is mainly concentrated on the shell of the shielding cylinder, with a maximum value of 0.338669 V·s/m^2. In the *z*-direction, the magnetic field is mainly concentrated outside the shielding cylinder, with an internal magnetic field of almost zero. In the *y*-direction, the magnetic field is mainly concentrated outside the shielding cylinder, with an internal magnetic field of almost zero. In the *x*-direction, the magnetic field is mainly concentrated outside the shielding cylinder, with an internal magnetic field of almost zero. Through simulation calculation, the internal magnetic field is 7×10^{-15} T~1.2×10^{-14} T, with extremely high uniformity.

5. Test verification

At present, the fluxgate magnetometer is a common instrument for measuring weak static magnetic fields and low-frequency vector magnetic fields. The relatively mature fluxgate magnetometer has a measurement range of -100000 nT to 100000 nT, and a resolution of 0.01 nT. However, the product materials measured in this system are less than 0.01 nT, so the fluxgate magnetometer is difficult to meet the requirements. Based on the efficient magnetic shielding system, this article has installed a zero field atomic magnetometer on the non-magnetic translation mechanism, sensitivity<15 fT/ √ Hz in 3-100 Hz band, size 15 mm×10 mm×5 mm, can exhibit extremely high sensitivity when the magnetic field background is small.

1) We open the cover of a section of the shielding cylinder and install the zero fields atomic magnetometer on the nonmagnetic translation mechanism inside the system;

2) We close the lid and run the zero fields atomic magnetometer to collect data;

3) We record the data.

The test results are shown in Figure 4.

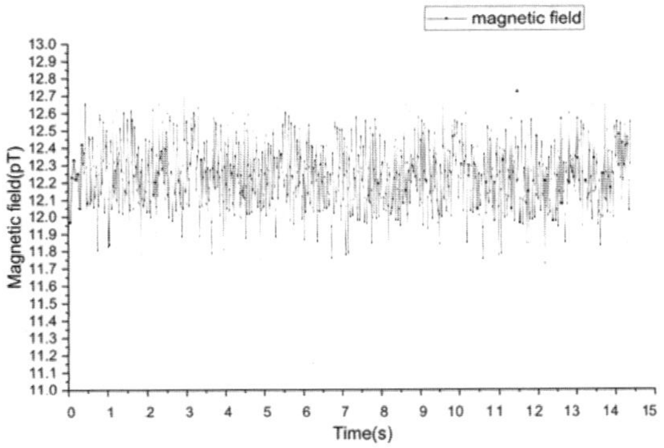

(a) Radial magnetic field fluctuation value

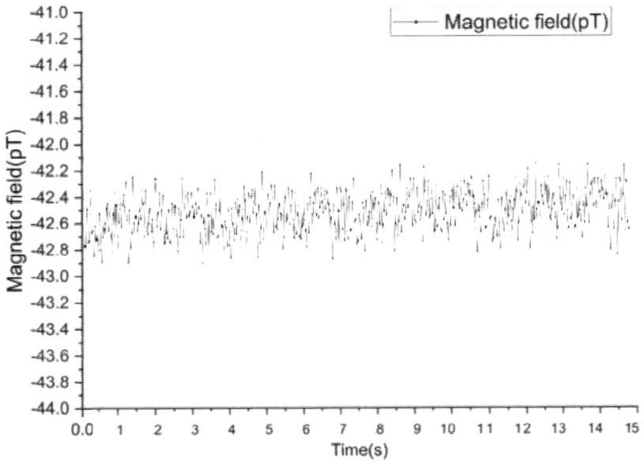

(b) Axial magnetic field fluctuation value

Figure 4. Test results

From the test results, it can be seen that the radial magnetic field is 11.8 pT to 12.8 pT, with a maximum fluctuation of 1 pT; the axial magnetic field is -42.2 pT to -43.0 pT, with a maximum fluctuation of 0.8 pT. The magnetic field value is stable and the shielding effect is good.

6. Conclusion

This article develops an extremely weak residual magnetic measurement system, with the effective space of the entire shielding system being φ 800 mm×1600 mm, and the magnetic field resolution is 0.001 nT. This system provides a strong guarantee for the accurate measurement of magnetic fields in aerospace, new material research, biomedical engineering, and other fields.

References

[1] K. L Liu, Z Heng, W. H Li, et al. The characteristics of geomagnetic field's gradient distribution and the magnetic force effects in near-earth space[J].Spacecraft Environment Engineering, 2012, 29(5):493-497.

[2] S. M. Acmillan, J. M. Quinn. The 2000 revision of the joint UK/US geomagnetic field models and an IGRF2000 candidate model. Earth, Planets and Space, 2000, 52:1149-1162.

[3] R. J. Gambino, M. M. Raja, S. Sampath, et al. Plasma Sprayed Thick Film Anisotropic Magneto Resistive (AMR) Sensors. IEEE Trans. Magn., 2002, 38(4): 656-659.

[4] M. Ilg, B. C. Chang, D. Hepner, et al. A Microcontroller Solution for AMR Magnetic Sensing in Flying Munitions Systems. Proceeding of the 2005 IEEE International Conference on Mechatronics, 2005, pp. 83-86.
[5] L. Liu, D. Zhang. Research progress of electromagnetic shielding materials[J]. Journal of Functional Materials, 2015, 46(3):3016-3019.
[6] K. Nemoto, M. Komori. A study of an active magnetic shielding method for the superconductive Maglev vehicle[J]. Physica C, 2010, 470:1777-1781.
[7] W. Y. Zhang, L. Shi, J. F. Zhang, et al. Research of Magnetic Shielding Efficiency of Multilayer Cylindrical Shield[J]. Journal of Hunan University (Natural Sciences), 2017, 44(8):84-90.
[8] F. F. Niu, X. N. Hu, C. Y. Chun, et al. Study of magnetic shielding system applied to superconducting gravity measurement[J]. Cryogenics & Superconductivity, 2023, 51(2):14-19.

Research on a kind of positioning anti-eccentric fixture for measuring thin wall gear with special big inner hole

Min He[1], Dongyan Cai[1], Hongbo Gao[1,2], Cui Li[1], Ke Zhang[1] and Rui Xing[1]

[1]National Institute of Measurement and Testing Technology, Chengdu, 610056, China

[2]Corresponding author's e-mail: 309963382@qq.com

Abstract: This paper mainly studies and designs the clamping tool which is installed in the measuring of big gears with big inner holes and thin walls. This paper analyzes the disadvantages of conventional clamping of large gears with special inner holes, and presents an example of the research and development of the Assembly of the mounting and positioning device of the special large inner hole gear based on the gear measuring instrument. At present, in the field of gear measurement, the measurement of large gears is always a technical problem. The technical ability of gear testing in our country is restricted by various factors. Most gear high-precision instruments and their accessories still depend on imports, which cannot be a good solution to the measurement of special gear installation and positioning problems from the economy and timeliness. Especially, large gears with large bores and thin walls face the problem of determining the position of the measuring device and the measured gears, that is, the measuring datum is difficult to determine [1]. This paper proposes how to use the existing gear measuring instrument to develop and use small accessories, so as to help the production enterprises and measuring institutions solve the problem of reference positioning in the actual production measurement process and determine the accuracy of special gear samples.

1. Introduction

In China, the accelerated in-depth development of the high-end manufacturing industry, machinery industry, aerospace, railway transportation, automobile industry, agricultural electricity mining, and other industrial core transmission components, all involve the key mechanism of gear. At present, the production and manufacturing of gears has become the largest industry in mechanical basic components in China. In 2016, the output value of the domestic gear industry reached 230 billion yuan, ranking first in the world [1]. In the field of gear measurement, contact-type gear measuring instruments are mainly used for measuring precision gears at home and abroad [2-3]. The problem facing the gear measuring instrument is how to determine the position between the measuring device and the measured gear. Especially for some special thin-walled and large internal holes, the installation and positioning of oversized gears on gear measuring instruments is a complex project [4]. Most gear high-precision instruments and their supporting installation components rely on imports, so it is difficult to solve the problem of a wide variety of different shapes of special gear clamping from the economy and timeliness. The following paper describes the current general gear clamping method, analyzes the drawbacks of the current clamping method for special big gears and focuses on explaining the clamping fixture components for thin-walled large internal hole gears.

2. Conventional clamping for gear measurement

At present, the clamping of measuring gears is generally adopted by mandrel with center clamping (Figure 1) or base chuck clamping (Figure 2).

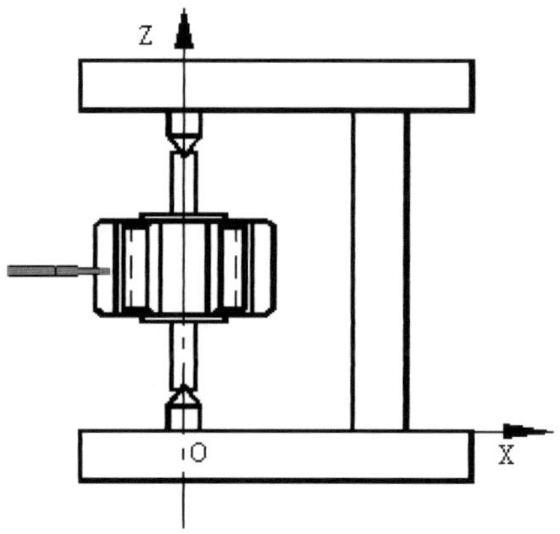

Figure 1. Mandrel with top clamping.

2.1 Mandrel with top clamping [5]

For thin gears with particularly large inner holes and large numbers of teeth, a non-standard gear with a tooth number of 580, a module of 1.02, a pressure angle of 30°, a tooth apex circle of 604 mm, a tooth root circle of 590 mm and a bore size of 580 mm is taken as an example. As shown in Figure 1, this gear requires a 580 mm diameter tapered mandrel to match. However, in actual production and use enterprises, non-standard gears involve various types, and each gear is matched with a mandrel. The coaxiality and straightness of the core shaft and the radial runout accuracy of the two ends are required to be within 2 μm, and the taper is 1:10,000 [6]. At present, it is difficult to process a 580 mm large-diameter high-precision mandrel and maintain a high-precision mandrel for a long time without corrosion and deformation. The preservation environment after processing needs to be dry and at 20°C constant temperature. Only a few constant temperature laboratories have harsh environmental conditions to maintain high-precision requirements. In order to maintain high precision and minimal deformation, large mandrels require the use of special metal materials. The commonly used materials for shaft parts include 35, 45, and 50 high-quality carbon steel, with 45 steel being the most widely used. Alloy steel can be used for heavy loads [6], axial dimensions, weight limitations, or certain special requirements, and chromium plating treatment can be applied to the surface of the core shaft. The density of carbon steel and alloy steel is about 7.8 tons per cubic meter, the mandrel mass of the diameter d is 580 mm and the busbar height L is 150 mm. The calculation is as follows.

$$\rho=7.8\ t/m^3=7.8\times10^{-3}\ g/mm^3$$

$$m=\pi\times r2\times L\times\rho=3.14\times2402\times150\times7.8\times10\text{-}3=211610g=211\ kg \quad (1)$$

At present, the main gear measuring instruments are from manufacturers such as Gleason in Germany, Mahr in Germany, and Klingelnberg Group in the United States. Most models of gear measuring instruments have a designed load-bearing capacity of less than 100 kg, which cannot withstand a core shaft with such a large mass of 211 kg. Large mass mandrels are prone to deformation and eccentricity of the instrument's center due to inertia during gear measurement rotation, resulting in damage to the instrument's center positioning system. At the same time, the axial eccentricity will cause the measurement error of gear radial runout to be amplified [7], and it is not feasible to install this special gear core shaft with the center fitting.

2.2 Base chuck clamping [8]

Figure 2 shows the large chucking gear. Because of the pressure brought by the weight of the large chuck to the turntable, especially in the measurement of gears with more than one tooth, the angle error of the turntable is caused. Taking 580 teeth as an example, the tooth profile and the tooth pitch are measured when the turntable rotates, and the turntable rotates 360°/580, that is 0.62069°. The angle deviation is easy to be caused by the increase of mass and inertia, and the uneven force on the contact tooth surface of the measuring needle is easy to wear and break [9] and finally leads to the uncertainty of the measurement results.

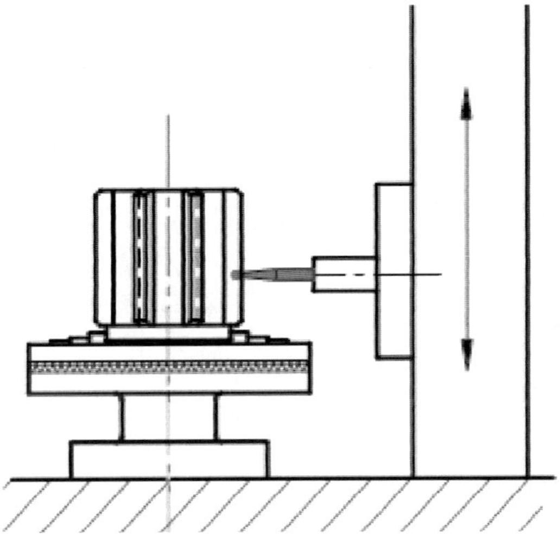

Figure 2. Base chuck clamping.

3. Design example of special gear fixture

The working turntable of the gear measuring instrument has three sets of positioning holes at an angle of 120°, and three assembly clamp components are developed. The whole gear clamping is realized by a turntable equipped with a laser measuring and positioning device. It can replace the traditional mandrel assembly method, which not only makes it less difficult to manufacture, but also has a small clamping contact surface, easy to process, easy to use, easy to disassemble and easy to store. It is not easy to cause deformation or damage due to motion interference with measuring instruments [10] and can be applied to different types of special gears.

3.1 Composition and working principle of a group of clamping components

A set of gear measuring clamping components includes a raised column (Figure 3), a t-shaped positioning rod (Figure 4), a pressure plate (Figure 5), a nut (Figure 6), and a positioning pallet (Figure 7).

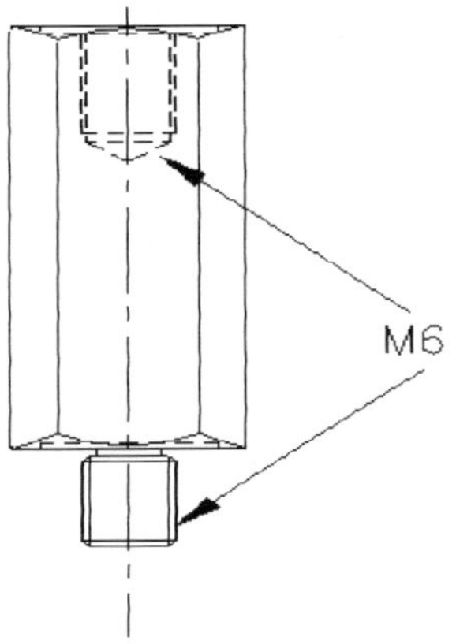

Figure 3. Raised column.

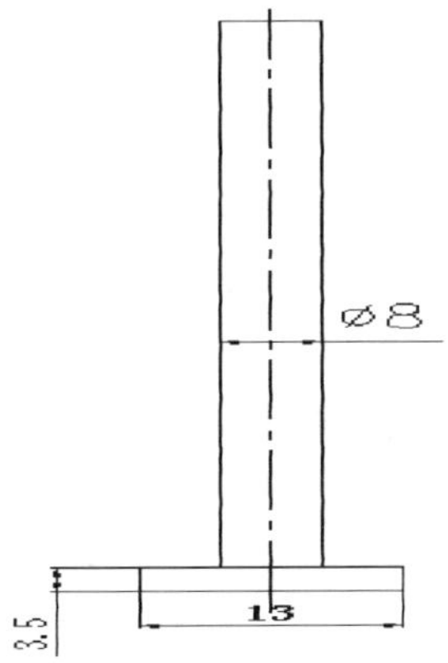

Figure 4. Positioning rod.

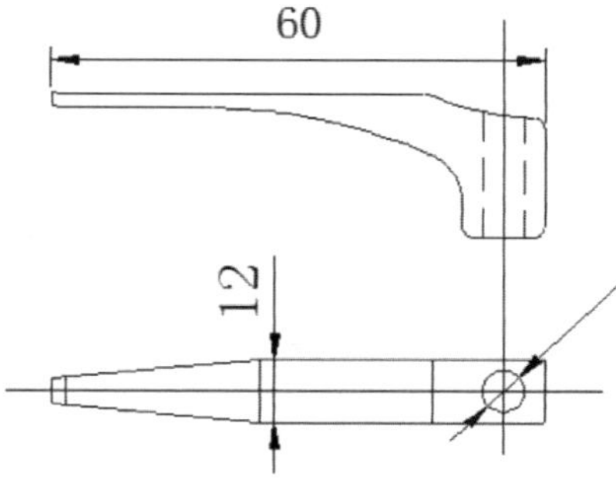

Figure 5. Pressure Plate.

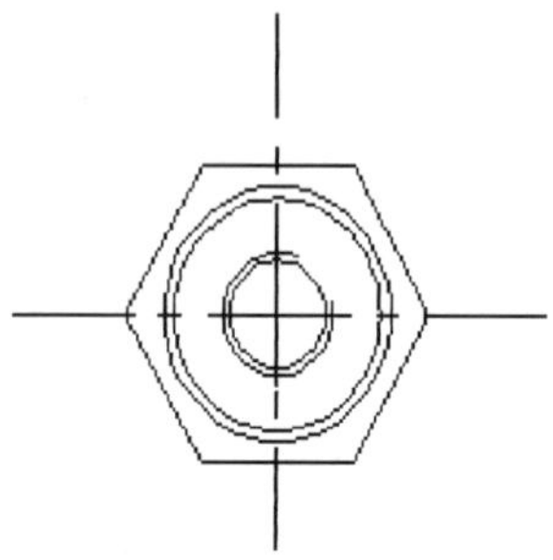

Figure 6. Nut.

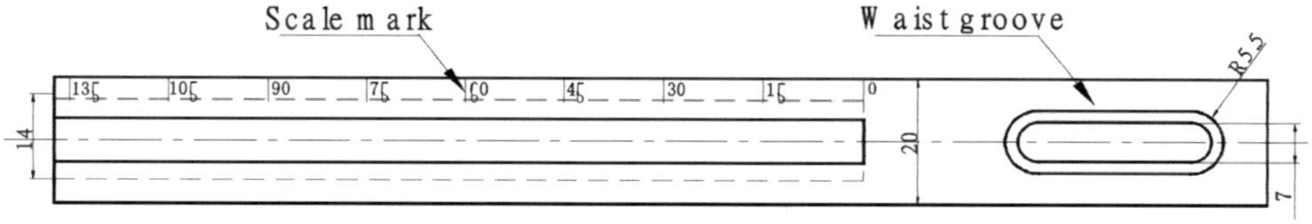

Figure 7. Positioning pallet.

A chute with a scale is processed along the length direction of the positioning plate, the positioning plate is provided with a connecting part, and a supporting surface is formed on one side of the positioning plate. The raised column and the connecting part are fixed with nuts so that the positioning plate is supported on the external platform. The T-shaped positioning rod can slide along the groove direction of the chute. The clamping plate is connected to the positioning rod, and a clamping surface is formed on one side of the clamping plate close to the positioning bracket. The bearing surface and the clamping surface are parallel to each other. The positioning rod is connected to the chute vertically through the sliding part. The positioning rod is arranged with a scale along its length direction. The clamping plate is rotationally connected with the positioning rod, and can be rotated around the axis of the positioning rod, so as to realize the contact of the inner or outer end face of the gear and adapt to the clamping of different scenes. The connection part is the positioning hole, one end of the pad high column is embedded with the positioning hole, and the embedded end of the pad high column is processed with a screw hole. The screw hole is matched with a locking screw, and the nut of the locking screw is used for pressing the hole wall of the positioning hole. The bearing surface, the circumferential surface of the positioning rod and the clamping surface form a clamping space for clamping the inner or outer edge of the gear. The assembly diagram of a group of fixture components is shown in Figure 8.

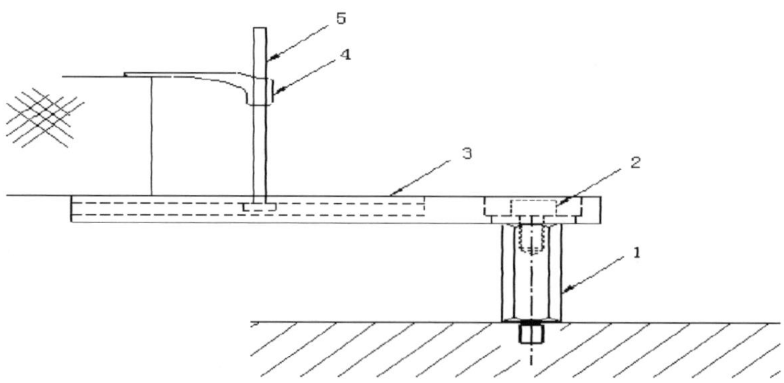

Figure 8. The assembly diagram of a group of fixture components.

3.2 Three assembly clamps on the turntable installation

As shown in Figure 9, the three assembly clamp components 100 are positioned in three directions of 120° on the gear measuring instrument turntable, with each fixed with the cushion height column respectively and the positioning plate fixed with locking screws.

4. Laser measuring device

The turntable 200 of the gear measuring instrument has a positioning center hole 201. The laser measuring device 300 is fixed in the center hole at the first installation station 201 (Figure 9). It has three sets of laser measuring heads 301 and a controling power supply. It is evenly distributed around the first installation station 201, and it is divided into 120° angles. Three laser beams match the radial line distribution of Turntable 200. In working condition, each set of the laser measuring head 301 can locate the position of the clamping assembly 100 in the corresponding direction. The emitted laser illuminates the clamping assembly 100 in the corresponding direction to form a positioning position. The position of the long pallet 100 is adjusted so that the laser beam is in the same position on the same part of the clamping assembly 100. The positioning plate distribution of the clamping assembly 100 is consistent with the radial line distribution of the turntable 200 to provide concentric clamping conditions for gear positioning.

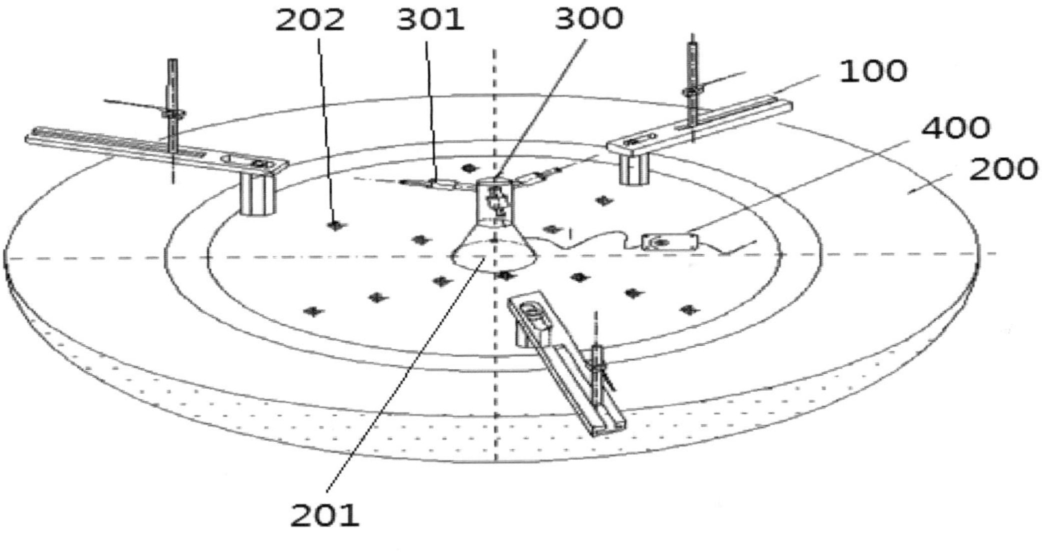

Figure 9. Measuring device.
100 - Fixture components, 200 - Turntable, 201 - Position center hole,
300 - Laser measuring device, 301 - Laser measuring head, 400 - Power source

5. New fixture installation measurement

The big gear is installed on the new fixture, and the beams of the three laser heads are directed to the right side of the long pallet. The left and right positions of the long pallet are adjusted so that the beams are aligned with the right side of the long pallet. The three data of placing the gear on the long tray scale are basically consistent. The pressure plate should be held down. At this time, the center of the gear basically coincides with the center of the turntable of the gear measuring instrument. Gear measurement is carried out, and the error diagram of tooth shape and tooth direction data is shown in Figure 10.

specification	No.of teeth	Normal module	Face width	Probe stylus diameter	Base diameter	Normal Press.angle	Helix angle
	527	1.002mm	10.000mm	1.0024mm	517.1899mm	20.0000°	17.5000°
Profile variation	Left/um						
	Fa	No.396	No.264	No.132		No.1	
		7.7	7.3	9.5		8.4	
	fHa	-8.8	-7.5	-10.0		-7.8	
	ffa	1.9	1.5	1.6		1.9	
	Right/um						
	Fa	No.396	No.264	No.132		No.1	
		5.4	4.8	4.3		6.3	
	fHa	-4.8	-4.7	-3.5		-6.0	
	ffa	2.5	1.5	2.0		1.7	

specification	No.of teeth	Normal module	Face width	Probe stylus diameter	Base diameter	Normal Press.angle	Helix angle
	527	1.002mm	10.000mm	1.0024mm	517.1899mm	20.0000°	17.5000°
Lead tooth alignment variation	Left/um						
	Fb	No.396	No.264	No.132		No.1	
		8.5	7.1	9.3		8.5	
	fHb	-7.1	-5.4	-7.4		-7.3	
	ffb	3.1	4.0	3.8		5.3	
	Right/um						
	Fb	No.396	No.264	No.132		No.1	
		3.9	4.8	5.0		3.9	
	fHb	-3.1	-1.3	-4.9		0.0	
	ffb	5.3	4.5	3.8		3.9	

Figure 10. Profile variation, lead-tooth alignment variation.

6. Advantages of the new fixture

After practical application, the new fixture has the following advantages:

- When the big gear is installed, as long as the outer edge of the gear is placed at the same scale as the long support plate, it has been ensured that the initial installation of the gear and the gear measuring instrument turntable has been basically concentric [10]. When the measuring gear is adjusted, the coarse adjustment can be skipped, and the centering accuracy is high.
- It can avoid the pressure brought by the large mandrel and the large chuck due to its own weight on the turntable, and it is not easy to cause deformation or damage due to movement interference with the measuring instrument [11].

7. Conclusion

This paper analyzes the infeasibility of conventional clamping for special big gear. A case study of installing and positioning device components based on gear measuring instruments is presented. The clamping assembly in this paper solves the problem that the gear measuring instrument cannot clamp the big gear with a large inner hole [12]. The fixture can be adapted to different types of gear installation positioning by forming a clamping space with the inner or outer edge of the gear. It can also save the manufacturing and preservation costs of custom mandrels.

References

[1] Zhang Z, Weng C.Z, Zeng J.F (2012) Gear testing technology, china machine press, Beijing, http://www.cmpbook.com.
[2] Shi Z.Y, Wang S.T, (2017) Gear industry meet the fourth opportunity, China Industry News, (6), www.cinn.cn.
[3] LOTZE W, HAERTIG F. (2001) 3D gear measurement by CMM.In:5th International Conference and Exhibition on Laser Metrology.
[4] Wang J.H (1996) The principle, characteristic and key technology of CNC gear testing center, Tool Engineering, Chengdu, https://www.youfabiao.com/gjjs/
[5] Cui Z.Q, Sun Z.G. (2007) Engineering material, China Machine Press, Beijing. http://www.cmpbook.com.
[6] LI R. (2006) Master Gear. JJG 1008-2006.Inspection and Quarantine of the People's Republic of China, Beijing
[7] Hu L.H, Lang Q.D (1987) Automotive gear and spline mapping, china communication press. Beijing, http://www.ccpcl.com.cn/
[8] Wang C, Xie L.X, Wen GH (2018) Optimization of machining process for hollow gear shaft center hole. Mechanical Engineer. Hei Longjiang.www.jxgcs.com
[9] Shan G.K, Li W.L, Zheng P. (2003) Measurement and compensation of eccentricity in the on-machine inspection of large gears, Journal of Mechanical Transmission, 48-50
[10] Xu X, (Xi`an Technological University), 2022, Position adjustment and measurement coordinate system establishment for side mounted gear measuring machine, DOI:10.27391/d.cnki.gxagu.2022.000643.
[11] Zhu W.G (2006) Research on principle and technology of in-position comprehensive measurement of large gear, Shenyang University of Technology, Beijing, http://www.doc88.com/p-0953990209849.html
[12] Guo T.T, Sun P.Y, Liu Wei, Zhao J (2020) Research on measuring method of tooth profile deviation of large gear, Tool Engineering, 64-67, DOI:10.3969/j.issn.1000-7008.2020.06.016

Design of a non-invasive blood pressure simulator based on oscillographic method

Yaohong Wang[1], Cheng Zeng[1], Meina Fang[1], Yu Yan[1,2], Zhonghao Wang[1] and Bowen Deng[1]

[1]Chongqing Academy of Metrology and Quality Inspection, Chongqing, 401123, China

[2]Corresponding author's e-mail: 1296278430@qq.com

Abstract. In this article, a non-invasive blood pressure simulation system based on the oscillographic method is proposed, and a piston-connecting rod-type variable air chamber is designed and installed in the system as the pulse waveform generation mechanism. It uses PWM constant current control and ultra-small subdivision technology to control the linear stepping motor to achieve stable and accurate displacement and to reproduce the pulse wave according to the blood pressure envelope, to simulate ambulatory blood pressure. In this article, a hardware circuit design scheme is also put forward, including a power supply circuit, pressure acquisition feedback circuit, stepping motor drive circuit, and so on. The experimental results show that the simulator can meet the requirements of JJF 1626-2017 *Calibration Specification for NIBP Simulators* and can be used as a calibration standard with efficiency and accuracy.

1. Introduction

Blood pressure is an important parameter of human vital signs and one of the most important indicators reflecting the hemodynamic state[1-2]. By measuring human blood pressure, which includes systolic and diastolic pressure, we can monitor the function of the human cardiovascular system and find abnormal conditions such as hypertension and hypotension. Blood pressure is an important index for the discovery, diagnosis, treatment, rehabilitation, and prevention of diseases[3-4]. Non-invasive blood pressure measurement is an indirect and non-invasive method using a cuff and pressure sensor[5]. Non-invasive automated sphygmomanometer based on oscillography which measures the oscillation waves in the blocked cuff, is widely used in medical institutions and families, because of its advantages such as easy operation and intuitive indication[6-7]. Non-invasive sphygmomanometer belongs to Class II medical devices, and the reliability of their metrological performance will directly affect the diagnosis results and treatment effects, while their malfunction can even endanger the health and life safety of patients[8]. JJG 692-2010 *Non-invasive Automated Sphygmomanometers* puts forward clear requirements for the measurement performance of both static and dynamic non-invasive sphygmomanometers. In this article, a non-invasive blood pressure simulation system is designed based on oscillography, which realizes the evaluation of measurement range, indication error, air tightness, overpressure protection, and other critical indexes of static pressure, through the standard pressure sensor and static pressure module. Meanwhile, the same system can also evaluate the dynamic blood pressure measurement performance of non-invasive sphygmomanometers by reproducing the human artery pulse wave through the blood pressure envelope.

2. Measurement principle

The oscillographic method is based on the principle of blood vessel unloading to realize blood pressure measurement, which means the extraction and analysis of human pulse waves. The inflatable cuff that reaches a certain pressure is used to block the blood vessel, and during the slow and uniform deflation process, certain means are used to record the pressure oscillation waves in the cuff. After the trend of gas pressure waves in the cuff is extracted and analyzed, systolic and diastolic blood pressure can be determined by a fixed ratio calculation method or mutation point criterion[9]. The principle of dynamic blood pressure simulation is derived from the process of measuring blood pressure using the oscillographic method, which involves reproducing the pulse wave corresponding to blood pressure through a certain executing mechanism. The shape and trend of pulse waves are closely related to the blood pressure envelope, and different blood pressure values have different blood pressure envelopes. The blood pressure envelope is a virtual line formed by connecting each pulse wave peak in a set of pulse waves. The blood pressure envelope can intuitively reflect the relationship between cuff pressure and pulse wave amplitude[10]. Dynamic blood pressure simulation requires the blood pressure simulator to restore the corresponding amplitude pulse wave at the corresponding cuff pressure based on the blood pressure envelope. The specific output waveform P (t) is represented by Equation (1):

$$\mathrm{P(t)} = \mathrm{P_c} + \varphi(\mathrm{DBP, SBP}) Y \mathrm{P_c} \quad (1)$$

where P_c is the pressure coefficient of the cuff, φ (DBP,SBP) is the pressure coefficient corresponding to the set systolic and diastolic blood pressure, T is the period of the pulse wave, and Y is the pulse waveform, which is represented by Equation (2):

$$Y = A\left[\sin(2\pi ft) + h\sin(4\pi ft + l)\right] \quad (2)$$

where A is the amplitude of the pulse wave, h is the position of the descending middle gorge, l is the position of the descending middle wave, and f is the frequency of the pulse wave.

3. System components

A linear actuator is used to compress the gas in the sealed chamber according to a certain pattern, generating a pressure pulse that is superimposed on the pressure waveform of the main chamber, to simulate the pulse wave picked up from the cuff. The blood pressure simulator is mainly composed of a microprocessor, pressure sensor, cylinder, piston, stepper motor, air pump, and solenoid valve. The microprocessor senses the static pressure of the pneumatic system and controls the stepper motor which is connected to the piston through a guide screw, and the piston is driven to move back and forth in the cylinder. The pressure of the entire gas circuit system including the cylinder will change because of the reciprocating motions of the piston, which will result in oscillation waves. The amplitude of the oscillation wave is determined by factors such as the amplitude of the piston motion, the static pressure of the gas circuit system, and the amount of gas. The cylinder piston group is equipped with a position sensor to indicate the actual position of the piston, to avoid errors caused by friction and position deviation. The pressure sensor is connected to the cylinder and the entire gas circuit to measure the gas pressure of the circuit. A direct current diaphragm pump is connected to the cylinder to inflate the entire gas circuit. The pump is equipped with additional components such as a filter with an inlet passage and a one-way valve with an outlet passage to improve inflation quality and gas tightness. There are two channels between the cylinder and the outside world. One is the front panel interface that connects to the non-invasive blood pressure device being tested, and the other is connected to the outside world through an electromagnetic valve as a vent. The block diagram of the noninvasive blood pressure simulation system is demonstrated in Figure 1.

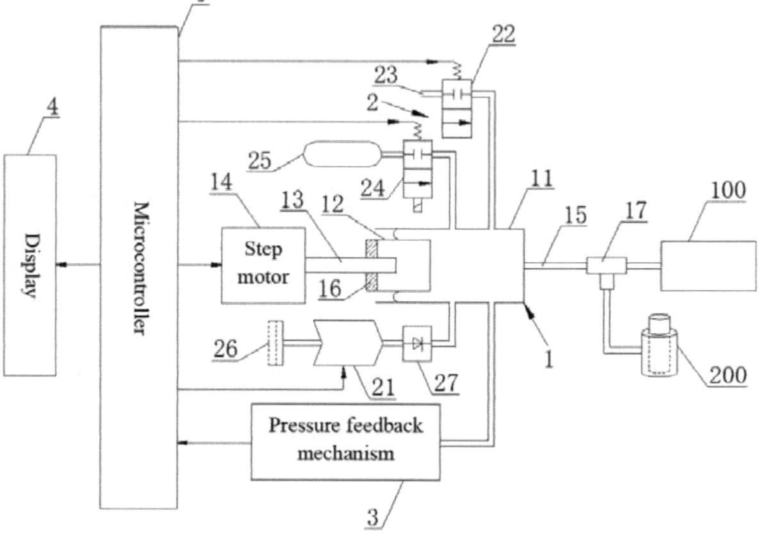

Figure 1. Block diagram of noninvasive blood pressure simulation system.

4. Hardware design

The hardware design of the simulation system mainly includes the design of the hardware circuit, and mechanical structure of the pulse wave generation platform and pneumatic device, which is used to generate pressure pulse waves and provide pneumatic pressure in static pressure measurement.

4.1. Hardware circuit design

The hardware platform of the non-invasive blood pressure simulation system consists of 8 circuits, which include a power supply, core control circuit, LCD circuit, pressure acquisition circuit, A/D conversion circuit, stepper motor drive circuit, pneumatic device drive circuit, and matrix keyboard drive circuit. The block diagram of the system hardware platform is shown in Figure 2.

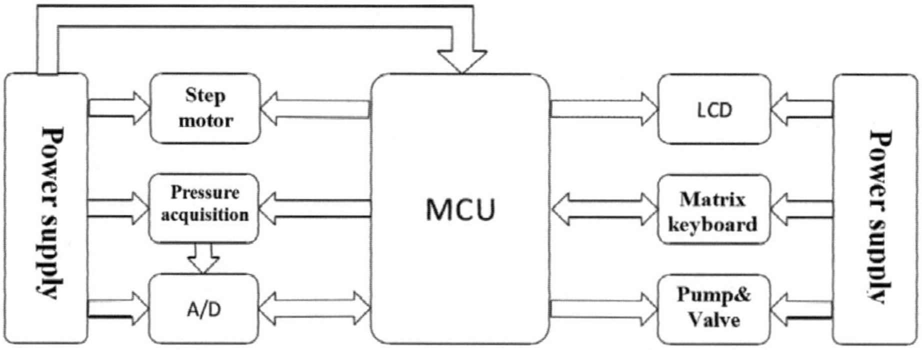

Figure 2. Block diagram of system hardware platform.

The boost conversion circuit and buck conversion circuit of the power supply are both designed with switch regulators, to realize high efficiency and power density with smaller size and heat dissipation area. LM5122 boost chip has a maximum input voltage of +65 V and a maximum output voltage of +100 V. The power supply of the system has multiple power networks, whose structure is relatively complex. The input voltage DC+12 V is converted into the required voltage at all levels by switching the power supply chip, and the ancestor capacitor and ceramic capacitor are connected in parallel at the input and output ends of the power network at all levels respectively, to play the role of filtering and decoupling. The control unit of the core control circuit adopts the STM32F103xE series microcontroller, with a main frequency of up to 72 MHz. It has a large number of powerful I/O ports and peripheral resources and is packaged in LQFP100 with a size of 14 x 14 mm. It is a compact and powerful microprocessor with rich hardware resources, such as 512 KB FLASH storage, a considerable number of timers, A/D conversion, and various communication interfaces. In addition, the external main power supply voltage of this controller is up to 4.0 V, while both the port pull and fill current are up to 150 mA.

4.2. Mechanical structure design

The design principle of the pulse wave generation platform is to use a linear stepper motor as the rhythm generator of the system, simulating the work of the human heart. By regularly squeezing the gas in the gas chamber, pressure waves are generated to simulate the pulse waves of a human body, achieving the goal of simulating blood pressure. In this article, the pulse wave generation platform of the non-invasive blood pressure simulation system is designed based on the oscillographic method, and the platform mainly consists of an air chamber, an air chamber fixing device, a piston with a guide screw, a linear stepper motor, and a stepper motor driver. The connection diagram of various devices on the pulse wave generation platform is shown in Figure 3.

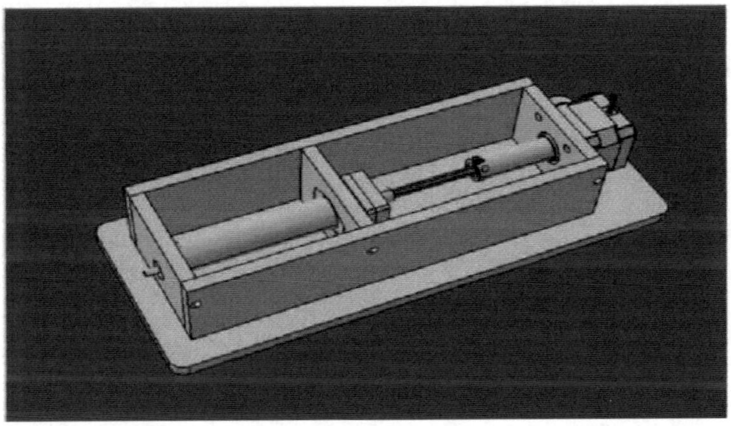

Figure 3. Connection diagram of pulse wave generation platform.

The volume of the gas chamber is not yet specified. The scheme adopted in the article is 20 cc, and the maximum simulated pressure is designed to be 300 mmHg. According to Boyle Mariotte's law, the product of pressure *P* and volume *V* of the gas chamber is a constant, which can determine the relationship between the stroke of the stepper motor, piston displacement, and bottom radius. The system adopts Ametek 43H4-2.33-A05 hybrid linear stepper motor, with a step angle of 1.8° and a step length of 0.05 mm. It is equipped with a high-performance subdivision driver and PWM constant current control to achieve small vibration and stable operation at low speed. The pneumatic device mainly consists of an air pump and an electromagnetic valve and can meet the requirements of static pressure testing: the pressure range is 0 to 40 kPa, and the air tightness is satisfying. The system uses a CJP37C series pressure pump manufactured by Xiamen Kunjin Electronics, which can provide a gas pressure of 50 kPa with air tightness of less than 3 mm Hg/min.

5. Experimental results and analysis

To evaluate the metrological performance (such as accuracy, repeatability, and stability) of the simulator, and to check whether the equipment can meet the requirement of JJF 1626-2017 *Calibration Specification for NIBP Simulators*, first, the simulator is connected to a standard pressure gauge and a metal container with a three-way valve to calibrate its static pressure simulation function, and the experimental results are shown in Table 1.

Table 1. Experimental results of static pressure of blood pressure simulator.

Standard pressure	Measurement value 1	Measurement value 2	Indication error
50.00 kPa	50.02 kPa	50.02 kPa	0.02 kPa
40.00 kPa	40.02 kPa	40.02	0.02 kPa
30.00 kPa	30.02 kPa	30.02 kPa	0.02 kPa
10.00 kPa	20.00 kPa	20.00 kPa	0.00 kPa
0.00 kPa	10.00 kPa	10.00 kPa	0.00 kPa

Secondly, the simulator is connected to a standard pressure gauge metal, a pressure converter, and a digital oscilloscope, then the simulator on adult mode is set, to calibrate its simulated pulse wave variability, and experimental results are shown in Table 2.

Table 2. Experimental results of simulated pulse wave variability.

Set value	Variability	Technical requirement
Systolic blood pressure 20.0 kPa	0.0%	≤10%
Average blood pressure 15.5 kPa	0.0%	≤10%
Diastolic blood pressure 13.3 kPa	0.0%	≤10%
Systolic blood pressure 8.0 kPa	0.0%	≤10%
Average blood pressure 5.3 kPa	0.0%	≤10%
Diastolic blood pressure 4.0 kPa	0.0%	≤10%

6. Conclusion

In the article, a non-invasive blood pressure simulation system is proposed based on the oscilloscope method, and a piston-connecting rod-type pulse blood pressure waveform generation mechanism with a variable air chamber is designed. PWM constant current control and ultra-small subdivision technology are used to control the direction and displacement of the linear stepper motor, and the pulse waves are replicated based on the blood pressure envelope. The experimental results show that the designed simulator can accurately simulate stable pulse waves, and metrological performance can fully meet the requirements of JJG 692-2010 *Non-invasive Automated Sphygmomanometer*, with high measurement accuracy and resolution. At the same time, it is easy to operate with excellent stability and repeatability. Further studies will focus on improving the anti-interference ability of the transmission line to avoid significant errors caused by signal interference during the working process.

Acknowledgments

This work was financially supported by the Scientific and Technological Project of Market Regulation Administration of Chongqing Municipality - Research and Application of Multi-parameter Vital Signs Simulator (Grant No. CQSJKJ2019011), and Special Project for Performance Incentive and Guidance of Scientific Research Institutions of Chongqing Municipality-Research and Implementation of Measurement Standard for Human Vital Signs (Grant No. cstc2020jxjl120012).

References

[1] Sun, Y., et al. (2018). Noninvasive blood pressure calibration of multi-parameter ECG monitor and its applications in quality control. Chinese Journal of Medical Physics, 35 (08): 951-955. doi: 10.3969/j.issn.1005-202X.2018.08.016.

[2] Jiang, W. J., et al. (2020). Quality Control and Data Analysis of Multi-parameter Monitor Non-invasive Blood Pressure Based on FLUKE Prosim8. China Medical Devices, 35 (11): 70-72. doi: 11.3969/j.issn.1674-1633.2020.11.020.

[3] Xu, J. J. (2016). Study on the Application of Support Vector Machine in Quality Control of Non-invasive Blood Pressure. China Medical Devices, 31 (11): 127-128, 139. doi: 10.3969/j.issn.1674-1633.2016.11.037.

[4] Li, H. L., et al. (2022). Development of wearable multi-physiological parameters monitoring equipment. Transducer and Microsystem Technologies. 41 (11): 73-76, 80. doi: 10.13873/J.1000-9787(2022)11-0073-04.

[5] Chu, Z. Q., et al. (2013). Research on the Performance Testing of Non-invasive Measurement of Blood Pressure (NIBP) and Problems of Multi-parameter Monitor. China Medical Equipment, 10 (7): 37-38. doi: 10.3969/J.issn.1672-8270.2013.07.013.

[6] Chen, Y. F. (2017). Discussion on performance testing techniques of non-invasive automated sphygmomanometer (oscillometry). Chinese Medical Equipment Journal, 38 (5): 132-134, 137. doi: 10.7687/j.issn1003-8868.2017.05.132.

[7] Cao, J. W. (2021). Progress and prospect of non-invasive blood pressure measurement based on pulse wave signal. Journal of Biomedical Engineering Research, 40 (2): 220-224. doi: 10.19529/j.cnki.1672-6278.2021.02.20.

[8] Wu, X. S., et al. (2017). Development of Simulation System of Noninvasive Blood-Pressure Meter. Clinical Medicine & Engineering, 24 (8): 37-38. doi: 10.3969/j.issn.1674-4659.2017.08.1038.

[9] Godana, A. A., et al. (2023). Bayesian longitudinal modeling of blood pressure measurements of hypertensive patients at Wachemo University Nigist Elleni Mohamed Memorial Teaching and Referral Hospital Hosanna, Southern Ethiopia. American Heart Journal, 9 (12): e22984. https://doi.org/10.1016/j.heliyon.2023.e22984.

[10] Koshy, A. N., et al. (2018). Accuracy of blinded clinician interpretation of single-lead smartphone electrocardiograms and a proposed clinical workflow. American Heart Journal, 205: 149-153. https://doi.org/10.1016/j.ahj.2018.08.001.

Design of power supply system for portable aviation inspection equipment

Yang Yang[1,2], Hao Ai[2,4], Shibo Li[1,3] and Guokai Xiao[2]

[1]Key Laboratory of Computing Power Network and Information Security, Ministry of Education, Shandong Computer Science Center (National Supercomputer Center in Jinan), Qilu University of Technology (Shandong Academy of Sciences), Jinan, China
[2]Shandong Standard Institute of Emerging Technologies and Innovations Co., Ltd., Jinan, China
[3]Shandong Provincial Key Laboratory of Computer Networks, Shandong Fundamental Research Center for Computer Science, Jinan, China

[4]Corresponding author's e-mail: aihao@sieti.cn

Abstract: Given that airborne electronic equipment may have a variety of wiring methods and may encounter a set of systems using multiple power supplies, it is necessary to take some measures on the circuit to ensure the normal work of the power supply and the post-circuit. A portable aviation detection equipment power supply system is designed in this paper. The system supports simultaneous access of two power supplies to identify and judge the input power supply and switch between the battery and the external power supply, effectively avoiding the security risks caused by battery backflow current. The battery power is monitored in real-time through the battery power detection circuit and will be reminded when the battery power is too low. Since the circuit uses a MOS tube to control the switching of the power supply, there is almost no voltage drop loss. In addition, the EMC protection circuit is set up at the power supply side to further improve the safety and reliability of the system.

1. Introduction

In the actual design, for the multi-power supply system, it is generally required that each power supply is independent of the other and does not interfere with each other to supply power to the product, but there will inevitably be multiple power supplies accessing the product at the same time. If certain measures are not taken in the circuit, it may lead to the post-stage circuit not working normally. And the battery side may also lead to certain security risks, such as backflow current, battery heating, and other conditions [1]. Based on this problem, an automatic power switch system is designed in this paper for airborne electronic equipment, which can identify and judge the type of access power supply, and automatically disconnect the battery connection when the external power supply is connected. This system aims to ensure that only one power supply supplies the equipment at the same time, avoids the safety risks caused by current backflow, and at the same time, ensures that the rear circuit will not lose power, which may cause serious consequences such as reset and restart, data loss, and even component damage. In addition, the voltage is smooth when switching. The system has the advantages of low power consumption, fast response, and good electromagnetic compatibility [2].

2. Principle of power switching system scheme

This scheme makes full use of the switching characteristics of the MOS tube to achieve the purpose of switching between the battery and external power supply. That is, for PMOS tubes, the gate-source voltage Vgs is less than a certain value, which is more suitable for high-end drive scenarios [3]. At the same time, the MOS tube has almost no voltage loss and is safer and more reliable. A classic circuit is shown below, as shown in Figure 1. Its circuit flow is as follows: when two power supplies are connected to the system at the same time since the external power supply has power, the Vgs of the PMOS is > 0, so the PMOS parasitic diode is cut off, and the rear load is powered by the external power supply; when the external power supply is disconnected, the PMOS is switched on, and the rear load is powered by the battery [4].

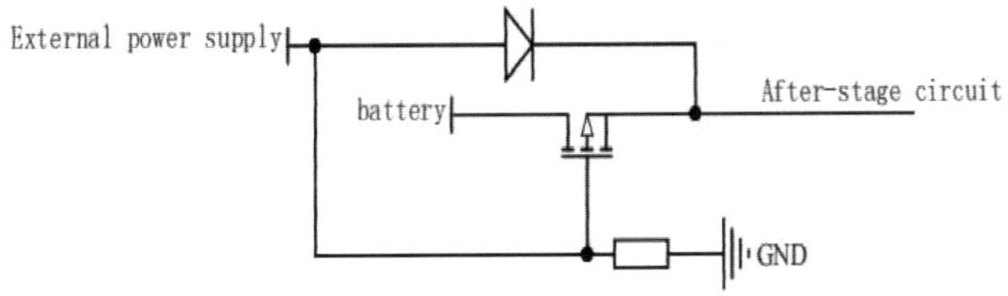

Figure 1. Power switching circuit of MOS tube.

Based on this circuit, appropriate improvement and optimization are made according to the actual situation, and a power switching system suitable for aviation airborne electronic equipment is designed. As shown in Figure 2, automatic and seamless power switching can be achieved after testing, and the voltage is stable and reliable when the battery and external power supply are powered separately [5]. When the two power supplies are connected at the same time, the external power supply is preferred for the power supply. At the same time, the battery power supply is cut off, forming a disconnect.

In aviation and airborne electronic equipment, a 28 V voltage power supply is commonly used to power various load equipment, and many chips only need 3 V or even lower operating voltage. In this system, a 3.6 V battery is designed to share the post-load circuit with a 28 V external power supply. When the battery is connected separately, Vg < Vs, the PMOS tube is on, and the battery supplies power to the rear circuit. When the 28 V external power supply is connected, the voltage is first regulated from 18 V to 5 V through the linear voltage regulator chip, and then from 5 V to 3.6 V. The 5 V voltage signal is connected to the gate of the MOS tube as a control signal to control the on-off of the MOS tube, to achieve the purpose of automatic switching between the battery and the external power supply. The 5 V voltage signal acts on the gate, while 3.6 V acts on the source, Vg > Vs, Vgs is about 1.4 V, the PMOS tube is off, and the connection between the battery and the circuit is cut off, to avoid the adverse effects such as battery damage caused by the external power supply for the battery to reverse charge [6].

The circuit also added EMC electromagnetic compatibility secondary protection circuit, to eliminate the power supply terminal voltage fluctuations and electrostatic impact on the circuit. A filter capacitor is added to the linear voltage regulator chip input and output, to further ensure the stability of the voltage.

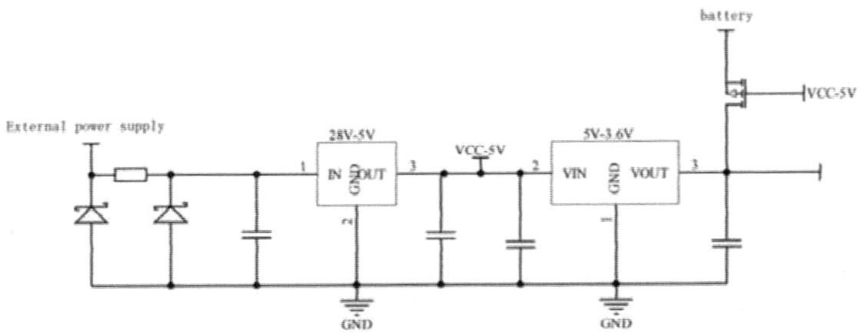

Figure 2. Battery and external power switching circuit.

In addition, the system also has power supply identification, battery remaining power detection, low power reminder, and other functions, as shown in Figure 3. When the I/O pin is high power, the system is powered by the external power supply. You can choose to enable some high-power functions or perform other processing. At the same time, the ADC voltage detection function can be used to detect the remaining battery power with a simple circuit. The diode and capacitor are used to achieve the purpose of decoupling filtering and protecting components, and the remaining battery power can be calculated according to the voltage change detected by the ADC pin of the MCU. At the same time, different methods can be used to warn when the battery power is low according to actual needs.

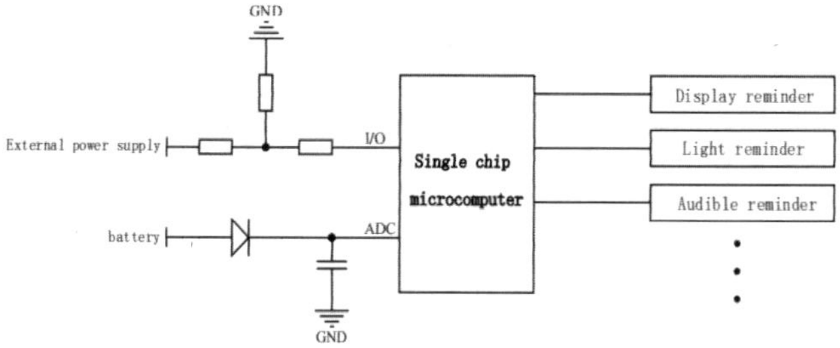

Figure 3. Power identification, power monitoring, and reminder.

3. Reliability test

To ensure that the whole power supply switching system meets the requirements, the oscilloscope is used to capture the voltage change in various cases and analyze whether it meets the expected requirements [7]. In Figures 4 - 7 and 10 - 11, the voltage unit of the oscilloscope is set to 5 V per cell, and the time unit is set to 4 ns per cell. The dashed line is used to draw the cell, and the solid line is used to represent the voltage reference line at 0 V. In Figure 8 and Figure 9, the oscilloscope voltage unit is set to 500 mV per cell, and the time unit is set to 400 ms per cell. The cells are also drawn with dotted lines, and the solid line is used to represent the voltage reference line at 0 V. This is because the voltage change time at the instant of power switching is about 200 ms, and the voltage amplitude change is about 200 mV. When observed in 5 V and 4 ns units, the voltage change cannot be clearly seen.

3.1. Batteries and external power supplies are supplied separately

3.1.1. The battery is powered separately. The test battery is a lithium battery with a nominal voltage of 3.6 V, and the actual voltage value is 3.68 V. The oscilloscope probe measures the battery voltage at the rear circuit, as shown in Figure 4.

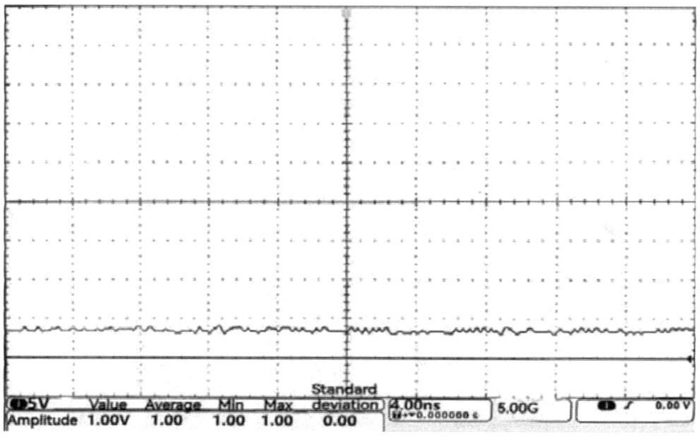

Figure 4. Battery-powered output.
(Units: 5 V and 4 ns per cell)

3.1.2. External power supply separately. The external power supply alone is tested, respectively measuring 28 V and 5 V, and the voltage of the three points of the rear circuit is tested.

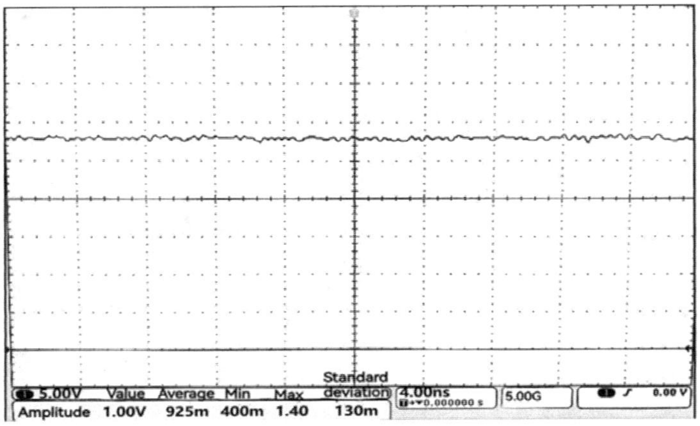

Figure 5. 28 V voltage of external power supply.
(Units: 5 V and 4 ns per cell)

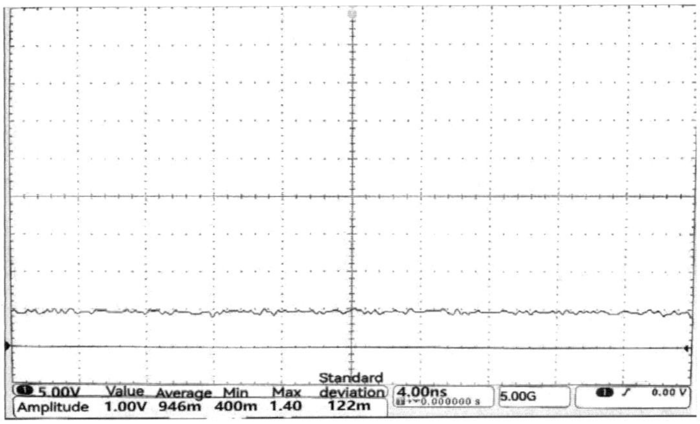

Figure 6. 5 V voltage of external power supply.
(Units: 5 V and 4 ns per cell)

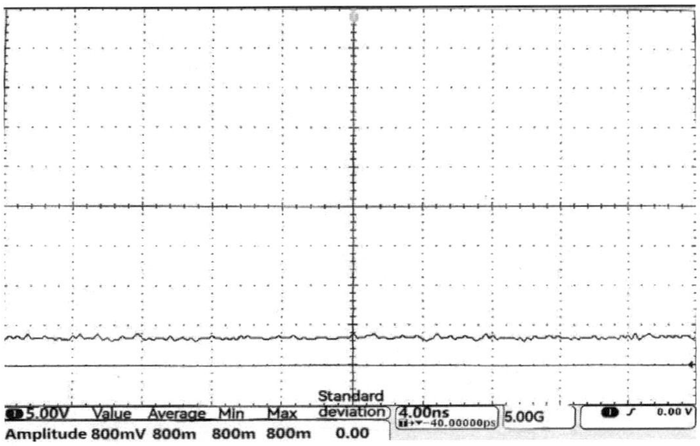

Figure 7. 3.6 V voltage of external power supply.
(Units: 5 V and 4 ns per cell)

3.2. Power switchover instant

The power switch moment includes four situations: connecting and disconnecting the external power supply when the battery is powered, and connecting and disconnecting the battery when the external power supply is powered. According to the circuit principle of this system, since the battery voltage is slightly higher than 3.6 V, connecting and disconnecting the external power supply when the battery is powered should produce smooth and stable voltage changes. When the external power supplies, regardless of whether the battery is connected, the voltage value should be stable, that is, when the external power supply is connected, the MOS tube will be controlled to disconnect the battery.

The oscilloscope is used to capture the waveform of the power switch moment to ensure that the power switch will not cause power failure and restart of the post-stage equipment, affecting the normal operation of the electronic equipment [8]. It can be found that the voltage change of the output to the rear circuit at the moment of power switching is smooth and slow, and the voltage is continuous. There is no power loss, and there is no relatively large voltage fluctuation, which meets the expected requirements of power supply reliability [9].

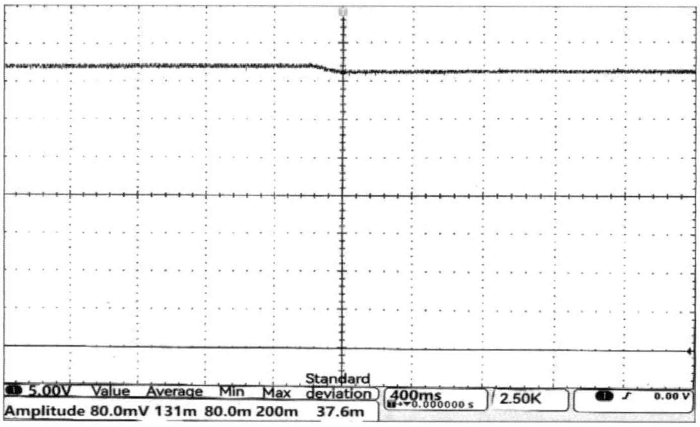

Figure 8. Instantaneous connection of the external power supply. (Units: 500 mV and 400 ms per cell)

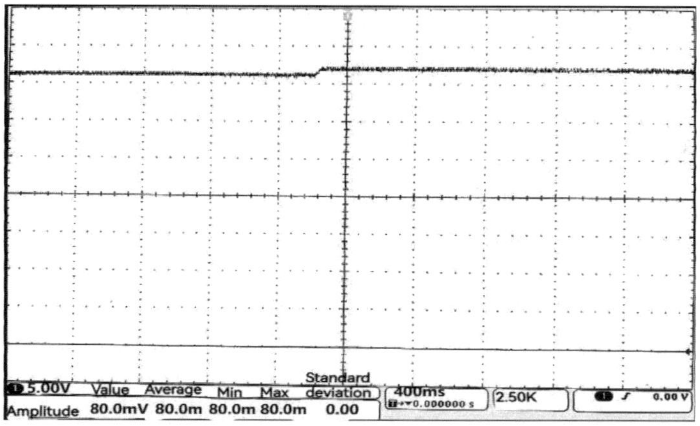

Figure 9. Instantaneous disconnect of external power supply. (Units: 500 mV and 400 ms per cell)

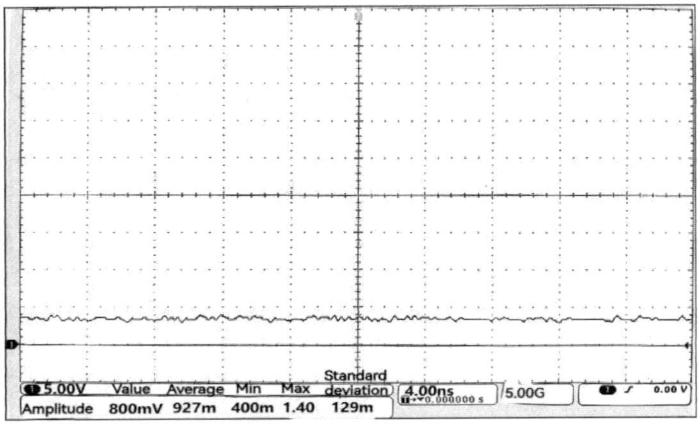

Figure 10. Instantaneous battery connection.
(Units: 5 V and 4 ns per cell)

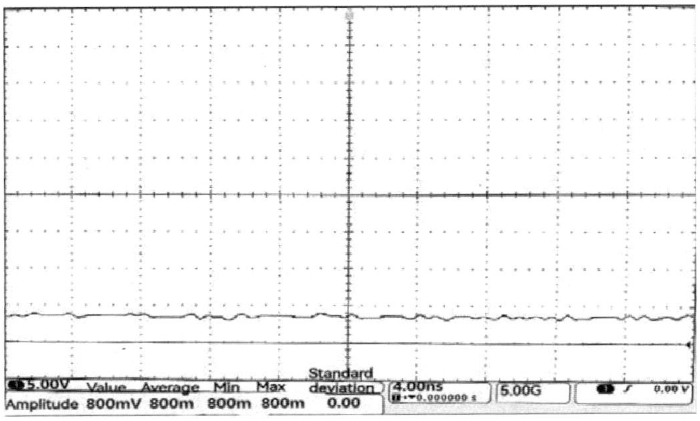

Figure 11. Battery disconnect moment.
(Units: 5 V and 4 ns per cell)

3.3. *Two power supplies are supplied at the same time*

To prevent the current from pouring back into the battery, the ammeter is used for testing. The specific method is to connect the ammeter in series to the battery end circuit. Provides 4.7mA of current to the circuit when the battery is connected alone, as shown in Figure 12. When the external power supply is connected, it can be observed that there is neither current output nor reverse current inflow at the battery end, to ensure the safety and reliability of the system, as shown in Figure 13.

Figure 12. Battery supply current.

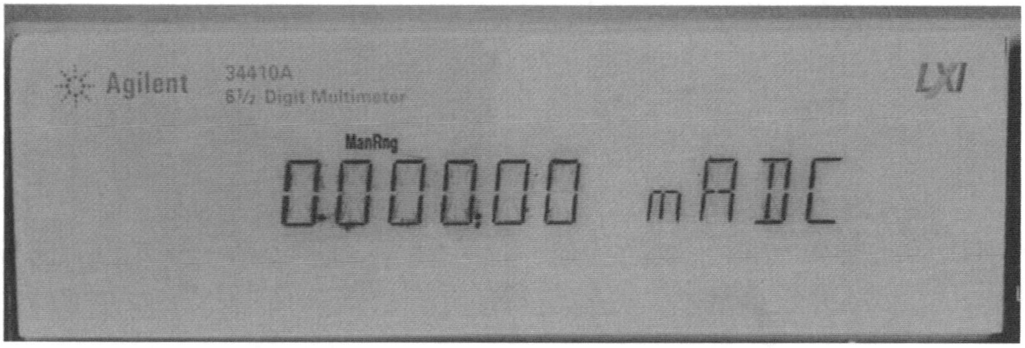

Figure 13. Battery current after the external power supply is connected.

4. System workflow

First of all, the power supply is identified and judged by the detection circuit (Figure 3) and enters different operating modes according to the judgment results. Due to the limited amount of power that the battery can store, the battery is usually used in low-power mode to avoid frequent battery replacement, only the necessary functions are retained. At the same time, DAC is used to detect the remaining power in the battery (Figure 3), and timely alarm reminds the need to replace the battery when the power is insufficient. When the power supply is switched or connected to the system at the same time, since there is no power consumption limit when using an external power supply, it will enter normal mode. The battery and the external power supply switching circuit (Figure 2) will automatically select the power supply to ensure that the current will not flow from the external power supply to the battery, causing safety hazards [10]. Figure 14 is the workflow block diagram of the power supply system of portable aviation inspection equipment.

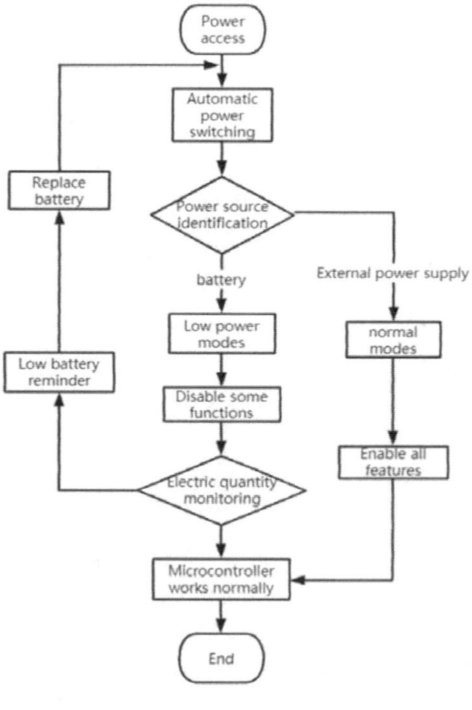

Figure 14. Block diagram of system workflow.

5. Conclusion

This paper introduces a portable power supply system for aviation inspection equipment, which supports automatic switching between two power supplies without losing power when switching, and has the advantages of battery power detection, low power reminder, power identification, and EMC circuit protection. It is mainly used in the power supply circuit of airborne electronic equipment. After testing, the system is stable and reliable, and the power supply voltage is stable, which has a certain promotion value and use value.

Acknowledgments

This work was funded by the Shandong major science and technology innovation project (2021JMRH0112A) and Qilu University of Technology (Shandong Academy of Sciences) Industry-University-Research collaborative innovation project (2021-CXY21), the Project of improving the innovation ability of small and medium-sized sci-tech enterprises in Shandong province (2022TSGC2167).

References

[1] Popoola A. I; Akinpelu E. O; Ewetumo T (2021) Development and Performance Evaluation of an Intelligent Electric Power Switching System. J. Journal of Trend in Scientific Research and Development. Volume 5, Issue 3.
[2] Adel Oubelaid; Nabil Taib; Toufik Rekioua (2022) Novel coordinated power sources switching strategy for transient performance enhancement of hybrid electric vehicles. J. Compel. Volume 41, Issue 5. 2022. PP 1880-1919.
[3] Jiang Xin-Jian. (2006) PLC control of single-load to Dual-power automatic switching, Electronic Glass Branch of Chinese Silicate Society. Proceedings of Electronic Glass Technology Exchange.
[4] E. Huang; C. Hill (1999) Optimising Trench MOS devices for power switching applications. IEE Colloquium Recent Advances in Power Devices1999. PP 2.
[5] Li Furong. (2021) Research on fast Switching Technology of Dual Power Supply, One Use, one backup System. J. Electrical Apparatus Industry, No.243 (02): PP 62-65.
[6] Yang Changcun. (2022) Design of an Intelligent Flexible Power Switching Device. J. Electronic Technology,51 (09): 316-317.
[7] Wang Wengui. (2015) Reliability Analysis of Dual Power Automatic Changeover Switch. J. Electronic World, No.482 (20): 150-151.
[8] Hu Jindong, He Feng. (2019) Research on Working Principle and Switching test of standby and Dual Power automatic changeover switches in substation. J. Gansu Science and Technology, 48 (03): 24-26+42.
[9] Huang Le. (2021) Design and discussion of intelligent detection System of Dual Power Automatic Changeover Switch (ATSE). J. Electronic testing, No. 465 (12): 8-10.
[10] Zhang Weiwei. (2015) Design of Intelligent Controller of Dual Power Automatic Transfer Switch. D. Shenyang University of Technology.

Hydraulic fault prediction of integrated transmission system based on LSTM

Xiongjun Liu [1,3], Yuwen Jin [1], Shuai Yao [1], Chuangbo Hao [1], Jinle Zhang [2] and Yi Xu [2]

[1] Beijing Jinghang Computation and Communication Research Institute, Beijing 100074, China
[2] China North Vehicle Research Institute, Beijing 100072, China

[3] Corresponding author's e-mail: liuxiongjun1.163.com

Abstract: Integrated drive-line fluids play a role in transmitting torque, providing control pressure, lubrication, and cleaning in the integrated drive-train. The increase of oil abrasive particles will accelerate the wear rate of rotating parts, while insufficient oil will lead to rapid ablation and gluing of rotating parts. According to statistics, more than 75% of hydraulic equipment failure is caused by oil pollution. Oil pollution leads to filter blockage which will cause fatal failure so that the entire transmission system cannot work. Therefore, it is extremely important to predict the fault and health management of the hydraulic system. This paper studies the oil fault of the integrated transmission system, proposes a data preprocessing method, and solves the problem of difficult extraction of deterioration index in the multi-channel strong noise signal of the hydraulic system through the information fusion algorithm and the mean smoothing algorithm. According to the abnormal data judgment criterion based on the sliding window, the identification of the early fault starting point of the hydraulic system is realized, which provides a basis for the start-up fault prediction and the division of normal and degraded states. A multi-channel fusion fault prediction algorithm for hydraulic systems based on a Long-Short Term Memory (LSTM) deep network is proposed, which realizes long-term and high-precision fault prediction of hydraulic systems. Finally, the oil failure prediction algorithm of the integrated transmission system is verified by using real vehicle fault data.

1. Introduction

In the integrated transmission hydraulic system, the coupling relationship among the pressure parameters is complex, and the failure degradation mechanism is complex, too. It is difficult to establish a failure model, and it is impossible to predict the failure through the model-based method [1-4]. A data-driven approach is digging mine equipment degradation information directly from historical data to predict deterioration trends and remaining service life. This method gives full play to the role of the data, which does not need to analyze its complex mechanical model and transmission relationship, reduces the dependence on the equipment degradation mechanism, and uses the data-driven fault prediction method to realize the fault prediction of the integrated transmission hydraulic system [5-6].

Data-driven online fault prediction generally only predicts a single channel, while an integrated transmission hydraulic system has many hydraulic parameters, and the integration of multiple hydraulic parameters can obtain more accurate fault prediction results. Long-Short Term Memory

(LSTM) deep network is used for long-sequence natural language processing and text processing, with strong long-sequence data processing capabilities. When the network structure is reasonably designed, its processing power increases with the increase of training data. The LSTM deep network fault prediction algorithm has been successfully used in the multi-parameter fault prediction of the NASA aero-engine simulation fault dataset. The network is used for comprehensive transmission hydraulic system fault prediction. High-precision fault prediction of hydraulic systems can be achieved by setting reasonable network depth, number of neurons at the network layer, and training duration.

In summary, the failure model cannot be constructed for the integrated transmission hydraulic system, therefore it is necessary to build a learning model based on historical data in the situation when offline training is possible. In this model, the multi-channel parameters of the hydraulic system will be integrated to train a powerful fault prediction model to realize the failure prediction of the integrated transmission hydraulic system.

2. Methods

2.1 Data preprocessing

2.1.1. Elimination of the outlier. This abnormal point is caused by the acquisition system, which is not the failure of the transmission system. The oil pressure stable degradation is a slow change process. The sudden failure of the hydraulic system should be through iterative comparison to eliminate the outlier, through the current moment data and the previous moment of the collection data for comparison. The previous moment set the zoom ratio. When the data fluctuation is large beyond the set range, it can be considered that the data point is a system abnormal point, and then we replace the current moment value with the previous moment sampling point.

$$x^t = \begin{cases} x^{t-1} & if\, x^t < \lambda x^{t-1} \\ x^t & if\, x^t > \lambda x^{t-1} \end{cases} \tag{1}$$

where λ is the scaling factor, and $\lambda = 0.8$.

2.1.2. Sliding window smoothing. Because the median value has a good smoothing effect on the data mutation value, the method of taking the median value of the data point in the sliding window is used to smooth the data, and the length of the sliding window is set to $slidewidth$.

$$x^t = median(x^{t:t+slidewidth}) \tag{2}$$

where $median()$ means to take the median value of the data series. We take $slidewidth = 100$, which takes a median point for every 100 sampling points.

2.1.3. Mean smoothing. We smooth the data by using the mean. The smoothing process is:

$$x^t = mean(x^{1:t}) \tag{3}$$

where $mean()$ represents the average processing from the moment of acquisition to this moment.

2.2 Hydraulic signal deterioration index normalization

To improve the versatility of model processing, the data of each channel is normalized, so that the distribution range of each channel data is consistent. In the process of predictive model training, each channel is equally important to the model parameters. Data normalization can improve the feature mining ability, prediction ability, and generalization ability of the model. We set the acquisition signal as $X_n^t, n = 1,...,N, t = T_0,...,T_t$, where N represents the sampling data of each channel to the period because the numerical range of each channel varies greatly. To improve the generalization ability of the model, the sampling value is first normalized, each channel is normalized separately, and all channels are normalized within the same range, as shown below:

$$X_i^{T_0:T_{0+t}} = \frac{X_i^{T_0:T_{0+t}} - \mu_i}{\sigma_i}, i = 1, \dots, N \tag{4}$$

where μ_i is the mean and σ_i is the variance of the i th channel.

2.3 Early fault detection

Correctly determining the starting point of faults is conducive to discovering early faults, reasonably segmenting training samples, and starting predictive models. The principle 3σ is often used to monitor anomalies. The sliding window 3σ principle is used to detect oil pressure abnormal data. When the deterioration index at a certain moment deviates from the variance of the previous eigenvalue at this moment L more than 3 times, it can be considered as an abnormal oil pressure parameter. The feature sequence is $\{X^t\}, t = T_0, \dots, T_t$.

$$\left\| x^{t+1} - \frac{1}{L}\sum_{i=t-L}^{t} x^i \right\| > 3 \times \sqrt{\frac{1}{L-1}\sum_{i=t-L}^{t}\left(x^i - \frac{1}{L}\sum_{i=t-L}^{t} x^i \right)} \tag{5}$$

2.4 Fault prediction algorithm of LSTM hydraulic system multi-channel fusion

2.4.1 LSTM algorithm architecture. LSTM increases the control logic of the forgetting gate, memory gate, and outputting gate compared with RNN. LSTM RNN replaces the hidden layer of traditional RNN through the memory unit, in which the correction coefficient represents the state through a node [7-10].

(1) Forget gate.

The first step of LSTM is to decide what information to keep and to discard. This is implemented by a Sigmoid function, while this function is called “forget gate”. Its input is x^t and h^{t-1}, and the output value is between 0~1, where 0 means all information is abandoned, and 1 means all information is retained. The forget gate can be expressed as:

$$c^t = f^t c^{t-1} + i^t tanh(W_{cx} X_n^t + W_{ch} h^{t-1} + b_c) \tag{6}$$

(2) Input gates and input nodes.

The next step is to choose to introduce new information and store it in the internal state. This step consists of two parts: (1) The Sigmoid function acts as an “input gate” to decide what information is updated; (2) The tanh function acts as a vector for the input node to generate a new state for updating the state. The input gate and input node equations can be expressed as:

$$\begin{aligned} i^t &= \sigma(W_{ix} X_n^t + W_{ih} h^{t-1} + W_{ic} c^{t-1} + b_i) \\ f^t &= \sigma(W_{fx} X_n^t + W_{fh} h^{t-1} + W_{fc} c^{t-1} + b_f) \end{aligned} \tag{7}$$

(3) Output gate

Finally, an “output gate” is formed by a Sigmoid function, which is used to determine the output information. The internal state is controlled in the range of -1~1 through the tanh function, and the output result is obtained by multiplying with the output gate, as shown below:

$$\begin{aligned} o^t &= \sigma(W_{ox} X_n^t + W_{oh} h^{t-1} + W_{oc} c^{t-1} + b_o) \\ h^t &= o^t tanh(c^t) \end{aligned} \tag{8}$$

2.4.2 Loss function. By constructing the objective function to train LSTM, we set the true remaining lifetime as:

$$R = \left\{ R_n^t \mid n = 1, \ldots, N; t = 1, \ldots, T_n \right\} \tag{9}$$

We take the mean square of the predicted value and the true value as the loss function, that is:

$$\text{Cost} = \frac{1}{N} \sum_{i=1}^{N} \left(R_i - h_i \right)^2 \tag{10}$$

As the objective function of the entire network architecture, the entire network is trained by backpropagation. The training process adopts the gradient descent optimization algorithm for backpropagation training and stops training when the training results converge or reach the maximum number of training rounds.

3. Experiments

This fault occurs in the real integrated transmission system. During the vehicle driving, the oil pressure before and after the control filter is imbalanced and then makes the hydraulic system fail. The raw data of hydraulic system failures is shown in Figure 1. The data sampling period is 50 ms, the sampling point is about 2.8×10^5, and the length of the collected data is 3.85 h.

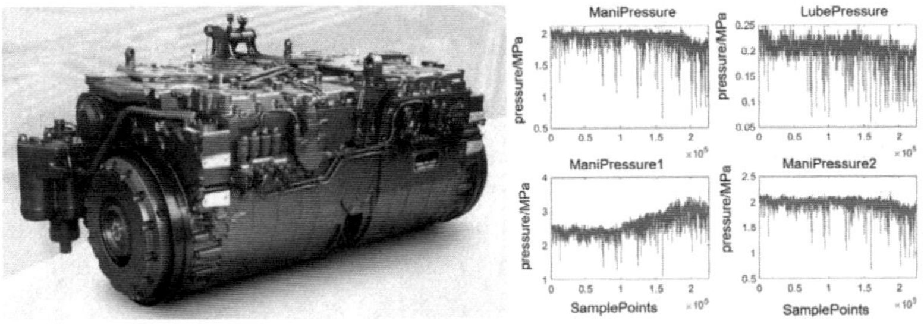

Figure 1. Integrated transmission system and raw data of hydraulic system failures.

In this paper, the deterioration degree index is first extracted, and the fault starting point is determined by 3σ principle. Finally, the remaining life prediction of the multi-channel hydraulic system is carried out by using LSTM.

3.1 Construction of dataset

The use of reasonable time intervals can reduce the amount of calculation and improve the prediction accuracy. For the sparse sampling rate data of 50 ms, we take a point every 600 points, which means the time interval between every two points is 30 s. After data preprocessing and 3σ principle fault point identification, the deterioration degree index is shown in the Figure 2, where the red dotted line is the fault starting point obtained by the 3σ principle judgment.

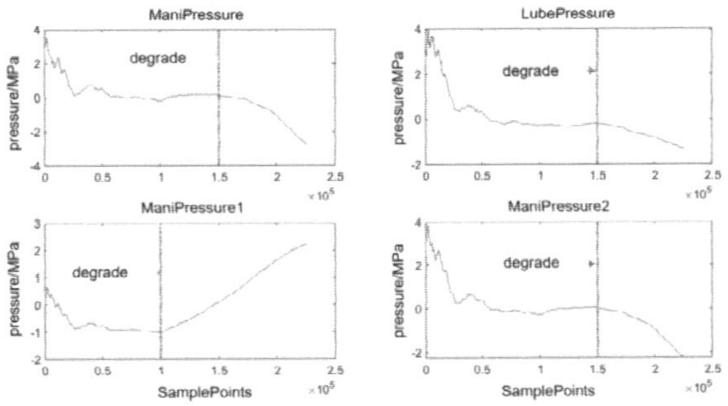

Figure 2. Early fault detection.

The fault prediction algorithm of the LSTM multi-channel integrated transmission system needs to construct a training set to train the model offline. Each training sample includes 4 channels, and each channel has 10 sampling points, constituting 4×10 input characteristics. Each sample corresponds to a residual life value. According to the 3σ principle, before the fault start time, RUL is constant. While the fault appears, RUL begins to be linear degradation. The remaining life value is obtained, as shown in the Figure 3. We input 4×10 oil pressure parameters to the LSTM RNN and the current oil pressure corresponding to the remaining life value will be predicted.

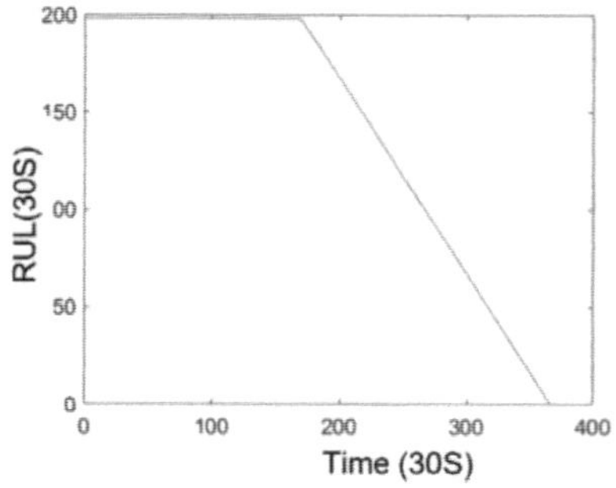

Figure 3. Remaining useful life (RUL).

3.2 Training process

In this paper, the LSTM RNN model is written by using the Pytorch framework. The LSTM RNN model has three layers. The number of neurons is set to 50, the Dropout is taken as 0.5, and the learning rate is set to 0.001. 80% of the sample points are randomly selected as the training data set, and the other 20% of the data not participating in the training is used to finally test the prediction accuracy of the training model.

The training process uses GPU for acceleration. The training platform is the same as the machine used in the second chapter of this article. The loss function value with the iterative process of its convergence curve is shown in the Figure 4. During the training process, the loss is rapidly reduced. After 100 rounds of training, the training process is completed.

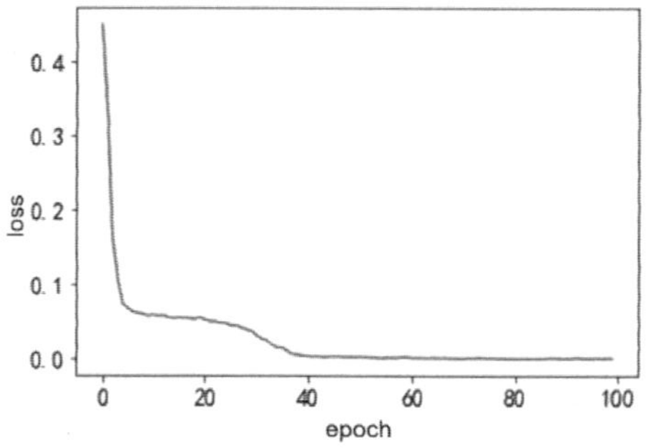

Figure 4. Loss function value curve.

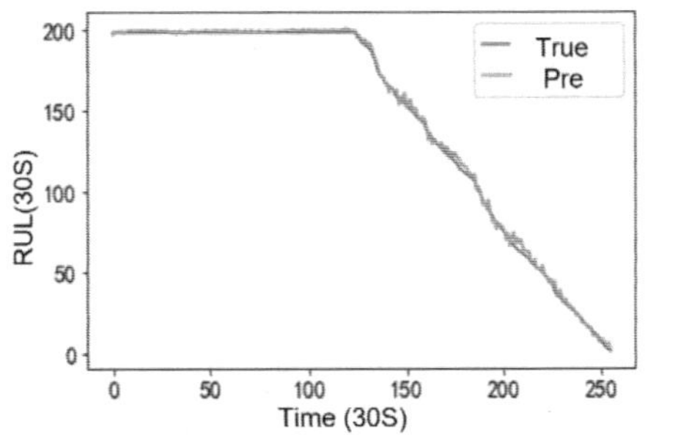

Figure 5. Prediction results of the training set.

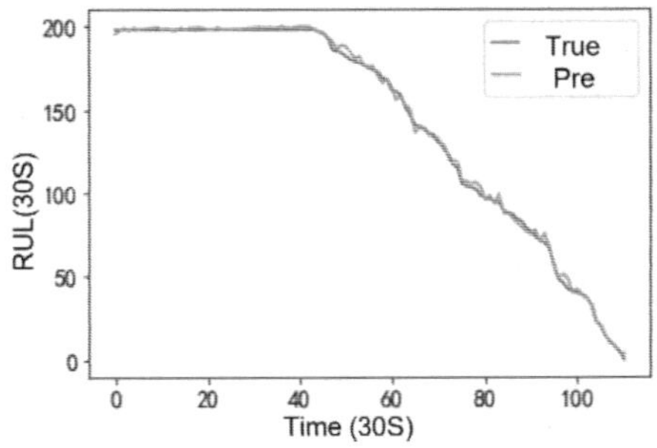

Figure 6. Prediction results of the test set.

The prediction results of the training set are shown in Figures 5 and 6. The predicted values of the training set are in good agreement with the real values. The trained model is used to test the test set to obtain prediction results and real value results, as shown in Figure 5. The test set also has a good prediction effect. From normal data to the beginning of degradation, the LSTM RNN fault prediction model can give an accurate remaining life value. Its prediction ability is not limited by the prediction time. From the degradation starting point, the remaining life can be obtained. There are 104 minutes from the final failure. LSTM RNN fault prediction model can dig deep into the degradation information contained in the oil pressure data, which is of great significance for the multi-channel residual life prediction of the integrated transmission hydraulic system.

SVR and gray prediction models are used to predict the oil pressure change trend of a single channel. The prediction accuracy of SVR and gray prediction models is significantly lower than that of LSTM, as shown in Figure 7 below. These two methods can only learn and predict for a single channel, which limits the model's modeling ability to the failure process.

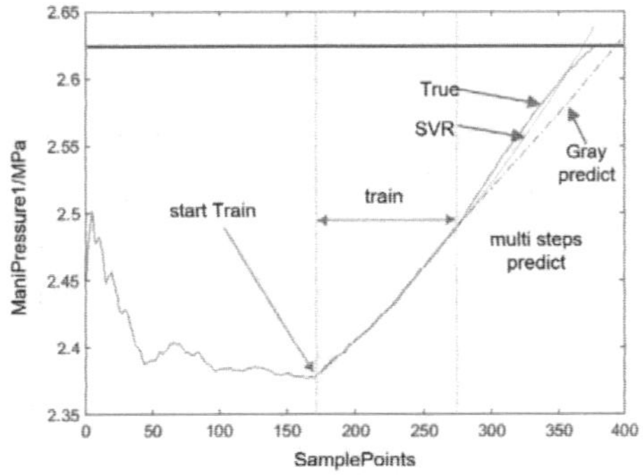

Figure 7. Prediction results based on Support Vector Machines (SVM).

4. Conclusion

In this paper, by establishing a multi-channel LSTM prediction model, online fault prediction and offline training are realized in the hydraulic system of the integrated transmission system. The prediction model is verified and analyzed through the actual fault data in the process of a real car sports car. LSTM has good prediction accuracy, benefiting from its deep learning architecture. The prediction ability of the LSTM model will be enhanced with the increase of sample data. Under the condition of sufficient data, the LSTM trained offline can achieve a longer prediction time and better prediction accuracy. The hydraulic fault prediction method of the integrated transmission system can better provide maintenance decisions.

Acknowledgments
This work was supported by the National Basic Scientific Research Program of China (Grant No. JCKY2019602B002, No. JCKY2020204C021 and No.NSFC52175074).

References

[1] Zheng Changsong, et al. Dynamic balance study of hydraulic system pollution control machine [J]. Tool & Hydraulics, 2005, 33 (8): 126–129.
[2] Li Jianlin, et al. Target pollution degree analysis of hydraulic and lubrication system of vehicle integrated transmission [C]. VII National Session Equipment Maintenance and Transformation and Proceedings of the 10th National Conference on Equipment Lubrication and Hydraulics, Hefei, 2011: 374–380.
[3] Zhang Jinle, et al. Research on oil contamination of integrated transmission hydraulic system [C]. The 9th National Equipment Lubrication and Hydraulics Proceedings of Academic Conferences, Yichang, Hubei, 2007: 111–114.
[4] Zheng Changsong, et al. Research and application of oil products to the life test of the bench of stereotyped products. China Mechanical Engineering, 2006, 17 (9): 975–978.
[5] Tang Decai. Reliability modeling and analysis of integrated transmission of armored vehicles [D]. Changsha: National University of Defense Technology, 2013.
[6] Gao Zhen, et al. Research on online monitoring sensor of integrated transmission oil metal abrasive particles [J]. Journal of Guangxi University (Natural Science Edition), 2017, 42 (2): 409–418.
[7] Srivastava N., Hinton G., Krizhevsky A., et al. Dropout: a simple way to prevent neural networks from overfitting [J]. The Journal of Machine Learning Research, 2014, 15 (1): 1, 929–1, 958.
[8] Ma M. and Mao Z. Deep-convolution-based LSTM network for remaining useful life prediction [J]. IEEE Transactions on Industrial Informatics, 2020, 17 (3): 1, 658-1, 667.
[9] Shi Z. and Chehade A. A dual-LSTM framework combining change point detection and remaining useful life prediction [J]. Reliability Engineering & System Safety, 2021, 205: 107, 257.
[10] Miao H., Li B., Sun C., et al. Joint learning of degradation assessment and RUL prediction for aero-engines via dual-task deep LSTM networks [J]. IEEE Transactions on Industrial Informatics, 2019, 15 (9): 5, 023-5, 032.

Journal of Physics: Conference Series **2724** (2024) 012017 doi:10.1088/1742-6596/2724/1/012017

Design and implementation of control software for radar antenna performance index test system

Bin Deng[1], Chaohua Liang[1,2], Jingjing Wang[1] and Wei Lv[1]

[1]Air Force Early Warning Academy, 288 Huangpu Street, Wuhan, Hubei Province, China

[2]Corresponding author's e-mail: 261624064@qq.com

Abstract. According to the actual requirements of radar antenna performance testing, focusing on the design of the instrument control software of the test system, in this paper, the functional requirement of the instrument control software of the system is analyzed, the basic framework of the software design is put forward, and the design ideas and implementation methods of the instrument and equipment management as well as the specific design methods of the instrument and equipment control are given. Practical applications have shown that the control software has advantages such as an intuitive interface, convenient operation, and good interactivity. Combined with system hardware, it can achieve fast and accurate measurement of radar antenna performance and has high practical value.

1. Introduction

The radar testing system is mainly used for testing the technical performance indicators and anti-interference performance of radar systems and has been widely used in various aspects such as radar development, production, and use. To improve the measurement accuracy and efficiency of radar testing systems, scholars in relevant fields have conducted in-depth research on the design of radar testing systems. For example, Gu and Liu analyzed the construction method of injection-type radar testing systems and conducted in-depth research on key technologies in response to the testing requirements of radar systems [1]. Zhou et al. proposed a new radar testing method based on simulation technology, which constructs a static combat electromagnetic environment and implements precise detection from terminal to terminal in the radar system [2]. Liu et al. proposed a design scheme for a Doppler navigation radar testing system based on semi-physical simulation, focusing on the hardware structure, software design, and semi-physical simulation system structure based on an echo simulator [3]. The above research results provide technical support for the construction of a radar antenna performance index testing system from a theoretical perspective.

The radar antenna performance testing system, abbreviated as the antenna performance testing system, is an automatic testing system designed specifically for rapid testing of radar antenna lobe width, sidelobe level, zero-depth, and other indexes [4-8]. A typical antenna performance testing system consists of two parts: hardware and software, as shown in Figure 1. Among them, the system hardware mainly consists of programmable measurement instruments, communication equipment, testing antennas, measurement control computers, and other components. Program-controlled measuring instruments mainly include the RF signal source, spectrum analyzer, digital storage oscilloscope, etc. The RF signal source is mainly used to generate test signals, the spectrum analyzer is mainly used to measure the antenna pattern of the received signal, and the oscilloscope is used to collect antenna azimuth data. The system software mainly includes instrument control software and data analysis software. Among them, the instrument control software mainly realizes the program control of RF signal source, spectrum analyzer, digital storage oscilloscope, and other measuring instruments, which is the key to system software design. Focusing on the design and implementation of the instrument control software of the antenna test system, in this paper, the Functional requirement of the instrument control software of the system is first analyzed, then the basic framework of the design of the instrument program control software on this basis is put forward, and finally, the specific program of the software design from the two aspects of instrument management and instrument control is given.

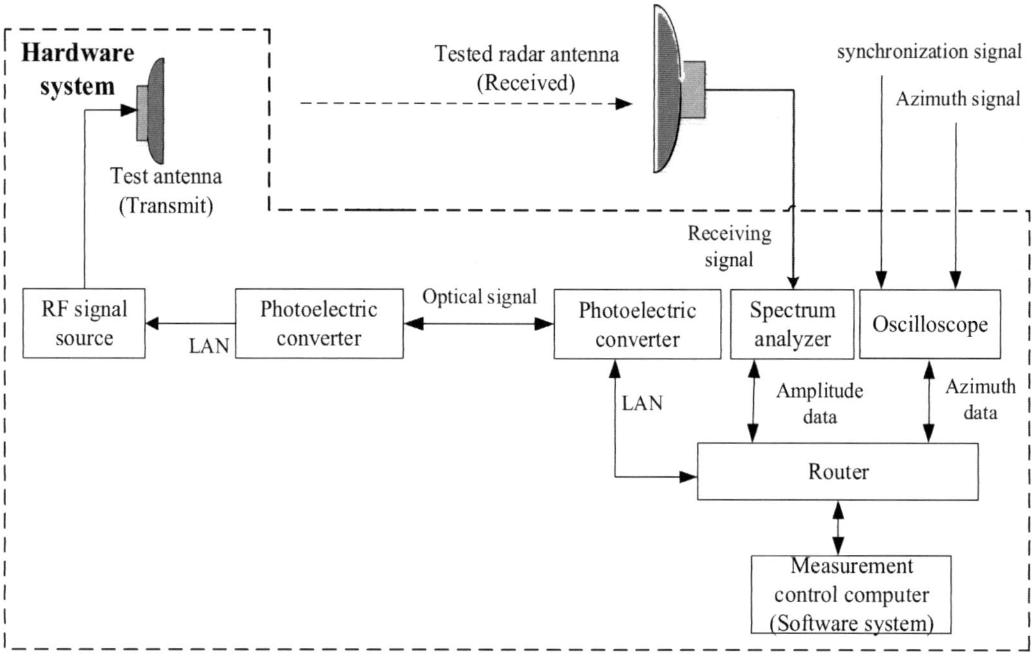

Figure 1. Basic composition diagram of the radar antenna receiving pattern testing system.

2. The Software Functional Requirement and Basic Architecture

2.1. Functional requirement of instrument control software

The software of the antenna performance testing system mainly includes instrument control software and data analysis software. Among them, the instrument control software mainly realizes the program control of RF signal source, spectrum analyzer, digital storage oscilloscope, and other measuring instruments. The data analysis software mainly analyzes the data collected by measuring instruments, generates the curve of the tested radar antenna pattern, and automatically generates an "antenna performance test report" based on the antenna performance test results and technical index requirements.

This paper focuses on the design and implementation of instrument control software, which mainly includes two parts: instrument and equipment management and instrument and equipment control. The specific functional requirement is as follows:

(1) Instrument and equipment management function: for commonly used electronic measuring instruments with typical communication interfaces such as serial port (RS232), GPIB, LAN, etc., the programmable computer can establish a test connection with them, automatically identify the device identity (manufacturer, instrument model, etc.), display the communication address, port number, and other status information of the instrument and equipment, and achieve management such as adding, deleting, and online status monitoring of the instrument and equipment.

(2) Instrument and equipment control function: first, the instrument parameter programming setting function. For RF signal sources, it is required to achieve programmable settings of signal frequency, output power, and other parameters. For the spectrum analyzer, it is required to be able to realize programmed settings of IF frequency, bandwidth, reference level, scanning time, and other parameters. For digital storage oscilloscopes, it is required to be able to achieve programmable settings of time sensitivity, vertical sensitivity, trigger source, and trigger electrical parameters. The second is the storage and display function of instrument data. For oscilloscopes, it is required to be able to read the signal waveform displayed by the instrument in real-time, store the waveform data on a programmable computer, and display it at the same time. For the spectrum analyzer, it is required to read the signal spectrum displayed by the instrument in real time, store the spectrum data in the programmable computer, and display it at the same time.

2.2. The basic architecture of instrument control software

To improve universality and portability, the system software adopts a virtual instrument software architecture, designed according to a three-layer structure of hardware layer, driver layer, and application layer [9-11]. This makes the testing system driver and hardware transparent and also makes the complex underlying bus control protocols and communication protocols more open. The system software design can be completed through simple secondary development based on the driver provided by the RF instrument. The basic architecture of the software system is shown in Figure 2. In the figure, the system software consists of three parts: hardware layer, driver layer, and application layer.

(1) Hardware layer. The spectrum analyzer, RF signal source, digital storage oscilloscope, and other measuring instruments in the system hardware communicate with the control computer remotely through the serial port (RS232), GPIB, LAN, and other typical communication protocols, to realize the remote control of the measuring equipment by the system software.

(2) Driver layer. It mainly includes a serial port to USB, GPIB to USB, and other communication interface drivers, as well as an RF signal source, spectrum analyzer, digital storage oscilloscope, and other measurement instrument drivers. The control computer controls the corresponding measuring equipment through driver layer-related programs.

(3) Application layer. According to the functional definition in software design, the application layer mainly includes two functional modules: instrument equipment management and instrument equipment control. Among them, the instrument and equipment management function module mainly realizes the online addition and deletion of instruments, the selection of serial port, GPIB, LAN, and other communication methods, the setting of GPIB address, IP address, and other communication addresses, and the setting of communication ports, as well as the display of the status information of measuring instruments, the communication between the control computer and measuring instruments (spectrum analyzer, RF signal source, oscilloscope, etc.), and the identification of measuring instruments. The instrument equipment control function module mainly realizes the setting of parameters such as measuring instrument signal frequency, output power, and reference level, as well as the reading and display of waveform data or spectrum data. Specifically, the parameter settings of the Spectrum analyzer mainly include center frequency, frequency span, starting frequency, ending frequency, amplitude reference level, scanning time, scanning mode, etc. The parameter settings of RF signal sources mainly include frequency, power, modulation switch, radio frequency switch, etc. The parameter settings of the oscilloscope mainly include horizontal sensitivity (time sensitivity), vertical sensitivity, trigger source, trigger electricity, etc.

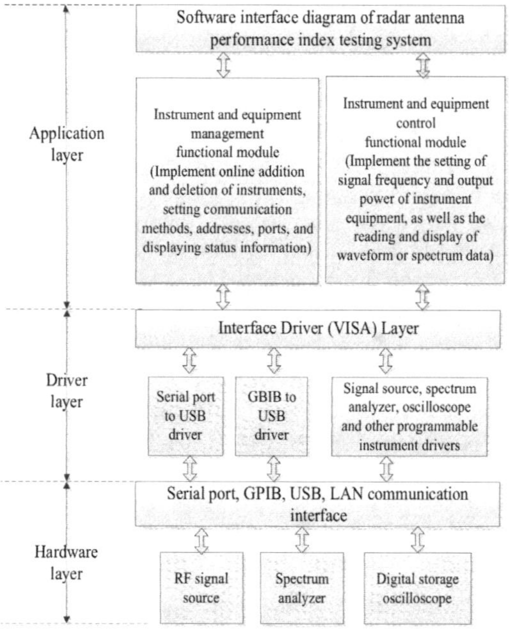

Figure 2. Instrument control Software architecture.

3. Implementation of instrument and equipment management module

The design and implementation of control software will be discussed from two aspects: instrument equipment management and instrument equipment control. Among them, the instruments and equipment include a signal source, spectrum analyzer, oscilloscope, etc, which is under control.

3.1. Functional requirement of instrument control software

The instrument equipment management module mainly implements functions such as online addition and deletion of instruments. The instrument addition function mainly involves the selection of communication methods, IP settings, and port settings, as well as the connection testing and display of status information for instrument equipment communication status. The instrument and equipment management interface designed based on the above functional definition is shown in Figure 3. In the figure, on the left is the “Tree structure of instrument and equipment resource management”, which mainly implements the functions of adding and deleting instruments and equipment. On the right is the “Instrument Equipment Communication Parameter Settings Bar”, which mainly implements functions such as communication mode selection, communication parameter settings, connection status testing, and status information display.

In the figure, when adding instruments in the instrument equipment management module, corresponding communication methods can be selected based on the specific communication interface of the measuring instrument. The communication interfaces that can be selected include a serial port (RS232), GPIB, LAN, and other methods. When serial communication mode is selected, communication parameter settings include port number (COM1, COM2, COM3), Baud, etc. When selecting the GPIB communication method, the communication parameter settings specifically include the GPIB address, port number, etc. When selecting the LAN communication method, the communication parameter settings specifically include IP address, port number, etc. After selecting the communication mode and setting the communication parameters of the instrument, the connection test of the communication status can be carried out. The status information of the instrument equipment will be displayed in the “Status Information Bar”.

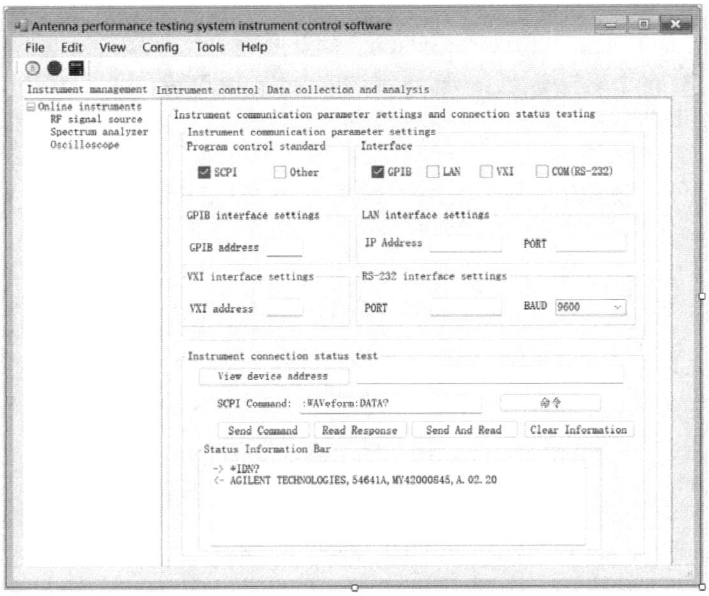

Figure 3. Interface Design of Instrument and Equipment Management Module.

3.2. Specific implementation

The core content of the instrument and equipment management module shown in Figure 3 mainly consists of two parts: one is the editing of the Tree structure of instruments and equipment. The second is the connection and status testing between computers and instrument equipment. The specific implementation of the above two functions will be discussed below.

3.2.1. Editing Tree structure of instruments and equipment. TreeView control in VisualStudio is mainly used for adding and deleting instruments and equipment. The Tree structure node of the number of instruments and equipment can be added through the Add method of the Nodes attribute of the control, and the Tree structure node of the number of instruments and equipment can be deleted through the Remove method of the Nodes attribute of the control. The Tree structure of instruments and equipment can be classified according to the communication interface, and the instruments and equipment linked online by the system or linked historically can be displayed according to the serial port (RS-232), GPIB, LAN, VXI, and other communication interfaces.

3.2.2. Connection and status testing between computers and instrument equipment. SCPI (Standard Commands for Programmable Instruments) is a standardized instrument programming language developed to solve programming problems for programmable instruments. It specifies the message structure and content between the PC and the instrument on the information exchange layer, suitable for the development of automatic measurement systems. At the same time, the Virtual Instrument Software Architecture (VISA) interface library provides rich functional instructions to establish the exchange of control instructions and test parameters between computers and instruments. The connection testing between the controller and instrument equipment can be achieved using VISA library functions and SCPI commands.

The basic procedure of the VISA control instrument is shown in Figure 4. First, the “Ivi. Visa. Interop. dll” Dynamic-link library is added to the reference of the control software project file. Then, the VISA resource manager handle (commonly used function is viOpenDefaultRM()) is open, and the instrument handle (commonly used function is viOpenf()) is open, that is, instrument control instructions through the iPrintf() function is sent through the SCIP command to make the instrument act or query the instrument status. Alternatively, the information returned by the instrument can be received through the viScanf() function and stored. It reads data from the output queue of the instrument and formats the resulting data in a certain string, thus, achieving control functions such as querying instrument model, setting instrument status, controlling instrument operation, reading measurement data, and processing instrument events.

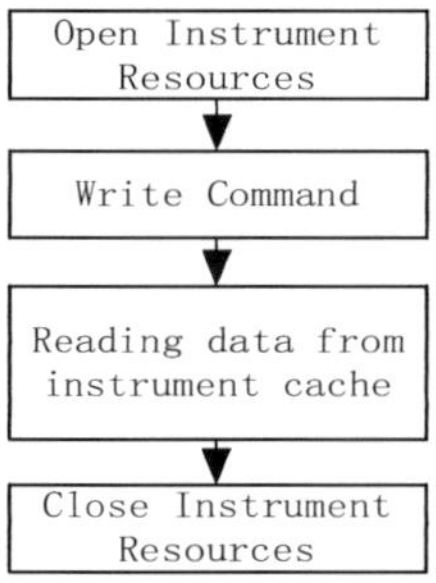

Figure 4. Basic Process of Instrument Control.

For example, by sending the "*IDN?" command, the instrument model can be queried, as shown in the status information column in Figure 3. After executing the "Query Instrument Signal" command, the return string is "AGILENTTECHNOLOGIES, 54641A, MY42000845, A.02.20". The string contains "54641A", which suppresses the model of the connected digital storage oscilloscope. From this, it can be confirmed that the model of the instrument being operated is correct and subsequent control can be carried out. This command is usually used as an initialization detection command to determine whether the communication hardware connection is normal. If the "*RST" command is sent, the factory settings of the instrument equipment can be restored. On a multi-instrument control platform, different SCPI control command streams can be selected based on the returned instrument information. Finally, the VISIA resource manager is released to handle through the viClose() function.

4. Implementation of Instrument Equipment Control Module

The design and implementation of control software will be discussed from two aspects: instrument equipment management and instrument equipment control. Among them, instrument and equipment control specifically include signal source, spectrum analyzer, oscilloscope, etc. In the design of instrument control software, the RF signal source only involves the setting of signal frequency, output power, and other parameters, while the spectrum analyzer and oscilloscope involve not only the setting of instrument parameters but also the reading and display of waveform data or spectrum data. The above three instrument programmable designs have certain similarities. Here we choose the RF signal source to specifically discuss the realization of its control function.

4.1. Function Definition and Interface Design

The parameters involved in the control of RF signal sources mainly include signal frequency, output power, modulation method, modulation type, modulation parameters, modulation switch, and RF switch. The modulation methods include external modulation and internal modulation. The modulation types include amplitude modulation, frequency modulation, and pulse modulation. The selection of modulation parameters depends on the modulation method and modulation type. In the internal modulation mode, when selecting the amplitude modulation type, the modulation parameter options include amplitude modulation frequency (kHz), amplitude modulation depth (%), etc. When selecting a frequency modulation type, its modulation parameter options include modulation frequency offset (kHz), etc. When selecting a pulse modulation type, its modulation parameters include pulse width (μs), Pulse repetition frequency (kHz), etc.

The RF signal source control interface designed based on the above functional definition is shown in Figure 5. Through the interaction columns of signal frequency, output power, signal form, modulation method, modulation parameters, modulation switch, RF switch, etc. in the figure, the setting of RF signal source-related parameters can be achieved. For example, In Figure 5, the signal frequency is set to 870 MHz, the output power is -10 dBm, and the modulation method is pulse modulation. The repetition period of the pulse signal is 2.5 ms, and the pulse width is 200 μs.

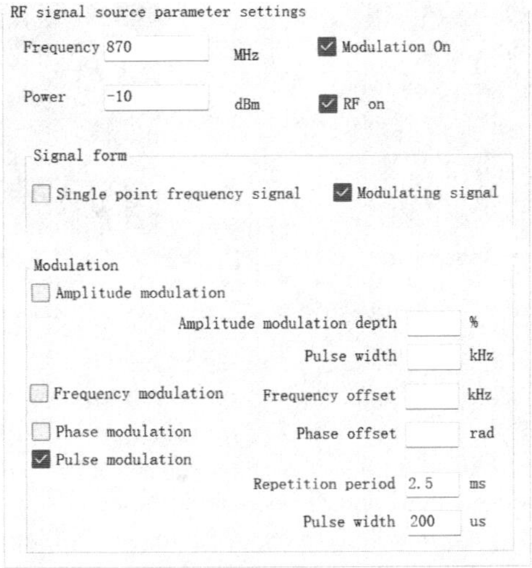

Figure 5. Basic Process of Instrument Control.

4.2. The specific implementation of functional code

4.2.1. RF signal source SCPI command. The RF signal source is mainly controlled by SCPI program control commands, which can be used to set the signal frequency, output power, modulation method, and related modulation parameters of the RF signal source.

4.2.2. The specific implementation of the program control function.

(1) The generation and verification of continuous wave signals:

The following is an example of generating a continuous wave signal with a frequency of 30 MHz and a power of -10 dBm to illustrate the implementation of its specific program control function. The corresponding SCPI commands are as follows:

① "FREQuency: CW30MHZ"//Set the signal waveform type to fixed-point frequency continuous wave and frequency to 30 Hz.

② "POWER-10DBM" //Set the power output of the signal signal to -10 dBm.

③ "OUTPut: STATe1" //Turn on the RF switch.

In software design, first, the VISA resource manager handle and instrument handle are open. Then, sending the SCIP commands from ① to ③ above can generate a continuous wave signal with corresponding frequency and power. The generated continuous wave signal can be tested and verified in a time domain and frequency domain respectively through an oscilloscope and spectrum analyzer. The test results are shown in Figure 6. Figure 6(a) shows the oscilloscope test results, which show a continuous wave waveform and a signal frequency of 30.12 MHz. Figure 6(b) shows the measurement results of a spectrum analyzer. The signal frequency is 30 MHz and the signal power level is -11.68 dBm. The measured signal power is 1.68 dB lower than the set value of the RF signal source (10 dBm), mainly due to the insertion loss of the test cable. The signal form, signal frequency, and output power set are compared to meet the signal setting requirements. (Note: the signal frequency displayed by the oscilloscope is not completely consistent with the signal frequency measured by the Spectrum analyzer, with a difference of 0.12 MHz, which is mainly caused by the error of the signal frequency measured by the oscilloscope).

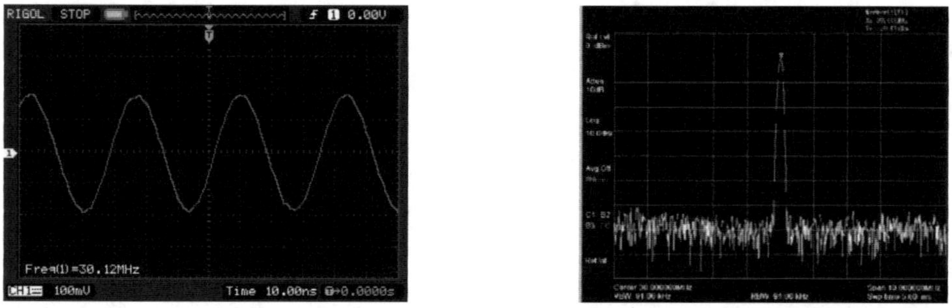

(a) Oscilloscope measurement results (b) Measurement results of Spectrum analyzer

Figure 6. Continuous wave signal measurement results.

(2) Generation and Verification of Pulse Modulation Signals:

Next, we generate a pulse width of 200 μs. The pulse modulation signal with a pulse repetition period of 2.5 ms (corresponding to a pulse repetition frequency of 400 Hz), a frequency of 870 MHz, and a power level of -10 dBm is used as an example to illustrate the implementation of its specific program control function. The corresponding SCPI commands are as follows:

① "FREQuency: CW 870 MHZ"//Set the signal waveform type to fixed-point frequency continuous wave and frequency to 870 MHz.

② "POWER -10 DBM" //Set the power output of the signal to -10 dBm.

③ "OUTPut: STATe 1" //Turn on the RF switch".

④ "OUTPut: Modulation: STATe 1" //Turn on the modulation switch.

⑤ "PULM: STATe 1" //Turn on the pulse output switch.

⑥ "PULM: SOURCE 1" //Set the pulse input to internal modulation.

⑦ "PULM: Internal: Period 2.5 ms" //Set the pulse repetition cycle.

⑧ "PULM: Internal: PWIDth 200 us" //Set the pulse width.

During software design, first, the VISA resource manager handle and instrument handle are open. Then, sending the above ① to ⑧ SCIP commands can generate the corresponding pulse modulation signal. The generated pulse modulation signal can be tested and verified in the time domain and frequency domain respectively through an oscilloscope and spectrum analyzer. The test results are shown in Figure 7. Figure 7(a) shows the oscilloscope test results, which display a pulse-modulated signal with a pulse repetition frequency of 403.2 Hz and a pulse width of 200 μs. Figure 7(b) shows the measurement results of the spectrum analyzer. The signal frequency is 870 MHz and the signal power level is -11.74 dBm. The measured signal power is 1.74 dB lower than the set value of the RF signal source (10 dBm), mainly due to the insertion loss of the test cable. The signal form, signal frequency, output power, pulse repetition frequency, pulse width, and other parameters set are compared to meet the signal setting requirements.

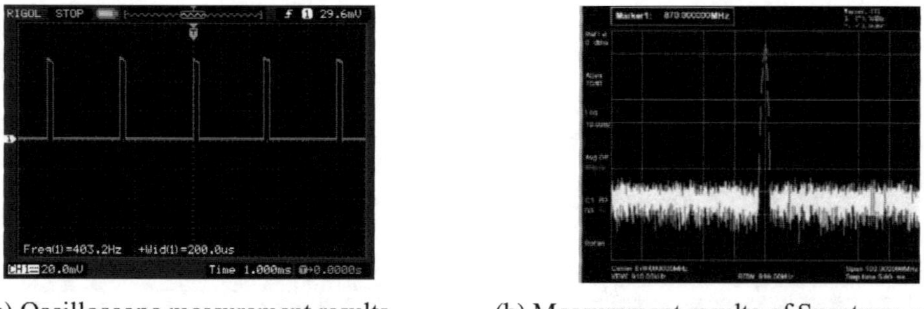

(a) Oscilloscope measurement results (b) Measurement results of Spectrum analyzer

Figure 7. Measurement Results of Pulse Modulation Signal.

5. Conclusion

The radar antenna performance index testing system is an automatic testing system designed specifically for radar antenna pattern testing. It is mainly used to quickly measure radar antenna lobe width, sidelobe level, zero-depth, and other parameters in actual environments. This paper focuses on the design of the instrument control software of the system. The functional requirement of the software is analyzed, the basic framework of software design is put forward, the design ideas and implementation methods of instrument and equipment management, and the specific design methods of instrument and equipment control are given. Practical applications have shown that the control software has advantages such as an intuitive interface, convenient operation, and good interactivity. Combined with system hardware, it can achieve fast and accurate measurement of radar antenna performance index, which has high practical value and good promotion prospects.

References

[1] Gu Zhenjie, Liu Yu. Research on the Construction Method of Injection Radar Test System [J]. Modern Defense Technology, 2016,44 (05): 155-160+200.

[2] Zhou Hongyu, Wang Jiuling, Feng Zhen, et al. Research on Building a Radar Testing Environment Based on Simulation Technology [J]. Modern Radar, 2014,36 (09): 79-82. DOI: 10.16592/j.cnki.1004-7859.2014.09.016.

[3] Liu Hao, Yu Jinsong, Zhang Haijie, et al. Design of Doppler radar testing system based on semi physical simulation [J]. Measurement and Control Technology, 2012,31 (10): 121 123+131. DOI: 10.19708/j.ckjs.2012.10.031.

[4] Deng Bin, Wang Jingjing, Lin Qiang, Xia Liang. Design and Implementation of a Radar Antenna Lobe Measurement System Platform [J]. Journal of Ordnance Equipment Engineering, 2019, (40): 15-19.

[5] Wen Qian. Research on Key Issues of the Measurement and Control Platform for Portable Radar Antenna System [D]. Nanchang Hangkong University, 2014.

[6] Song Ping. Research on millimeter wave radar antenna test system [D]. Shenyang Aerospace University, 2015.

[7] Huang Yan. Design and implementation of the master control software of the antenna near-field tester system and DBF monitoring control software [D]. Yantai University, 2017.

[8] Du Junjie. Design and implementation of coherent millimeter wave radar echo simulation software [D]. University of Electronic Science and Technology of China, 2013.

[9] Liu Yi Design and Implementation of RF Front end Component of Real time USB Spectrum analyzer [D] Chengdu: University of Electronic Science and Technology of China, 2020.

[10] Deng Zhiwei A Solution for Remote Control and Management System of Intelligent Instruments [D] Beijing: Beijing Jiaotong University, 2011.

[11] Chen Zhu Design of oscilloscope instrument driver and program control software based on LAN mode [D] Chengdu: University of Electronic Science and Technology of China, 2020.

Structure design and deformation analysis of double hydraulic cylinder deep-sea pressure simulator

Jianmei Wu [1, 2] and Yang Li [1]

[1]School of Mechanical Engineering, Anhui Sanlian University, Hefei 230601, China

[2]Corresponding author's e-mail: E-mail: 819504383@qq.com

Abstract: Because of the particularity of the deep-sea environment and the difficulty of on-site observation and research, it is particularly important to establish a deep-sea simulation system for research. Based on the high measurement accuracy requirement of deep-sea simulation equipment, a design scheme of a deep-sea pressure simulation device with a double hydraulic cylinder that can measure small water pressure changes is proposed. Firstly, the working parameters and pressure principle of the design are introduced, and then the deformation of the deep-sea pressure simulation device is analyzed. The mathematical model of piston feed and pressure without considering the deformation of the sealed pressure cylinder and the mathematical model of piston feed and pressure considering the deformation of the sealed pressure cylinder are established respectively. Finally, the relationship curve is analyzed by using origin software. The results show that when the deep-sea pressure simulator is used to simulate the underwater environment, it is necessary to consider not only the influence of liquid compression on the pressure but also the influence of deformation on the pressure.

1. Introduction

The reliability of underwater instruments and equipment must be tested and determined by experimental devices that simulate underwater environments. With the further development of underwater resources, higher requirements are put forward for the accuracy of deep-sea simulation experimental devices [1-4]. Therefore, it is of great significance to design a deep-sea pressure simulation device with high accuracy. In [5-6], hydraulic pumps and electro-hydraulic servo valves are used for pressure control. However, due to the control accuracy of the electro-hydraulic servo valve and the fluctuation of motor speed, it was difficult to achieve relatively high precision pressure control. In [7- 8], the liquid volume in the piston feed compression simulator is used for pressure control. The control accuracy of this method was high, but its pressure control range was limited, which made it difficult to achieve high-precision pressure control under a large pressure environment, so its application range was narrow. To simulate the underwater pressure environment with high precision, we should not only consider the high-pressure environment but also consider the precision control under the high-pressure environment. Based on the above factors, a pressure control method of a double hydraulic cylinder was proposed based on the compressibility of liquid [9-10]. The pressurization scheme was realized by controlling the piston feed of the main and auxiliary hydraulic cylinders, in which the piston feed of the main pressurization system could make the pressure device reach the target pressure value range quickly. The piston feed of the auxiliary pressurization system could accurately control the small pressure, to achieve the goal of simulating the deep-sea pressure environment with high precision.

2. Mechanism design

To reduce the error caused by pressure fluctuation, a deep-sea pressure simulator with a double hydraulic cylinder is developed, which can measure the small water pressure change. The device can realize high-precision deep-sea environment simulation. The deep-sea pressure simulation device is shown in Figure 1. The device mainly consists of three parts: a sealed pressure cylinder for simulating an underwater environment, a feed rate control system for controlling pressure changes, and a support for supporting the simulation device. The sealed pressure cylinder comprises a cylinder cover, cylinder body, pressure gauge, and bolt group. The control system includes Motor 1, Reducer 1, ball screw 1, main hydraulic cylinder, Motor 2, Reducer 2, ball screw 2, and auxiliary hydraulic cylinder.

Figure 1. Deep-sea pressure simulator.
1-Sealed pressure cylinder; 2-Motor 1; 3-Reducer 1; 4-Ball screw 1; 5-Main hydraulic cylinder; 6-Support; 7-Motor 2; 8-Reducer 2; 9-Ball screw 2; 10-Auxiliary hydraulic cylinder.

Table 1. Main technical parameters of deep-sea pressure simulator.

Parameters	Value
Effective volume of pressure cylinder (V'/mm^3)	Φ270×235
The wall thickness of pressure cylinder (t_1/mm)	25
The thickness of pressure cylinder cover (t_2/mm)	50
Maximum working pressure (P/MPa)	5.0
Simulated depth (H/m)	0~500
Pressure control accuracy (S/FS)	0.02%

The deep-sea pressure simulator can simulate the ocean depth of 0-500 m, and its main technical parameters are shown in Table 1. The device adopts a double hydraulic cylinder pressure system, which can realize the precise pressure requirement of 0-5.0 MPa through the coordinated feed of the main and auxiliary pressure system pistons. The structure block diagram of the pressure control system of the high-precision pressure simulation device is shown in Figure 2. The control object of the control system is the feed amount of the piston, and the purpose of control is to make the pressure inside the pressure cylinder reach the corresponding depth value of the marine environment, to simulate the depth of the marine environment to the greatest extent. The principle of its control is when the piston is fed, the pressure signal is converted into a digital signal through A/D conversion and is received by the data acquisition system. The current feed rate is calculated by detecting the control algorithm module of the data feed rate and is compared with the set feed rate. After calculation, the control signal is output. In Figure 2, the dotted boxes are the main components of the main and auxiliary pressurizing systems. The cross-sectional area of the piston of hydraulic cylinder 1 of the main pressurizing system is larger than that of the hydraulic cylinder 2 of the auxiliary pressurizing system. During the operation of the main pressurizing system, the internal pressure of the pressure device can quickly reach the target pressure range, while the auxiliary pressurizing system can accurately control the small pressure. Thus, the goal of simulating the deep-sea pressure environment with high precision is achieved.

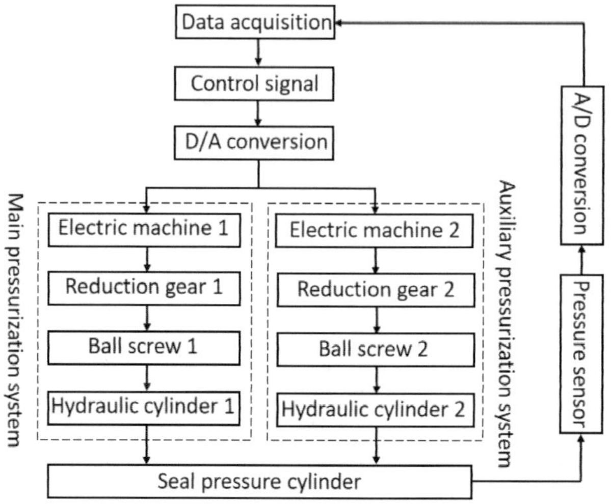

Figure 2. Block diagram of pressure control system structure.

3. Deformation analysis

The deformation of the sealed pressure cylinder, pipeline, and bolt will all lead to the reduction of the internal pressure of the whole device. However, due to the slight influence of pipeline deformation and bolt deformation, only the deformation of sealed pressure cylinders is studied [11].

Firstly, the assembly model of the sealed pressure cylinder is imported into ANSYS software, and then a 5.0 MPa pressure load is added. Finally, the results can be obtained through analysis, as shown in Figure 3.

Figure 3 shows the deformation analysis result of the sealed pressure cylinder under the pressure of 5.0 MPa. The deformation of the pressure gauge is not required to be considered in the analysis, so only the deformation result of the pressure cylinder is observed. According to the analysis of Figure 3 (a), the maximum deformation of the pressure cylinder is located in the center area of the cylinder cover, which is the orange area shown in Figure 3 (a). The maximum deformation is 0.041627 mm, and the average deformation of the lid is about 0.0258 mm. The maximum deformation of the cylinder is located in the inner wall of the cylinder, and it is difficult to directly observe the result of its deformation in the left figure, so it is necessary to pull the cylinder out separately to analyze the result, as shown in Figure 3 (b). It is concluded that the maximum deformation of the cylinder is located in the middle area of the inner wall of the cylinder, that is, the red area shown in Figure 3 (b), and the maximum deformation is 0.015694 mm. The average deformation of the cylinder is about 0.0114 mm.

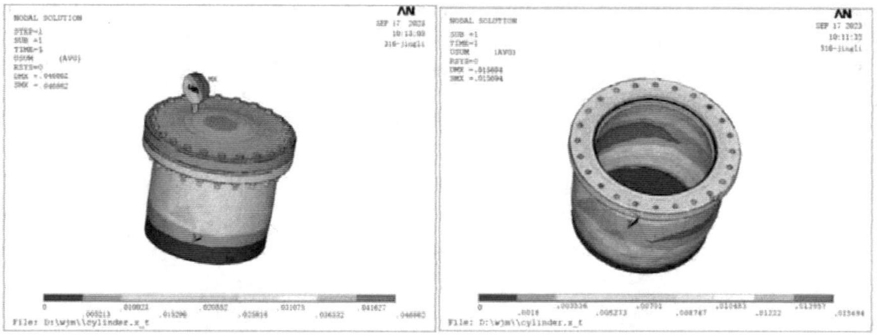

(a) (b)

Figure 3. Deformation analysis of pressure cylinder.

4. Results and analysis

4.1 The relationship between piston feed and pressure without considering the deformation

The main parameters of the main and auxiliary pressurization systems are shown in Table 2.

Table 2. The main parameters of the main and auxiliary pressurization system.

Parameters	Main pressurization system	Auxiliary pressurization system
Piston area (A/mm^2)	314	12.6
Maximum stroke (L/mm)	160	150
Coefficient of liquid compression (k/MPa^{-1})	5.9×10^{-4}	
Initial liquid volume (V/mm^3)	1.32×10^{7}	

The piston feed of the main and auxiliary pressure systems can compress the liquid and thus play the role of pressure. According to the compressibility conditions of the liquid, it can be obtained:

$$\Delta V' = kVP = V_L \tag{1}$$

where k is the coefficient of liquid compression (MPa^{-1}); V is the initial liquid volume (mm^3); $\Delta V'$ is the volume compression of liquid (mm^3) under the action of Pressure P; V_L is the corresponding volume of the piston (mm^3) when it is fed a certain stroke, and $V_L = L_1A_1 + L_2A_2$, in which $L_1$and L_2 refer to the feed stroke of the main and auxiliary hydraulic cylinder pistons (mm), and $A_1$and A_2 refer to the cross-sectional area of the main and auxiliary hydraulic cylinder pistons (mm^2).

Thus, the relationship between piston feed and pressure is obtained.

$$P = \frac{A_1}{kV}L_1 + \frac{A_2}{kV}L_2 \tag{2}$$

The relationship curve is shown in Figure 4.

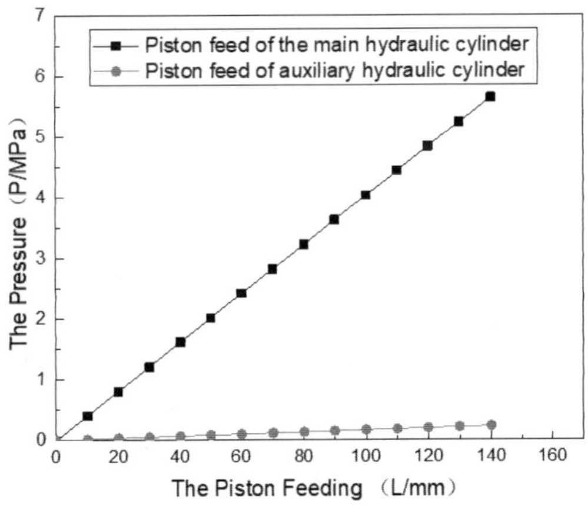

Figure 4. The relation between piston feeding and pressure without considering deformation.

As shown in Figure 4, the black box line is the relationship curve between the feed rate and pressure when the piston of the main hydraulic cylinder is fed, and the red dot line is the relationship curve between the feed rate and pressure when the piston of the auxiliary hydraulic cylinder is fed. It can be seen from the figure that the change of pressure inside the pressure cylinder caused by the piston feed of the auxiliary hydraulic cylinder is an order of magnitude smaller than that of the piston feed of the main hydraulic cylinder, and the relationship between the piston feed and pressure is approximately linear in the range of 0-140 mm, so accurate pressure control can be carried out through this relationship curve.

4.2 The relationship between piston feed and pressure when considering deformation

Volume before deformation:

$$V = \pi r^2 h \tag{3}$$

where r is the inside diameter of the pressure cylinder (mm); h is the inside height of the pressure cylinder (mm).

Volume after deformation:

$$\boldsymbol{V' = \pi \int_0^{h+\Delta h} f^2(h) dh} \tag{4}$$

where $f(h)$ is the inside diameter of the pressure cylinder at different heights (mm); Δh is the elongation of the pressure cylinder cap (mm).

Thus, the total deformation is:

$$\Delta V = \pi \int_0^{h+\Delta h} f^2(h) dh - \pi r^2 h \tag{5}$$

Because the volume deformation of the sealed pressure cylinder is small, the deformation of the sealed pressure cylinder can be simplified into axial and radial deformation. Thus, the deformation of the simplified model ΔV is as follows:

$$\Delta V = \pi (r + \Delta r)^2 (h + \Delta h) - \pi r^2 h \tag{6}$$

where Δr is the amount of increase in the inside diameter of the pressure cylinder under a certain pressure (mm); Δh is the amount of increase in the inside height of the pressure cylinder under a certain pressure (mm).

Through the above deformation analysis, it has been obtained that under the pressure environment of 5.0 MPa, the average deformation of the cylinder body and the average deformation of the cylinder cover are about 0.0114 mm and 0.0258 mm respectively. Therefore, axial deformation: $\Delta h = 0.0258\ mm$; radial deformation: $\Delta r = 0.0114\ mm$. The deformation of the sealed pressure cylinder is calculated by Formula (6): $\Delta V \approx 3,749\ mm^3$.

The above analysis is the deformation of the pressure cylinder under 5.0 MPa pressure. According to the same steps, the deformation analysis of the assembly of the pressure cylinder by ANSYS analysis software can be analyzed within the range of 0-5.0 MPa, and the average axial deformation Δh and radial deformation Δr of pressure cylinder can be also analyzed. The greater the pressure is, the greater the deformation ΔV of the pressure cylinder is. Therefore, to obtain the target pressure, we should not only consider the effect of liquid compression on the pressure but also consider the reverse effect of the deformation of the pressure cylinder on the pressure. Therefore, it can be concluded that the deformation of the pressure cylinder should be considered to simulate the deeper underwater pressure environment.

At this time, the formula of liquid volume compression should be revised as follows:

$$\Delta V' = V_L - \Delta V \tag{7}$$

The corrected relationship between piston feed and pressure is thus obtained.

$$P = \frac{A_1}{kV} L_1 + \frac{A_2}{kV} L_2 - \frac{\Delta V}{kV} \tag{8}$$

The relationship curve is shown in Figure 5.

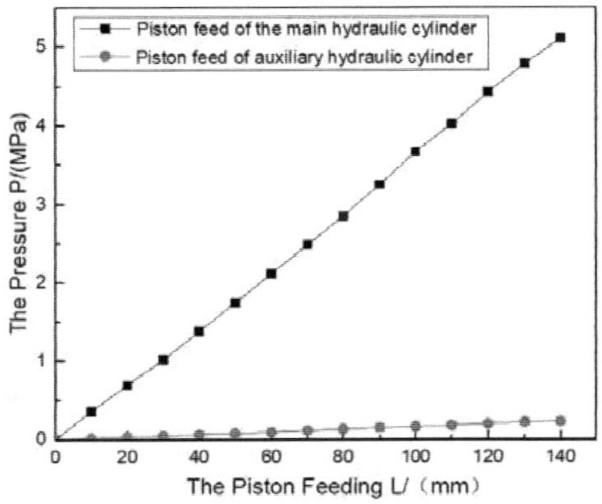

Figure 5. The relation between piston feeding and pressure with considering deformation.

It can be seen from Figure 5 that when considering the deformation of the pressure cylinder, the pressure of the piston of the main hydraulic cylinder can reach 5.0 MPa only when the piston of the auxiliary hydraulic cylinder is fed to more than 135 mm, and the pressure value changes slightly when the piston of the auxiliary hydraulic cylinder is fed. Compared with Figure 4, when pressurizing to the same target pressure value, the piston needs more feed considering the deformation of the pressure cylinder. Therefore, the previous conclusion is verified: when using the deep-sea pressure simulator to simulate the deep-sea environment, it is necessary to consider the reverse influence of the deformation of the pressure cylinder on the pressure.

5. Conclusion

To improve the simulation accuracy of the deep-sea pressure simulator, based on the compressibility of the liquid, a design scheme of a double-hydraulic cylinder deep-sea pressure simulator that can measure small water pressure changes is proposed. The pressurization scheme is realized through the coordinated feed of the main and auxiliary hydraulic cylinder pistons, in which the main hydraulic cylinder piston can make the pressure device reach the target pressure value range quickly while the auxiliary hydraulic cylinder piston can accurately control the small pressure, to achieve the goal of simulating the deep-sea pressure environment with high precision. This provides a new solution for the development of a high-precision deep-sea simulator.

Through the deformation analysis of the deep-sea pressure simulator, it is concluded that the deformation ΔV of the pressure cylinder is also different under different pressures within the pressure range of 0-5.0 MPa. Therefore, to obtain the target pressure value, it is necessary to consider not only the influence of liquid compression on the internal pressure of the deep-sea pressure simulator but also the reverse influence of the deformation of the deep-sea pressure simulator on the internal pressure. However, as the above analysis mainly stays in theory and there are some errors, the subsequent work needs to develop experimental equipment to further verify and optimize the control method, to obtain higher dynamic pressure simulation accuracy.

Acknowledgments

The paper is funded by the Key Project of Natural Science of Anhui Province (2022AH051981), the University's Excellent and Top-notch Talent Cultivation Funding Project of Anhui Province (gxyq2022141), and the Key Project of Natural Science of Anhui Sanlian College (KJZD2023006).

References

[1] Yang H., Chen C. J., Zhou J. R., et al. (2019). Design and Simulation of Hydraulic Loading Control for Deep-sea Ultrahigh-Pressure Simulator. Machinery Design and Manufacture, (05): 52-55+59. DOI: 10.19356/j.cnki.1001-3997.2019.05.013.

[2] Merckelbach L. M., Briggs R. D., Smeed D. A., et al. (2008). Current measurements from autonomous underwater gliders. In: IEEE/OES/CMTC 9th Working Conference on Current Measurement Technology, United States. pp. 61-67.
[3] Gong S. F., Cheng D., and Dong P. (2018). Research on deep-sea environment simulation test technology. Ship Science and Technology, 40 (23): 49-52. https://kns.cnki.net/kcms2/article/abstract?v=lVku9qtM8H90Q_1oRSHC7772Qw_ZD8_LPodVaO7cU1j1xWMv_LyfGF_hdP154u3VGFDp2gk6H1-c5Byo640J4Uor_eb_pcBTUcfgQu5gwnDkd7md-RmoCsOHM0Ft5s1khl5wPBfJYuM=&uniplatform=NZKPT&language=CHS.
[4] Wood J. D., Terray E., Strong B., et al. (2005). Measurement of deep-water ocean waves from a subsurface mooring. In: Institute of Electrical and Electronics Engineers Computer Society. Brest. pp: 166-171.
[5] Hou J. W. (2005). Research on electro-hydraulic proportional pressure control technology of deep-sea extreme environment simulation platform, Zhejiang University. https://kns.cnki.net/kcms2/article/abstract?v=lVku9qtM8H8K7_fljiGauMaFrqtA5wqcLkVRCV0hGS3rJ2fgA--CMnOsV1F-Lhyx4OCLJB2SW6nFVRr_d3Sdd9zNDPoyHP8r8KWfB0tCAOlsMkOs9qU8kr56r96vqQae&uniplatform=NZKPT&language=CHS.
[6] Geng L. P. (2012). Development of Pressure Control Test System for Large Pressure Cylinder, Huazhong University of Science and Technology. https://kns.cnki.net/kcms2/article/abstract?v=lVku9qtM8H-23PeX1PsvNNZgi5vDV7UmB3WDfnp73PxweK-jwOlXHH_io89b4IUOrgPStVekLo08rxNK371xjo4IKIC5bqa3n1f4sKsTDu2RYBXe_cQnCjCsF-eTUOeifbIzTcusGk8=&uniplatform=NZKPT&language=CHS.
[7] Luo S., Li W. J., and Zhang J. X. (2007). Research on High Precision Pressure Control System Based on Changing Volume. Chinese Hydraulics & Pneumatics, (05): 18-19. https://kns.cnki.net/kcms2/article/abstract?v=lVku9qtM8H8ZxDxv4LAfwcQ2RKQmG___5F62FS0hlpS7dcKohwF7PGJJB5ZQBRgQf-DzihtzkgGqA1-VY2YqSvhfMfq1-_SXv4px2p5-GYmfx5UTgv89r8heQoAkPxfF&uniplatform=NZKPT&language=CHS.
[8] Tao S. J. (2013). Research on Pressure Cylinder Pressure Control System Based on Volume Change, Huazhong University of Science and Technology. https://kns.cnki.net/kcms2/article/abstract?v=lVku9qtM8H8J61lK8qei_MYO4UAYdlkeKtjB5qDOgl-uFh_IF3EJXwlDkNbji3YI5UT-EJaDs9d2mgVevbf32_wFSNpbyvcP6rgSPS6TKumn9figkyl4IqHDi_lfBVQ6yyn5i4doQus=&uniplatform=NZKPT&language=CHS.
[9] Wang H. Y., Hu S. S., Liu Z. S., et al. (2021). Double piston high precision deep-sea pressure simulation method, Acta Metrologica Sinica, 42 (02): 199-203. https://kns.cnki.net/kcms2/article/abstract?v=lVku9qtM8H_mHbMD3qOZKAV3BJeMJfNKnvUKvIZLLziR783ffe6G-nioYCGWSNWY2hhTRLBgqu2glu7i_rOuDpw0aUtfW1BdIzOnzcMnRqMvO8pbLB4stAOtxm7uOeFyAr-xz8uxl4Q=&uniplatform=NZKPT&language=CHS.
[10] Yang Y. C. (2017). Research on Automatic Calibration Method of Piston Pressure Gauge. Acta Metrologica Sinica, 38 (06): 708-712. https://kns.cnki.net/kcms2/article/abstract?v=lVku9qtM8H_rDhfv1FnO93D-BgG3oo0Su-zsfBt3yJFkelfomVSG9iAXnErwAdDcUpb1nUNM-qeXE325u6i2L-8dZKmv3mwg_2PMUt4_LK9jU6eD3fO7BYUMqDGYZTGWdpAuf00_BxY=&uniplatform=NZKPT&language=CHS.
[11] Wu J. M. (2015). Research on deep-sea pressure simulation test device, Hefei University of Technology. https://kns.cnki.net/kcms2/article/abstract?v=lVku9qtM8H91KinCq-2XTkxDbZ9fj2zQ9pk3jRO9aJXZWcQ7lY3RLmnf-Tfd5pXFguYhgSXSeC0-wBGcFS4psfFvhl6lFrbG13e7YR5grlgjvTpW7ttGs5bJBTUiEUJy2CQRnr0d81E=&uniplatform=NZKPT&language=CHS.

Global sensitivity analysis of fatigue life based on structural health monitoring

Shiyu Liu[1, 2]

[1]COMAC Shanghai Aircraft Design & Research Institute, Shanghai, China

[2]Corresponding author's e-mail: liushiyu@comac.cc

Abstract: Fatigue cracks are one of the most common causes of structure failure. Structural health monitoring (SHM) has shown important practical significance in ensuring the safety of civil infrastructures by providing continuous, accurate, and reliable monitoring at low cost. Given that data provided by SHM involves various types of uncertainty, this study presents the interval-random hybrid model for fatigue life based on fatigue crack growth theory. By introducing global sensitivity indices, the influences of different variables on the variance of structural fatigue life are analyzed. To enhance computational efficiency, the method is proposed by combining the space partition with unscented transformation (SP-UT).

1. Introduction

In aerospace, bridges, and other important infrastructures intended for long-term service, repeated cyclic loads can result in fatigue crack growth at the location of existing initial cracks. These cracks then continue to expand, ultimately leading to fatigue damage. Fatigue and cracks are the main causes of structure failure [1]. Real-time and accurate monitoring of structural fatigue cracks during operation is necessary [2]. The massive data collected by the SHM system include numerous factors that impact structural fatigue life [3]. A broad range of uncertainties including randomness and fuzziness exists and must be effectively characterized and described. Mechanical models can be used to accurately predict fatigue damage life and perform reliability sensitivity analyses. These measures can help inform decision-making and logistical maintenance planning to ensure optimal structural protection. The history of research on structural fatigue reliability is relatively short, and many studies have focused on constructing fatigue reliability models [4-5]. Some studies introduced sensitivity analysis to optimize the design of structure fatigue life, which is not yet well developed [6-7], mainly focusing on local sensitivity analysis and lack of consideration for various mixed uncertainty situations. Variance-based sensitivity indices provide a global perspective on the output performance variance [8]. In this paper, variance-based sensitivity indices are introduced to analyze the important measure of various engineering uncertainties for the fatigue life of structures. Considering the diverse and uncertain nature of data from the SHM system, the interval-random hybrid model for fatigue life based on fatigue crack growth theory is developed. In addition, the corresponding indices and efficient and robust algorithms are proposed.

2. Variance-based global sensitivity indices

2.1. Sobol's indices

Sobol's indices are the most classic and widely used indices to determine the average effect of input variables on the output response when they are fixed at a specific point. Numerous scholars have proposed variance-based importance measures, and this paper utilizes the commonly recognized Sobol's indices for research purposes. Sobol's indices are based on the high-dimensional model decomposition equation, because the structural limit state function $Y=g(\boldsymbol{X})$, $\boldsymbol{X}=\{X_1,\ldots,X_n\}$ is an *n*-dimensional random independent variable, and then Y and $V(Y)$ can be obtained as follows [9]:

$$Y=g_0+\sum_{i=1}^{n}g_i(X_i)+\sum_{i\neq j}g_{ij}(X_i,X_j)+\ldots+g_{12\ldots n}(X_1,X_2,\ldots,X_n) \tag{1}$$

$$V(Y)=\sum_{i=1}^{n}V_i+\sum_{i<j}V_{i,j}+\sum_{i<j<k}V_{i,j,k}\cdots+V_{1,2,\ldots,n} \tag{2}$$

Based on the above decomposition equation, S_i for the contribution of the input variable X_i to $V(Y)$ is obtained as:

$$S_i=\frac{V_i}{V(Y)}=\frac{V\left[E(Y\mid X_i)\right]}{V(Y)} \tag{3}$$

where $V(Y)$ is the total variance of the model output; $V\left[E(Y\mid X_i)\right]$ is the average reduction in the variance of the model output when the variable X_i is fixed over its entire domain of distribution. If S_i is larger, it means that the uncertainty in X_i contributes more to the variance of the output.

2.2. Variance-based sensitivity indices for interval-random model

Variables with large data dispersion are defined as interval variables. The model contains random and interval variables $M=g(\mathbf{X},\boldsymbol{Y})$ and $\boldsymbol{X}=(X_1,X_2,\cdots,X_n)$, which are independent random variables with probability density function $f_{X_i}(x_i)(i=1,2,\cdots,n)$ and $\boldsymbol{Y}=(Y_1,Y_2,\ldots,Y_m)$. Thus, the output response M has interval-random mixed uncertainty, and the variance of M is $V_M=\left[V_M^l,V_M^u\right]$. By setting $X_i{=}x_i^*$ or $Y_j{=}y_j^*$, the conditional variance of M is $V_{M|X_i=x_i^*}=\left[V^l_{M|X_i=x_i^*},V^u_{M|X_i=x_i^*}\right]$ or $V_{M|Y_j=y_j^*}=\left[V^l_{M|Y_j=y_j^*},V^u_{M|Y_j=y_j^*}\right]$.

Based on the distance measure proposed by Liu [10], the impact of the variable on variance is positively correlated with the distance between the unconditional and conditional variance. $\delta_{X_i}^{V_M}$ and $\delta_{Y_j}^{V_M}$ reflect the size of the contribution of X_i and Y_i to the uncertainty of the output variance, respectively. $\delta_{X_i}^{V_M}$ and $\delta_{Y_j}^{V_M}$ can be obtained as follows:

$$\delta_{X_i}^{V_M}=E_{X_i}\left[d^2\left(V_M,V_{M|X_i=x_i^*}\right)\right]=\int_{-\infty}^{+\infty}d^2\left(V_M,V_{M|X_i=x_i^*}\right)f_{X_i}(x_i)dx_i \tag{4}$$

$$\delta_{Y_j}^{V_M}=\varphi^c_{V_M^{Y_j}}=\frac{\varphi^u_{V_M^{Y_j}}+\varphi^l_{V_M^{Y_j}}}{2} \tag{5}$$

$$d^2\left(V_M, V_{M|X_i=x_i^*}\right)=\left[\left(\frac{V_M^l+V_M^u}{2}\right)-\left(\frac{V_{M|X_i=x_i^*}^l+V_{M|X_i=x_i^*}^u}{2}\right)\right]^2+\frac{1}{12}\left[\left(V_M^u-V_M^l\right)-\left(V_{M|X_i=x_i^*}^u-V_{M|X_i=x_i^*}^l\right)\right]^2 \tag{6}$$

$$\varphi_{V_M^{Y_j}}\left(y_j^*\right)=d^2\left(V_M, V_{M|Y_j=y_j^*}\right)=\left[\left(\frac{V_M^l+V_M^u}{2}\right)-\left(\frac{V_{M|Y_j=y_j^*}^l+V_{M|Y_j=y_j^*}^u}{2}\right)\right]^2+\frac{1}{12}\left[\left(V_M^u-V_M^l\right)-\left(V_{M|Y_j=y_j^*}^u-V_{M|Y_j=y_j^*}^l\right)\right]^2 \tag{7}$$

3. Efficient method for variance-based global sensitivity indices

Yun et al. [11] proposed a highly efficient method by combining space-partition (SP) and unscented transformation (UT) for solving variance-based global sensitivity indices. By reducing the nonlinearity in the response function by SP and utilizing the strong capability of the UT to explore the probability, all the indices for each order can be estimated with a set of UT samples. The computation process is as follows:

Step 1: We divide the entire integration space into $\prod_{i=1}^{n} N_i (i=1,\ldots,n)$ subspaces with equal probability by the density function $f_{\mathbf{X}}(\boldsymbol{x})$ of the random variable $\boldsymbol{X}$ and apply the UT and optimization toolbox to solve for the unconditional variance of the output response $V_M=\left[V_M^l, V_M^u\right]$.

Step 2: M Gaussian points $(x_{i1},\ldots,x_{iM})$ and their corresponding weights $(w_{i1},\ldots,w_{iM})$ are generated based on the sampling density function $f_{X_i}(x_i)$ of the one-dimensional input variable X_i.

Step 3: We fix the variable X_i at $x_{il}(l=1,\ldots,M)$ and divide the entire integration space of $\boldsymbol{X}_{\sim i}$ into $\prod_{\substack{j=1 \\ j\neq i}}^{n} N_j (j=1,\ldots,i-1,i+1,\ldots,n)$ subspaces with an equal probability according to sampling density $f_{\mathbf{X}_{\sim i}}(\mathbf{x}_{\sim i})$. We apply the UT and optimization toolbox to solve the conditional variance $V_{M|X_i=x_{il}}=\left[V_{M|X_i=x_{il}}^l, V_{M|X_i=x_{il}}^u\right](l=1,\ldots,M)$ in each subspace. $\delta_{X_i}^{V_M}$ can be estimated by Equation (4) and Equation (6).

Step 4: We fix the variable $Y_j (j=1,\ldots,m)$ at the interval equidistant point $y_j^{(k)}(k=1,\ldots,N_f)$, and the conditional variance $V_{M|Y_j=y_j^{(k)}}=\left[V_{M|Y_j=y_j^{(k)}}^l, V_{M|Y_j=y_j^{(k)}}^u\right]\left(k=1,\ldots,N_f\right)$ can be obtained. With $V_M=\left[V_M^l, V_M^u\right]$ from Step 1, $\delta_{Y_j}^{V_M}$ can be computed by Equation (5) and Equation (7).

Based on the above computation process, the computational cost of the SP-UT method to obtain $\delta_{X_i}^{V_M}$ and $\delta_{Y_j}^{V_M}$ is $M\times 2\times(n-1)\times\prod_{\substack{j=1 \\ j\neq i}}^{n} N_j$ and $m\times N_f\times 2\times n\times\prod_{i=1}^{n} N_i$, respectively.

4. Fatigue life sensitivity analysis based on SHM

Runyang Bridge is a super large bridge composed of a cable-stayed bridge and a suspension bridge. The suspension bridge has a main span of 1, 490 m. The bridge uses oorthotropic steel decks (OSDs) which are prone to fatigue cracking, especially at the welded connection points between the top plate and U-rib. Thus, developing the SHM system on Runyang Bridge is critical. Figure 1 shows that fiber grating strain sensors are employed to detect the strain at the top plate and the U-rib weld [12].

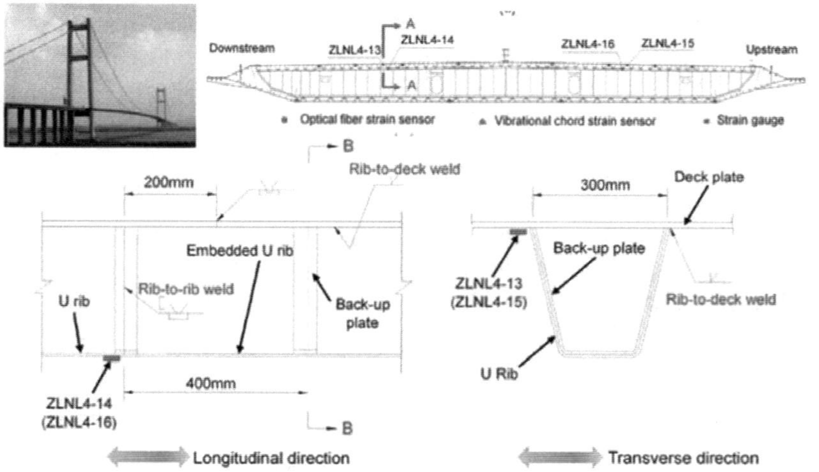

Figure 1. SHM system on Runyang Bridge.

The fatigue life model based on crack growth theory has the following function:

$$t = \left[365 \cdot ADTT \cdot C \cdot S_{re}^{m} \cdot \left(1 + (n-1) \cdot \alpha / 2\right) \right]^{-1} \cdot \int_{a_0}^{a_c} \frac{da}{\left(Y(a) \cdot \sqrt{\pi a} \right)^{m}} \quad (8)$$

where $ADTT$ is average daily load cycles; S_{re} is daily equivalent stress range; C is fatigue property coefficient; m is the inverse slope of the log-log curves; n is bridge service year; α is the linear coefficient of traffic growth; a_0 and a_c are initial and critical crack length, respectively; $Y(a)$ is the dimensionless function, which can be expressed as:

$$Y(a) = 1.12 \cdot \left[\frac{2w}{\pi a} \cdot \tan\left(\frac{\pi a}{2w} \right) \right]^{\frac{1}{2}} \quad (9)$$

According to Deng et al. [13], S_{re} and $ADTT$ are usually treated as a random distribution to fit the probability density function based on the long-term monitoring data, as shown in Figure 2, which reveals that there are large errors in the fitting results. Thus, this paper sets S_{re} and $ADTT$ as interval variables with the basic change interval being $[2.6, 8.7]Mpa$ and $[3053, 9413]$ respectively. Other parameters are considered random variables and the detailed parameter information can be found in Table 1.

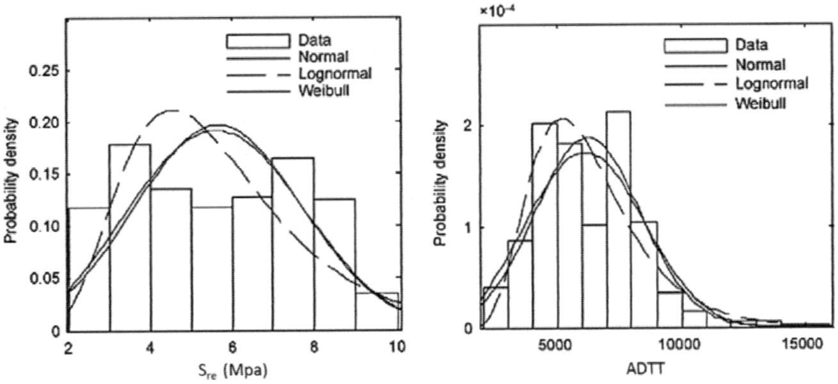

Figure 2. Bin histograms and fitted PDFs of S_{re} and $ADTT$ [13].

Table 1. Detail fatigue parameter information of RSB [13].

Random parameter (lognormal)	Location	Scale	Constant parameter	
C, MPa$^{-3}\cdot m^{-1/2}$	6.89×10^{-12}	0.45	α =2%	w =0.014 m
a_0, m	0.0005	0.5	m=3	a_c=0.007 m

The global sensitivity indices of random variables and interval variables for the fatigue life variance are given in Figure 3 and Figure 4, respectively. The fatigue property coefficient C is more important than the initial crack length a_0, and the daily equivalent stress range S_{re} is more crucial than average daily load cycles $ADTT$, which offers a theoretical foundation to guide actual engineering.

From Figure 3 and Figure 4, the results obtained by SP-UT coincide with those of the MCS method, indicating the accuracy of SP-UT in calculating the variance-based global sensitivity indices. Table 2 gives the computation number of SP-UT and MCS, which shows that SP-UT is much more efficient than MCS.

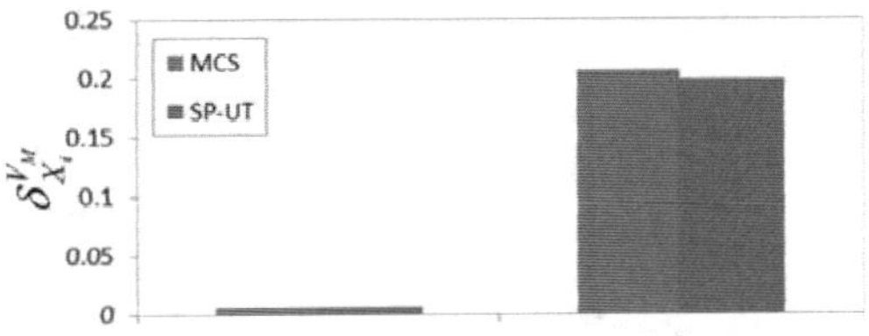

Figure 3. $\delta_{X_i}^{V_M}$ for fatigue life of random variables.

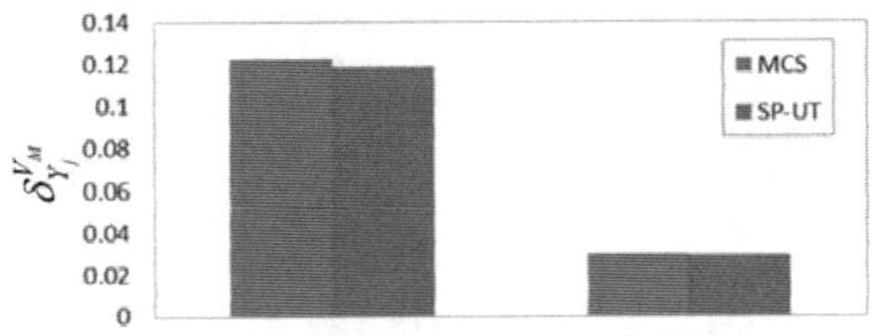

Figure 4. $\delta_{Y_j}^{V_M}$ for fatigue life of interval variables.

Table 2. Comparison of the computation number between two methods.

Method	$\delta_{X_i}^{V_M}$	$\delta_{Y_j}^{V_M}$
MCS	$n\times N_r^2=2\times\left(10^6\right)^2=2\times10^{12}$	$m\times N_f\times N_r=2\times10\times10^6=2\times10^7$
SP-UT	$M\times2\times(n-1)\times\prod_{j=1, j\neq i}^{n}N_j=5\times2\times1\times10=100$	$m\times N_f\times2\times n\times\prod_{i=1}^{n}N_i=2\times10\times2\times2\times10^2=8000$

5. Conclusion

Drawing upon measured health monitoring data, this research presents a variance-based global sensitivity analysis for fatigue life based on crack growth theory. Given the challenge of accurately fitting uncertainty data obtained from health monitoring with stochastic probabilistic models, determining the magnitude or boundaries of their unknowns is comparatively straightforward. Thus, this paper proposes the interval-random variance-based global sensitivity indices and uses the SP-UT method to enhance computational efficiency. Furthermore, the significance of various variables fatigue life is analyzed to provide a theoretical foundation for practical engineering.

References

[1] Xuan F., Zhu M., and Wang G. (2021). Retrospect and Prospect on Century-long Research of Structural Fatigue. *Journal of Mechanical Engineering*, 6. pp. 26-51.
[2] Fang Y. and Kong F. (2022). Comparison of Structural Probabilistic and Non-probabilistic Reliability Computational Methods under Big Data Condition. *Structural Durability & Health Monitoring*, 16. pp. 129-143.
[3] Fang F., Zheng H., Wang Y., and Qiu L. (2021). Mechanical Structural Health Monitoring: A Review. *Journal of Mechanical Engineering*, 57. pp. 269-292.
[4] Zhang H., Liu Y., Dend Y., and Feng D. (2021). Fatigue reliability assessment of orthotropic steel deck weld details considering the correlation between variables. *Journal of Vibration and Shock*, 4. pp. 105-113.
[5] Alshoaibi A., Ghazwani M., and Hakami M. (2021). Fatigue life and reliability assessment of metal structures. *Engineering Solid Mechanics*, 9. pp.13-22.
[6] Hu B., Huang X., and Du S. (2023). Reliability Sensitivity Analysis of the Fatigue Life of Wind Turbine Bearings. *Journal of Northeastern University (Natural Science)*, 8. pp. 1, 128-1, 135.
[7] Huang X., Wang P., Li H., and Zhang Z. (2021). Fatigue reliability and sensitivity analysis of turbine disk with fuzzy failure status. *Journal of Northwestern Polytechnical University*, 6. pp. 1, 312-1, 319.
[8] Ivan I. (2021). Risk of estimators for Sobol's sensitivity indices based on metamodels. *Electronic Journal of Statistics*, 15. pp. 235-281.
[9] Sobol I. and Myshetskaya E. (2008). Monte Carlo estimators for small sensitivity indices. *Mcma* 13, pp. 455-465.
[10] Liu H. (2004). Ranking fuzzy numbers based on a distance measure. *Journal of Shandong University*, 39. pp. 30-36.
[11] Yun W., Liu Z., and Zhao L. (2017). An efficient method for variance-based global sensitivity analysis by combination of space-partition and unscented transformation. *Journal of National University of Defense Technology*, 39. pp. 166-173.
[12] Deng Y., Li A., and Feng D. (2018). Fatigue Reliability Assessment for Orthotropic Steel Decks Based on Long-Term Strain Monitoring. *Sensors*, 18. pp.181.
[13] Deng Y., Ding Y., Li A., and Lin R. (2012). Fracture fatigue reliability of welded details in bridge steel box girders. *Science China Technological Sciences*, 10. pp. 122-128.

Stress detection method of pipeline based on vibration characteristics

Jinzhou Li [1], Xiaoben Liu [1], Tong Shi [1], Hong Zhang [1,3], Hao Wang[1], Shengyi Sha[2] and Mingzhe Bu[2]

[1]Beijing Key Laboratory of Urban Oil and Gas Distribution Technology, MOE Key Laboratory of Petroleum Engineering, National Engineering Research Center for Pipeline Safety, China University of Petroleum, Beijing, China
[2]PipeChina Engineering Technology Innovation Co. Ltd., China

[3]Corresponding author's e-mail: hzhang@cup.edu.cn

Abstract: Stress concentration during the operation of long-distance transmission pipelines is an inevitable problem. In this paper, a pipeline stress detection method based on vibration characteristics is proposed. We establish a finite element model for knock detection and determine the law of vibration frequency and pipeline stress. The reliability of the model is verified by the knock detection experiments. It is found that the natural frequency stress of the vibration of the sample shows a close correlation under the knocking effect. The study results provide an innovative process for pipeline stress detection.

1. Introduction

During the operation of long-distance transmission pipelines, the stress concentration in the local area of the pipeline is caused by geological disasters and improper installation processes. At present, there have been many fracture accidents caused by changes in pipeline structure [1]. The non-destructive testing method is a significant technique for assessing and ensuring the safety of pipelines [2].

At present, the non-destructive testing methods for pipeline stress are mainly classified as X-ray diffraction method, Barkhausen noise method, IMU inertial navigation method, etc. The free vibration characteristics of the structure can reflect the stress state of the structure to some extent. Zhang et al. measured the boom load through the change of natural frequency, and the error was about 5% [3]. Tang and Gao studied the rectangular plate with residual stress by using the method of pulse excitation and found that the greater the residual stress is, the greater the change in natural frequency is [4]. Liao et al. studied the structural modal of the parts with residual stress and found that the existence of residual stress will increase the natural frequency [5].

There have been a lot of studies at home and abroad based on the non-destructive testing methods of vibration characteristics [6-8]. However, no scholars have studied pipeline steel at present, so it is difficult to apply the knock detection method to pipeline stress measurement. The material of the sample selected in this study is X80 pipeline steel, and the rattle vibration model of plate specimens under different stresses is established. The relationship between stress and vibration frequency of plate specimens is studied, and the reliability of the simulation model is proved by experiments, to provide support for the engineering application of the rattle detection method.

2. Vibration theory

2.1. Natural frequency solution

The differential equation of motion of an undamped free vibration system is expressed as [9]:

$$M\ddot{a}(t)+Ka(t)=0 \tag{1}$$

where M represents the mass matrix of the system; $\ddot{a}(t)$ represents the second derivative of node displacement concerning time; K represents the stiffness matrix of the system; $a(t)$ is the displacement array of nodes.

Its solution can be expressed in the following form:

$$a=\phi\sin\omega(t-t_0) \tag{2}$$

where ϕ is a vector; ω represents the vibration frequency; t represents the variables of time; t_0 represents the constant of time.

If Equation (1) is substituted into Equation (2), it will become a problem of solving eigenvalues.

$$K\phi-\omega^2M\phi=0 \tag{3}$$

2.2. Tap detection principle

When a structural system is excited, it vibrates at a specific frequency, which is called the natural frequency of the structure. For undamped single-degree-of-freedom systems, the tapping natural frequency is calculated, as shown in Equation (4).

$$f_n=\frac{1}{2\pi}\left(\frac{k}{m}\right)^{1/2} \tag{4}$$

where k is the stiffness; m represents the mass (unit: kg); f_n represents the n-th natural frequency of the structure (unit: Hz).

The natural frequency of the structure is affected by the stiffness and mass. In the knock test of the structural system, the multi-order natural frequency can usually be obtained [10]. The stiffness is different, the natural frequency is different, and the vibration signal generated by the knock is different. This is the principle of judging whether there is stress concentration in the specimen by the vibration generated by the knock. In this study, the sample was subjected to free vibration by tapping excitation, and the natural frequency was obtained by collecting the vibration signal.

3. Numerical simulation

3.1. Model establishment

The knock vibration model of the X80 plate sample was established by COMSOL, and the vibration of the sample under different stress states was analyzed. The modeling process is as follows:

(1) Modeling and meshing: The model was constructed based on an actual pipeline sample, and the grid was divided by sweeping (seen in Figure 1);

(2) Material properties: According to the actual situation, the selected material is X80 pipeline steel, E is 210 GPa, v is 0.3, and ρ is 7, 860 kg/m^3;

(3) Boundary conditions: The left side of the sample is subject to fixed constraints as boundary conditions. The specimens are loaded to different stress states (0-160 MPa, every 20 MPa) by applying different X-direction displacements to the right side of the model. On this basis, the transient force is applied to the middle position of the XY plane of the model to realize the simulation of knocking vibration;

(4) Solution and post-processing: The model was solved transiently, and a probe was set in the middle of the model to record and store the vibration signal of the model.

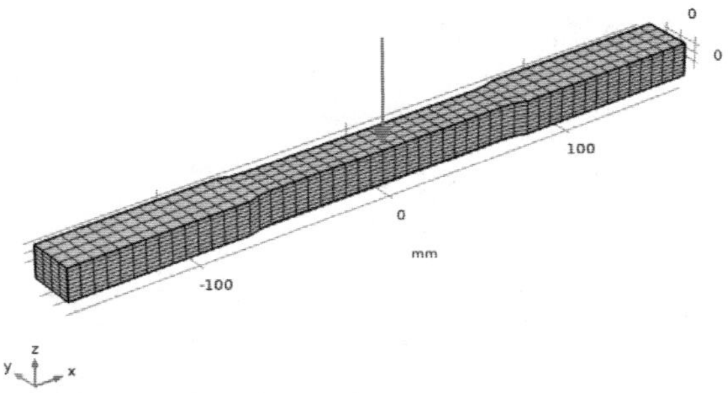

Figure 1. Finite element model of X80 plate specimen.

3.2. Simulation results and analysis

The vibration of the sample under different tensile stress states is studied. Firstly, different stress states are obtained by changing the boundary conditions. On this basis, the transient knocking vibration simulation is carried out in the middle of the model, and the vibration signal is obtained (seen in Figure 2).

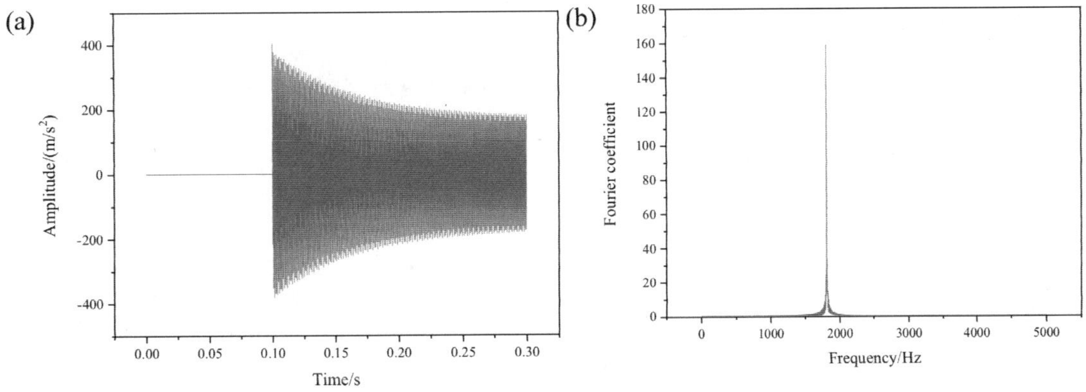

Figure 2. (a) Time domain graph (simulation); (b) Frequency domain graph (simulation).

With the increase of time, the amplitude of vibration will decrease, and there is a certain attenuation phenomenon. There is an obvious peak in the spectrum, and the frequency corresponding to the peak represents the natural frequency of the structure.

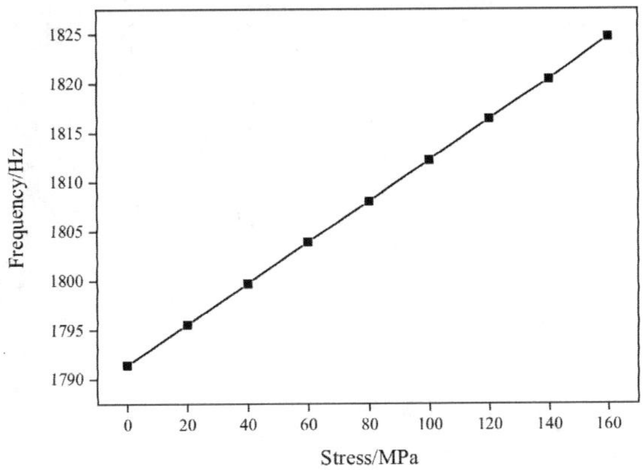

Figure 3. Relationship between tensile stress and dominant frequency of vibration (simulation).

The main frequency of vibration is positively correlated with the tensile stress in Figure 3. This is because the stress state changes the stiffness of the structure, so that the vibration frequency changes, indicating that the natural frequency can reflect the stress state of the structure.

4. Experimental verification

4.1. Sample and experimental procedure

The sample was designed and sampled axially from the real X80 pipe. As seen in Figure 4, R is the radius of curvature.

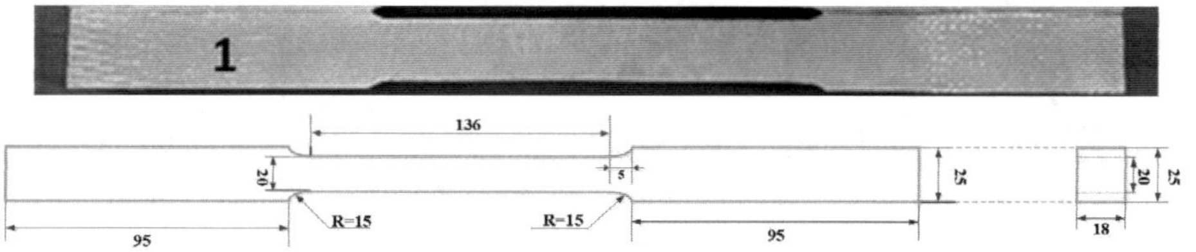

Figure 4. Sample size (mm).

The knocking vibration experiment of the X80 base metal sample was carried out to obtain the vibration signals under different load conditions. The difference in knocking vibration characteristics under different axial tensile stress conditions was analyzed. Figure 5 shows the experimental test system. The experimental steps are as follows:

(1) We install the pneumatic stretching fixture of the stretching machine as required and set up the limit device;

(2) The tensile machine clamps the samples and adjusts the tensile machine to make the sample force zero;

(3) We connect the vibration sensor with the signal amplifier and the storage device (oscilloscope) and place them in a suitable position;

(4) The samples were loaded with different loads by starting the tensile machine. The loading speed was 1 mm/min, and the constant load was 10 s after each increase of the load;

(5) Knocking experiments were carried out on the samples under different stress states, and the vibration data were stored;

(6) We import the data into the computer for analysis.

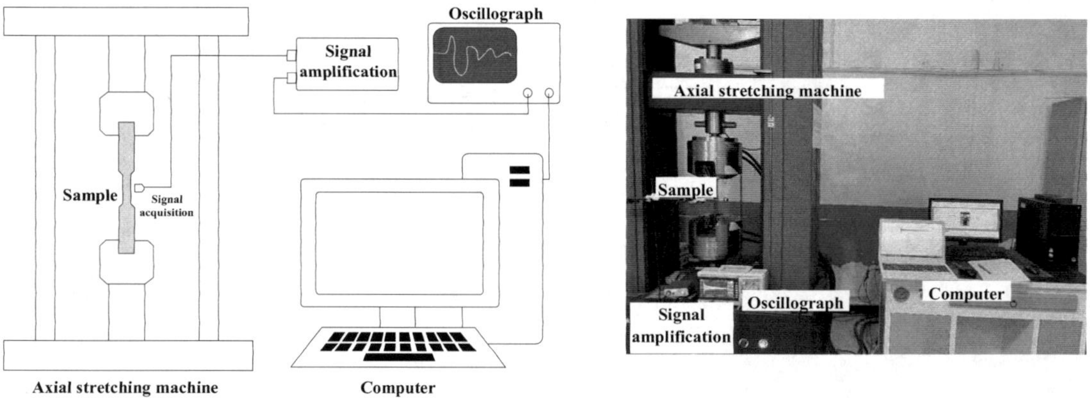

Figure 5. Test system.

4.2. Analysis and verification of experimental results

The knocking vibration experiment was carried out for the base metal under different axial tensile stress states. The tensile stress value was the same as that of the numerical simulation.

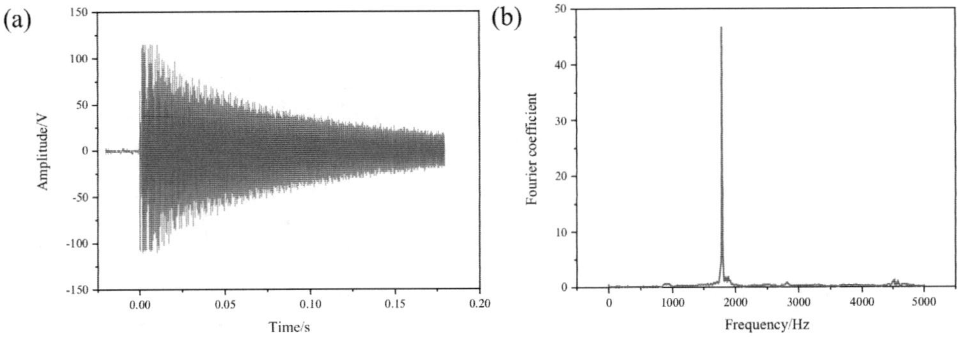

Figure 6. (a) Time domain graph (experiment); (b) Frequency domain graph (experiment).

The vibration amplitude will decrease with time in Figure 6. This is because during the vibration of the specimen, the amplitude of the vibration will be reduced due to the influence of air and other damping. There is also a dominant frequency in the spectrogram. The experimental results are similar to the simulation results. To further analyze the regularity of the experimental results, the main frequency of vibration under different tensile stress is obtained.

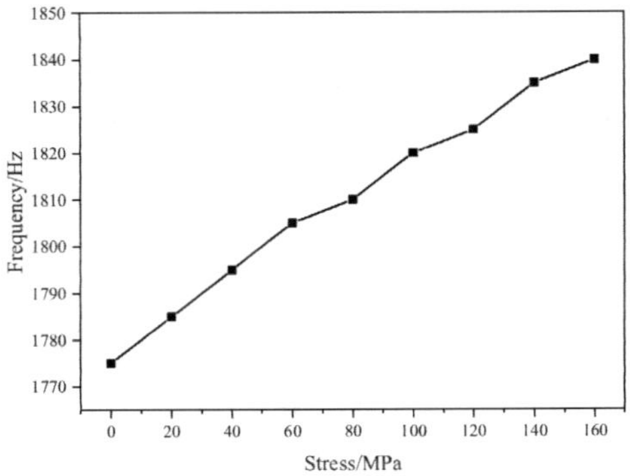

Figure 7. Relationship between tensile stress and dominant frequency of vibration (experiment).

The experimental results are very similar to the finite element simulation results in Figure 7. The main frequency of the sample vibration is positively correlated with the tensile stress, but there are some differences in the value, as seen in Table 1.

Table 1. Vibration results under various stress states.

Categories	Results								
Stress (MPa)	0	20	40	60	80	100	120	140	160
Simulation results (Hz)	1, 791	1, 795	1, 799	1, 803	1, 808	1, 812	1, 816	1, 820	1, 824
Experimental results (Hz)	1, 775	1, 785	1, 795	1, 805	1, 810	1, 820	1, 825	1, 835	1, 840
Error (%)	0.92	0.59	0.26	0.06	0.11	0.43	0.47	0.8	0.83

Under various stress states, the difference between the simulation results and the experimental result is small, and the dominant frequencies of both are in the range of 1, 775-1, 840 Hz. After comparison, the error is less than 1%. The reasons for the differences may be as follows: First, the model is relatively simplified compared with the experiment; Second, environmental factors have some influence; Third, equipment accuracy also has an influence. From the comparison of the results, both the change trend and the absolute value error are relatively small, which can prove the feasibility of the tapping vibration method to obtain pipeline stress states. It is a reliable method for non-destructive testing and evaluation of small sample stress.

5. Conclusion
This study came up with a pipeline tap detection means based on vibration characteristics. The reliability of the finite element method is verified by the experimental method. The feasibility of the knock detection stress method is illustrated. According to the results of the tapping vibration detection method, there is a quantitative relationship between the natural frequency of the vibration and the tensile stress. It can evaluate the situation of pipe stress and can be used as an effective means to estimate the safety status of pipelines in active service.

Acknowledgments
① National Key R&D Program of China (Grant No. 2022YFC3070100);
② Natural Science Foundation of China (Grant No. 52004314);
③ Young Elite Scientists Sponsorship Program by Beijing Association for Science and Technology (Grant No. BYESS2023261);
③ The Science and Technology Research Project of PipeChina (Grant No. WZXGL202105);
④ The Science and Technology Research Project of PipeChina (Grant No. WZXGL202104);
⑥ Science Foundation of China University of Petroleum, Beijing (No. 2462023BJRC005).

References
[1] Yuan Q. W., Qiu G. H., Tian S. S., and Shen J. C. (2013). Research on microwave detection of crack defects in medium and long-distance pressure pipelines. *Journal of Electronic Measurement and Instrumentation* 1-12. http://kns.cnki.net/kcms/detail/11.2488.TN.20231010.1336.032.html.
[2] Cheng Y. F. (2020). Some technical considerations for ensuring the long-term safe operation of the China-Russia Eastern Gas Pipeline. *Oil and Gas Storage and Transportation* 39 (1) 1-8. DOI: 10.6047/j.issn.1000-8241.2020.01.001.
[3] Zhang G. C., Zhang D. Q., Chen Z. K., Zheng W. G., Ma G. X., and Meng J. (2008). Research on the force test method of boiler hangers based on the natural frequency method. *Thermal Power Generation* 37 (10) 39-42. DOI: 10.19666/j.rlfd.2008.10.011.
[4] Tang G. and Gao Y. Y. (1999). Discussion on the influence of welding residual stress on the natural frequency of rectangular thin plates with four sides simply supported *J. of Xiangtan Normal University: Social Science Edition* 3 77-80. DOI: CNKI: SUN: XAFB. 0.1999-03-020.
[5] Liao D. H., Cheng A. G., and Xie H. C. (2010). Modal analysis and optimization of car body structure considering stamping residual stress and thickness variation. *Vibration and Shock* 29 (3) 30-35. DOI: 10.3969/j.issn.1000-3835.2010.03.008.
[6] Liu Z. G., Li Y. J., Cao Y. Y., and Zhao L. Z. (2022). Research on Knock Detection Defect Recognition Based on EM. *Aeronautical Computing Technique* 52 (02): 53-57. DOI: 10.3969/j.issn.1671-654X.2022.02.013.
[7] Naotoshi Y. (2023). Hammering sound of concrete with defects and spalling risk. *Tunnelling and Underground Space Technology* 131, 104, 789, 0886-7, 798. https://doi.org/10.1016/j.tust.2022.104789.
[8] Duffour P., Morbidini M., and Cawley P. (2006). A study of the vibroacoustic modulation technique for the detection of cracks in metals. *The Journal of the Acoustical Society of America* 119 (3): 1, 463. DOI: 10.1121/1.2161429.
[9] Mao K., Zhu X. P., Liu J., Liu Y. S., and Hu W. P. (2006). Finite element analysis of the natural frequency of windshield of a certain type of aircraft *J. of Xi'an Institute of Technology* 26 (2) 146-149. DOI: 10.3969/j.issn.1673-9965.2006.02.011.
[10] Chao J. M., Dong J., and Wang X. F. (2020). Damage Identification Test and Finite Element Simulation of Timber Connections. *Building Structures* 50 (S2) 648-653. DOI: 10.26914/c.cnkihy.2020.075859.

IOP Publishing
2724 (2024) 012021 doi:10.1088/1742-6596/2724/1/012021

Design of differential probe with harmonic magnetic field and its data process method

Lin Yang[1], Xinhua Wang[1,2], Tao Sun [1], Naixiang Hu [1] and Zisheng Guo[1]

[1] Faculty of Materials and Manufacturing, Beijing University of Technology, Beijing, 100124, China

[2]Corresponding author's e-mail: wxhemma2005@163.com

Abstracts: To identify the condition of steel pipelines beyond the line of sight during operation, this study presents a design methodology for a differential probe, along with a harmonic magnetic field excitation technique and its corresponding data processing methodology. The coaxial arrangement of the excitation coils in a planar configuration generates a potent induced magnetic field, ensuring a sufficient standoff distance. Under the influence of harmonic excitation, the magnetic permeability of the pipeline surface decreases, resulting in an increased detection depth. Periodic alterations in the magnetic permeability of the pipeline surface enhance the amplitude of defects. Adjusting the position of the induction coil minimizes interference from the excitation magnetic field, thereby obtaining a pure induced magnetic field. Differential processing is applied to the two induction coils to counteract interference from the background magnetic field. Finally, the effectiveness and reliability of this probe are validated through three experiments: pipeline positioning, weld seam inspection, and damage detection.

1. Introduction

The transportation of pipelines is favored for its cost-effectiveness and economic advantages. However, due to the presence of corrosive and acidic substances in the transported media and exposure to harsh working conditions characterized by high pressure and humidity, pipelines inevitably develop various forms of defects. Failing to promptly detect and repair pipeline damage often leads to catastrophic consequences. Therefore, conducting regular inspections on pipelines is of paramount importance.

Typically, pipelines, in addition to being coated with anti-corrosion layers, may require an insulation layer (either for thermal retention or cooling) or burial underground [1]. Presently, such pipelines are often subjected to manual inspections and visual examinations. However, these inspection methods exhibit a degree of uncertainty and randomness, making it challenging to achieve a scientific evaluation. To address this, Wang et al. [2] have devised a method for crafting flexible probes coupled with harmonic magnetic field detection technology. This approach effectively focuses the excitation magnetic field and penetrates the pipeline's outer insulation layer, enabling the detection of defects such as cracks, grooves, and through-holes at a standoff distance of up to 100 mm. Yang et al. [3] designed a three-phase array probe with a zero-flux position for placing TMR sensors to counteract interference from the excitation magnetic field, thereby enhancing probe lift-off distance. Zhao et al. [4] have introduced an array-focusing detector to increase the energy density of the excitation magnetic field on the pipeline. Hu et al. [5] improved the accuracy of large-area corrosion measurement by adding Mn-Zn ferrite magnetic cores to the receiving probes. Ona et al. [6] have optimized the distance between the transmitting and receiving coils through numerical analysis and experimental testing to enhance the probe's lift height.

Nevertheless, detecting defects in pipelines at high lift-off poses a tricky problem [7]: small damage characteristic signals are often too weak to be effectively captured in the acquired induced magnetic field signals. Differential sensors, as a design approach for probes, can effectively address this issue. Ona et al. [8] have designed an improved pick-up coil structure by combining it with a Maxwell bridge as a magnetic field measurement sensor, further enhancing the probe's standoff distance. Zhang et al. [9] have introduced a planar three-phase array probe, in which the output of the probe is only proportional to the size of the damage.

Reliable damage target identification algorithms are equally crucial. Existing target extraction algorithms, include Empirical Mode Decomposition [10], Variational Mode Decomposition [11], Independent Component Analysis [12], and Stochastic Resonance [13], employ time-frequency analysis tools combined with filtering to separate and extract target signals. These algorithms are effective for noise reduction. However, due to the shared time-frequency characteristics between background magnetic fields (the induced magnetic field at non-defective locations and the excitation magnetic field) and target magnetic fields (variations in the induced magnetic field at pipeline defect locations), these methods cannot eliminate the background field from the measured signal. This paper presents a harmonic magnetic field excitation method and a differential probe fabrication method to achieve non-contact detection of pipeline damage. The harmonic magnetic field excitation method systematically alters the magnetic permeability of the pipeline surface, enhancing the amplitude of damage targets. Two coaxially arranged excitation coils in a planar configuration provide a sufficiently strong excitation magnetic field to achieve a significant lift-off distance. The pick-up coil is positioned at a zero-flux location to mitigate interference from the excitation magnetic field. By matching damping resistors, the induction coil operates in a critical damping state to reduce noise interference. Combining a differential amplifier ensures that the system output is solely related to abnormal changes in the pipeline's induced magnetic field. Simultaneously, a data processing method is proposed to denoise, target frequency extraction, and defect characteristic identification.

The remaining sections of this paper are outlined as follows: Section II provides an introduction to the detection method, encompassing the principles of harmonic magnetic field detection and the operational principles of the probe. Section III delves into the fabrication of the probe, the method for matching damping resistors, and the experimental setup. Section IV, in conjunction with experimental data, elucidates denoising, target frequency extraction, and damage characteristic identification methods. Experimental verification is also conducted in this section. Finally, Section V summarizes this paper.

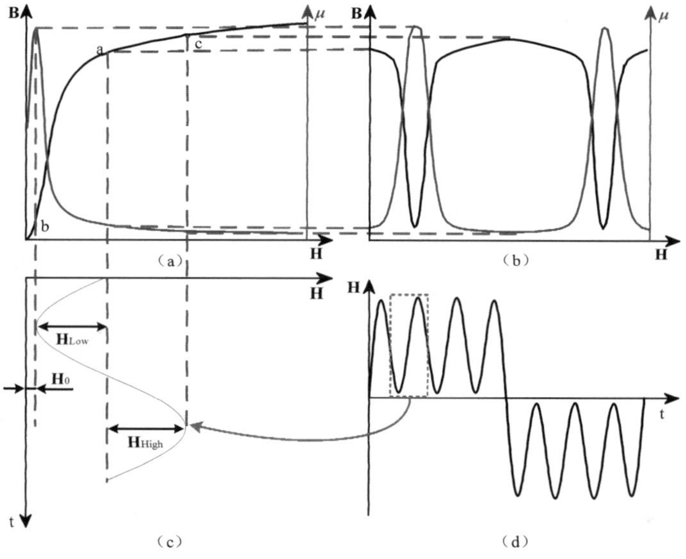

Figure 1. The detection principles of harmonic magnetic field.

2. Detection methods

2.1 Harmonic excitation sources

As elucidated in reference [14], the high magnetic permeability of steel pipes is a crucial factor leading to the under-detection of small defects. High magnetic permeability generates a strong induced magnetic field, and the magnetic field variation at the defect location causes minimal disturbance to the measured magnetic field. Consequently, short yet deep defects are challenging to detect. In light of this issue, we have improved our harmonic excitation source based on previous work, represented as follows:

$$\begin{cases} f(t) = A\sum_{i=1}^{n-1}\frac{f_i(t)}{2i-1} + B\,f_n(t) \\ f_i(t) = \sin\left[(2i-1)2\pi f\,t\right] \end{cases} \tag{1}$$

where A represents the fundamental amplitude, B signifies the peak harmonic amplitude, n denotes the number of harmonics, and f represents the fundamental frequency of the harmonics.

In this work, where n is 4 and f is 100 Hz, the excitation waveform is illustrated in Figure 1(d). Under the influence of the high-frequency magnetic field, the magnetic field strength varies as a→b→a→c→a, as depicted in Figure 1(a). The magnetic permeability of the pipeline continuously changes in response to the high-frequency magnetic field, as shown in Figure 1(b). When the amplitude of the excitation field reaches point c in Figure 1(a) (H_0+H_{Lwo}+H_{High}), the permeability of the pipeline is minimized.

2.2 Probe design

Theoretically, it should be feasible to separately measure the excitation magnetic field and then subtract it from the measurement results obtained from the test sample. However, the change in the induced magnetic field caused by damage is minute and requires amplification to ensure an adequate resolution. The presence of a large excitation magnetic field restricts the dynamic range of the signal amplifier, limiting the capability to identify the extent of damage.

The pick-up coil indirectly measures the magnetic field by detecting the rate of change of magnetic flux and is not affected by the static magnetic field. Placing the pick-up coil at the zero-flux position of the excitation coil can eliminate the excitation magnetic field. Based on this approach, a differential detection probe has been designed. The probe comprises an excitation coil, an adjusting coil, and two pick-up coils, as illustrated in Figure 2. The excitation coil and the adjusting coil share the same winding direction. The excitation coil generates the primary magnetic field for detection, while the adjusting coil provides an auxiliary excitation field for magnetic field focusing. Two pick-up coils are symmetrically arranged on the circumference of the excitation coil, measuring changes in magnetic flux density along the z-axis direction.

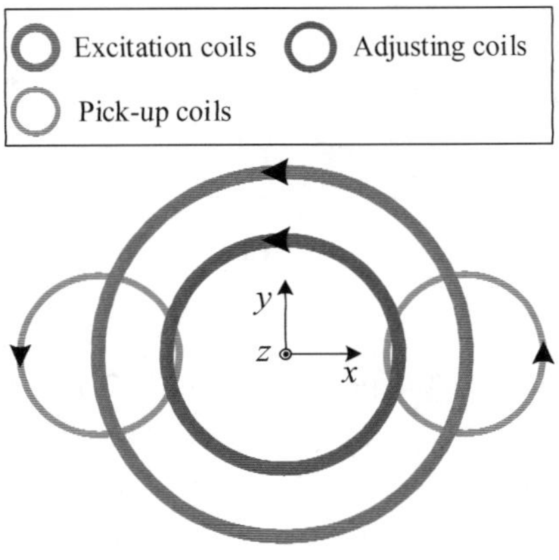

Figure 2. The structure diagram of the detection probe.

In practice, it is not possible to eliminate the excitation magnetic field. Thus the magnetic flux in each pick-up coil is the vector sum of the induced magnetic field and the excitation magnetic field and can be represented as follows:

$$\Phi_P = \oiint \left(\mathbf{B}_{EC} + \mathbf{B}_{exc} + \mathbf{B}_{adjI} + \mathbf{B}_{adjII}\right) \cdot \mathbf{n}\, dS_P \tag{2}$$

where Φ_P is the magnetic flux of the pick-up coils, **n** is the pick-up coil measurement direction vector, S_P is the pick-up coil area, and $\mathbf{B}_{EC}$, $\mathbf{B}_{exc}$, **and** $\mathbf{B}_{adj}$ are the induced magnetic field, excitation magnetic field, and adjusting magnetic field in the pick-up coils, respectively. Then, the electric potential of a pick-up coil is

$$E_P = n_p \cdot \frac{d\Phi_P}{dt} \tag{3}$$

where n_p is the turns of a pick-up coil. There are two design approaches for the two induction coils. One is adding the outputs of the two sensors within the same group, doubling the sensitivity of the sensors. The other is the differential of the outputs between the two coils. The essence of pipeline damage detection is to highlight the magnetic field variations at the damage location as effectively as possible. The differential design approach is better suited for removing interference from the background magnetic field and expanding the dynamic range of the subsequent circuitry. Therefore, the probe adopts a differential design approach, and its output is represented in Eq. (4).

$$v = E_{PI} - E_{PII} \tag{4}$$

where E_{PI} and E_{PII} are the electric potentials of two pick-up coils, respectively.

3. Probe fabrication and experimental setup

3.1 Probe fabrication and testing

The framework of the probe is a 3D-printed model, has grooves, and guides in the excitation coil frame and the pick-up coil frame to facilitate the adjustment of the relative positions between the excitation coil and the pick-up coil. The excitation coil and tuning coil are connected in series with the same winding direction, as depicted in Figure 3(a). The compensation coil has an inner diameter of 300 mm and a total of 140 turns, while the excitation coil has an inner diameter of 350 mm and a total of 240 turns.

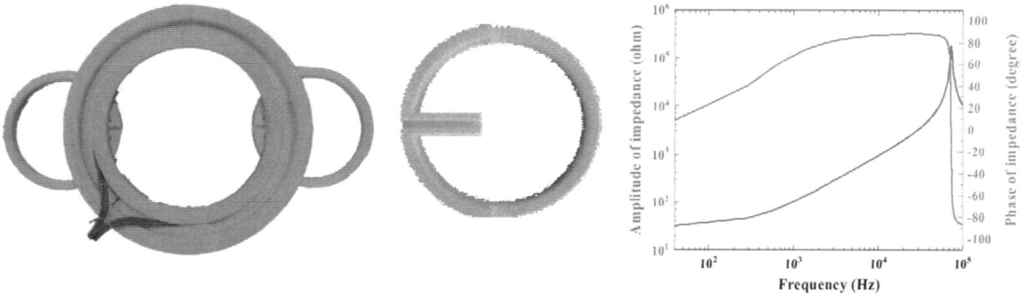

(a) (b) (c)

Figure 3. Detection probe and its impedance diagram

The two pick-up coils are symmetrically arranged near the zero-flux position with a radius of 435 mm. Figure 3(b) represents one of the pick-up coils, and each pick-up coil has an inner diameter of 200 mm with a total of 200 turns and a resistance of 32.7 ohms. Figure 3(c) displays the frequency response curve of the pick-up coil, with the first resonant frequency at 72.992 kHz. From this, it can be inferred that the inductance value of the pick-up coil is 196.992 mH, and the capacitance value is 0.24 nF.

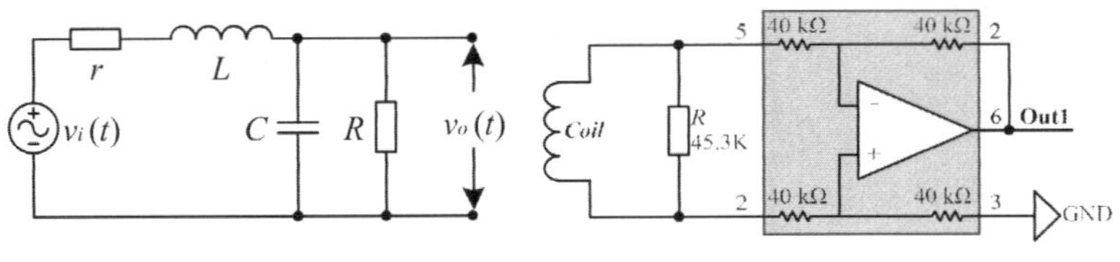

Figure 4. Induction coil equivalent circuit and its matching circuit.

3.2 Damping resistance matching

The pick-up coil can be equivalently modeled as an RLC circuit, as illustrated in Figure 4(a). Its mathematical model takes the form of a second-order differential equation, as shown in Eq. (5). In this equation, ξ represents the damping coefficient, as indicated in Eq. (6), and ω is the resonant frequency of the circuit. When ξ is greater than 1, the pick-up coil is in an overdamped state, resulting in a sustained phase lag in the induced voltage signal. When ξ is less than 1, the system becomes unstable, leading to increased coil output noise. Therefore, to ensure that the pick-up coil operates at critical damping, the matching resistance R is calculated to be 45.557 kΩ using Eq. (7). Figure 9(b) depicts the modulation circuit for the pick-up coil, where the matching resistance R is connected in parallel with the pick-up coil. The induced voltage of each pick-up coil is obtained by the differential amplifier AD8276.

$$\frac{d^2v_o}{dt^2} + 2\xi\omega\frac{dv_o}{dt} + \omega^2 v_o = \frac{v}{LC} \tag{5}$$

$$\xi = \frac{RrC + L}{2\sqrt{LCR(R+r)}} \tag{6}$$

$$R = \frac{L}{rC + 2\sqrt{LC}} \tag{7}$$

3.3 Experimental instruments

The schematic diagram of the experimental setup, as shown in Figure 5, comprises a harmonic excitation source, a detection probe, signal modulation circuitry, a data acquisition system, and a PC. Within the harmonic excitation source, the signal generator module generates four sets of sinusoidal reference signals with frequencies f, 3f, 5f, and 7f. These reference signals are synthesized into a harmonic reference signal through harmonic modulation circuitry, where the fundamental amplitude A is set to 1, and the highest harmonic amplitude B is 0.75. The reference signal is then amplified by a power amplification module to drive the coil, generating the harmonic magnetic field. The data acquisition system adjusts the frequency of the harmonic excitation magnetic field by controlling the frequency.

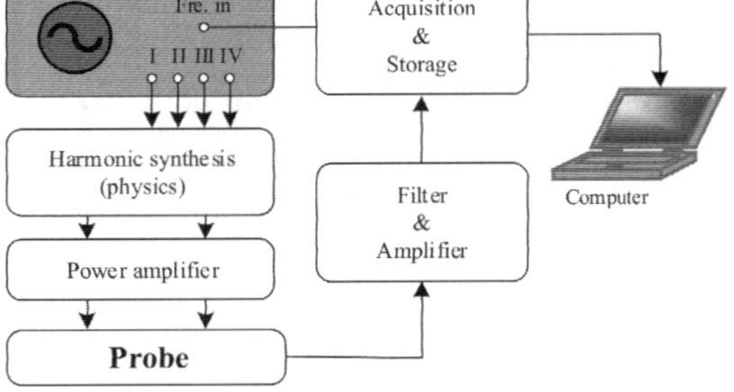

Figure 5. Schematic diagram of the testing instrument.

 2724 (2024) 012021 doi:10.1088/1742-6596/2724/1/012021

The signal modulation circuit is shown in Figure 6. Two pick-up coil voltage signals are connected to the modulation circuit via shielded twisted pairs. The differential signals are differentially amplified using the differential instrumentation amplifier AD620 and compensated using a voltage follower constructed with the operational amplifier AD548 to suppress common-mode interference induced by the transmission of weak signals. They are then further amplified by AD4895. The total amplification factor of the signal modulation circuit is 340x and includes a low-pass filter and a high-pass filter. The low-pass filter is utilized to eliminate high-frequency noise signals, with a cutoff frequency of 123 Hz. The high-pass filter serves to remove DC bias and unnecessary harmonic components, with a cutoff frequency of 103 Hz. Finally, the differential-induced voltage signals are digitized synchronously by the analog-to-digital converter.

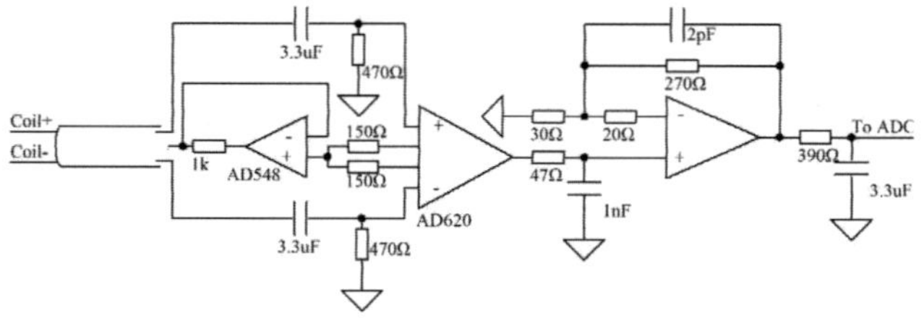

Figure 6. Schematic diagram of constant current source.

4. Tests and results

4.1 Signal analysis

Due to circuit bias in the operational amplifier circuit and the incomplete removal of unnecessary harmonic components by the band-pass filter, original data exhibits deviations and significant noise. We employ an integrated data preprocessing algorithm to enhance the signal-to-noise ratio for accurate damage target identification. This preprocessing algorithm includes the elimination of even-order harmonic and the extraction of the highest-order harmonic components, as depicted in Figure 7(a).

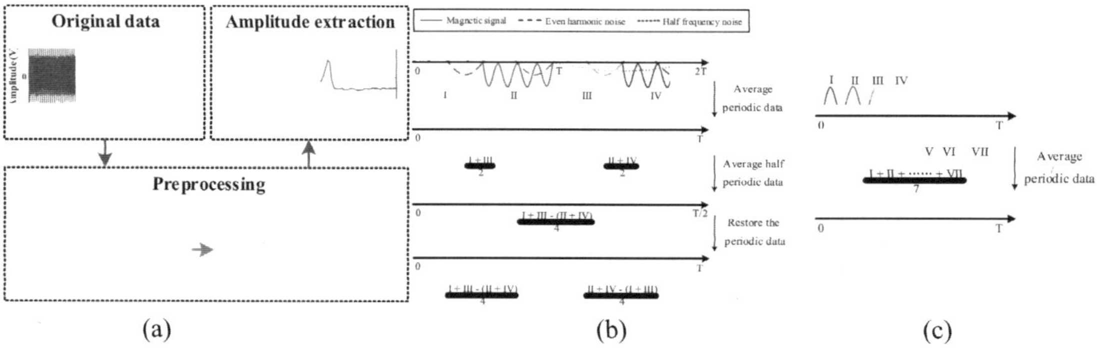

(a) (b) (c)

Figure 7. Signal processing flow.

1) Removal of even-order harmonic: As depicted in Figure 7(b), the process begins by summing the data over a period and taking the average. Then, the data from the latter half of a cycle is reversed within one cycle and averaged with the data from the first half of the cycle. Finally, the half-cycle data is restored to the full-cycle data.

2) Extraction of the highest-order harmonic component: As illustrated in Figure 7(c), the process begins by determining the number N of the highest-order harmonic component within one harmonic cycle. Subsequently, the harmonic signals within one cycle are divided into N groups. Finally, the data corresponding to each group is accumulated and averaged to extract the highest-order harmonic signal.

3) Orthogonal phase-sensitive detection technique: First, two orthogonal reference signals are constructed for the target frequency f harmonic and denoted as

$$\begin{cases} y(t) = \sin 2\pi ft \\ y'(t) = \cos 2\pi ft \end{cases} \tag{8}$$

where *f* is the frequency of the target harmonic. The cross-correlation function between the differential-induced voltage signal and the reference signals can be represented as

$$\begin{cases} R_{sy} = \frac{1}{T}\int_0^T s(t)y(t)dt = \frac{\lambda}{2}\cos\psi \\ R_{sy'} = \frac{1}{T}\int_0^T s(t)y'(t)dt = -\frac{\lambda}{2}\sin\psi \end{cases} \tag{9}$$

where $T = 1/f$ is the period of the target harmonic, λ is the amplitude of the differential signal, and ψ is the phase angle of the differential signal. The variation in the amplitude λ of the target harmonic component can be calculated as

$$\lambda = 2 \cdot \sqrt{R_{sy}^2 + R_{sy'}^2} \tag{10}$$

Using this method, the amplitude of the pipeline's induced magnetic field can be calculated. This allows for the determination of the position of the characteristic magnetic field signal associated with damage during the detection process.

4.2 Experimental research

In this section, the effectiveness of the proposed probe and data processing methods were validated through two laboratory test cases: pipeline routing and weld seam inspection, as well as an engineering demonstration case: pipeline defect detection.

Case 1. Pipeline positioning experiment

The test pipeline consists of two 45# steel pipes with an outer diameter of 76 mm and a wall thickness of 3 mm, as shown in Figure 8(a). The two pipes are placed parallel to each other with a spacing of 300 mm. The total detection distance is 250 cm, and the lift-off height h is 450 mm, approximately 6 times the pipe diameter. During detection, the probe is moved horizontally along the detecting guideline while the magnetic field data is collected, and then the collected data is processed (Figure 7). The results, as shown in Figure 8(b), reveal clear amplitude variations in the induced voltage of the differential probe, indicating effective pipeline localization. However, it is not possible to determine the number of pipes and the distance between parallel pipes based solely on amplitude information. Therefore, gradient processing is applied to the amplitude data, and the results are shown in Figure 8(c). The gradient-processed results indicate that there are two parallel pipes, and the spacing between them is approximately 303 mm, as determined from the distance between two zero-crossing points of the oscillation.

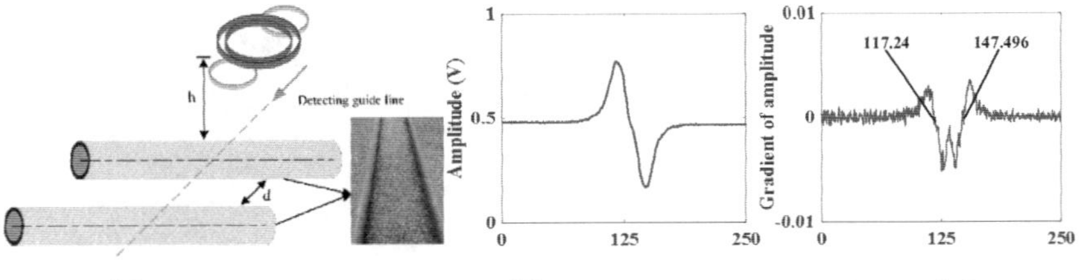

(a) (b) (c)

Figure 8. Pipeline positioning experiment and results.

Case 2. Welding detection experiments

The test pipeline and process are the same as the localization experiment. There is a circumferential weld seam at the center of the pipe. The probe's centerline is kept directly above the pipe, and the probe is moved horizontally along a detection guideline for weld seam inspection, as shown in Figure 9(a).

The total distance of the detection path is 300 cm, and the lift-off height h is 300 mm, approximately 4 times the pipe diameter. The detection results are displayed in Figure 9(b), where a significant amplitude is observed around the 140 cm position, deviating approximately 9 cm from the actual weld seam position. The reasons for this positioning error include: 1) Data preprocessing algorithms causing the loss of front-end data; 2) Inconsistent movement speed while manually moving the probe.

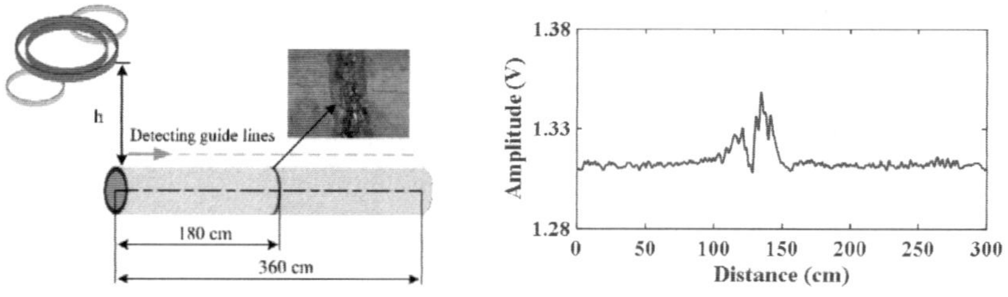

(a) (b)

Figure 9. Weld detection experiment and results.

Case 3. Defect detection experiment

A pipeline test platform was established in the suburbs, as shown in Figure 10(a). The test platform includes two 20# steel pipes with an outer diameter of 300 mm and a wall thickness of 10 mm. One of the pipes has a 100 mm × 4 mm × 3 mm groove with slight corrosion (100 mm × 100 mm), and a 40 mm × 40 mm × 3 mm square wear-thinning area. The probe's centerline is maintained directly above the damaged section of the pipeline, and the probe is moved horizontally along the guideline for defect detection. The total distance of the detection path is 800 cm, and the lift-off height h is 600 mm, approximately 2 times the pipe diameter. The results, as shown in Figure 10(b), reveal a significant decrease in probe amplitude output at 200 cm and 600 cm, corresponding to the locations of the damage.

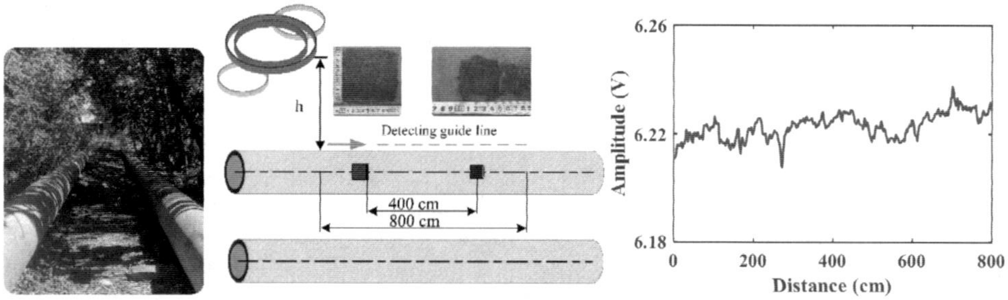

(a) (b)

Figure 10. Pipeline defect detection experiment and results.

5. Conclusion

This work proposed a method for detecting external pipeline damage using a differential probe fabrication technique and harmonic magnetic field excitation. The probe provides a sufficiently strong excitation magnetic field. Two pick-up coils are placed near the zero magnetic flux position, and a differential circuit and matching resistors are designed to reduce noise and background magnetic field interference. The improved harmonic magnetic field excitation method helps enhance the identification of damage targets. Test results demonstrate that this method can be effectively used for pipeline localization, weld seam detection, and pipeline wall thickness loss detection. In summary, this work provides an effective approach for pipeline damage detection, with the potential for practical engineering applications.

Journal of Physics: Conference Series **2724** (2024) 012021 doi:10.1088/1742-6596/2724/1/012021

References

[1] Wang X, Yang L, Sun T, et al. (2023) A review of development and application of out-of-pipe detection technology without removing cladding[J]. Measurement, 113249. https://doi.org/10.1016/j.measurement.2023.113249.

[2] Wang X, Gu Y, Chen Y, et al. (2020) A new technology for steel pipeline damage detecting without removing cladding[J]. Measurement, 159: 107700. https:// doi.org/10.1016/j.measurement.2020.107700.

[3] Yang L, Wang X, Sun T, et al. (2022) A technology of full perimeter inspection for steel pipeline without removing cladding[J]. Measurement, 190: 110746. https://doi.org/10.1016/j.measurement.2022.110746.

[4] Zhao Y, Wang X, Sun T, et al. (2021) Non-contact harmonic magnetic field detection for parallel steel pipeline localization and defects recognition[J]. Measurement, 180: 109534. https://doi.org/10.1016/j.measurement.2021.109534.

[5] Hu B, Yu R, Liu J. (2016) Experimental study on the corrosion testing of a buried metal pipeline by transient electromagnetic method[J], Anti-Corros. Methods Mater. 63 (4) 262–268. https://doi.org/10.1108/ACMM-10-2014-1444.

[6] Ona D I, Tian G Y, Sutthaweekul R, et al. (2019) Design and optimisation of mutual inductance based pulsed eddy current probe[J]. Measurement, 2019, 144: 402-409. https://doi.org/10.1016/j.measurement.2019.04.091.

[7] Zhang N, Ye C, Peng L, et al. (2019) Novel array eddy current sensor with three-phase excitation[J]. IEEE Sensors Journal, 19(18): 7896-7905. https://doi.org/10.1109/JSEN.2019.2919661.

[8] Ona D I, Tian G Y, Naqvi S M. (2019) Investigation of signal conditioning for Tx-Rx PEC probe at high lift-off using a modified Maxwell's bridge[J]. IEEE Sensors Journal, 20(5): 2560-2569. https://doi.org/10.1109/JSEN.2019.2951287.

[9] Zhang N, Ye Ch, Peng L, et al. (2021) Eddy current probe with three-phase excitation and integrated array tunnel magnetoresistance sensors[J]. IEEE Transactions on industrial electronics.68(6):5325-5336, https://doi.org/ 10.1109/TIE.2020.2989704.

[10] Huang Y, Wu D, Zhang Z, et al. (2017) EMD Based Pulsed TIG Welding Process Porosity Defect Detection and Defect Diagnosis using GA-SVM, J. Mater. Process. Technol. 239 92–102. https://doi.org/10.1016/j.jmatprotec.2016.07.015.

[11] Xiao Q, Li J, Sun J, et al. (2018) Natural-gas pipeline leak location using variational mode decomposition analysis and cross-time–frequency spectrum, Measurement 124, 163–17. https://doi.org/10.1016/j.measurement.2018.04.030.

[12] J. Ahmad, A. Akula, R. Mulaveesala, et al. (2019) An independent component analysis based approach for frequency modulated thermal wave imaging for subsurface defect detection in steel sample, Infrared Phys. Technol. 98, 45–54. https://doi.org/10.1016/j.infrared.2019.02.006.

[13] Wan C, Pan M, Zhang Q, et al. (2018) Magnetic Anomaly Detection Based on Stochastic Resonance, Sens. Actuat. A 278, 11–17. https://doi.org/10.1016/j.sna.2018.05.009.

[14] Karpenko O, Efremov A, Ye C, et al. (2020) Multi-frequency fusion algorithm for detection of defects under fasteners with EC-GMR probe data. NDT & E International, 110: 102227. https://doi.org/10.1016/j.ndteint.2020.102227.

Rapid determination of impurities in carbon dioxide for food by gas chromatography

Delong Hu[1], Tingting Ren[1], Xuemei Yang[1,2], Xianghong Li[1] and Qian Zou[1]

[1]Chongqing Academy of Metrology and Quality Inspection, China

[2]Corresponding author's e-mail: 876536427@qq.com

Abstract. This article describes the experimental principle and method of rapid determination of carbon monoxide, methane, propane, total hydrocarbon, methanol, ethylene oxide, acetaldehyde, vinyl chloride, benzene, and other common impurities in carbon dioxide by gas chromatograph. By using food additive carbon dioxide as a sample, the rapid analysis conditions, such as chromatographic column, valve switching time, and column box temperature control, were optimized. The retention time of impurity components in food additive carbon dioxide was determined using the external standard method. The relative standard uncertainty of CH_4, C_3H_8, CO, CH_3OH, CH_2CH_2O, CH_3CHO, CH_2CHCl, and C_6H_6 analysis is within 3% by using the established analytical method proposed in this article, which is equivalent to the national standard GB 1886.228-2016. The linear correlation coefficient r is greater than 0.9995, and the linear error is not greater than 2.0, indicating excellent linearity. In the meantime, in the case of 6 consecutive sample injections, the detection time is shortened by nearly three times. The experimental results show that the analytical method established in this article is suitable for the rapid determination of impurities in carbon dioxide for food, which can improve the accuracy of experimental results and work efficiency.

1. Introduction

Carbon dioxide is an important substance widely used in carbonated beverages, beer, and food preservation, which is closely related to daily life. The food industry of China has a huge demand for nearly 2 million tons of carbon dioxide every year, which is significantly increasing year by year. Therefore, it is necessary to detect harmful impurities in carbon dioxide used in food. However, at present, the relative testing method and traceability system have not been established in China. Large companies such as Coca-Cola require carbon dioxide purchased in China to be sent to foreign institutions for testing, indicating that there is still space for improvement in this field in China. Currently, food-grade carbon dioxide in China is mainly a by-product of chemical products, which is the carbon dioxide gas source. National food safety standard GB 1886.228-2016 *Food additive - Carbon dioxide* was officially implemented in 2017, which specifies detection indicators and methods. However, some detection methods proposed in the standard are relatively cumbersome, require complex types of equipment, lack traceability systems, and have high analytical uncertainty. Therefore, this article established a rapid detection method for carbon dioxide in food additives, which can quickly and accurately determine the concentration of carbon dioxide in food, solving the problems present in the national standard.

The carbon dioxide used in food contains impurities such as carbon monoxide, methane, propane, total hydrocarbons, methanol, ethylene oxide, acetaldehyde, vinyl chloride, and benzene, all of which

 2724 (2024) 012022 doi:10.1088/1742-6596/2724/1/012022

but carbon monoxide are hydrocarbons with a large number of hydrogen atoms, and carbon monoxide can be converted into methane through catalytic hydrogenation. Therefore, GC - FID (Flame Ionization Detector) is used for the detection of impurities in this article [1-2]. Due to the differences in the use of separation columns for each component in GC (Gas Chromatography), in order to achieve the goal of simultaneous detection of target impurity components in a single injection, separation channels are added in the original single flow chromatography system. Based on the difference in peak time of each channel, separation components in different channels respond separately on the FID detector. By adding devices such as automatic switching pneumatic valves and resistance columns, unnecessary and longer retained components in each separation channel are cut and blown back without affecting the peak detection of the target component, saving chromatographic separation and detection time [3]. In order to further improve sensitivity and repeatability, the chromatographic conditions were optimized by increasing quantitative ring volume [4].

2. Experimental method

2.1. Experimental principle of GC - FID

When organic compounds burn in a hydrogen flame, chemical ionization reactions occur, producing positive and negative ions. There is a potential difference between the emitter and collector of the detector, and ion pairs form an ion flow between the two electrodes. The electron flow collected by the collector generates a voltage signal through the high resistance load, which is amplified by a light microcurrent amplifier and transported to the recorder to obtain the corresponding chromatographic peaks of the substance [5-6].

2.2. Experimental instrument

Modify type SC-3000 GC, and equip it with glass microsphere column SCBWQ (φ3×2 m, (60~80) mesh), carbon molecular sieve filled column TDX-01 (φ 3×1 m, (60~80) mesh), resistance column OV-101(φ 3×1 m, (60~80) mesh), capillary column KB-PLOT Q (30 m×0.53 mm×20 μm), and Double FID detector. The internal configuration diagram of GC after renovation is shown in Figure 1.

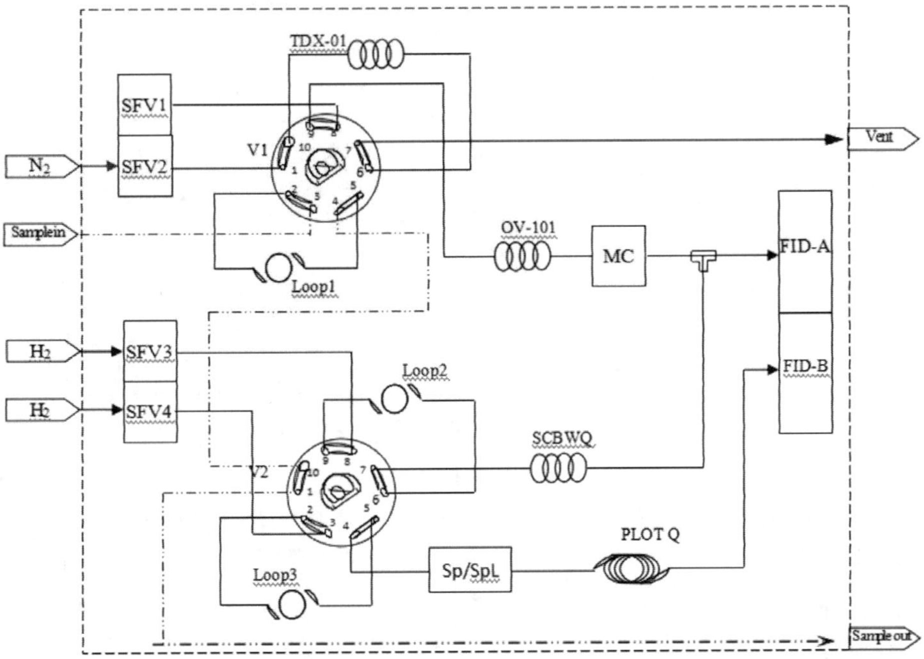

Figure 1. The internal configuration diagram of GC after renovation.

2.3. Experimental conditions

In order to achieve the optimal experimental conditions for analyzing impurities in carbon dioxide used as a food additive, the initial temperature of the columns is adjusted to 80°C, the temperature is kept for 5 minutes, and then the column temperature is elevated to 120°C at the rate of 10°C/min. The temperature is kept for 5 minutes, and the size of the sample ring is set at 1 mL. The detailed specific experimental conditions are shown in Table 1.

Table 1. Detailed specific conditions.

Impurities	Detector	Chromatographic column	Flow rate (mL/min)	Split ratio
C_3H_8 (5~50) μmol/mol	FID-A	SCBWQ φ3×2 m	N_2: 25	None
CO, CH_4 (5~50) μmol/mol	Methanator + FID-A	TDX-01 φ3×1 m; OV-101 φ3×1 m	N_2: 25	None
C_2H_4O, C_3H_8, CH_4O, C_2H_4O, C_2H_3Cl, C_6H_6 (1~5) μmol/mol	FID-B	KB-PLOT Q 30 m×0.53 mm×20 μm	N_2: 3	1:2

2.4. Experimental process

The pneumatic valve V2 is switched. The sample gas is set to respectively enter the quantitative Loop 1, Loop 2, and Loop 3 through the 3-2 and 5-4 channels of V1, as well as the 10-9, 6-5, and 2-1 channels of V2, to complete the injection. Carrier gas 2 (SFV2) brings the sample from Loop 1 into the TDX-01 column. CO and CH_4 components are first separated by the chromatographic column, and CO is converted into CH_4 [7-8] through a methanator, then CH_4 is detected by the FID-A detector, and V1 is activated to return to the initial position. The remaining components are purged and emptied by carrier gas 1 (SFV1), preventing the components with a long retention time in the TDX-01 column from remaining on the column and a large amount of CO_2 entering methanation. At the same time, V2 is started, and carrier gas 4 (SFV4) will bring the sample from Loop 3 into the KB-PLOT Q column through the injection port. CH_4, C_3H_8, CH_2CH_2O, CH_3OH, CH_3CHO, and CH_2CHCl [9-11] will be pre-separated by the KB-PLOT Q column before entering the FTD-B detector for detection. Carrier gas 3 (SFV3) will bring the sample from Loop 2 into the SCBWQ column for separation, and total volatile hydrocarbons (THC) are detected by the FID-A detector. CO, CH_4, and THC are displayed in the FID-A chromatogram, while CH_4, C_3H_8, CH_2CH_2O, CH_3OH, CH_3CHO, CH_2CHCl, and C_6H_6 are sequentially displayed in the FID-B chromatogram. CH_4 can be quantified in both the FID-A and FID-B chromatograms, which can be used as a tool for validation and comparison.

3. Experimental results

3.1. Separation order of impurity components in food additive carbon dioxide

The chromatographic columns used in the experiment are all non-polar or low-polar, and the components of smaller relative molecular weight and lower boiling point flow out earlier. If the boiling points of the components are the same or similar, saturated alkanes flow out first. Components with complex isomerism flow out earlier when saturated hydrocarbons are with the same carbon number. Unsaturated hydrocarbons with double bonds in the middle peak earlier than those without. In order to fully determine the peak time of each component, the external standard method was used to determine the retention time of impurity components in food additive carbon dioxide, and the gas chromatography analysis spectrums are shown in Figures 2 and 3. Figure 2 shows the channel FID-A detection results, including volatile THC (0.503 min), CO (1.105 min), and CH_4 (1.817 min). Figure 3 shows the channel FID-B detection results, including volatile CH_4 (1.105 min), C_3H_8 (2.618 min), CH_2CH_2O (2.999 min), CH_3OH (3.465 min), CH_3CHO (3.629 min), CH_2CHCl (4.465 min), C_6H_6 (8.487 min). The concentrations of all components are 5 μmol/mol.

2724 (2024) 012022 doi:10.1088/1742-6596/2724/1/012022

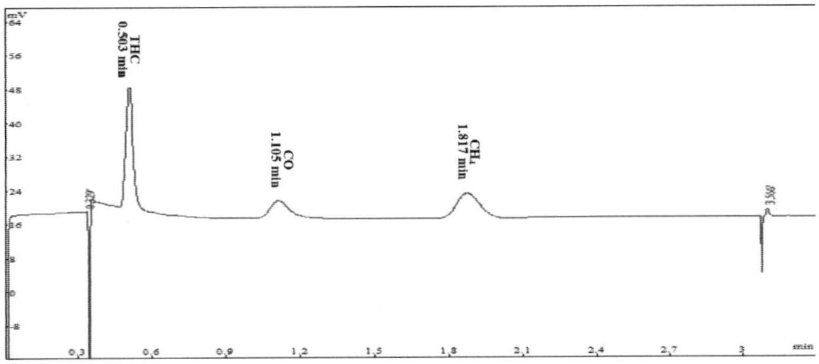

Figure 2. The gas chromatography analysis spectrum of channel FID-A.

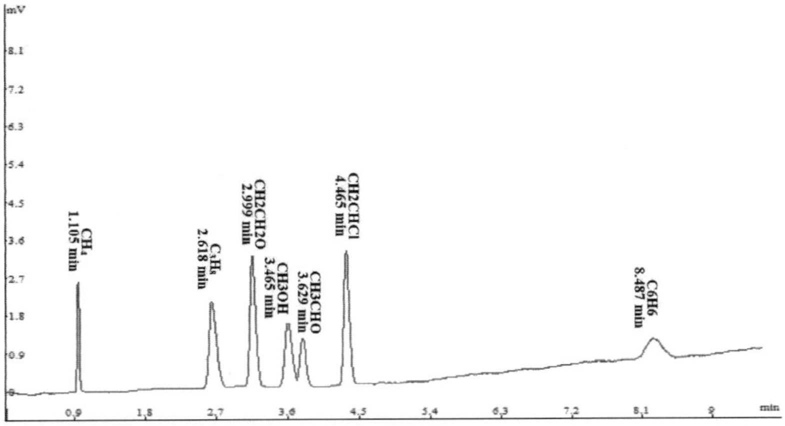

Figure 3. The gas chromatography analysis spectrum of channel FID-B.

3.2. Linearity and detection limit

Under the optimal analysis conditions elaborated in Section 2.3, CH_4, CO, and C_3H_8 standard gases with concentrations of 5.00 μmol/mol, 25.0 μmol/mol, and 50.0 μmol/mol, as well as CH_2CH_2O, CH_3OH, CH_3CHO, CH_2CHCl, and C_6H_6 standard gases with concentration of 1.00 μmol/mol, 2.00 μmol/mol, and 5.00 μmol/mol, are injected into the modified SC-3000 GC to perform gas analysis. The concentration of standard gas is made as the x-axis and the chromatographic peak area is made as the y-axis for linear regression. The detection limit is 3 times the baseline noise of the mass of the corresponding impurity component. The experimental results of linearity and detection limit are shown in Table 2.

Table 2. The experimental results of linearity and detection limit.

Component	Linear regression	Correlation coefficient	Detection limit (ng)
CH_4	$y = 936.676x - 115.115$	0.9999	2.5
CO	$y = 8932.920x + 958.435$	0.9999	2.4
C_3H_8	$y = 768.856x - 86.110$	0.9999	3.2
CH_2CH_2O	$y = 151.377x + 1.637$	0.9998	5.0
CH_3OH	$y = 9200.171x + 3.458$	0.9998	6.7
CH_3CHO	$y = 179.033x - 2.745$	0.9998	7.2
CH_2CHCl	$y = 213.127x + 2.895$	0.9998	6.3
C_6H_6	$y = 60.004x + 0.520$	0.9998	7.6

3.3. Repeatability

Repeatability describes the divergence degree of measurement data. Usually, relative standard deviation (RSV) represents repeatability. The experimental results of repeatability are shown in Table 3.

Table 3. The experimental results of repeatability.

Impurity	Concentration (μmol/mol)	Peak area (μV•s)						RSV
		1	2	3	4	5	6	(%)
CH_4	5.00	4516.5	4534.3	4512.6	4523.7	4513.2	4563.3	0.43
	25.0	23267.8	23178.4	23192.1	22987.5	23191	23055.3	0.45
	50.0	46860.7	46920.8	46764.5	46851.9	46765.2	46642.8	0.21
CO	5.00	45304.6	45623.1	45262.7	45261.6	45384.2	45751.3	0.46
	25.0	227221.3	225352.6	226453.5	227953.1	225634.2	226646.4	0.43
	50.0	446348.5	443361.6	448464.8	448120.6	447374.9	445284.9	0.44
C_3H_8	5.00	3783.7	3757.4	3781.7	3775.0	3743.1	3762.1	0.42
	25.0	19014.8	19003.9	18983.7	18827.2	18971.9	18879.8	0.41
	50.0	38470.6	38645.8	38531.5	38428.9	38386.4	38236.8	0.36
CH_2CH_2O	1.00	150.9	146	147.8	151.6	153.4	144.7	2.4
	2.00	300.4	298.3	298.2	309.0	305.8	302.6	1.5
	5.00	766.6	751.7	748.9	740.4	773.0	756.1	1.6
CH_3OH	1.00	206.9	201.9	202.7	196.7	198.7	195.2	2.2
	2.00	400.5	404.1	404.2	411.6	409.9	393.3	1.7
	5.00	984.91	1000.01	1027.64	1018.97	994.26	1009.68	1.6
CH_3CHO	1.00	175.7	174.9	172.8	183.6	178.2	180.8	2.3
	2.00	366.9	360.1	368.6	354.8	356.6	358.2	1.6
	5.00	886.9	919.6	889.9	894.0	885.3	892.5	1.5
CH_2CHCl	1.00	218.9	214.7	215.1	210.6	206.4	206.7	2.4
	2.00	435.0	442.3	437.6	435.9	431.7	423.0	1.6
	5.00	1048.3	1060.6	1070.3	1100	1061.8	1050.7	1.8
C_6H_6	1.00	57.8	57.6	60.7	58.9	60.8	59.1	2.4
	2.00	122.5	122.1	124.3	117.3	122.8	120.7	2.0
	5.00	296.5	295.2	295.8	298.0	297.2	309.7	1.9

4. Conclusion

This article modified an SC-3000 GC to determine the impurities in food additive carbon dioxide. The modified GC is equipped with 2 valves, 4 columns, 3 quantitative tubes, and 2 FID detectors that can automatically merge to achieve rapid and accurate analysis of various impurities in food additive carbon dioxide. The retention time of impurity components in food additive carbon dioxide was determined using the external standard method, which effectively achieved qualitative analysis. The established analytical method and optimal experimental conditions can quickly and accurately analyze the impurity content of THC, CO, CH_4, C_3H_8, CH_2CH_2O, CH_3OH, CH_3CHO, CH_2CHCl, and C_6H_6 through only one injection. The analytical method has good linearity and high precision, which is consistent with the national standard GB1886.228-2016 *Food additive - Carbon dioxide*, but with more than 3 times efficiency. The experimental results show that the method presented in this article is suitable for the determination of impurities in food additive carbon dioxide, providing a new and rapid analysis method for impurity analysis in food additive carbon dioxide.

References

[1] Xu, D. M. (2022). Determination of total non-methane hydrocarbon by auto sampling gas chromatography with Summa Canister. Fujian Analysis & Testing, 31(6): 27-29. doi: 10.3969/j.issn.1009-8143.2022.06.06.

[2] Liu, W. M., et al. (2013). NMHC analysis by gas chromatography with single column. Analytical Instrumentation, 02: 23-26. doi: 10.3969/j.issn.1001-232x.2013.02.006.

[3] Liu, Z. S., et al. (2021). High precision temperature controller and its parameter tuning are realized by using the principle of time delay filter. Electronic Design Engineering, 29(8): 106-110, 114. doi: 10.14022/j. issn1674-6236.2021.08.023.

[4] Xing, Q. Q., et al. (2023). Application process of gas chromatography-mass spectrometry in food analysis. Food & Machinery, 39(6): 234-240. doi: 10.13652/j.spjx.1003.5788.2022.80787.

[5] Liu, H., et al. (2022). Analysis of trace amount of CO and CO_2 in polymer grade propylene by gas chromatography. Polyester Industry, 35(4): 36-40. doi: 10.3969/j.issn.1008-8261.2022. 04. 010.

[6] Lu, W., et al. (2022). Simultaneous determination of methanol and ethanol in mouthwash water by gas chromatography. Detergent & Cosmetics, 10: 23-27. doi: 10.3969/j.issn.1006-7264. 2022.10.005.

[7] Fu, M. Y., et al. (2014). Gas chromatographic analysis of trace gas impurities in carbon tetrafluoride. Chemical Propellants & Polymeric Materials, 12(1): 93-96. doi: 10.16572/j. issn1672-2191.2014.01.005.

[8] Gui, Y. Y. (2020). Analysis of trace carbon monoxide and carbon dioxide by gas chromatography. Analytical Instrumentation, 01: 89-92. doi: 10.3969/j.issn.1001-232x.2020.01.019.

[9] Shan, J., et al. (2018). Method research of analyzing trace gas impurity using gas chromatography. Journal of Astronautic Metrology and Measurement, 38(4): 86-89. doi: 10.12060/j.issn.1000-7202.2018.04.16.

[10] Li, J. X., et al. (2023). Improvement of the experiment for determining the composition and content of alkane mixture by gas chromatography. Chemical Enterprise Management, 30: 55-61. doi: 10.19900/j.cnki.ISSN1008-4800.2023.30.013.

[11] Hu, G. (2014). Analysis of trace hydrocarbon impurities in propylene by gas chromatography. Refine and Chemical Industry, 25: 50-52,64. doi: 10.16049/j.cnki.lyyhg.2014.01.002.

Feature optimization combined with UPerNet-Twins model for eucalyptus extraction from Sentinel-2A image

Wenchao Huang [1], Bo Wei [1, 2,4] and Chang Wei [3]

[1]College of Geomatics and Geoinformation, Guilin University of Technology, Guilin 541004, China
[2]Guangxi University Key Laboratory of Ecological Spatiotemporal Big Data Perception Service, Guilin University of Technology, Guilin 541004, China
[3]Guangxi Land and Resources Planning and Design Group Co., Ltd, Nanning 530022, China

[4]Corresponding author's e-mail: superweibo@glut.edu.cn

Abstract. To improve the extraction accuracy of eucalyptus from Sentinel-2A image, two key factors of feature construction and extraction model are considered. The original band spectrum, custom vegetation index, red edge spectral index, and texture features are obtained from the image. The Relief F-PSO-SVM model is used to screen out the best feature subset. A UPerNet-Twins combination model is used to realize the high-precision extraction of eucalyptus for the study area. The experiments show that the original spectrum plays a significant role in the extraction of eucalyptus. In addition, the texture features, vegetation index, and red edge spectral index are helpful to the extraction of eucalyptus, which are increased by about 5.84%, 3.81%, and 3.00%, respectively, compared with the IOU of the original spectrum. Moreover, the IOU of the optimal feature subset obtained by feature selection is increased by about 2.22% compared with the original feature set. The UPerNet-Twins combination model has the highest extraction accuracy of eucalyptus with an IOU of 0.8496 compared with DeeplabV3+, PspNet-UNet, FCN, UperNet, and UperNet-Vision Transformers models.

1. Introduction

Eucalyptus spp. (scientific name: Eucalyptus spp.) is the collective name of the genus Eucalyptus, the genus Eucalyptus and the genus Myrtaceae. It can be used in pulp and paper, wood-based panels, and construction industries[1]. Through the automatic or semi-automatic identification or extraction of remote sensing images, the acquisition and monitoring of key information such as eucalyptus distribution, area, and health status can be achieved. The extraction range, accuracy, timeliness, and human resource consumption are greatly superior to manual extraction[2]. Based on the GF-6 image, Wang et al. In [3] constructed a vegetation index and red edge index and used a decision tree as an extraction model to extract eucalyptus forest information in Liuzhou area. In [4], the GF-2 image was selected to extract eucalyptus by spectral and texture features and vegetation index. These studies have shown that it is not very effective to extract eucalyptus only by relying on a single original spectral feature. It is necessary to combine vegetation index, red edge index, texture, and other features on the basis of the original spectrum. Due to the 'dimension disaster', it is necessary to optimize the features to obtain features that are more helpful for extraction[5]. In recent years, feature optimization has promoted the improvement of remote sensing image classification and information extraction results[6].

Hao et al. In [7] used the Relief F feature selection method and decision tree classification to extract wetland information and obtained 86.9% classification accuracy. The Relief F-PSO-SVM feature selection method was proposed in [8], and the mixed features were extracted based on this method, which achieved the highest accuracy in land use classification.

In addition to feature selection, the extraction model is another key factor affecting remote sensing image classification and information extraction. In recent years, UPerNet[9] and Twins[10] models have been proposed in deep learning models. This paper combines the advantages of the UPerNet, and Twins models, recombines the UPerNet-Twins model, and considers the importance of feature construction to the performance of the model. The Relief F-PSO-SVM method is used to select the best feature subset, and the extraction of eucalyptus from Sentinel-2A image in the study area is discussed.

2. Research Area and Research Methods

2.1. Overview of the study area

Due to the suitable geographical location, natural conditions, and climatic conditions, eucalyptus in Guangxi has formed the industrialization and scale of plantations. According to the data from 2020, the planting area of eucalyptus is more than 30 million acres, accounting for more than half of the total area of eucalyptus in China. The selected study area is located in Nanning, Guangxi, with a longitude range of 107°47′53″ to 107°58′41″ and a latitude range of 23°6′8″ to 23°55′3″, with a total area of about 2.5 square kilometers (Figure 1). After resampling of Sentinel-2A image data of the study area, the spectral resolution is set to 10 m, and only 10 bands (B2, B3, B4, B5, B6, B7, B8, B8a, B11, B12) are retained.

2.2. Multidimensional dataset construction

Because the single original spectral characteristics of other tall trees, such as pine and mahogany, are similar to the characteristics of eucalyptus, the original spectrum, vegetation index, red edge index, and extracted texture features are used to improve the extraction accuracy of eucalyptus. Nine vegetation indexes (NDVI, RVI, DVI, EVI, GSAVI, IPVI, NormR1, NormR2, NDeI) and six red edge indexes (NDVIre1, NDVIre2, NDVIre3, NDre1, NDre2, CIre) are obtained by using the original band weighting operation. Gray-Level Co-occurrence Matrix (GLCM) is used to extract texture features. After principal component analysis, the first two images were selected for texture feature extraction, and a total of 16 texture features were extracted.

2.3. Relief F-PSO-SVM feature selection method

Feature selection refers to the process of selecting the best feature subset in the feature set. Feature selection methods are generally divided into three types: Filter, Wrapper, and Embedded. The Relief F algorithm used in this paper belongs to the filtering type, and the PSO-SVM combination model belongs to the embedded type. The combination of the Filter-Wrapper algorithm can not only reduce the computational complexity and be very friendly to high-dimensional data sets but also take into account the interaction between features. This can result in the highest superiority of the selected feature subset.

2.4. Eucalyptus extraction feature combination scheme

In terms of extracting the feature set of eucalyptus, six feature combination schemes were designed based on the original feature set (Table 1). In order to more accurately explore the role of the red edge index, the red edge band (B5, B6, B7) is not included in the original band spectral feature scheme but is classified into the red edge index.

Table 1. Feature Combination Schemes.

Number	Characteristic variable set
Scheme 1	Original spectrum
Scheme 2	Scheme 1+vegetation index
Scheme 3	Scheme 1+red edge index
Scheme 4	Scheme 1+texture feature
Scheme 5	Pre-optimal feature set
Scheme 6	Best feature subset

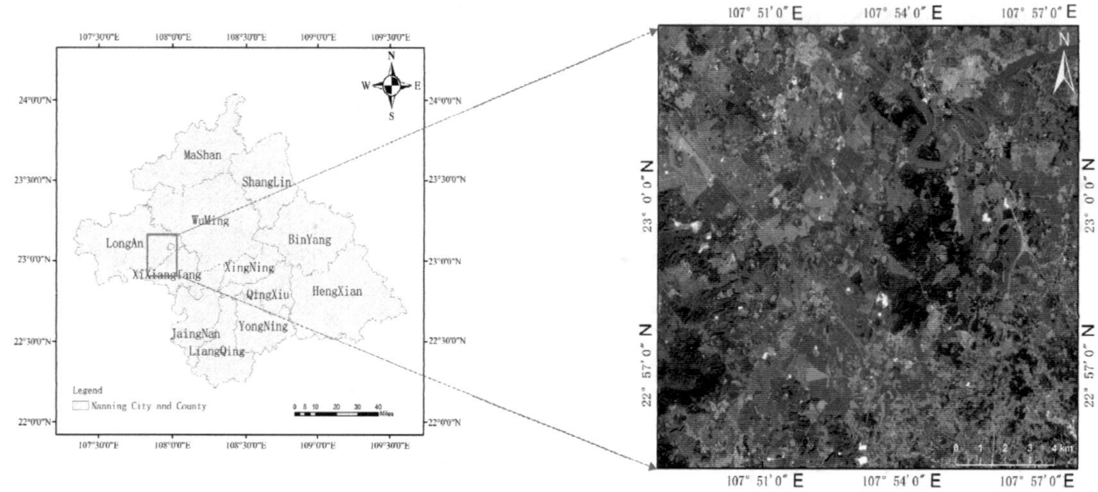

Figure 1. Location of Study Area.

2.5. UPerNet-Twins combination model

In this paper, UPerNet is combined with the Twins model to form the UPerNet-Twins model (Figure 2), which is mainly composed of a feature pyramid and pyramid pooling model. The left feature extraction module is replaced by the Twins model. The Twins model has two kinds of visual converter architectures. In this paper, the Twins-PCPVT model is used as the image semantic segmentation model. The process of image classification by the UPerNet-Twins model through feature extraction is as follows. (1) A remote sensing image with the size of $H \times W \times C$ is input, where H, W, and C represent the length, width, and number of input channels of the input image, respectively; (2) Through the feature extraction of Twins module, four feature maps of different sizes are obtained respectively. The minimum feature map (a4) fuses the global average pooling layer of four feature maps of different sizes through the pyramid pooling model to obtain the feature map b4; (3) The feature map b4 is upsampled three times and fused with the corresponding size of the Twins model to obtain three feature maps (b3, b2, b1); (4) The feature maps b1, b2, b3, and b4 are magnified to the original feature size and fused by bilinear difference method; (5) The 3×3 convolution kernel is used to classify the pixels of the feature fusion image, and the final classification result is obtained by restoring the size to $H \times W$ by the bilinear difference method.

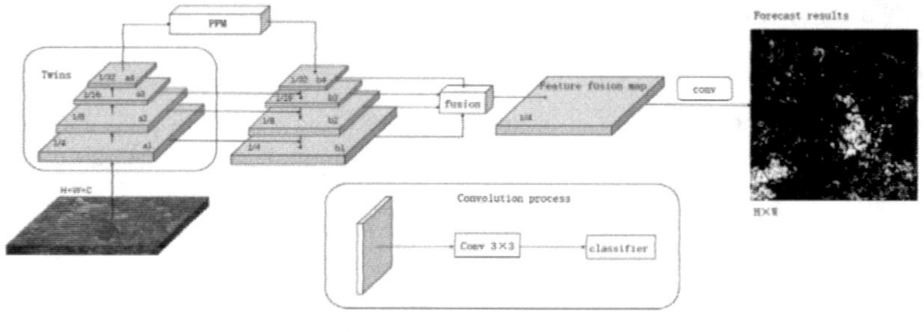

Figure 2. UPerNet-Twins Model Structure Diagram.

3. Results and Analysis

3.1. Best feature subset

Firstly, the Relief F algorithm was used to screen the 41 features (10 original spectral features, 9 vegetation indices, 6 red edge indices, and 16 texture features). The importance of the features was ranked according to their weight (Figure 3). The greater the weight is, the greater the correlation with the eucalyptus category is. Initially, 23 features with larger weights are retained, and the initial feature subset is obtained. Then, PSO-SVM is used to re-screen the initial feature subset, and the particle swarm achieves convergence and finally obtains the best feature subset with a feature number of 13 (Table 2).

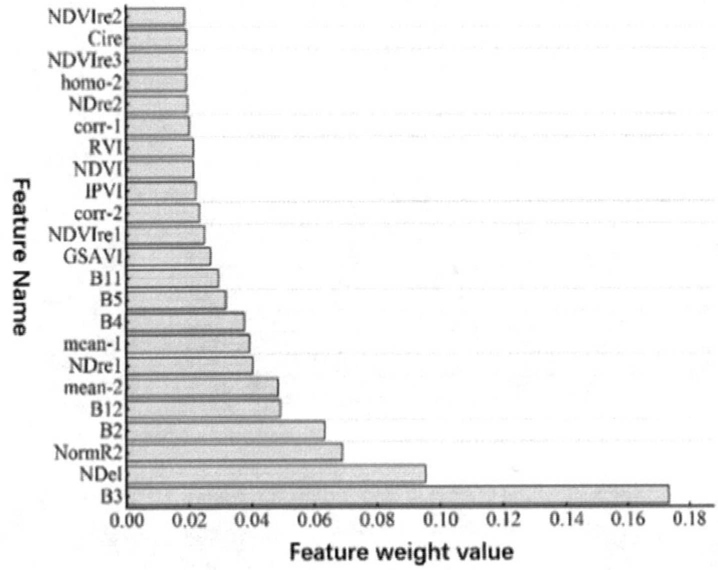

Figure 3. Relief F Algorithm Weight Alignment.

Table 2. Optimal Feature Subset.

Category	Feature name
Original spectrum	B2, B3, B4, B5, B12
Vegetation index	NDVI, NDeI, NormR2
Red edge index	NDVIre1, NDre1
Texture feature	mean-1, mean-2, homo-2

3.2. Comparison of Eucalyptus extraction results

3.2.1. Comparison of different feature combination results under the UPerNet-Twins model. Based on the UPerNet-Twins model, eucalyptus was extracted according to the six feature combination schemes in Table 1. The results are shown in Table 3. Table 3 shows that although scheme 1 has the lowest interaction using the original spectral features alone, it can still achieve a certain extraction accuracy. Hence, there will still be five original spectral features retained in the best feature subset scheme 6. The vegetation index, red edge index, and texture feature were added separately in schemes 2 ~ 4. The results showed that the extraction effect of scheme 4 with the texture feature was the best, and the interaction ratio reached 84.18%. Although scheme 5 contains all the features, it is not simple that the more features can show the better extraction effect. Scheme 6 has only 13 features, which is far less than scheme 5. However, its extraction accuracy is the best among the 6 schemes.

Displays the original images and labels of the study area and the prediction maps of the six schemes (Figure 4). By comparing different contrast schemes with the original image labels, it can be seen that using the best feature subset has the best effect. It can be seen from the prediction maps of schemes 1 to 3 that in areas where eucalyptus trees are sparsely distributed, there are a large number of missing and wrong mentions. The main reason is that the resolution is limited and there is a mixture of pixels in some small areas. Schemes 4 to 5 greatly reduced the phenomenon of missing extractions, but the phenomenon of unclear and ambiguous eucalyptus extraction still appeared in many places. Scheme 6 has been significantly improved, and it is very close to the original image label in all aspects.

Table 3. Comparison of Extraction Accuracy of Eucalyptus with Different Feature Combinations.

Scheme	Accuracy (%)	Recall (%)	F1-Score (%)	IOU (%)
Scheme 1	91.59	84.41	87.86	78.34
Scheme 2	92.16	88.31	90.19	82.15
Scheme 3	91.64	87.84	89.7	81.34
Scheme 4	92.00	90.84	91.41	84.18
Scheme 5	92.32	88.85	90.55	82.74
Scheme 6	92.37	91.38	91.87	84.96

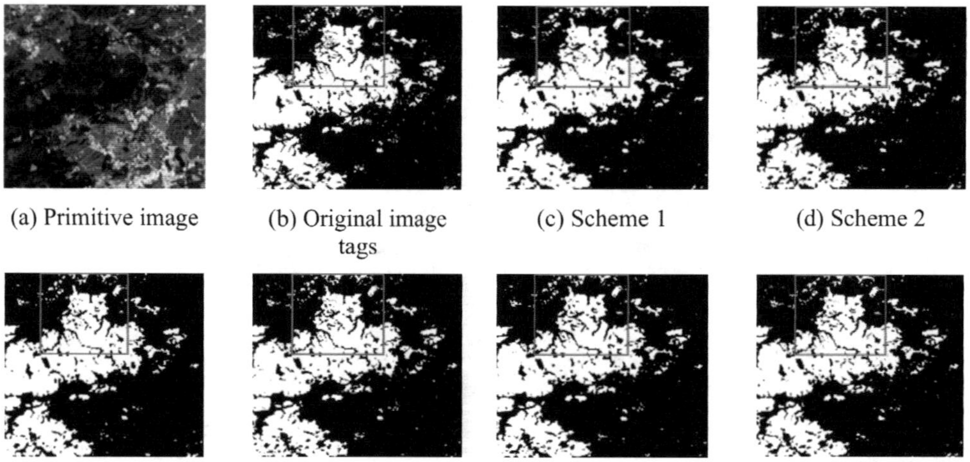

(a) Primitive image (b) Original image tags (c) Scheme 1 (d) Scheme 2

(e) Scheme 3 (f) Scheme 4 (g) Scheme 5 (h) Scheme 6

Figure 4. Prediction Map of Eucalyptus Extraction with Different Feature Combination Schemes.

3.2.2. Comparison of different model results under the best feature subset. By comparing the extraction accuracy of eucalyptus in different models (Table 4), it can be seen that the UPerNet-Twins model is the best, followed by the PSPNet-UNet model, the UperNet model, the DeepLabV3 + model, the UperNet-Vision Transformers model, and the FCN model. Under the same optimal feature subset, the maximum difference in extraction accuracy IOU is about 15%, and the average IOU difference is about 8%.

Table 4. Comparison of Extraction Accuracy of Eucalyptus from Different Models.

Scheme	Accuracy (%)	Recall (%)	F1-Score (%)	IOU (%)
DeeplabV3+	93.31	81.91	87.24	77.37
PspNet-UNet	97.16	83.15	89.61	81.18
FCN	88.04	76.85	82.06	69.58
UPerNet	91.99	86.47	89.15	80.42
UPerNet-Vision Transformers	84.49	82.95	83.71	71.99
UPerNet-Twins	92.37	91.38	91.87	84.96

4. Conclusions

The Relief F-PSO-SVM model was used to optimize the features of the feature set obtained from the Sentinel-2A image of the study area, and the best feature subset was obtained. Eucalyptus extraction and comparison experiments were carried out under the combined UPerNet-Twins model in this paper. The main conclusions are as follows.

(1) The Relief F-PSO-SVM model is suitable for feature selection. 13 feature subsets are selected from the original 41 features, with the highest extraction accuracy and effect can be obtained by using relatively few features. This shows the practicability of feature selection.

(2) The visual attention model shows superiority over the classical convolutional neural network in various visual tasks, but it still needs to be optimized and improved in efficiency. Compared with DeeplabV3+, PspNet-UNet, FCN, UperNet, and UperNet-Vision Transformers models, the IOU of the new combined UPerNet-Twins model is increased by at least about 3.8 % for eucalyptus extraction based on the optimal features.

Although eucalyptus extraction has achieved good accuracy in this paper, further research is still needed on multiple numbers of mixed species, a better feature optimization method, and a finer neural network model.

Acknowledgments
This work was supported in part by the National Natural Science Foundation of China under Grant 41461085, in part by the Foundation of Guangxi Key Laboratory of Spatial Information and Geomatics under Grant 16-380-25-04, and in part by the Doctoral Foundation of Guilin University of Technology under Grant 1996015.

References

[1] DZIKITI, S., GUSH, M.B., LE MAITRE, D.C., et al. (2016) Quantifying potential water savings from clearing invasive alien Eucalyptus camaldulensis using in situ and high resolution remote sensing data in the Berg River Catchment, Western Cape, South Africa. Forest Ecology and Management., 361: 69-80. DOI: 10.1016/j.foreco.2015.11.009.

[2] QIN, J.T., LIU, S.N., ZHANG, L.Q., et al. (2023) Estimating Eucalyptus spp. volume with single tree crown data extracted from UAV remote sensing image. Journal of Northeast Forestry University., 51: 96-102. DOI: 10.13759/j.cnki.dlxb.2023.06.005.

[3] WANG, Z.Y., REN, C., LIANG, Y.J., et al. (2021) Object-oriented Eucalyptus plantation forest information extraction based on the red-edge feature of GF-6. Bulletin of Surveying and Mapping., 531: 6-11. DOI: 10.13474/j.cnki.11-2246.2021.0167.

[4] Liang, W.H., LIU, J.K., ZHANG, W., et al. (2017) Extracting Eucalyptus information using GF-2 images based on an object-oriented method. Journal of Zhejiang A&F University., 34: 721-729. DOI: 10.13466/j.cnki.lyzygl.2017.06.010.

[5] ZHANG, X.Y., FENG, X.Z., JIANG, H. (2009) Feature space optimization for object-based classification. National Remote Sensing Bulletin., 13: 664-669. DOI:10.11834/jrs.20090409.

[6] CHEN, J., LI, H., LIU, Y.F., et al. (2023) Remote sensing recognition of agricultural crops based on Sentinel-2 data with multi-feature optimization. http://kns.cnki.net/kcms/detail/10.1759.P.20230531.0953.006.html.

[7] HAO, Y.F., MAN, W.D., WANG, J.H., et al. (2021) Wetland information extraction based on Relief-F algorithm and decision tree method. Journal of Liaoning Technical University (Natural Science)., 40: 225-233. https://fxky.cbpt.cnki.net/WKD/WebPublication/index.aspx?mid=fxky.

[8] XIAO, Y., JIANG, Q.G., WANG, B., et al. (2016) Object based land-use classification based on hybrid feature selection method of combining Relief F and PSO. Transactions of the Chinese Society of Agricultural Engineering., 32: 211-216. DOI: 10.11975/j.issn.1002-6819.2016.04.030.

[9] WANG, Z.H., LI, J., TAN, Z.L., et al. (2023) Swin-UperNet: A semantic segmentation model for Mangroves and Spartina alterniflora Loisel based on UperNet. Electronics., 12: 1111-1125. DOI:10.3390/ELECTRONICS12051111.

[10] CHU, X., TIAN, Z., WANG, Y., et al. (2021) Twins: Revisiting the design of spatial attention in Vision Transformers. IEEE Computer Vision and Pattern Recognition. Colorado Springs.20881-20890. https://ui.adsabs.harvard.edu/abs/2021arXiv210413840C/abstract.

Real-time analysis of star sensor attitude accuracy

Jingneng Fu[1,2,3,4,5], Wen Luo[1,2,3,5] and Hui Zhang[1,2,3]

[1]Institute of Optics and Electronics, Chinese Academy of Sciences, Chengdu 610209, China
[2]Key Laboratory of Science and Technology on Space Optoelectronic Precision Measurement, Chinese Academy of Sciences, Chengdu 610209, China
[3]University of Chinese Academy of Sciences, Beijing 100049, China
[4]Youth Innovation Promotion Association, Chinese Academy of Sciences, Beijing 100029, China

[5]Corresponding author's e-mail: fujingneng@ioe.ac.cn (Jingneng Fu); wxwdstr@163.com (Wen Luo)

Abstract. In the conventional methods of star sensor accuracy evaluation, neither the accuracy calculation formula nor the Monte Carlo method can comprehensively reflect the attitude measurement accuracy of star sensors in real-time. In this paper, a real-time analysis and calculation model for the attitude measurement accuracy of star sensors is proposed. Firstly, the basic attitude measurement model of the star sensor is established through the pinhole imaging model. Moreover, a moving frame of star sensor attitude measurement is established to avoid the defects of attitude representation by the Euler angle in the star sensor accuracy analysis. Then, combined with the error transfer of implicit overdetermined equations proposed in this paper, the error transfers of starspot extraction random error, star catalog random error, and intrinsic parameter system error are provided. Finally, simulation and field experiments are conducted to verify the accuracy of the proposed error analysis theory. The experimental results show that the attitude measurement accuracy of star sensors can be accurately estimated by using single-frame data of guide stars.

1. Introduction

The vast number of vehicles operating in outer space has expanded the range of human activity and cognition. In order to accomplish specific tasks like earth imaging, satellite communication, space observation, and interstellar travel, orbit and attitude accuracy should be ensured. As an attitude sensor, the star sensor is characterized by high accuracy, low power consumption, long life, strong anti-interference, and independent navigation without reliance on other systems. The star sensor is widely used in aeronautics and aerospace and has become the mainstream direction of attitude sensor research.

Accuracy is the most critical performance indicator of the star sensor. There are two types of methods to calculate the accuracy of star sensors; one is the theoretical calculation formula, and the other is the Monte Carlo algorithm. In the research of star sensors, there are many formulas to represent star sensor accuracy, including accuracy of attitude angle [1], optical axis pointing accuracy [2,3], noise equivalent angle [4], and accuracy of starspot extraction [5,6]. However, these are only representations of partly influencing factors of attitude accuracy. As such, they cannot estimate star sensor accuracy in real-time. For example, the accuracy of attitude angle [1], proposed by Abreu, is widely used in the design of star sensors. Attitude angle accuracy is an approximate estimate of the impact of starspot extraction error on the accuracy of attitude angle. Therefore, attitude angle accuracy cannot be estimated when the star sensor is aiming at a specific sky. Other factors, such as star catalog error and the intrinsic parameter error of the star sensor, cannot evaluate the impact on attitude measurement. The Monte Carlo algorithm [7] is a direct method to calculate star sensor accuracy. However, thousands of simulation experiments are required. This does not meet the requirements of real-time evaluation of the accuracy of the star sensor.

This paper aims to comprehensively evaluate the influence of various factors on the measurement accuracy of star sensors in real-time. Firstly, the attitude measurement model of the star sensor based on the moving frame of the Euler angle is proposed. As such, the attitude accuracy dependence on the coordinate system and the singularity of the Euler angle are avoided. Moreover, the error transfer of implicit overdetermined equations is derived and applied to the accuracy analysis of star sensor attitude. Finally, the validity of the analysis model of attitude accuracy is confirmed by both simulation and field data of the star sensors. Once the star sensor completes attitude capture, the corresponding attitude accuracy on each frame of the star image can be synchronously estimated. Therefore, the proposed accuracy analysis of the star sensor attitude can be classified as real-time.

This paper is organized as follows. In Section 2, the attitude measurement model of the star sensor based on the moving frame is established according to the pinhole imaging model. In Section 3, the error transfer of implicit overdetermined equations is proposed. Section 4 refers to the experiments that confirm the error transfer model's validity through star sensor simulation and field data. Lastly, Section 5 concludes the paper.

2. Attitude measurement model of star sensors

2.1 Basic model

Firstly, the celestial coordinate system and the star sensor coordinate system are established, which are respectively expressed as $O - X_I Y_I Z_I$ and $O - X_C Y_C Z_C$.

Then, the yaw angle (optical axis pointing to right ascension), pitch angle (optical axis pointing to declination), and roll angle, which are the attitude angles of the star sensor, are defined as (α, δ, ϕ) [8]. Among these, yaw angle α is the included angle between the axis OX_I and the projection of axis OZ_C on plane $X_I O Y_I$, where the counterclockwise direction of axis OX_I is forward direction; pitch angle δ is the included angle between the axis OZ_C and the projection of axis OZ_C on plane $X_I O Z_I$, where the counterclockwise direction of projection is the forward direction; roll angle ϕ is the included angle between the axis OY_C and the projection of axis OZ_I on plane $X_C O Y_C$, where the clockwise direction of the projection is the forward direction.

According to the above definition of attitude angles of star sensor, the attitude transformation matrix [9] $\boldsymbol{M}_{IS}(\alpha, \delta, \phi)$ from the star sensor coordinate system (*S* system) to the celestial coordinate system (*I* system) is obtained.

It is assumed that the coordinates of *N* starspots on the image plane are obtained as $\{(x_i, y_i) | 1 \leq i \leq N\}$ by processing the star image captured by the star sensor. Moreover, the coordinates of the corresponding guide stars in the celestial coordinate system are $\{(\alpha_i, \delta_i) | 1 \leq i \leq N\}$, where (α_i, δ_i) respectively stand for the right ascension and the declination of i th guide star. The attitude measurement of the star sensor is presented according to the pinhole imaging model as follows:

$$\boldsymbol{v}_i = \boldsymbol{M}_{IS} \boldsymbol{w}_i, 1 \leq i \leq N, \tag{1}$$

where the unit direction vector of the guide star in the celestial coordinate system is expressed as:

$$\boldsymbol{v}_i = [\cos\alpha_i \cos\delta_i \quad \sin\alpha_i \cos\delta_i \quad \sin\delta_i]^T, \tag{2}$$

The unit direction vector of the guide star in the star sensor coordinate system can be written as:

$$\boldsymbol{w}_i = \frac{1}{r_i} [-(x_i - x_o) \quad -(y_i - y_o) \quad f]^T, \tag{3}$$

where $r_i = \sqrt{(x_i - x_o)^2 + (y_i - y_o)^2 + f^2}$ is the normalization parameter, and (x_o, y_o) and f are the principal point and the focal length of the star sensor, respectively.

Finally, the following equation is obtained based on Equation (1):

$$\boldsymbol{L}(\boldsymbol{X}, \boldsymbol{Y}) = \begin{bmatrix} \boldsymbol{M}_{IS}\boldsymbol{w}_1 - \boldsymbol{v}_1 \\ \boldsymbol{M}_{IS}\boldsymbol{w}_2 - \boldsymbol{v}_2 \\ \cdots\cdots \\ \boldsymbol{M}_{IS}\boldsymbol{w}_N - \boldsymbol{v}_N \end{bmatrix} = \boldsymbol{0}, \tag{4}$$

where $\boldsymbol{X} = [x_1, y_1, \dots, x_N, y_N; \alpha_1, \delta_1, \dots, \alpha_N, \delta_N; x_o, y_o, f]^T$ is the influential factor of attitude measurement and $\boldsymbol{Y} = [\alpha, \delta, \phi]^T$ is the attitude angle of the star sensor.

2.2 Moving Frame Model

Two problems regarding the Euler angle must be addressed to describe the attitude accuracy of the star sensor. The first one is the singularity of the Euler angle representing attitude, i.e., when $\delta = \pm\frac{\pi}{2}$, α, and ϕ are not unique. The other more critical problem is that the same longitudinal arc lengths s_α and s_δ correspond to different arc angles θ. In other words, the derivative of arc length with respect to arc angle is inconsistent:

$$\frac{ds_\alpha}{d\theta} = \cos\delta \frac{ds_\delta}{d\theta} = 1. \tag{5}$$

Theoretically, when the distribution of guide stars is determined in the field of view, the attitude accuracy of the star sensor is independent of the star sensor attitude. However, the derivative of latitudinal arc length with respect to the arc angle shown in Equation (5) is dependent on latitude. In other words, the yaw angle error in Equation (4) increases with the latitude for the same distribution of guide stars. Therefore, the proper definition of star sensor attitude angle is necessary to characterize the star sensor's attitude accuracy correctly.

According to Equation (5), the aforementioned problem disappears when the optical axis points towards the equatorial plane of the celestial sphere. The space determined by $\boldsymbol{Y} = [\alpha, \delta, \phi]^T$ is a non-Euclidean space. However, its local space contains Euclidean geometric properties in the form of a three-dimensional manifold [10]. To ensure Euclidean geometric properties of the Euler angle, $\boldsymbol{Y} = [\alpha, \delta, \phi]^T$ needs to establish a set of local bases that change with the optical axis (or a moving frame). Therefore, in this paper, it is proposed that the reference attitude matrix $\boldsymbol{M}_{IS}^{Ref}$ and local attitude matrix $\boldsymbol{M}_{IS}^{Loc}$ are jointly employed to represent the attitude:

$$\boldsymbol{M}_{IS} = \boldsymbol{M}_{IS}^{Ref}\boldsymbol{M}_{IS}^{Loc}, \tag{6}$$

where reference attitude matrix $\boldsymbol{M}_{IS}^{Ref}$ is the theoretical attitude matrix of the star sensor. Since the true value of $\boldsymbol{M}_{IS}$ is typically unknown, the estimated value of attitude is usually used as a representation of $\boldsymbol{M}_{IS}^{Ref} = \widehat{\boldsymbol{M}}_{IS}$.

To avoid confusion, the attitude angle corresponding to the local attitude matrix $\boldsymbol{M}_{IS}^{Loc}$ is still represented by $\boldsymbol{Y} = [\alpha, \delta, \phi]^T$, i.e.:

$$\boldsymbol{M}_{IS}^{Loc}(\alpha, \delta, \phi) = [\boldsymbol{R}_Z(\phi)\boldsymbol{R}_Y(\delta)\boldsymbol{R}_X(\alpha)]^T, \tag{7}$$

where $\boldsymbol{M}_{IS}^{Loc}$ is represented by the 1-2-3 Euler rotation. $\boldsymbol{R}_X(\theta)$, $\boldsymbol{R}_Y(\theta)$ and $\boldsymbol{R}_Z(\theta)$ denote the basic transformation matrix of OX axis, OY axis, and OZ axis rotating by an angle of θ, respectively.

In addition, a similar problem exists in the coordinate $\{(\alpha_i, \delta_i) | 1 \le i \le N\}$ of the guide star in the celestial coordinate system. Similarly, the local rotation matrix is defined as $\{\boldsymbol{R}_{SI}^{Loc}(\alpha_i, \delta_i) | 1 \le i \le N\}$, where $\boldsymbol{R}_{SI}^{Loc}(\alpha_i, \delta_i) = \boldsymbol{R}_Y(\delta_i)\boldsymbol{R}_X(\alpha_i)$, and α_i and δ_i represent the yaw angle and pitch angle of the unit direction vector of the guide star in the local celestial coordinate system.

Finally, Equation (4) is modified into:

$$\boldsymbol{L}(\boldsymbol{X},\boldsymbol{Y})=\begin{bmatrix}\boldsymbol{M}_{IS}^{Ref}\boldsymbol{M}_{IS}^{Loc}(\alpha,\delta,\phi)\boldsymbol{w}_1-\boldsymbol{R}_{SI}^{Loc}(\alpha_1,\delta_1)\boldsymbol{v}_1\\ \boldsymbol{M}_{IS}^{Ref}\boldsymbol{M}_{IS}^{Loc}(\alpha,\delta,\phi)\boldsymbol{w}_2-\boldsymbol{R}_{SI}^{Loc}(\alpha_2,\delta_2)\boldsymbol{v}_2\\ \cdots\cdots\\ \boldsymbol{M}_{IS}^{Ref}\boldsymbol{M}_{IS}^{Loc}(\alpha,\delta,\phi)\boldsymbol{w}_N-\boldsymbol{R}_{SI}^{Loc}(\alpha_N,\delta_N)\boldsymbol{v}_N\end{bmatrix}=\boldsymbol{0}. \tag{8}$$

The attitude angle $\boldsymbol{Y}=[\alpha,\delta,\phi]^T$ of the star sensor is always in the proximity of the zero element under the representation of the moving frame. Therefore, the dependency of attitude accuracy on the coordinate system selection is avoided. Moreover, the reference attitude matrix $\boldsymbol{M}_{IS}^{Ref}=\widehat{\boldsymbol{M}}_{IS}$ does not need to solve the corresponding Euler angle. Thus, the Euler angle singularity problem is resolved.

The factors affecting the attitude measurement accuracy of the star sensor include the extraction error of starspots, the star catalog error, and the intrinsic parameter error. The extraction error of starspots and the star catalog error are random errors, while the intrinsic parameter error is a system error.

Due to conjugating object space light and image space light, the angular position error of the guide star in the star catalog is equivalent to the extraction error of starspots on the image plane. The angular position error of a guide star in the star catalog is less than 0.01" [11]. On the other hand, the angular error of starspot extraction is more than one order of magnitude higher. Therefore, compared with the extraction error of the starspot, the angular position error of a guide star in the star catalog can be ignored and will not be further discussed in this paper. Equation (8) is thus simplified as:

$$\boldsymbol{L}(\boldsymbol{X},\boldsymbol{Y})=\begin{bmatrix}\boldsymbol{M}_{IS}^{Loc}(\alpha,\delta,\phi)\boldsymbol{w}_1-\widehat{\boldsymbol{M}}_{IS}^T\boldsymbol{v}_1\\ \boldsymbol{M}_{IS}^{Loc}(\alpha,\delta,\phi)\boldsymbol{w}_2-\widehat{\boldsymbol{M}}_{IS}^T\boldsymbol{v}_2\\ \cdots\cdots\\ \boldsymbol{M}_{IS}^{Loc}(\alpha,\delta,\phi)\boldsymbol{w}_N-\widehat{\boldsymbol{M}}_{IS}^T\boldsymbol{v}_N\end{bmatrix}=\boldsymbol{0}. \tag{9}$$

3. Error Transfer of Implicit Overdetermined Equations

We make the implicit overdetermined equations be:

$$\boldsymbol{L}(\boldsymbol{X},\boldsymbol{Y})=\boldsymbol{0}, \tag{10}$$

where $\boldsymbol{X}$ and $\boldsymbol{Y}$ are independent and dependent variables, respectively.

We take the differential expansion of Equation (10) with respect to $(\boldsymbol{X},\boldsymbol{Y})$:

$$\boldsymbol{C_X}\delta\boldsymbol{X}+\boldsymbol{C_Y}\delta\boldsymbol{Y}=\boldsymbol{0}, \tag{11}$$

where $\delta\boldsymbol{X}$ and $\delta\boldsymbol{Y}$ are small quantities with respect to $\boldsymbol{X}$ and $\boldsymbol{Y}$, respectively.

A generalized inverse solution is used as follows:

$$\delta\boldsymbol{Y}=-\left(\boldsymbol{C}_Y^T\boldsymbol{C}_Y\right)^{-1}\left(\boldsymbol{C}_Y^T\boldsymbol{C}_X\right)\delta\boldsymbol{X}, \tag{12}$$

where $\boldsymbol{C}_X=\frac{\partial\boldsymbol{L}}{\partial\boldsymbol{X}}$ and $\boldsymbol{C}_Y=\frac{\partial\boldsymbol{L}}{\partial\boldsymbol{Y}}$ are sensitivity matrices of $\boldsymbol{L}$ with respect to $\boldsymbol{X}$ and $\boldsymbol{Y}$, respectively.

The mean vector represents the system error in Equation (12) due to the definite error transfer relation:

$$\boldsymbol{E}[\delta\boldsymbol{Y}]=-\boldsymbol{E}\left[\left(\boldsymbol{C}_Y^T\boldsymbol{C}_Y\right)^{-1}\left(\boldsymbol{C}_Y^T\boldsymbol{C}_X\right)\delta\boldsymbol{X}\right]. \tag{13}$$

Since it is impossible to obtain true values of $(\boldsymbol{X},\boldsymbol{Y})$, estimated value $(\widehat{\boldsymbol{X}},\widehat{\boldsymbol{Y}})$ are used for approximate representation:

$$\Delta\boldsymbol{Y}=-\left(\boldsymbol{C}_{\widehat{Y}}^T\boldsymbol{C}_{\widehat{Y}}\right)^{-1}\left(\boldsymbol{C}_{\widehat{Y}}^T\boldsymbol{C}_{\widehat{X}}\right)\Delta\boldsymbol{X}, \tag{14}$$

where $\boldsymbol{C}_{\widehat{X}}=\left.\frac{\partial\boldsymbol{L}}{\partial\boldsymbol{X}}\right|_{\boldsymbol{X}=\widehat{\boldsymbol{X}}}$ and $\boldsymbol{C}_{\widehat{Y}}=\left.\frac{\partial\boldsymbol{L}}{\partial\boldsymbol{Y}}\right|_{\boldsymbol{Y}=\widehat{\boldsymbol{Y}}}$ are sensitivity matrices of function $\boldsymbol{L}$ with respect to $\boldsymbol{X}$ and $\boldsymbol{Y}$ at the estimated value $(\widehat{\boldsymbol{X}},\widehat{\boldsymbol{Y}})$, and $\Delta\boldsymbol{X}$ and $\Delta\boldsymbol{Y}=\left[\Delta_\alpha,\Delta_\delta,\Delta_\phi\right]^T$ are system errors of influential factors and attitude measurement, respectively.

Due to its random distribution, the random error is represented by the covariance matrix of Equation (12):

$$\boldsymbol{E}[(\delta \boldsymbol{Y})^T \delta \boldsymbol{Y}] = \boldsymbol{E}\left[\left(\boldsymbol{C}_Y^T \boldsymbol{C}_Y\right)^{-1}\left(\boldsymbol{C}_Y^T \boldsymbol{C}_X\right)\left((\delta \boldsymbol{X})^T \delta \boldsymbol{X}\right)\left(\boldsymbol{C}_X^T \boldsymbol{C}_Y\right)\left(\boldsymbol{C}_Y^T \boldsymbol{C}_Y\right)^{-1}\right]. \tag{15}$$

Similarly, since the true value of $(\boldsymbol{X}, \boldsymbol{Y})$ cannot be obtained, corresponding estimated values $(\widehat{\boldsymbol{X}}, \widehat{\boldsymbol{Y}})$ are used for the following approximate representation:

$$\boldsymbol{U}_Y = \left(\boldsymbol{C}_{\hat{Y}}^T \boldsymbol{C}_{\hat{Y}}\right)^{-1}\left(\boldsymbol{C}_{\hat{Y}}^T \boldsymbol{C}_{\hat{X}}\right)\boldsymbol{U}_X\left(\boldsymbol{C}_{\hat{X}}^T \boldsymbol{C}_{\hat{Y}}\right)\left(\boldsymbol{C}_{\hat{Y}}^T \boldsymbol{C}_{\hat{Y}}\right)^{-1}, \tag{16}$$

where $\boldsymbol{U}_X = \boldsymbol{E}[(\delta \boldsymbol{X})^T \delta \boldsymbol{X}]$ denotes the covariance matrix of independent variable error, and $\boldsymbol{U}_Y = \boldsymbol{E}[(\delta \boldsymbol{Y})^T \delta \boldsymbol{Y}]$ denotes the covariance matrix of dependent variable error. The diagonal elements of both reflect the random noise level.

Since Equation (10) represents implicit overdetermined equations, Equation (14) and Equation (16) are different from the error transfer of implicit well-determined equations [12] provided by the Joint Committee for Guides in Metrology (JCGM). They can be considered as more general error transfer formulas.

4. Experimental Analysis

4.1 Simulation Experiment

A major trend of star sensor development [13] is low cost and miniaturization, which means smaller optical apertures and fewer guide stars. In the most extreme cases, only two guide stars are detected in the field of view. Then, an important issue is the distribution of the two guide stars in the field of view which can lead to high attitude accuracy. In this section, the answer is directly addressed through single-frame data analysis, and the theoretical results are compared by thousands of Monte Carlo calculations.

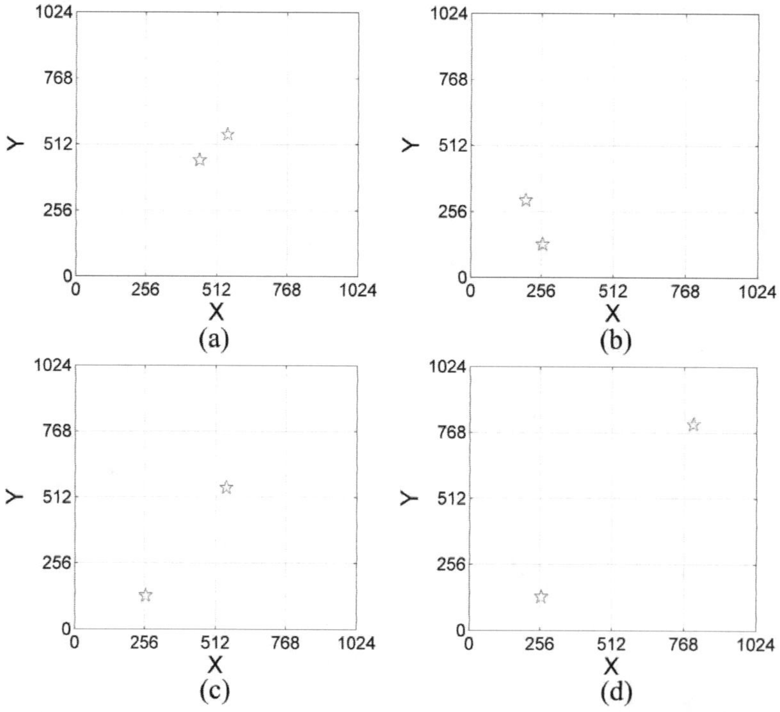

Figure 1. Different distributions of two guide stars in the field of view.

Table 1. Relevant experimental parameters of a star sensor.

Field of view	Focal length (mm)	Pixel size (μm)	Image resolution	Principal point (x_o, y_o)	Extraction accuracy of starspots σ_{XY}
20°×20°	31.9407	11	1024×1024	(512, 512)	0.1 pixel

The parameters of the star sensor are provided in Table 1. Then, the distribution of two stars in the field of view is divided into four types, as shown in Figure 1 and Table 2.

(1)Two stars near the center of the field of view;

(2)Two stars near the edge of the field of view in the same region;

(3)A single star near the center of the field of view, the other one near the edge;

(4)Two stars near the edge of the field of view in different regions.

Table 2. Experimental parameters associated with attitude.

Experiment case	Theoretical coordinates of star image points	Coordinates of guide stars	Theoretical attitude angle (α, δ, ϕ)
1	(450, 450) (550, 550)	(50.0316°, 80.3041°) (65.7234°, 79.6752°)	(60°, 80°,120°)
2	(200, 300) (260, 130)	(15.2452°, 80.5187°) (16.2617°, 77.0100°)	(60°, 80°,120°)
3	(550, 550) (260, 130)	(65.7234°, 79.6752°) (16.2617°, 77.0100°)	(60°, 80°,120°)
4	(800, 800) (260, 130)	(92.9039°, 75.7004°) (16.2617°, 77.0100°)	(60°, 80°,120°)

Each experiment case is simulated $K = 1000$ times by Monte Carlo to estimate the actual attitude angle error. The coordinates of starspots with extraction error, parameters of the star sensor shown in Table 1, coordinates of guide stars shown in Table 2, and the Singular Value Decomposition (SVD) algorithm [14] are all employed to solve the attitude matrix of the star sensor $\boldsymbol{M}_{IS}$. Theoretical attitude parameters in Table 2 are placed into the reference matrix of the star sensor $\boldsymbol{M}_{IC}^{Ref}$, and Equation (6) is used to calculate the local attitude matrix of the star sensor $\boldsymbol{M}_{IS}^{Loc} = \left(\boldsymbol{M}_{IS}^{Ref}\right)^T \boldsymbol{M}_{IS}$. The attitude angle $\left(\delta_{\alpha,k}, \delta_{\delta,k}, \delta_{\phi,k}\right)$ corresponding to $\boldsymbol{M}_{IS}^{Loc}(1 \le k \le K)$ is solved by employing Equation (17):

$$\begin{cases} \delta_\alpha = \operatorname{asin}\left(\frac{-\boldsymbol{M}_{IS}^{Loc}(2,3)}{\sqrt{1-\left(\boldsymbol{M}_{IS}^{Loc}(1,3)\right)^2}}\right) \\ \delta_\delta = \operatorname{asin}\left(\boldsymbol{M}_{IS}^{Loc}(1,3)\right) \\ \delta_\phi = \operatorname{asin}\left(\frac{-\boldsymbol{M}_{IS}^{Loc}(1,2)}{\sqrt{1-\left(\boldsymbol{M}_{IS}^{Loc}(1,3)\right)^2}}\right) \end{cases} . \tag{17}$$

The attitude errors estimated by Monte Carlo are obtained according to Equation (18):

$$\begin{cases} \sigma_{\alpha}^{(MC)} = \sqrt{\frac{1}{K}\sum_k \delta_{\alpha,k}^2} \\ \sigma_{\delta}^{(MC)} = \sqrt{\frac{1}{K}\sum_k \delta_{\delta,k}^2}. \\ \sigma_{\phi}^{(MC)} = \sqrt{\frac{1}{K}\sum_k \delta_{\phi,k}^2} \end{cases} \tag{18}$$

It is incorrect to perform the Monte Carlo algorithm by directly obtaining the attitude angle via Equation (4) and calculating the attitude measurement error according to the difference between the true values of the attitude angle. As analyzed in Section 2, the Euler angle does not belong to the Euclidean space category, and Euclidean distance is not defined in the global space. Consequently, the incorrect Monte Carlo algorithm is meaningless for representing the star sensor accuracy.

The coordinates of single-frame starspots with extraction error, the parameters of the star sensor in Table 1, and the coordinates of guide stars in Table 2 are substituted into Equation (16). Thus, the covariance matrix of random errors in attitude measurement $\boldsymbol{U_Y}$ is obtained. The measurement errors of attitude angles estimated by using single-frame data are denoted by:

$$\left(\sigma_{\alpha}^{(Est)}, \sigma_{\delta}^{(Est)}, \sigma_{\phi}^{(Est)}\right) = \left(\sqrt{\boldsymbol{U_Y}(1,1)}, \sqrt{\boldsymbol{U_Y}(2,2)}, \sqrt{\boldsymbol{U_Y}(3,3)}\right). \tag{19}$$

Table 3. Comparison of attitude measurement error among various methods in the simulation experiment.

Experiment case	Accuracy of attitude measurement $(\sigma_\alpha, \sigma_\delta, \sigma_\phi)$(“)		
	Monte Carlo method (Equation (18))	Proposed method (Equation (19))	Abreu’s method (Ref. 1)
1	(5.0, 5.0, 206.3)	(5.1, 5.1, 206.3)	(5.0, 5.0, 112.4)
2	(16.3, 16.9, 158.1)	(16.2, 17.0, 159.3)	(5.0, 5.0, 112.4)
3	(5.4, 6.0, 58.6)	(5.4, 6.0, 57.2)	(5.0, 5.0, 112.4)
4	(5.3,5.0,33.4)	(5.0,5.0,33.9)	(5.0,5.0,112.4)

For comparison purposes, Abreu’s attitude measurement accuracy evaluation of star sensor [1] is employed. This method can be expressed as $\left(\sigma_{\alpha}^{(Abreu)}, \sigma_{\delta}^{(Abreu)}, \sigma_{\phi}^{(Abreu)}\right)$.

As shown in Table 3, it can be observed that the results obtained by the proposed method are relatively close to the Monte Carlo results, which proves the effectiveness of the algorithm. The assumption of uniform distribution of guide stars is not valid for a few stars. Moreover, the attitude measurement accuracy of the star sensor depends on the distribution of guide stars in the sky. Abreu’s method would result in a large deviation in the estimation of attitude measurement accuracy of the star sensor, especially in the roll angle accuracy estimation.

When only two guide stars are present in the field of view and are distributed on both sides of the principal point, the yaw angle and the pitch angle can be obtained with relatively high accuracy. Furthermore, the accuracy of the roll angle increases with the distance between the guide stars.

After analyzing the extraction error influence of starspot on attitude measurement accuracy, the influence of system errors of principal point and focal length on attitude measurement accuracy is further investigated. It is assumed that the calibrated values of the principal point and the focal length of the star sensor are (513, 511) and 31.9000 mm, respectively. The corresponding system errors of the intrinsic parameter are (1, -1) pixels and 0.0407 mm, which are the attitude measurement errors induced by the estimation of system errors of the intrinsic parameter by substituting 70.3”, -70.3”, and -260” into Equation (14). The parameters of the case 4 experiment shown in Table 2 are taken as input, and the calculation result of the proposed method by using single-frame data is obtained as $\left(\Delta_{\alpha}^{(Est)},\Delta_{\delta}^{(Est)},\Delta_{\phi}^{(Est)}\right)$ = (-66.9", -69.6", -0.1"). Based on the star sensor parameters shown in Table 1, the coordinates of the two guide stars, and the theoretical coordinates of starspots in Table 2, the attitude matrix of the star sensor $\boldsymbol{M}_{IS}$ is solved by employing the SVD algorithm. Theoretical attitude parameters in Table 2 are substituted into the reference matrix of the star sensor $\boldsymbol{M}_{IS}^{Ref}$. Equation (6) is used to calculate the local attitude matrix of the star sensor $\boldsymbol{M}_{IS}^{Loc} = \left(\boldsymbol{M}_{IS}^{Ref}\right)^T \boldsymbol{M}_{IS}$, and the attitude angles corresponding to $\boldsymbol{M}_{IS}^{Loc}$ are obtained as $\left(\Delta_{\alpha}^{(MC)},\Delta_{\delta}^{(MC)},\Delta_{\phi}^{(MC)}\right)$ =(-66.7", -69.5", -0.1"). Without changing other parameters, the system error of focal length is set to 0 mm, and recalculation results by using single-frame data are obtained as $\left(\Delta_{\alpha}^{(Est)},\Delta_{\delta}^{(Est)},\Delta_{\phi}^{(Est)}\right)$ =(-70.8",-71.2",-0.2"); $\left(\Delta_{\alpha}^{(MC)},\Delta_{\delta}^{(MC)},\Delta_{\phi}^{(MC)}\right)$ =(-70.8", -71.2", -0.2"). Compared with the calculation results of the Monte Carlo method, it is found that the analysis theory of the proposed method well reflects the influence of the system error of intrinsic parameters on attitude measurement.

In general, the star sensor image can be approximated in paraxial terms, i.e., $|x_i - x_o|, |y_i - y_o| \ll f$. Then, the following expressions can be written:

$$\begin{cases} \dfrac{\partial \boldsymbol{w}_i}{\partial x_o} = \dfrac{1}{f}[1 \quad O(\varepsilon^2) \quad O(\varepsilon)]^T \\ \dfrac{\partial \boldsymbol{w}_i}{\partial y_o} = \dfrac{1}{f}[O(\varepsilon^2) \quad 1 \quad O(\varepsilon)]^T \\ \dfrac{\partial \boldsymbol{w}_i}{\partial f} = \dfrac{1}{f}[O(\varepsilon) \quad O(\varepsilon) \quad O(\varepsilon^2)]^T \end{cases}, \tag{20}$$

where ε is a small quantity associated with the field of view angle and uppercase $O(*)$ denotes equivalent order. Then:

$$\begin{aligned} \Delta \boldsymbol{w}_i &= \frac{\partial \boldsymbol{w}_i}{\partial x_o}\Delta_{x_o} + \frac{\partial \boldsymbol{w}_i}{\partial y_o}\Delta_{y_o} + \frac{\partial \boldsymbol{w}_i}{\partial f}\Delta_f \\ &\approx \frac{1}{f}[\Delta_{x_o} \quad \Delta_{y_o} \quad \Delta_{x_o}O(\varepsilon) + \Delta_{y_o}O(\varepsilon)], \end{aligned} \tag{21}$$

where $\left(\Delta_{x_o},\Delta_{y_o}\right)$ and Δ_f are system errors of principal point and focal length, respectively.

Equations (20) and (21) show that the effect of the focal length system error on the unit direction vector $\boldsymbol{w}_i$ in the star sensor coordinates can be ignored in comparison with the effect of the principal point system error on the unit direction vector in the star sensor coordinates. In other words, once the star map matching is completed, the system error of focal length has less influence on the accuracy of attitude measurement than that of the principal point. This observation is consistent with the simulation results.

4.2 Field Experiment

In this paper, the error analysis model of star sensor attitude measurement is verified via the field experiment data. Before the experiment, the intrinsic parameters, including the principal point, the focal length, and the distortion of the star sensor, are calibrated, as shown in Table 4. The star sensor is fixed on a three-dimensional rotating platform. In contrast, the zero position of the *Z*-axis of the rotating platform coincides with the optical axis of the star sensor. The star sensor extracts starspot coordinates on the image plane, guides star tracking, matches star maps, and communicates with the host computer. After matching the star map, the extracted coordinates of starspots and the corresponding guide stars are output to the data acquisition computer at the frequency of 8 Hz via the communication interface of the star sensor. Each frame can output up to five guide stars. When the number of the extracted guide star exceeds five, the truncation output contains the first five tracked guide stars.

Table 4. Relevant parameters of the star sensor used in field experiments.

Detector size (pixel)	Extraction accuracy of starspots (1σ)	The principal point in the *X* direction	The principal point in the *Y* direction	Focal length (mm)	Field of view (°)	Sensitivity (Mv)
1536×1536	0.1 pixel	752	773	23.9087	Φ = 20	6.0

In the field experiments, the star sensor points at the zenith and the rotating platform are static. The earth is equivalent to a rotating platform moving slowly and uniformly. A total of $K = 5\,000$ frames of data are collected. The trajectory formed by the superposition of extracted coordinates of star image points is shown in Figure 2(a).

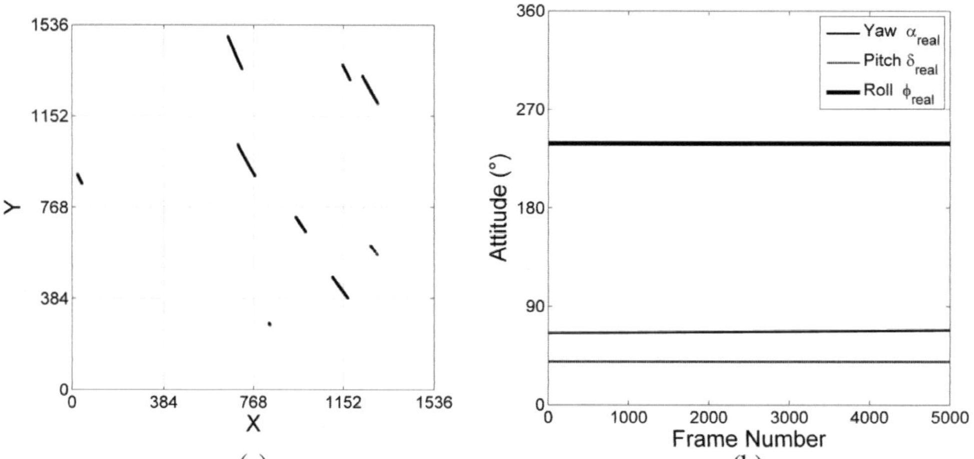

(a) (b)

Figure 2. Field experiments: (a) trajectory formed by the superposition of starspots on the image plane, and (b) calculated attitude of the star sensor.

Since the 3-axis rotating platform is in a static state relative to the ground, the earth is equivalent to a rotating platform subjected to a uniform motion. By combining guide stars and star catalog, the attitude angles $(\alpha_{real}, \delta_{real}, \phi_{real})$, as shown in Figure 2(b) and the attitude matrix $\boldsymbol{M}_{IS}$ of the star sensor are estimated. The attitude angle of a star sensor can be obtained by fitting estimation as the theoretical value $(\alpha_{ref}, \delta_{ref}, \phi_{ref})$. Then, the reference matrix of a star sensor $\boldsymbol{M}_{IS}^{Ref}$ is calculated. The local attitude matrix of the star sensor $\boldsymbol{M}_{IS}^{Loc} = \left(\boldsymbol{M}_{IS}^{Ref}\right)^T \boldsymbol{M}_{IS}$ is calculated according to Equation (6). Furthermore, as shown in Figure 3, the attitude angles $(\delta_{\alpha,k}, \delta_{\delta,k}, \delta_{\phi,k})(1 \le k \le K)$ corresponding to $\boldsymbol{M}_{IS}^{Loc}$ are calculated according to Equation (17). Furthermore, statistical analysis is performed on the measurement errors of local tracking window attitude by Equation (18), and $\left(\sigma_\alpha^{(MC)}, \sigma_\delta^{(MC)}, \sigma_\phi^{(MC)}\right)$ = (1.7”,1.5”,15.9”) is obtained.

After calibrating intrinsic parameters in a star sensor, the measured starspot extraction accuracy is approximately equal to $\sigma = 0.1$ pixel obtained by statistical analysis of an angular distance error. For independence assumption of angle measurement errors in $\boldsymbol{X}$ and $\boldsymbol{Y}$ directions in the image plane coordinates, $\sigma_{XY} = \frac{1}{\sqrt{2}}\sigma = 3.3''$. Then, the covariance matrix of random error of star image points extraction is equal to $\boldsymbol{U}_{\boldsymbol{X}} = f^2\sigma_{XY}^2\boldsymbol{I}_{2N\times 2N}$.

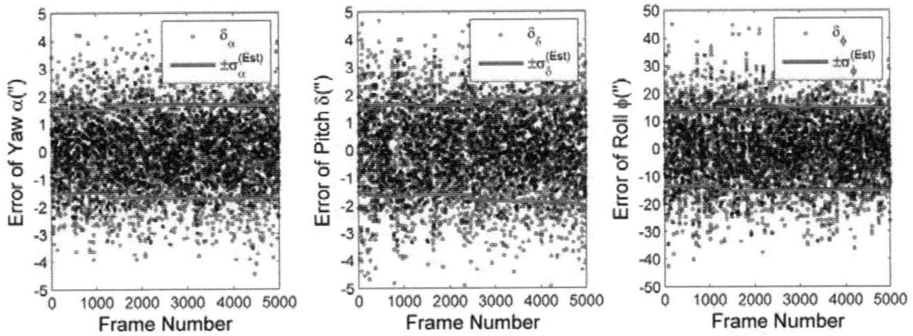

(a) (b) (c)

Figure 3. Positive and negative mean square deviation of attitude measurement errors $(\delta_\alpha, \delta_\delta, \delta_\phi)$ obtained by Monte Carlo method and by the proposed error analysis: (a) yaw angle, (b) pitch angle, and (c) roll angle.

The star sensor attitude angle calculates the attitude measurement accuracy of single star extraction error $(\alpha_{real}, \delta_{real}, \phi_{real})$ and the relevant parameters of the star sensor are listed in Table 4. By substituting them into Equation (16) and Equation (19), the measurement errors of the attitude angles of single-frame data$\left(\sigma_\alpha^{(Est)}, \sigma_\delta^{(Est)}, \sigma_\phi^{(Est)}\right)$ are obtained, as shown in Figure 3, with the corresponding averaged values of (1.6", 1.7", 14.7"). The results are relatively close to the measured values, with the difference in measurement accuracy of the roll angle being 1.2" and the relative error being roughly 8%. When using the Abreu method [1], the evaluated measurement accuracy of star sensor attitude is obtained as $\left(\sigma_\alpha^{(Abreu)}, \sigma_\delta^{(Abreu)}, \sigma_\phi^{(Abreu)}\right)$ = (1.5", 1.5", 13.6"). The difference in measurement accuracy of the roll angle is 2.3", and the relative error is approximately equal to 15%. Moreover, with an increase in the number of guide stars, their uniform distribution is gradually achieved, and the measurement accuracy of the pitch and yaw angles obtained by Abreu's method is gradually improved. However, for a relatively large number of guide stars, the roll angle calculation accuracy of Abreu's method is still lower than the proposed method. Therefore, it is necessary to correct Abreu's method.

To verify the system error of intrinsic parameter to attitude measurement accuracy, the calibrated values of the principal point and the focal length of the star sensor are assumed as (753, 772) and 23.9050 mm, respectively. Then, calculate the attitude matrix $\boldsymbol{M}_{IS}$ for data matching to a star map. Moreover, according to the calibrated theoretical values of the principal point and the focal length of a field star sensor, which are (752,773) and 23.9087 mm, the attitude, i.e., the reference matrix $\boldsymbol{M}_{IS}^{Ref}$ of a star sensor, is calculated for data matching a star map. Equation (6) is used to calculate the local attitude matrix of the star sensor $\boldsymbol{M}_{IS}^{Loc} = \left(\boldsymbol{M}_{IS}^{Ref}\right)^T \boldsymbol{M}_{IS}$. According to Equation (17), the attitude measurement errors $\left(\Delta_{\alpha,k}^{(MC)}, \Delta_{\delta,k}^{(MC)}, \Delta_{\phi,k}^{(MC)}\right)(1 \le k \le K)$ corresponding to $\boldsymbol{M}_{IS}^{Loc}$ is solved (black solid lines shown in Figure 4). Furthermore, the average values are obtained as (-48.8", -45.8", -2.8"). The calibrated values of the principal point and the focal length of the indoor star sensor are (753,772) and 23.9000 mm, respectively. The system errors of intrinsic parameters are (1.0, -1.0) pixels and -0.0037 mm. The attitude measurement errors $\left(\Delta_{\alpha,k}^{(Est)}, \Delta_{\delta,k}^{(Est)}, \Delta_{\phi,k}^{(Est)}\right)(1 \le k \le K)$ (Red solid lines are shown in Figure 4) induced by the system errors of intrinsic parameters are estimated by Equation (14) for 47", -47", and -32". Due to the small variation of guide star distribution in the field of view, the average errors are obtained as (-45.7", -48.2", -2.7"). Compared with the Monte Carlo, a minor difference exists between the two values. On the one hand, the aforementioned proves the effectiveness of the system error analysis method for intrinsic parameters. On the other hand, it is shown that the Euler angle has manifold characteristics, and the local space of the Euler angle is equivalent to a Euclidean space.

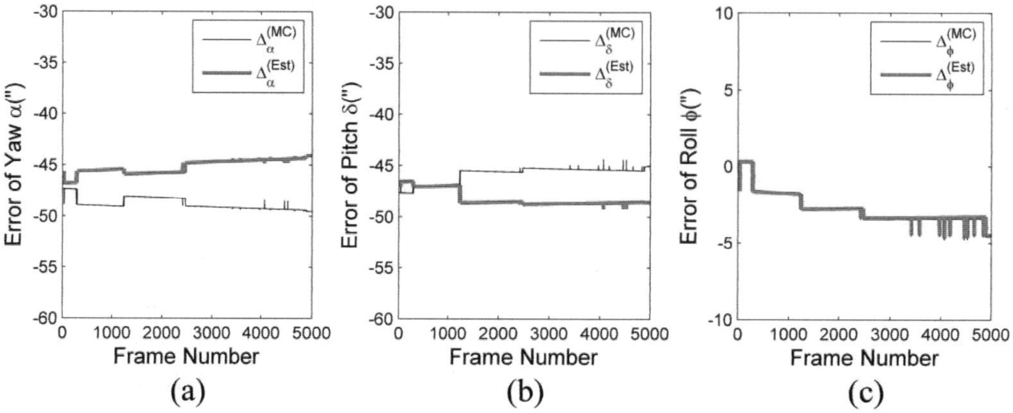

Figure 4. In the field experiments, when the systematic errors of intrinsic parameters are set as (1.0,-1.0) pixels and -0.0037mm, the system errors of attitude angle obtained by the Monte Carlo (solid black lines) and the proposed method (solid red lines). Here, (a), (b), and (c) represent error analysis results of yaw, pitch, and roll, respectively.

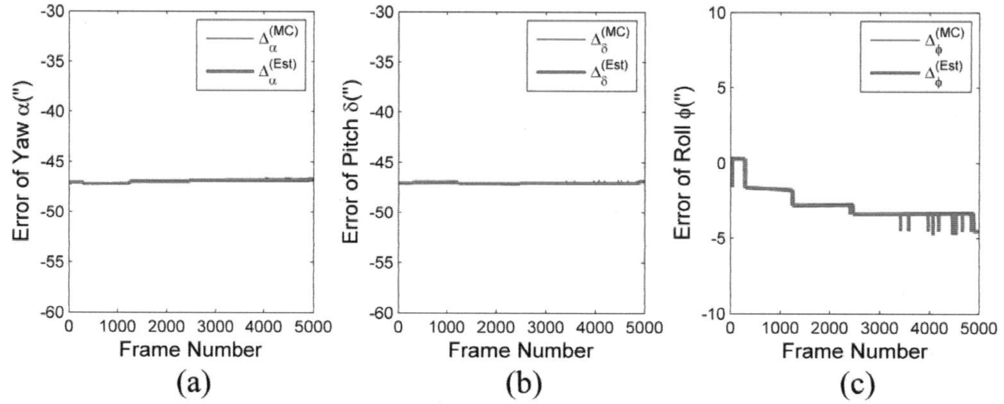

Figure 5. In the field experiments, when the systematic errors of intrinsic parameters are set as (1.0,-1.0) pixels and 0mm, the system errors of attitude angle obtained by the Monte Carlo (solid black lines) and the proposed method (solid red lines). Here, (a), (b), and (c) represent error analysis results of yaw, pitch, and roll, respectively.

Furthermore, according to the proposed method, the intrinsic parameters in the star sensor without error in focal length equal to (752, 773) and 23.9087 mm are substituted to calculate $\left(\Delta_{\alpha,k}^{(MC)}, \Delta_{\delta,k}^{(MC)}, \Delta_{\phi,k}^{(MC)}\right)$ and $\left(\Delta_{\alpha,k}^{(Est)}, \Delta_{\delta,k}^{(Est)}, \Delta_{\phi,k}^{(Est)}\right)$, as shown in Figure 5. Respective averaging results are (-47.0", -47.0", -2.7") and (-46.9", -47.0", -2.7"), i.e., an insignificant difference is observed between the two. Combined with the conducted experimental investigations, it can be concluded that the measurement system error of attitude angle caused by the principal point error is higher than the one caused by the focal length error.

Regarding the system errors of intrinsic parameters, the contribution of principal points to attitude measurement system error is much larger than that of a single-star extraction error to attitude measurement random error. The deviation caused by the system error of the principal point is constant, which can be eliminated by the calibration of the installation attitude matrix. However, it is difficult to eliminate the random error of attitude measurement caused by the starspot extraction error.

Finally, the verification results of field experiments are summarized in Table 5.

Table 5. Comparison of attitude measurement errors among different methods in field experiments.

			Yaw	Pitch	Roll
The contribution of starspot extraction error to attitude measurement error (1σ)	Monte Carlo method		1.7”	1.5”	15.9”
	Proposed method		1.6”	1.7”	14.7”
	Abreu’s method		1.5”	1.5”	13.6”
The contribution of intrinsic parameter error to attitude measurement error	With error in focal length	Monte Carlo method	-48.8”	-45.8”	-2.8”
		Proposed method	-45.7”	-48.2”	-2.7”
	Without error in focal length	Monte Carlo method	-47.0”	-47.0”	-2.7”
		Proposed method	-46.9”	-47.0”	-2.7”

5. Conclusions

In this paper, the Euler angle was selected to intuitively represent the attitude of a star sensor. By employing the moving frame, the defects of attitude representation by the Euler angle in the accuracy analysis of star sensors were avoided, including the dependence of attitude accuracy on the selection of the coordinate system and the singularity of the Euler angle. The error transfer of implicit overdetermined equations was derived, and the attitude accuracy analysis of a star sensor was established, including the random and system error transfer models. The error transfer model of a star sensor was analyzed from three aspects: random error of starspot extraction, random error of star catalog, and system error of intrinsic parameter.

Conducted experimental analysis shows that the contribution of the random error of the star catalog to the attitude measurement of a star sensor can be ignored. Regarding the influences of system errors of intrinsic parameters on the attitude measurement accuracy of a star sensor, the system error of focal length contributes less to attitude measurement than the principal point.

The most significant difference between this paper and the previous models for attitude error or accuracy analysis of a star sensor is that the accuracy analysis proposed in this paper is a real-time error transfer model. After the star sensor completes the attitude capture, the corresponding attitude accuracy on each frame of the star image can be synchronously estimated.

Acknowledgments

The authors would like to thank their colleagues working with them in the department of the Institute of Optics and Electronics at the Chinese Academy of Sciences. The authors also would like to thank the anonymous reviewers for their very constructive comments and suggestions.

References

[1] Abreu, R., “Stellar attitude determination accuracy with multiple-star-tracking advanced star tracker,” *Proceedings of SPIE* - The International Society for Optical Engineering. 1949, 216-227 (1993).

[2] Eisenman, A. R., and CC Liebe. “The advancing state-of-the-art in second generation star trackers,” in *Aerospace Conference*, pp. 111-118, IEEE, Snowmass, USA (1998).

[3] Zhang, X., J. Xiang, and Y. Zhang, “ Error analysis of line-of-sight measurement for video satellite,” *Advances in Mechanical Engineering*. 9(11), 1-9(2017).

[4] Liebe C C, “Accuracy performance of star trackers - a tutorial,” *IEEE Transactions on Aerospace Electronic Systems*. 38(2), 587-599 (2002).

[5] Hancock, B. R., et al. "CMOS active pixel sensor specific performance effects on star tracker/imager position accuracy." *Functional Integration of Opto-electro-mechanical Devices & Systems* International Society for Optics and Photonics, 2001.
[6] Hui, et al. "Minimum variance unbiased subpixel centroid estimation of point image limited by photon shot noise," *Journal of the Optical Society of America. A, Optics, image science, and vision.* 27(9), 2038-2045 (2010).
[7] Sun, Ting, F. Xing, and Z. You. "Optical system error analysis and calibration method of high-accuracy star trackers," *Sensors*, 13(4), 4598-4623 (2013).
[8] Zhang, H., et al. "Error analysis of spaced-based celestial positioning for space object," *Journal of Astronautics*, 31(5), 1345-1351 (2010).
[9] Markley F L, Crassidis J L, *Fundamentals of Spacecraft Attitude Determination and Control*, 1st ed., Springer, New York (2014).
[10] Hempel, J., *Intersection Calculus on Surfaces with Applications to 3-Manifolds*, 1st ed., American Mathematical Society, Providence (1983).
[11] SAO Star Catalog. https://heasarc.gsfc.nasa.gov/W3Browse/star-catalog/sao.html, 2004-11-08/2022-03-01.
[12] Iso, E. *Guide to the Expression of Uncertainty in Measurement*, 1st ed., Joint Committee for Guides in Metrology, (2008).
[13] Ho, K. "A survey of algorithms for star identification with low-cost star trackers." *Acta Astronautica*, 73,156-163.(2012).
[14] Pal, M., and M. S. Bhat, " Least square estimation of spacecraft attitude along with star camera parameters," *IFAC Proceedings Volumes*. 47(1), 20-25 (2014).

A robust estimation method of GNSS/IMU fusion kalman filter

Yanyan Pu [1, 2] and Shihuan Liu [1]

[1]CCCC Lvjian (Xiamen) Technology Co., Ltd., Xiamen 361000, China

[2] Corresponding author's e-mail: 869695594@qq.com

Abstract. Aiming at the traditional gross error detection method of GNSS/IMU fusion Kalman filter, which simply retains or eliminates abnormal observation data and does not make full use of useful observation information, a new robust estimation method of Kalman filter is proposed. Based on the chi-square statistical hypothesis test, this method constructs continuous change measurement using weight coefficient, fully excavates measurable innovation between normal value and abnormal value, and establishes a new measurement update equation of Kalman filter robust estimation. Finally, through GNSS/IMU integrated navigation simulation, the advantages of the new Kalman filter are verified: no parameter adjustment is required, and the statistical error fluctuation is smaller than that of the traditional adaptive filter.

1. Preface

Kalman Filter (KF) is a kind of filter that uses a random linear system. The method of optimal estimation is widely used in many fields, such as navigation guidance and control, communication, and image processing.

Theoretically, KF can obtain the optimal state estimation only when the structural parameters and noise statistical characteristics parameters of the stochastic linear system are accurately known. If the measured statistical characteristics do not match the system modeling parameters, the accuracy of KF state estimation will decline, and even filter divergence may occur in serious cases. The chi-square test (C2T) is a traditional measurement fault detection method [1]. On the premise of accurate state space modeling, chi-square statistics can be constructed through the innovation of KF and its mean square error matrix to detect whether the measurement is abnormal. If the measurement is normal, the measurement update will be carried out. Otherwise, if the measurement is abnormal, the measurement update will be abandoned. However, it is often difficult to accurately determine the system noise parameters in practical applications, which makes the threshold setting of chi-square detection statistics a major problem. If the chi-square detection threshold is set too large, abnormal measurements may be introduced into the filter to reduce the accuracy of filter estimation; If the threshold setting is too small, some normal measurements may be excluded, which will reduce the measurement utilization and the accuracy of state estimation. The root cause of this problem lies in the binary measurement anomaly judgment principle of traditional chi-square detection. The measurement is either judged as normal or abnormal, and there is no other intermediate situation.

Concerning Sage Husa Adaptive KF (SHAKF) or robust adaptive filtering method [2-5], measurement conditions can at least be more finely divided into "normal/trust", "suspicious/partial trust" and "abnormal/discard" categories. The size of the measurement noise variance matrix can be modified or given different measurement utilization weights for different measurement categories. The weight can be directly calculated according to the KF innovation size to achieve a seamless and continuous transition of weight size. However, the ideas of the two adaptive filtering methods are not completely the same. By analyzing the characteristics of binary utilization of innovation weight in traditional chi-square detection, this paper gives a continuous utilization method of innovation weight, improves the binary value of weight to continuous value, and generalizes it, thus obtaining a multi-component chi-square detection method for multi-dimensional measurement. In this paper, the proposed new method is called Soft Chi-square Test KF (SC2TKF), while the traditional method can be called Hard Chi-square Test KF. Obviously, the new method is more universal, and the traditional hard chi-square test can be regarded as a special case of the new soft chi-square test method. At last, the KF simulation of the soft chi-square detection method is carried out by using the inertial navigation/satellite navigation combination, and the results show that the new method has better filtering accuracy and stability than Sage Husa adaptive filtering.

2. Traditional measurement fault chi-square detection principle

The state space model of stochastic linear systems is written as[6-8]

$$\begin{cases} \boldsymbol{X}_k = \boldsymbol{\Phi}_{k/k-1}\boldsymbol{X}_{k-1} + \boldsymbol{\Gamma}_{k-1}\boldsymbol{W}_{k-1} \\ \boldsymbol{Z}_k = \boldsymbol{H}_k\boldsymbol{X}_k + \boldsymbol{V}_k \end{cases} \tag{1}$$

where $\boldsymbol{X}_{k\,/\,n\times 1}$ is the state vector of dimension, $\boldsymbol{Z}_{k\,/\,m\times 1}$ is the dimensional measurement vector; $\boldsymbol{\Phi}_{k/k-1}, \boldsymbol{\Gamma}_{k-1}, \boldsymbol{H}_k$ are the known system structure parameters, respectively, called $n\times n$. One step state transition matrix of the dimension $n\times l$ of dimension system noise allocation matrix and $m\times n$ dimensional measurement matrix is depicted; $\boldsymbol{W}_{k-1\,/\,l\times 1}$ is the system noise vector of dimension, and $\boldsymbol{V}_{k\,/\,m\times 1}$ is the measurement noise vector of dimension, both of which are zero mean Gaussian white noise vector sequences, and they are mutually uncorrelated. The following basic assumptions of Kalman filtering about noise are met.

$$\begin{cases} \mathrm{E}[\boldsymbol{W}_k] = \boldsymbol{0}, & \mathrm{E}[\boldsymbol{W}_k\boldsymbol{W}_j^{\mathrm{T}}] = \boldsymbol{Q}_k\delta_{kj} \\ \mathrm{E}[\boldsymbol{V}_k] = \boldsymbol{0}, & \mathrm{E}[\boldsymbol{V}_k\boldsymbol{V}_j^{\mathrm{T}}] = \boldsymbol{R}_k\delta_{kj} \\ \mathrm{E}[\boldsymbol{W}_k\boldsymbol{V}_j^{\mathrm{T}}] = \boldsymbol{0} \end{cases} \tag{2}$$

Among them, $\boldsymbol{Q}_k$ is the semi-positive definite; $\boldsymbol{R}_k$ is the positive definite; δ_{kj} is the Kronecker function.

The Kalman filtering formula is:

$$\begin{cases} \hat{\boldsymbol{X}}_{k/k-1} = \boldsymbol{\Phi}_{k/k-1}\hat{\boldsymbol{X}}_{k-1} \\ \boldsymbol{P}_{k/k-1} = \boldsymbol{\Phi}_{k/k-1}\boldsymbol{P}_{k-1}\boldsymbol{\Phi}_{k/k-1}^{\mathrm{T}} + \boldsymbol{\Gamma}_{k-1}\boldsymbol{Q}_{k-1}\boldsymbol{\Gamma}_{k-1}^{\mathrm{T}} \\ \boldsymbol{r}_k = \boldsymbol{Z}_k - \boldsymbol{H}_k\hat{\boldsymbol{X}}_{k/k-1} \\ \boldsymbol{A}_k = \boldsymbol{H}_k\boldsymbol{P}_{k/k-1}\boldsymbol{H}_k^{\mathrm{T}} + \boldsymbol{R}_k \\ \boldsymbol{K}_k = \boldsymbol{P}_{k/k-1}\boldsymbol{H}_k^{\mathrm{T}}\boldsymbol{A}_k^{-1} \\ \hat{\boldsymbol{X}}_k = \hat{\boldsymbol{X}}_{k/k-1} + \boldsymbol{K}_k\boldsymbol{r}_k \\ \boldsymbol{P}_k = (\boldsymbol{I} - \boldsymbol{K}_k\boldsymbol{H}_k)\boldsymbol{P}_{k/k-1} \end{cases} \tag{3}$$

where $\hat{\boldsymbol{X}}_{k/k-1}$ and $\boldsymbol{P}_{k/k-1}$ represent the state one-step prediction and its mean square error matrix respectively; $\boldsymbol{r}_k$ and $\boldsymbol{A}_k$ respectively represent the measurement prediction error (i.e., innovation) and its mean square error matrix. Theoretically, innovation sequence $\boldsymbol{r}_k$ is the zero mean white noise; $\boldsymbol{K}_k$ is the filter gain; $\hat{\boldsymbol{X}}_k$ and $\boldsymbol{P}_k$ represent the state estimation and its mean square error matrix, respectively.

Statistics can be constructed from measurement innovation and its mean square error matrix [1].

$$\lambda_k = \boldsymbol{r}_k^{\mathrm{T}} \boldsymbol{A}_k^{-1} \boldsymbol{r}_k \tag{4}$$

where λ_k obeys the degree of freedom as m chi-square distribution, i.e., $\lambda_k \sim \chi^2(m)$. Under normal measurement conditions, the statistics value of λ_k should be relatively small; If the measurement is abnormal, λ_k will become larger. Whether the measurement is normal or not is generally based on a selected threshold T_{Dm} as the judgment threshold, namely,

$$\begin{cases} \lambda_k \le T_{Dm} & \textit{Measurement is normal}\left(\textit{denoted by } \chi_k = 1\right) \\ \lambda_k > T_{Dm} & \textit{Measurement is anomaly}\left(\textit{denoted by } \chi_k = 0\right) \end{cases}. \tag{5}$$

This is the chi-square detection principle of the Kalman filter for measuring faults. Statistics can be calculated during filtering λ_k. According to its size, it is monitored whether the measurement is abnormal in real-time, and then it is decided whether to update the measurement. The measurement and its mean square error matrix update formula in Formula (3) can be rewritten as:

$$\begin{cases} \hat{\boldsymbol{X}}_k = \hat{\boldsymbol{X}}_{k/k-1} + \sqrt{\chi_k}\boldsymbol{K}_k \boldsymbol{r}_k \\ \boldsymbol{P}_k = (\boldsymbol{I} - \chi_k \boldsymbol{K}_k \boldsymbol{H}_k)\boldsymbol{P}_{k/k-1} \end{cases}. \tag{6}$$

Obviously, when $\chi_k=1$, $\lambda_k \le T_{Dm}$. Normal Kalman filtering measurement update is performed. And when $\chi_k=0$, $\lambda_k > T_{Dm}$. The measurement update is abandoned to isolate the abnormal measurement.

3. A new robust Kalman filter estimation method

The precondition for the effectiveness of the traditional chi-square detection method is that the filtering system model is accurate. In practical applications such as inertial navigation/satellite navigation integrated navigation, the system structural parameter modeling of Kalman filtering is relatively accurate and stable, while the noise parameters are difficult to be fully accurately modeled due to the change of motion state, imperfect modeling, system aging or operating environment, which will lead to the chi-square detection threshold T_{Dm}. It is difficult to determine accurately in strict accordance with theoretical methods. If the threshold value is set too high, the probability of fault detection will be reduced, and there is a risk that more outliers will be introduced into the filtering measurement, which will increase the filtering error; If the threshold is set too small, the probability of false alarm will increase. The frequent false alarm will reduce measurement utilization, and the reduction of measurement correction will also reduce the accuracy of filter estimation. Of course, for high-precision inertial navigation systems, more than 50% of satellite navigation measurements are abandoned randomly. For example, even if the satellite navigation measurement is reduced from 1 Hz to 0.1 Hz, it has little impact on the integrated navigation performance; However, for low-precision inertial navigation systems, the change of satellite navigation utilization rate from 1 Hz to 0.5 Hz may have a significant impact on the combination accuracy.

The result of the traditional chi-square detection method is binary, not 0 is 1, and there is no intermediate state. In radar target detection and other fields, this method mainly uses "yes/no" as the indicator, and its application is very reasonable. However, for integrated navigation and other occasions, there may be a large number of intermediate states (suspicious) between normal (trusted) and abnormal (discarded) measurement information, so it is not appropriate to simply use chi-square to detect binary results. This also goes against the commonly used measurement adaptive filtering and robust filtering methods. For example, in Sage Husa measurement adaptive Kalman filtering, the size of the measurement noise variance matrix is automatically adjusted by the size of the measurement innovation, which is equivalent to the comprehensive utilization of all measurement information. Compared with the normal measurement, the measurement weight is reduced when the measurement innovation is medium (suspicious) and is very small when the innovation is abnormal. If it tends to zero, it is equivalent to abandonment. According to the innovation utilization characteristics of adaptive filtering, Formula (6) can be improved to the following:

$$\begin{cases} \hat{\boldsymbol{X}}_k = \hat{\boldsymbol{X}}_{k/k-1} + \sqrt{\rho_k}\boldsymbol{K}_k\boldsymbol{r}_k \\ \boldsymbol{P}_k = (\boldsymbol{I} - \rho_k\boldsymbol{K}_k\boldsymbol{H}_k)\boldsymbol{P}_{k/k-1} \end{cases} \tag{7}$$

$$\rho_k = \begin{cases} 1 & \lambda_k \le T_{Dm} \\ T_{Dm} / \lambda_k & \lambda_k > T_{Dm} \end{cases} \tag{8}$$

Obviously, when the equivalent measurement innovation is abnormal, the equivalent filter gain in Formula (7) is $\sqrt{\rho_k}\boldsymbol{K}_k$. The utilization of innovation will gradually decline with the increase of innovation, which is a soft decline mode, so it is called the soft chi-square detection method in this paper. Compared with Formula (7), the method of binary hard decline (from 1 to 0) of innovation utilization rate in the traditional chi-square detection Formula (6) can be called the hard chi-square detection method.

It is worth noting that if the measurement in Formula (1) is multidimensional, the traditional chi-square detection regards multidimensional measurement as a whole. As long as any component is abnormal, the chi-square detection method will also reduce the innovation utilization rate of other normal measurement components at the same time. This processing method is not very reasonable. To avoid this defect, it is necessary to do chi-square detection on each measurement component one by one. If, in the Kalman filtering model, the measurement components are uncorrelated, that is, the measurement noise mean square error matrix $\boldsymbol{R}_k$ is a diagonal matrix, the chi-square detection of each sequential measurement update is equivalent to the processing of a single measurement component one by one after the sequential filtering processing is adopted. If $\boldsymbol{R}_k$ is a nondiagonal matrix, for the new interest $\boldsymbol{r}_k$ of $i\,(i = 1, 2, \cdots, m)$ components, $A_{k(ii)}$ is the innovation mean square error matrix $\boldsymbol{A}_k$ of i, which is i Column diagonal elements and recorded statistics.

$$\lambda_{k(i)} = r_{k(i)}^{\mathrm{T}} A_{k(ii)}^{-1} r_{k(i)} = r_{k(i)}^2 / A_{k(ii)} \tag{9}$$

Each statistic $\lambda_{k(i)}$ is subject to a chi-square distribution with the degree of freedom of 1, that is $\lambda_{k(i)} \sim \chi^2(1)$.

Concerning Formula (7), for the promotion and establishment of $\lambda_{k(i)}$, the constructed Kalman filter measurement update method is as follows:

$$\begin{cases} \boldsymbol{K}_k = \boldsymbol{P}_{k/k-1}\boldsymbol{H}_k^{\mathrm{T}}\sqrt{\boldsymbol{\rho}_k}\boldsymbol{A}_k^{-1} \\ \hat{\boldsymbol{X}}_k = \hat{\boldsymbol{X}}_{k/k-1} + \boldsymbol{K}_k\boldsymbol{r}_k \\ \boldsymbol{P}_k = (\boldsymbol{I} - \boldsymbol{K}_k\sqrt{\boldsymbol{\rho}_k^{\mathrm{T}}}\boldsymbol{H}_k)\boldsymbol{P}_{k/k-1} \end{cases} \tag{10}$$

$$\sqrt{\boldsymbol{\rho}_k} = \mathrm{diag}\left(\sqrt{\rho_{k(1)}} \;\; \sqrt{\rho_{k(2)}} \;\; \cdots \;\; \sqrt{\rho_{k(m)}}\right) \tag{11}$$

$$\rho_{k(i)} = \begin{cases} 1 & \lambda_{k(i)} \le T_{D1} \\ T_{D1} / \lambda_{k(i)} & \lambda_{k(i)} > T_{D1} \end{cases} \tag{12}$$

In Formula (10), the filter gain $\boldsymbol{K}_k$ is modified. The purpose is to ensure the subsequent mean square error matrix $\boldsymbol{P}_k$ Symmetry of the calculation of Formula (7) is ρ_k special case when all components are equal. In Formula (12), the parameter T_{D1} is a chi-square statistic with a degree of freedom of 1 $\lambda_{k(i)}$. The threshold value is set for fault detection. The significance level is 0.05 $T_{D1} \approx 3.8$.

4. Simulation verification

4.1 Inertial navigation/satellite navigation integrated navigation model

The inertial navigation system/satellite navigation system loose integrated navigation system is taken as an example for simulation experiment verification. The system status is 15 dimensions, and the measurement is 3 dimensions. The specific modeling is as follows:

$$\begin{cases} \dot{\boldsymbol{X}} = \boldsymbol{F}\boldsymbol{X} + \boldsymbol{G}\boldsymbol{W}^b \\ \boldsymbol{Z} = \tilde{\boldsymbol{p}}_{\text{INS}} - \tilde{\boldsymbol{p}}_{\text{GNSS}} = \boldsymbol{H}\boldsymbol{X} + \boldsymbol{V} \end{cases} \tag{13}$$

$$\boldsymbol{X} = \begin{bmatrix} \boldsymbol{\phi}^{\mathrm{T}} & (\delta \boldsymbol{v}^n)^{\mathrm{T}} & (\delta \boldsymbol{p})^{\mathrm{T}} & (\boldsymbol{\varepsilon}^b)^{\mathrm{T}} & (\boldsymbol{\nabla}^b)^{\mathrm{T}} \end{bmatrix}^{\mathrm{T}}$$

$$\boldsymbol{F} = \begin{bmatrix} \boldsymbol{M}_{aa} & \boldsymbol{M}_{av} & \boldsymbol{M}_{ap} & -\boldsymbol{C}_b^n & \boldsymbol{0}_{3\times3} \\ \boldsymbol{M}_{va} & \boldsymbol{M}_{vv} & \boldsymbol{M}_{vp} & \boldsymbol{0}_{3\times3} & \boldsymbol{C}_b^n \\ \boldsymbol{0}_{3\times3} & \boldsymbol{M}_{pv} & \boldsymbol{M}_{pp} & \boldsymbol{0}_{3\times3} & \boldsymbol{0}_{3\times3} \\ & & \boldsymbol{0}_{6\times15} & & \end{bmatrix}, \quad \boldsymbol{G} = \begin{bmatrix} -\boldsymbol{C}_b^n & \boldsymbol{0}_{3\times3} \\ \boldsymbol{0}_{3\times3} & \boldsymbol{C}_b^n \\ \multicolumn{2}{c}{\boldsymbol{0}_{9\times6}} \end{bmatrix}, \quad \boldsymbol{W}^b = \begin{bmatrix} \boldsymbol{w}_{\mathrm{g}}^b \\ \boldsymbol{w}_{\mathrm{a}}^b \end{bmatrix}$$

$$\boldsymbol{H} = \begin{bmatrix} \boldsymbol{0}_{3\times3} & \boldsymbol{0}_{3\times3} & \mathbf{I}_{3\times3} & \boldsymbol{0}_{3\times3} & \boldsymbol{0}_{3\times3} \end{bmatrix}$$

$\boldsymbol{\phi}^{\mathrm{T}}, \delta\boldsymbol{v}^n, \delta\boldsymbol{p}, \boldsymbol{\varepsilon}^b, \boldsymbol{\nabla}^b$ represent inertial navigation misalignment angle, velocity error, position error, gyro random constant drift, accelerometer random constant deviation, respectively; $\boldsymbol{w}_{\mathrm{g}}^b$ and $\boldsymbol{w}_{\mathrm{a}}^b$ are the gyro white noise and accelerometer white noise; $\boldsymbol{V}$ is the measuring white noise for the position of satellite guidance receiver; $\boldsymbol{C}_b^n$ is the inertial navigation attitude matrix; n represents the "East North Sky" navigation coordinate system; b represents the "right front top" carrier coordinate system; $\tilde{\boldsymbol{p}}_{\text{INS}}, \tilde{\boldsymbol{p}}_{\text{GNSS}}$ are the position output of inertial navigation system and satellite navigation system, including latitude, longitude, and altitude; In [6], there are specific expressions for $\boldsymbol{M}_{**}$.

4.2 Simulation conditions

The PSINS toolbox software is used to simulate a section of a vehicle driving track, which includes stationary, acceleration and deceleration, uniform speed, turning, climbing, and other stages. The total simulation time is about 1000 s, and the driving mileage is about 7.5 km. The driving speed is shown in Figure 1, and the two-dimensional plan of the track's relative position change is shown in Figure 2. The mark "★" in the figure is the starting point of vehicle driving.

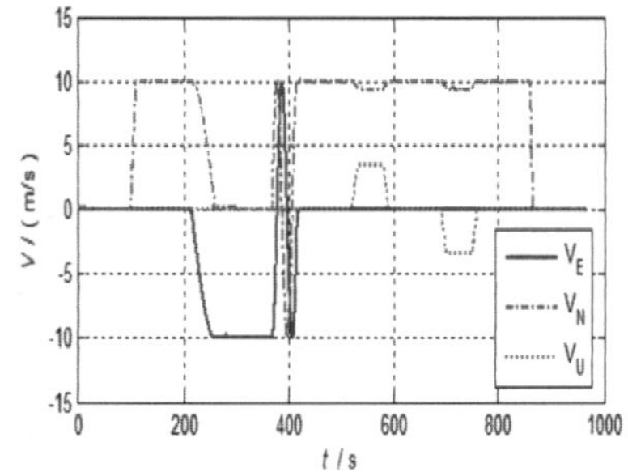

Figure 1. Driving speed of vehicle.

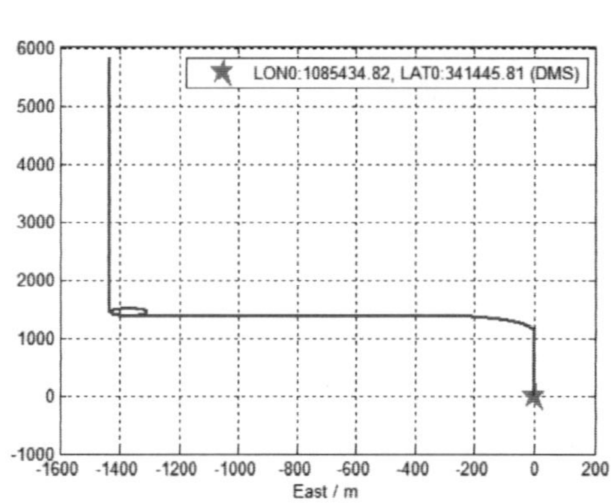

Figure 2. Plane track of vehicle driving.

Inertial sensor data and satellite navigation positioning data are generated according to trajectory simulation, in which inertial data is sampled at 100 Hz, and satellite navigation measurement is conducted at 1 Hz. Then, inertial navigation solution and inertial navigation/satellite navigation integrated navigation Kalman filter solution are carried out. Various errors injected in the simulation process are listed in Table 1.

Table 1. Simulation error settings.

Error name (unit)	Numerical value
Gyro random constant drift ($°/\mathrm{h}$)	0.03
Gyro random walk ($°/\sqrt{\mathrm{h}}$)	0.001
Accelerometer random constant deviation (μg)	100
Accelerometer random walk ($\mu g/\sqrt{\mathrm{Hz}}$)	5
Initial horizontal misalignment angle ($'$)	0.5
Initial azimuth misalignment angle ($'$)	20
Initial speed error (m/s)	0.1
Initial position error (m)	1

In particular, the position measurement error of the satellite navigation system (three components of latitude and longitude) is set as Gaussian white noise with zero mean value, and the variance in the normal period is $(1\mathrm{m})^2$, except for the following two periods: (1) The noise variance between 200 s and 400 s is set to $(10\mathrm{m})^2$; (2) It is set as pollution noise between 600 s and 800 s. Taking the satellite navigation altitude measurement error as an example, it is expressed as

$$\delta H \sim \begin{cases} \mathrm{N}(0,(1\mathrm{m})^2) & \text{w.p.} \quad 0.9 \\ \mathrm{N}(0,(100\mathrm{m})^2) & \text{w.p.} \quad 0.1 \end{cases} \tag{14}$$

where "w.p." means "with probability". Formula (14) represents noise δH . Under normal circumstances, the variance is approximately $(1\mathrm{m})^2$, 90% of the Gaussian distribution is regarded as normal noise; Under abnormal conditions, the variance is $(100\mathrm{m})^2$, 10% of the Gaussian distribution is regarded as outlier noise.

An example of a group of height measurement error simulations is shown in Figure 3. It can be seen from the figure that the overall fluctuation of height error becomes larger between 200 s and 400 s, and the simulated noise variance changes; However, individual error amplitude jumps greatly between 600 s and 800 s, simulating small probability outlier jumps.

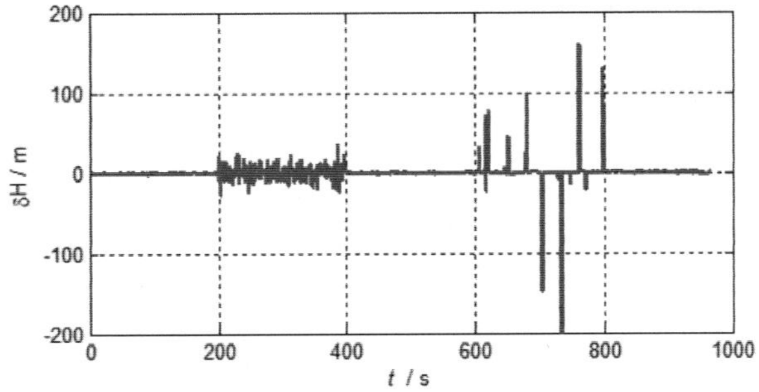

Figure 3. Height measurement error of satellite guidance.

4.3 Simulation results

The results of attitude misalignment angle, velocity error, and position error of inertial navigation/satellite navigation integrated navigation are shown in Figure 4. It can be seen from the figure that the error of integrated navigation increases with the measurement noise of satellite navigation between 200 s and 400 s, and the magnitude of this error and the variance of the whole range of the satellite navigation error are set to $(10\mathrm{m})^2$. The results are consistent. Between 600 s and 800 s, the state estimation is not affected by the outliers of satellite conductance measurement noise, and the filter has a good outlier isolation effect, with only a small variance $(1\mathrm{m})^2$. Like the measurement noise, it always maintains a high filtering accuracy.

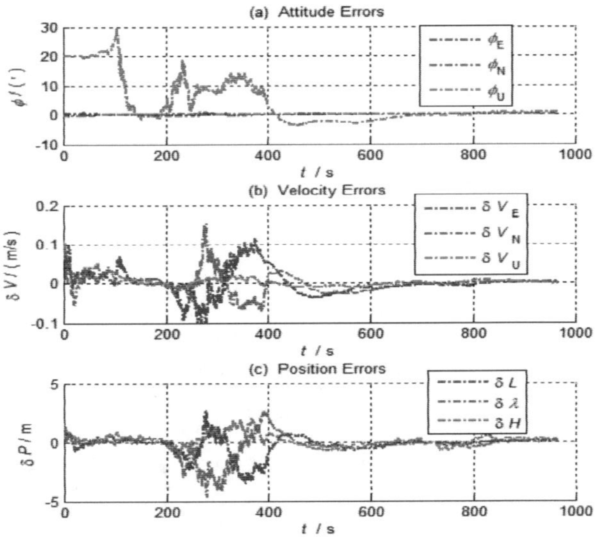

Figure 4. Estimation error based on soft chi-square detection.

In addition, the paper also made a comparative simulation of the new method (SC2TKF) and the traditional Sage Husa measurement noise adaptive filter (SHAKF), carried out 50 Monte Carlo simulations, and counted the navigation parameter errors. The results showed that SC2TKF was more stable, of which the RMS (Root Mean Square) error statistics of latitude error were shown in Figure 5, and other errors were similar, so they would not be listed one by one. It is worth mentioning that there is also a fading factor in SHAKF that needs to be set carefully (b=0.5). If the setting of the fading factor is not appropriate, it will have a greater negative impact on the filtering results. In a word, the new method does not need to set any parameters and has the advantages of convenient use and strong adaptability.

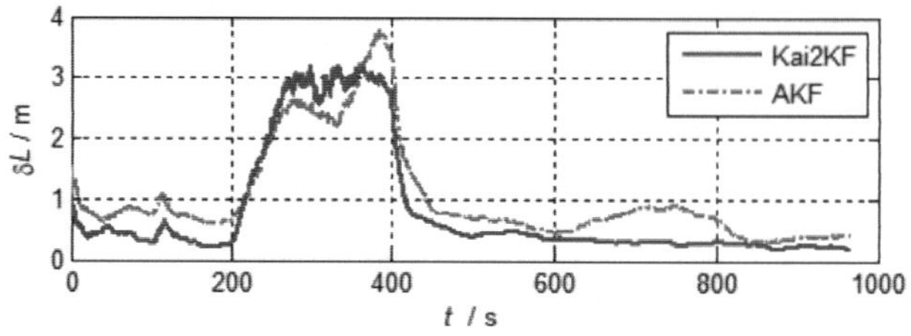

Figure 5. Comparison of latitude errors of different filtering methods.

5. Conclusion

In the actual integrated navigation system, in addition to the classical Sage Husa adaptive filtering based on measurement noise adaptation which is more practical, other adaptive methods and theories often have ideal assumptions or cumbersome parameter settings, which lack feasibility and are difficult to apply to complex high-dimensional systems, so there are few reports of practical applications. This paper proposes a new quality control method for the GNSS/IMU Kalman filter. It can effectively perform chi-square statistical hypothesis tests on each measurement component one by one, eliminate abnormal observations, retain useful information, avoid traditional overall tests, eliminate all measurement information, and effectively improve the stability of GNSS/IMU fusion positioning.

Acknowledgments

This paper is financially supported by the research on key technology of high precision positioning and detection of multi-sensor combination based on GNSS/IMU/odometer (Grant No.CQSJKJ2022037).

References

[1] Qin Y.Y, Zhang H.Y, Wang S.H. (2015) Kalman filter and integrated navigation principle [M]3rd EditionXi'an: Northwest Polytechnical University Press, 2015.
[2] Sage A P, Husa G W. (1969) Adaptive filtering with unknown prior statistics [C]//Proceedings of Joint Automatic Control Conference, Boulder Colorado, 1969: 760-769.
[3] Yan G.M, Deng Y, (2020) Review of practical Kalman filtering technology in traditional integrated navigation [J]Navigation, positioning and time service, 2020,7 (2): 50-64
[4] Yang Y.X (2017) Adaptive dynamic navigation positioning [M]2nd EditionBeijing: Surveying and Mapping Press, 2017.
[5] Zhang Q.Q, Zhao Lo, Zhou J.H, (2020) Application of robust adaptive fractional filtering algorithm in PPP/INS integrated navigation [J]Navigation, positioning and time service, 2020,7 (05): 73-81.
[6] Yan G.M, Weng J, (2019) Strapdown Inertial Navigation Algorithm and Integrated Navigation Principle [M]Xi'an: Northwest Polytechnical University Press, 2019.
[7] Erfianti R,Asfihani T & Suhandri H F.(2023).GNSS/IMU Sensor Fusion Performance Comparison of a Car Localization in Urban Environment Using Extended Kalman Filter. IOP Conference Series: Earth and Environmental Science(1). doi:10.1088/1755-1315/1127/1/012006.
[8] Kaczmarek Adrian,Rohm Witold,Klingbeil Lasse & Tchórzewski Janusz.(2022).Experimental 2D extended Kalman filter sensor fusion for low-cost GNSS/IMU/Odometers precise positioning system. Measurement. doi:10.1016/J.MEASUREMENT.2022.110963.

Application of calibration-free wavelength modulated spectral absorption technique in temperature measurement under ultra-short optical path

Yunhao Ren[1], Feng Gao[1], Xiaoming Duan[1], Yu Cai[1], Yachao Li[1] and Yihui Zhou[1,2]

[1]Dalian University of Technology, Dalian 116000, China

[2]Corresponding author's e-mail: zflower@dlut.edu.cn

Abstract: The measurement of gas temperature parameters is of great significance in scientific research and industrial production. Wavelength modulation technology, as a non-contact optical detection means, has the characteristics of strong anti-interference ability and is widely used in various complex and harsh environments. The accuracy of temperature measurement is affected by short optical paths and weak absorption. In this paper, a calibration-free wavelength modulated ultra-short optical path measurement system is established, and the selection rules of spectral lines under the ultra-short optical path and weak absorption are proposed. The measurement results of wavelength modulation without calibration are in good agreement with those of thermocouple temperature measurement, and the relative error is less than 10%.

1. Introduction

Tunable diode laser absorption spectroscopy (TDLAS) has the advantages of non-intrusive measurement, high sensitivity, multi-parameter synchronous measurement, etc. Therefore it is widely used in the measurement of flow field temperature [1] without interfering with the original flow form and getting the real description of objectives. Absorption spectroscopy includes two methods, namely the direct absorption spectroscopy (DAS) and the wavelength modulation spectroscopy (WMS). Wavelength modulation technology can greatly suppress the low-frequency noise in the absorbed signal because of the superposition of high-frequency signal on the low-frequency signal, which makes wavelength modulation suitable for the poor flow field environment with weak absorption.

In wavelength modulation technology, the absorbed signal and the linear function are coupled to each other, and there is no complete decoupling method at present. Therefore, when using WMS for temperature detection, it is generally necessary to calibrate the demodulated harmonic signal [2]. In the case of temperature measurement in a bad environment, direct calibration is difficult, so researchers have done a lot of work on wavelength modulation without calibration. Li et al. developed a fast response coordinated diode laser absorption sensor and measured the temperature in a shock tube with an optical path L=15.24 cm using the peak inversion flow field parameter method. The measured results were consistent with the theoretical value, and the relative error was less than 1.2% [3]. Sun et al. proposed a new calibration-free full-linear fitting wavelength modulated absorption spectrum temperature measurement method, which compares simulated normalized second harmonic signals with measured results to determine the properties of gases [4]. Goldenstein et al. used the calibration-free method proposed by K SUN to measure the temperature, velocity, and static pressure of the gas in the Stanford expansion tube with optical path L=7.5 cm. The result showed that the residual difference between the simulated harmonic signal and the measured signal was less than 2.5%. The temperature, pressure, and velocity measured by the experiment are consistent within the range of 5%, 3.8%, and 1.5% predicted by the theory [5].

In the above experiments measured by calibration-free wavelength modulation technology, the measured optical path is large, and the optical path is not interfered with by other substances such as glass. The accuracy of temperature measurement by wavelength modulation technology is affected by the intensity of the absorbed signal, which declines as the optical path decreases for the same spectral line. When the laser passes through the glass, absorption will be reduced. Therefore, more attention needs to be paid to the selection of spectral lines. In this paper, the selection criteria of spectral lines under short optical path and weak absorption are systematically studied, and the calibration-free measurement system at room temperature is verified.

2. The basic principle of calibration-free wavelength modulation technology for temperature measurement

Wavelength modulation technology refers to the laser in the low-frequency sawtooth signal and the high-frequency sine signal driven by the laser. The light intensity signal to be measured by the flow field is accepted by the detector and then demodulated through the lock amplifier to absorb signals at a specific frequency. From these absorbed signals, information such as the flow field temperature and constituent concentrations can be obtained. The specific formula has been elaborated in many literatures and will not be described too much here [6, 7]. This section mainly describes the specific steps of uncalibrated wavelength modulation.

The relationship between the intensity of the incident laser $I_0(t)$ and the intensity of the absorbed laser $I_t(t)$ can be described by the Beer-Lambert law:

$$I_t(t) = I_0(t)\exp\left[-\sum_j A_j\, \emptyset_j(v(t), T, P, \chi)\right] \tag{1}$$

where $I_0(t)$ represents the incident laser intensity signal, $\emptyset_j$ represents the linear function of absorbed component j, $v(t)$ represents the change of laser wave number with time, and A_j refers to the integral absorbance of absorbed component j, which can be expressed as:

$$A_j = \int_{-\infty}^{\infty} \alpha(v)dv = \int_0^L S_j(T)P_j dl \tag{2}$$

Figure 1 shows the solution of the wavelength modulation temperature without calibration is mainly divided into the following steps [8]: (1) The relationship between the actual absorbed laser intensity and time $I_t^M(t)$ is measured. (2) The relationship between the wave number of the laser $v(t)$ and the change of intensity with time $I_0(t)$ is measured. (3) The intensity of the transmitted laser light is simulated according to the Beer-Lambert law. (4) The actual laser intensity and the simulated laser intensity are fed into the same lock-in amplifier to obtain the harmonic signal. (5) The simulated harmonic signal is approximated to the actual by iterating the free parameters to find the temperature *T*.

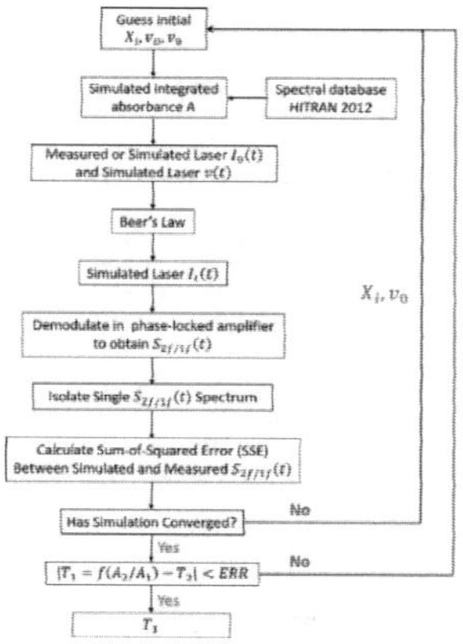

Figure 1. WMS temperature measurement calibration free flowchart.

3. Selection of spectral lines

The most important step in WMS-2f/1f temperature measurement is to select the appropriate spectral line. Zhou X et al. described the line selection criteria for temperature measurement using wavelength modulation technology [9,10,11]. In this section, we briefly discuss the principle of line selection for measuring the temperature of the flow field in the case of a short optical path through quartz glass.

When selecting the spectral line, we must first ensure that the laser can have enough absorption after passing through the glass, and then select the spectral line with lower cost under the premise of ensuring strong absorption. At the same time to avoid the interference of adjacent spectral lines, the two spectral lines should have a large enough low-state energy level difference to ensure a high-temperature measurement sensitivity. When there is no way to ensure sufficient absorption and high-temperature measurement sensitivity at the same time, priority should be given to ensuring sufficient absorption while trying to get closer to high sensitivity.

Based on the above rules, a pair of absorption lines were selected from the HITRAN2012 database, as shown in Table 1.

Table 1. Absorption line parameters.

Wavelength (nm)	$S\ (cm^{-1}.atm^{-1})$	$E\ (cm^{-1})$	$\gamma_{\text{air}}\ (cm^{-1})$	$\gamma_{\text{self}}\ (cm^{-1})$
1392.5	0.3719	136.7617	0.0997	0.434
1359.5	0.2740	325.3479	0.0775	0.391

4. H_2O absorption temperature measurement experiment

4.1 Experimental system

The experimental equipment is shown in Figure 2, and the real image of the experiment is shown in Figure 3. The measuring system includes a signal generator, two DFB lasers, a laser mounting base, two pieces of high-purity quartz glass, two photodetectors, N_2 purge device, germanium etalon, wavemeter, and data acquisition system.

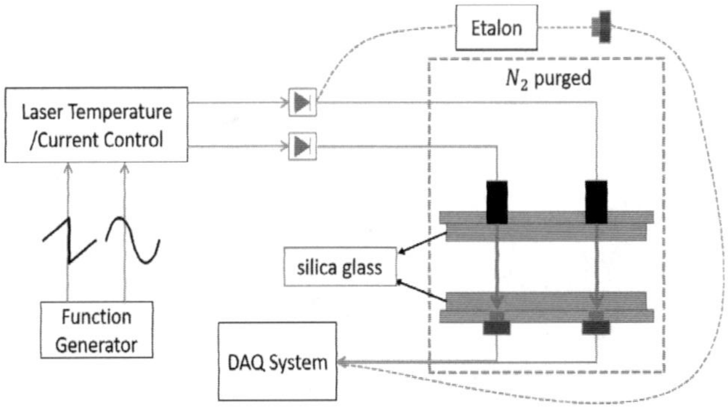

Figure 2. Schematic of TDLAS measurement device.

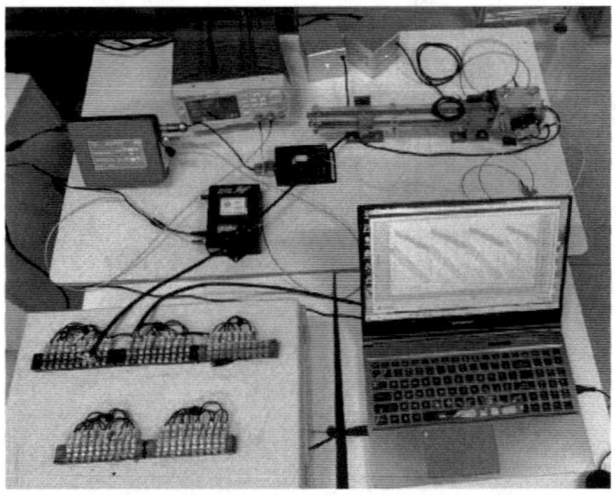

Figure 3. Real image of the experiment.

The two DFB lasers operating near 1392.5 nm and 1359.5 nm are installed on the laser mounting mount and the flow field temperature is monitored with a thermocouple. By adjusting the drive seat output center current and measuring with a wavemeter, the laser center wavelength is adjusted to 1392.534 nm and 1359.512 nm respectively, which are located near the center of the water molecule absorption line. The signal generator outputs a sawtooth signal and a modulation signal to the laser drive seat. The modulation signal selects the normalized second harmonic peak maximum corresponding to the modulation signal, and the size of the scan signal should be able to make the second harmonic shape completely display a cycle. The signal size of the signal generator input to the laser controller is shown in Table 2.

Table 2. Load to laser controller signal size.

Wavelength (nm)	Scan amplitude $a_s\ (cm^{-1})$	Scan rate $f_s(Hz)$	Modulation amplitude $a_m(cm^{-1})$	Modulation frequency $f_m(kHz)$
1392.5	2.20	200	0.24206	20
1359.5	2.25	200	0.30	20

4.2 Analysis and discussion of experimental results

The initial laser intensity of the laser $I_0(t)$ is measured without absorption by N_2 purging. The change of laser wavenumber with time $v(t)$ is measured by the standard and the wavemeter. First of all, the wavelength of the laser at the initial time needs to be determined by the wavelength meter, and it is converted into the wavenumber. Secondly, the interference signal was measured by using germanium etalon, and the relative wavenumber change relationship with time was obtained by using the program developed by Li et al [12]. The specific processing steps can be referred to Li et al.'s work and Du et al.'s work [12,13].

The actual absorbed signal and the analog absorbed signal are passed into the same phase-locked amplifier program for demodulation. The parameter Settings of the phase-locked amplifier are shown in the table. After three iterations, the normalized second harmonics obtained by the final demodulation are shown in Figure 4. It can be seen that the simulated and actual normalized second harmonics of the two lasers fit very well.

The measurement results of wavelength modulation temperature without calibration are shown in Figure 5. The temperature measured by the thermocouple is 296 K, and the experimental results are basically the same as those measured by the thermocouple, with a relative error of less than 5%. The experimental results show that the spectral line pairs of 1392.5 nm and 1359.5 nm are feasible for the measurement of temperature under an ultra-short optical path.

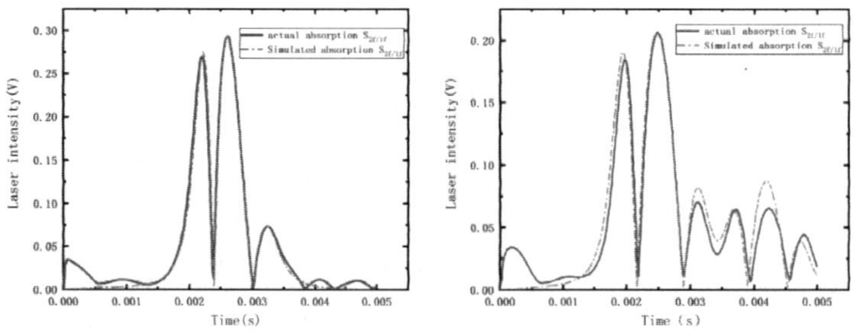

(a) (b)

Figure 4. Comparison between simulated absorption normalized second harmonic and actual situation.

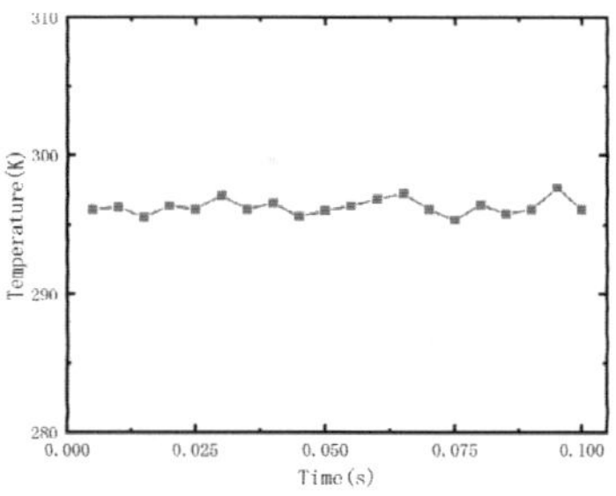

Figure 5. WMS temperature measurement results.

5. Conclusion

The temperature measuring platform was built by using the wavelength modulation technique of a laser passing through quartz glass under the background of an ultra-short optical path. Weak absorption caused by short optical range and glass can be improved by selecting spectral lines with strong line strength, strong absorption, and high thermometric sensitivity, while interference from neighboring spectral lines should be avoided. When there is a contradiction with the traditional spectral line selection rules, the premise should be to ensure sufficient absorption. Taking H_2O absorption as an example, the temperature of the laboratory environment was measured using 1392.5

nm and 1359.5 nm lasers, and the relative error was less than 5% compared with the thermocouple temperature measurement results. The experimental results show that the surface calibration wavelength modulation technique has the advantages of strong reliability and high precision in the measurement of the ultra-short optical temperature range. The next step is to apply this system to the temperature measurement of an ultra-short optical path oscillating tube (similar to a shock tube).

Acknowledgments
This research was supported by the National Key Research and Development Program of China (2018YFA0704601).

References

[1] Li, R., Li, F., Lin, X., et al. (2023) Linear calibration-free wavelength modulation spectroscopy. Microwave and Optical Technology Letters., 65, 5: 1024-1030. https://doi.org/10.1002/mop.33063.
[2] Wang, K.Y., Shao, J., Shao, L.G., et al. (2021) A pressure-calibration method of wavelength modulation spectroscopy in sealed microbial growth environment. Chinese Physics B., 30, 5: 054203. DOI:10.1088/1674-1056/abda32.
[3] Li, H., Farooq, A., Jeffries, J.B, et al. (2007) Near-infrared diode laser absorption sensor for rapid measurements of temperature and water vapor in a shock tube. Applied Physics B., 89: 407-416. DOI:10.1007/s00340-007-2781-9.
[4] Sun, K., Chao, X., Sur, R., et al. (2013) Analysis of calibration-free wavelength-scanned wavelength modulation spectroscopy for practical gas sensing using tunable diode lasers. Measurement Science and Technology., 24, 12: 125203. DOI:10.1088/0957-0233/24/12/125203.
[5] Goldenstein, C.S., Strand, C.L., Schultz, I.A., et al. (2014) Fitting of calibration-free scanned-wavelength-modulation spectroscopy spectra for determination of gas properties and absorption line shapes. Applied optics., 53, 3: 356-367. DOI:10.1364/AO.53.000356.
[6] Li, R., Li, F., Lin, X., et al. (2023) Linear calibration-free wavelength modulation spectroscopy. Microwave and Optical Technology Letters., 65, 5: 1024-1030. https://doi.org/10.1002/mop.33063.
[7] Zang, Y.P., Xu, Z.Y., Xia, H.H., et al. (2020) Method for measuring high temperature spectral line parameters based on calibration-free wavelength modulation technology. DOI:10.3788/CJL202047.1011001.
[8] Sun, K. (2013) Utilization of multiple harmonics of wavelength modulation absorption spectroscopy for practical gas sensing. Stanford University. http://purl.stanford.edu/rb361sv6972.
[9] Zhou, X., Jeffries, J.B., Hanson, R.K. (2005) Development of a fast temperature sensor for combustion gases using a single tunable diode laser. Applied Physics B., 81:711-722. DOI:10.1007/s00340-005-1934-y.
[10] Wang, W.F., Yang, B., Liu, H.F., et al. (2023) A multiline fitting method for measuring ethylene concentration based on WMS-2f/1f. Scientific Reports, 13, 1: 15302. DOI:10.1038/s41598-023-42398-2.
[11] Zhao, R., Zhou, B., Zhang, J., et al. (2023) Rapid online tomograph in non-uniform complex combustion fields based on laser absorption spectroscopy. Experimental Thermal and Fluid Science, 147: 110930. https://doi.org/10.1016/j.expthermflusci.2023.110930.
[12] Li, Z., Wang, Z., Mével, R., et al. (2023) A utility for characterising laser diode wavelength-to-time response for wavelength modulation spectroscopy application. Applied Physics B, 129, 1: 5. DOI:10.1007/s00340-022-07940-6.
[13] Du, Y., Peng, Z., Ding, Y. (2020) A high-accurate and universal method to characterize the relative wavelength response (RWR) in wavelength modulation spectroscopy (WMS). Optics Express, 28, 3: 3482-3494. DOI: 10.1364/OE.383216 DOI:10.1364/OE.383216.

Experimental research on instrumentation for measuring automotive exhaust particle concentrations

Yuanhao Li [1,3], Guangze Li [2], Liuyong Chang [2] and Quan Zhou [1]

[1]School of Mechanical and Electrical Engineering, China Jiliang University, Hangzhou 310018, China
[2]Zhongfa Aviation Institute of Beihang University, Hangzhou 311115, China

[3]Corresponding author's e-mail: 1416906503@qq.com

Abstract: Due to the absence of domestic automotive exhaust measurement methods, this paper, based on the principle of condensation nucleus growth counting, has developed a nanoparticle counter suitable for automotive exhaust detection. The paper incorporates the peak value counting method and conducts comparative experiments between different counting methods of the nanoparticle counter and the results obtained by electrostatic counting, as well as parallel experiments with a TSI CPC. The following results were obtained: (1) Compared to the traditional rising edge counting method, the peak value counting method significantly improves counting accuracy. (2) When the concentration is below 12, 500 P/cc, the peak value counting method can count more particle signals. (3) In the testing of complex particle clusters in automotive exhaust, the nanoparticle counter can be used for measuring automotive exhaust emissions.

1. Introduction

High-temperature combustion sources are the primary contributors to atmospheric PM2.5 pollution, with the vast majority of these particles being in the nanoscale range for example nanoparticle concentrations in vehicle exhaust exceed 90% [1]. These nanoparticles can penetrate the bloodstream and readily adsorb polluting gases in the atmosphere, forming secondary aerosols, and can also carry viruses and bacteria, posing significant health hazards to humans. For the latest ultra-low emissions from vehicles, aircraft, power plants, and other sources [2-4], assessing the harm of exhaust particulates solely based on mass concentration is insufficient. Even though ultrafine particles represent a small proportion by mass, their potential for harm is substantial. In comparison to new vehicles, in-use vehicles emit higher levels of particulate number, making an assessment of exhaust emissions based solely on particulate mass concentration inadequate. It is essential to incorporate the measurement of particulate number concentration. Currently, several countries such as the Netherlands and Finland have initiated the use of particulate number (PN) measurement technology for in-use vehicles' exhaust inspections. In China, there is an urgent need to establish relevant standards and develop measurement equipment and instruments [5].

Pollutants in the exhaust emissions from vehicles include harmful gases (such as carbon dioxide, nitrogen oxides, carbon monoxide, and volatile organic compounds) and particulate matter (such as suspended particles and fine particles) [6, 7]. These pollutants not only directly affect air quality but can also contribute to environmental issues such as climate change, acid rain, and photochemical smog [3]. Furthermore, automotive exhaust emissions pose potential health threats to human beings. When

 2724 (2024) 012027 doi:10.1088/1742-6596/2724/1/012027

harmful gases and particulate matter in vehicle exhaust enter the human respiratory system, they may lead to respiratory diseases, cardiovascular diseases, allergies, and even cancer. Fine particles, particularly those with a diameter of less than 2.5 micrometers, have high penetration capabilities, can deeply enter the lungs, and be absorbed into the bloodstream, causing more significant health effects [4, 8-10]. Presently, detection technologies for gaseous pollutants in vehicle exhaust are relatively mature and include laser gas analyzers for measuring carbon dioxide (CO2) and nitrogen oxide (NOx) concentrations [11, 12], as well as opacity meters [13]. However, with the increasing awareness of the harm of particulate emissions from vehicles [14, 15], governments and environmental agencies have imposed more stringent regulations and standards for vehicle emissions, along with higher testing requirements, making particulate matter detection increasingly important. There has been a significant shift in recent years from traditional particle mass (PM) measurements to particle number (PN) measurements for particulate detection [16-18], with a particular focus on the quantity and size distribution of fine particles [18, 19]. Therefore, the independent development of equipment suitable for measuring particulate number concentration in in-use vehicles in China is of great significance.

2. Working principle of the nanoparticle counter

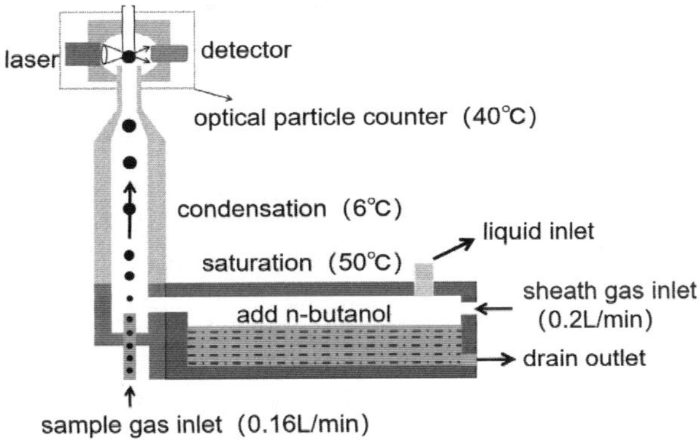

Figure 1. Schematic representation of the nanoparticle counter instrument.

This paper presents a nanoparticle counter designed for automotive exhaust detection based on the condensation nucleus growth counting principle, as shown in Figure 1. The essential components of the CPC (Condensation Particle Counter) include a saturation stage, a condensation stage, and an optical counting unit. The saturation stage operates at a temperature of 50°C and is used to heat isobutyl alcohol to create a stable vapor source. Particles undergo condensation and growth in the condensation stage, which operates at a temperature of 6°C. When the particles to be measured enter the condensation stage, the supersaturated isobutyl alcohol vapor condenses on the particle surfaces due to the temperature difference. This increases particle diameter, transforming them from the nanoscale to submicron or micron levels. Subsequently, the particles enter the optical counting unit, where they are illuminated by a laser and the scattered signal is collected and converted into electrical pulses by photodiodes, ultimately enabling particle counting.

3. Signal correction work

The traditional counting method relies on counting based on the rising edge of the signal, as illustrated in Figure 2. Using a set threshold, it counts an event when the signal transitions from below the threshold to above it. Consequently, this rising edge counting method may miss counts when dealing with multiple pulse signals within a certain range. This situation arises because after the particle concentration increases, particles overlap when passing through the optical path of the optical counter, causing the received optical signal to not fall back to the bottom noise. At the same time, overlapping particles will also interact with each other to generate more scattered light, resulting in more peak signals, which to some extent leads to technical difficulties in the case of particle overlap.

To address this issue, the peak value counting method is introduced, as depicted in Figure 3. The peak value counting method is based on counting the peaks of the pulse signals. By identifying local maxima in the existing signal acquisition frequency, it calculates the wave peaks and counts them, ensuring accurate frequency identification for the current signal. Therefore, the peak value counting method does not suffer from counting losses and effectively enhances the accuracy of the counting results.

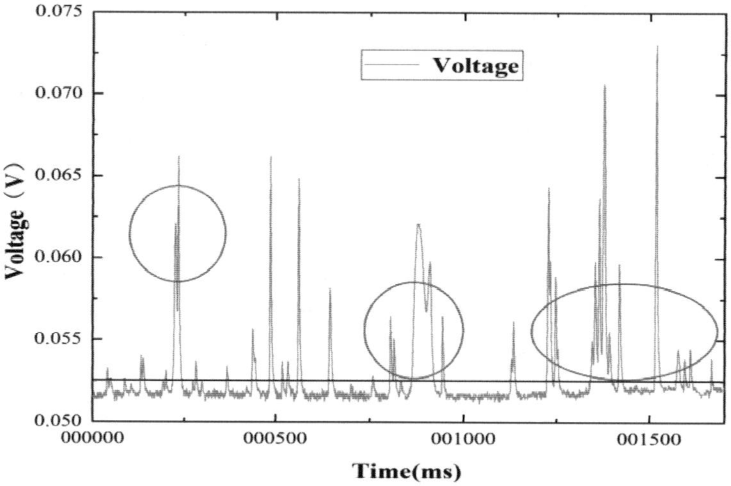

Figure 2. Rising edge counting method.

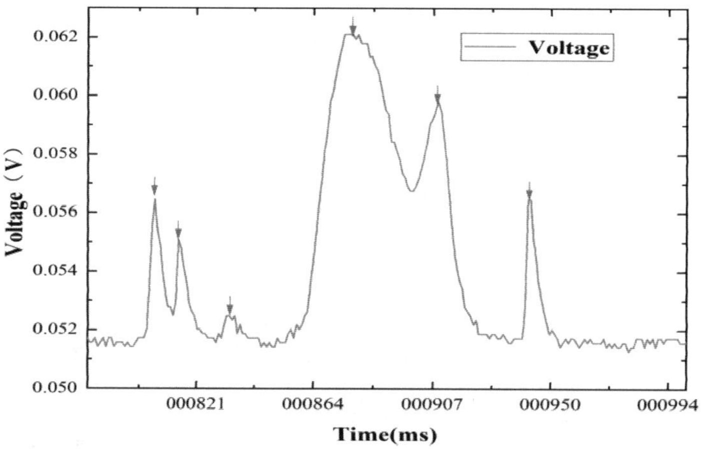

Figure 3. Maximum counting method.

4. Experimental testing and validation

As shown in Figure 4, to validate the improved nanoparticle counter, comparative experiments were conducted at different particle concentrations using a self-developed nanoparticle counter and a TSI electrostatic counter. The experimental setup included an aerosol generator for generating stable particle sources, a Differential Mobility Analyzer (DMA) used for particle size selection, the Electrostatic counter (ELT) used as the reference instrument for comparison, and the self-developed nanoparticle counter. In the experiments, different concentrations of 50 nm sodium chloride nanoparticles were generated by adjusting the sheath-to-aerosol flow ratio of the DMA and the mass concentration of the sodium chloride solution. The counting results of the self-developed nanoparticle counter and the electrostatic counter were simultaneously tested and compared.

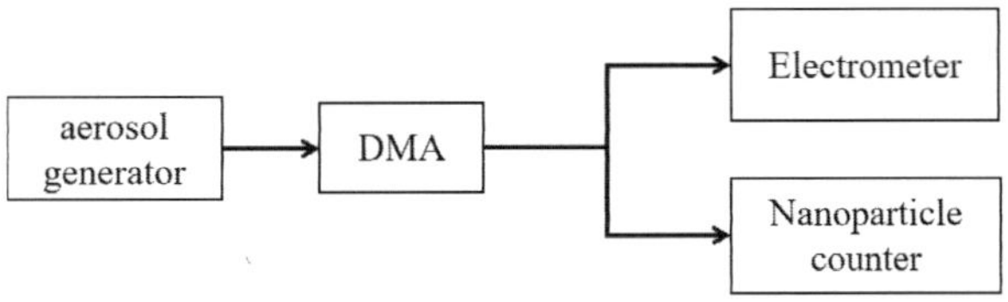

Figure 4. Schematic diagram of nanoparticle counter and electrometer testing experiment.

As depicted in Figure 5, the experimental results clearly demonstrate that, at concentrations exceeding 10000 P/cc, the self-developed nanoparticle counter using the rising edge counting method yields significantly lower counts compared to the electrostatic counter (ELT). This discrepancy arises because, at higher particle concentrations, a greater number of particles trigger high plateau signals, producing multiple consecutive peaks that cannot be accurately counted using a single rising edge detection. The introduction of the peak value counting method effectively corrected the counting issue. As shown in the graph, the counting results obtained after employing the peak value counting method closely align with those of the electrostatic counter.

After implementing the peak value counting method, the counting results of the self-developed nanoparticle counter and the electrostatic counter exhibited a more consistent trend over time and demonstrated a positive correlation. It is worth noting that when the particle number concentration falls within the range of 12500 to 15750 P/cc, the measurement results of the self-developed nanoparticle counter and the electrostatic counter are nearly identical. At lower concentrations, the nanoparticle counter produces measurements higher than the electrostatic counter, while at concentrations above 15700 P/cc, the nanoparticle counter yields measurements lower than those of the electrostatic counter.

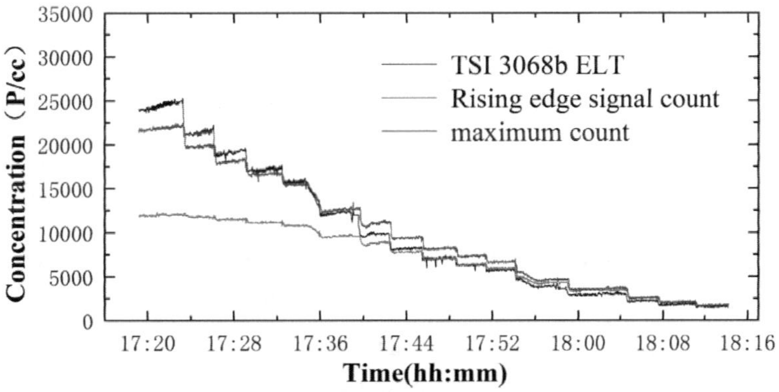

Figure 5. Comparison of different counting methods.

To verify the detection performance of this system on complex particle clusters in automotive exhaust, the exhaust was diluted to a concentration below 20, 000 P/cc using a dilution system. The self-developed nanoparticle counter and TSI Model 3775 CPC were simultaneously tested. During a 2-minute test, the vehicle's engine speed was varied from a low speed of 700 r/s to a high speed of 4000 r/s, then reduced back to low speed, with 30-second intervals maintained at 4000 r/s and 2000 r/s. The test results, as shown in Figure 6, indicate that the self-developed BH CPC exhibits good consistency with the TSI 3775 CPC.

It still has good consistency between high-speed and low-speed changes, and the response speed of the self-developed nanoparticle counter to concentration changes is similar to that of TSI CPC. Therefore, it is believed that the nanoparticle counter, with the help of a diluter, can measure the number and concentration of automotive exhaust particles.

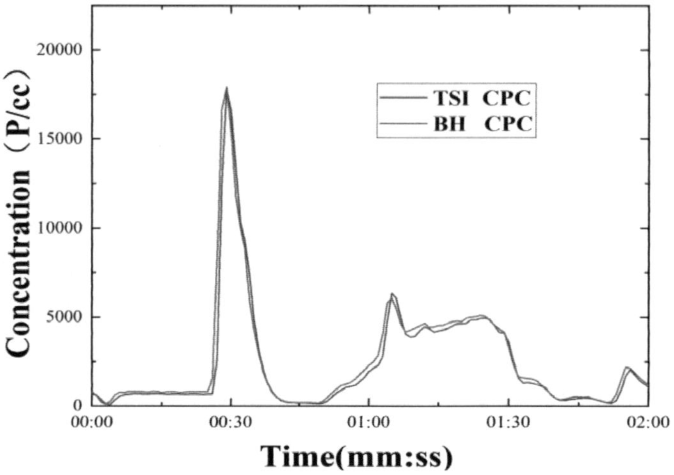

Figure 6. Nanoparticle Counter Automotive Exhaust Gas Detection Results.

5. Conclusion

Based on the principle of condensation nucleus growth counting, this paper has developed a nanoparticle counter for automotive exhaust detection, achieving the following results:

(1) Compared to the traditional rising edge counting method, the peak value counting method significantly improves counting accuracy.

(2) When the concentration is below 12, 500 P/cc, the peak value counting method can count more particle signals.

(3) In testing complex particle clusters in automotive exhaust, the self-developed BH CPC (Condensation Particle Counter) results closely match the counts obtained with the TSI CPC, indicating that the BH CPC can be used for measuring automotive exhaust emissions.

References

[1] A. S M, A. D, H. S. Particulate material 2.5 change in the city using EURO4 urban buses technology case study: Arak City, Iran [J]. International Journal of Environmental Science and Technology, 2022, 19 (10). https://doi.org/10.1007/s13762-022-04205-9.

[2] FOWLER D, BRIMBLECOMBE P, BURROWS J, et al. A chronology of global air quality [J]. Philosophical Transactions of the Royal Society A, 2020, 378(2183): 20190314. https://doi.org/10.1098/rsta.2019.0314.

[3] HEI. State of Global Air 2020 [M]. HEI Boston. 2020. https://www.healtheffects.org/annou-ncements/state-global-air-2020-reports-air-pollutions-impact neonatal mortality.

[4] POPE III C A, COLEMAN N, POND Z A, et al. Fine particulate air pollution and human mortality: 25+ years of cohort studies [J]. Environmental research, 2020, 18(3):10-24. https://doi.org/10.1016/j.envres.2019.108924.

[5] Chen Zirong, On standardizing motor vehicle exhaust emission testing and strengthening automobile pollutant emission management [J]. Heilongjiang Environmental Journal, 2023, 36(06): 52-54. https://www.cnki.com.cn/Article/CJFDTOTAL-HLJO202306018.htm.

[6] Zhang Quansheng, Study on Pollution of Urban Automobile Exhaust Emissions and Its Prevention and Control Measures [J]. Energy and energy conservation, 2021, 11(1): 63-64. doi:10.16643/j.cnki.14-1360/td.2021.11.024.

[7] SHRIVASTAVA M, NGUYEN A, ZHENG Z, et al. Kinetics of soot oxidation by NO2 [J]. Environmental Science & Technology,2010,44(12):796-801. https://doi.org/10.1021/es903672y.

[8] NAKAI S, JST C, NITTA H, et al. Respiratory health associated with exposure to automobile exhaust. III. Results of a cross-sectional study in 1987, and repeated pulmonary function tests from 1987 to 1990 [J]. Archives of Environmental Health: An International Journal, 1999, 54(1): 26-33.https://doi.org/10.1080/00039899909602233.
[9] STURM R. Theoretical deposition of diesel exhaust particles in the respiratory tract of children [J]. J Public Health Emerg, 2019, 3(1):2-3. doi: 10.21037/jphe.2019.09.01.
[10] KONUR O. Cancer Caused by Diesel Fuel Exhaust Emissions: A Review of the Research [J]. Petrodiesel Fuels, 2021, 1(1):75-92. https://doi.org/10.1201/9780367456252.
[11] Liu Yi, Wang Changchun, Yuan Yibing, Application and Improvement of Laser Gas Analyzer in Exhaust Gas Monitoring and Detecting [J]. Automation in Petro-Chemical Industry, 2023, 59(02):86-87+90. https://www.cnki.com.cn/Article/CJFDTOTAL-LYHG202302021.htm.
[12] LI Xiang, LI Ling-xiao, LI Han-guang. Stability Analysis of Laser Dust Particle Counter [J]. Telecom Power Technology, 2018, 35(09): 123-125. DOI: 10.19399/j.cnki.tpt.2018.09.047.
[13] Teng Qin, Liu Yiming, Xia Shunli, Design of a Sampling Type Smoke Opacimeter [J]. Car Engine, 2002,06(1):25-29.
https://kns.cnki.net/kcms2/article/abstract?v=IkZuKgafC3Wajoh1L6KAaOOeMkrK7KN75we5MX2srsHhiHjaxvQSAtIFVypTJ3jVdw6Age18rAXXIG0c4zi8AQi_D72BHMIPCH6GffjnAsPXhtrDwUYSMLp-fmCU3x3Z07If9EiTBw0=&uniplatform=NZKPT&language=CHS.
[14] ANWAR M N, SHABBIR M, TAHIR E, et al. Emerging challenges of air pollution and particulate matter in China, India, and Pakistan and mitigating solutions [J]. Journal of Hazardous Materials, 2021, 16(1): 25-26. https://doi.org/10.1016/j.jhazmat.2021.125851.
[15] ALEMAYEHU Y A, ASFAW S L, TERFIE T A. Exposure to urban particulate matter and its association with human health risks [J]. Environmental Science and Pollution Research, 2020, 27:491-506. https://link.springer.com/article/10.1007/s11356-020-09132-1.
[16] GIECHASKIEL B, MELAS A, MARTINI G, et al. Overview of vehicle exhaust particle number regulations [J]. Processes, 2021, 9(22): 12-16. https://doi.org/10.3390/pr9122216.
[17] Peng Min.research on influence factors of particulate matter emission detection from gasoline engine.[D]
https://d.wanfangdata.com.cn/thesis/ChJUaGVzaXNOZXdTMjAyMzEyMjISCUQwMTMzOTQ4MhoIOWloM3B1dmo%3D
[18] Wang Zemin. Analysis of Automobile Exhaust Emission Detection Technology and Treatment Measures[J].TimesCars,2023,1(1):7-9.
https://xueshu.baidu.com/usercenter/paper/show?paperid=1e0b0060n45s06k0q74m0tp0qw738136&site=xueshu_se.
[19] Xu Sheng, Analysis of excess air coefficient measurement deviations in automobile exhaust gas detection [J]. Equipment management and maintenance, 2020, 1(5): 108-109. DOI: 10.16621/j.cnki.issn1001-0599.2020.08.57.

Evaluation of measurement uncertainty of drop hammer impact test for composite materials

Li Lei [1,2] **and Zhang Yao** [1, 2]

[1]Shanghai Aircraft Design and Research Institute, Shanghai 201210, China

[2]Corresponding author's e-mail: sculilei@126.com (Li Lei); zhangyao1@comac.cc (Zhang Yao)

Abstract. Future aircraft structures are usually dominated by composites, and one of the key defects and forms of damage to composite structures is the impact damage caused by low-energy (especially low-speed) foreign matter. This kind of low-energy impact test is usually completed by the drop hammer testing machine, so accurately assessing the impact test of the falling hammer completed in accordance with the standard test method, and the reliability of the permissible value of the impact performance of the objective reaction composite is of great significance to the future aircraft structure design. Taking a standard composite test piece as an example, considering the key factors in the whole process of impact test, a mathematical model is constructed, and the measurement uncertainty of different configuration composite test pieces of different configurations is calculated according to the test data, which provides a basis for aircraft structural design.

1. Introduction

Future aircraft structures are usually dominated by composite materials. And one of the key defects and damage forms of composite structures is impact damage caused by low-energy (especially low-speed) foreign objects [1-2]. This low-energy impact test is usually performed by a drop-weight testing machine. The allowable value of composite impact performance obtained by the drop weight testing machine is an important reference index for aircraft structural design, so it is very important to accurately obtain the allowable value of composite impact performance through testing. The accuracy of the impact test measurement results is closely related to the measurement method and measurement equipment, and the correct evaluation of its uncertainty can compare the reliability of the allowable value of composite impact performance objectively, which is of great significance to future aircraft structure design.

Zhang Xiaojuan and Zhang Boping from Northwestern Polytechnical University have made useful explorations on the relationship between impact energy and pit depth and gave the corresponding reference formula [3-4], and Serge Abrate from Southern Illinois University in the United States also gave the relationship between impact force and pit depth according to Hertz's contact law [5].

In this paper, the process of testing the uncertainty of the system by a specific impact test is used to check the accuracy of the allowable value of the impact performance of the composite material and reduce the evaluation workload and improve the accuracy.

2. Experiments and calculations

The impact resistance of composite materials is evaluated according to JJF1059.1-2012 "Evaluation and Representation of Measurement Uncertainty" [6].

1) Test method. According to ASTM D7136 "Standard Test Method for Measuring the Damage Impedance of Fiber Reinforced Polymer Matrix Composites to Drop Weight Impact Events".

2) Environmental conditions. The test method requires a room temperature of 10 ℃ ~ 35 ℃ and relative humidity ≤ 85%. The room temperature of this test was 20 ℃ and the humidity was 79%.

3) Test equipment: Instron CEAST 9350 drop weight testing machine:0.59J ≦ impact energy ≦ 757J.

4) Test piece. The test piece is made of medium-mold high-strength carbon fiber reinforced M21C epoxy resin unidirectional with prepreg, the material specification is CMS-CP-309: Type 34, Class 3, Class 194, grade M21C/34%/UD194/IMA/6.4-3, and the nominal thickness of the single layer of the laminate after curing is 0.187 mm. All test pieces mentioned in this document are manufactured using automatic wire laying and autoclave curing auxiliary processes. The padding used for the test is shown in Table 1, and the test piece configuration and testing machine are shown in Figure 1.

Table 1 List of laminate lay-up for pit test of damage characteristics of composite fuselage.

Ply Code	Number of Plies	Nominal thickness (mm)	Ply order
1201	12	2.244	[+45/-45/0/90/+45/-45] s
1202	12	2.244	[+45/-45_2/90/+45/0] s
1203	12	2.244	[+45/-45/+45/90/-45/0] s

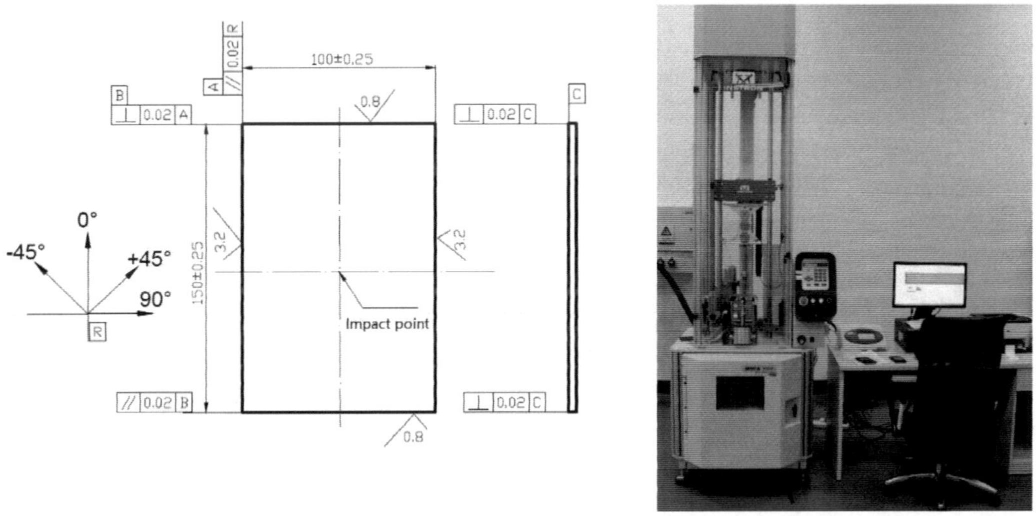

Figure 1. Pit test piece and testing machine of laminate damage characteristics.

3. Sources of uncertainty and mathematical models

The working principle of the drop weight impact test is to lift the corresponding mass of the heavy hammer to a certain height, do free fall movement, convert the gravitational potential energy of the drop weight into kinetic energy, and impact the test piece at a certain speed, so as to verify the impact resistance of the test piece, the test piece is washed out of a certain pit, and there is a certain relationship between the impact energy and the depth of the pit.

The uncertainty of the impact test results includes 1. the uncertainty introduced by repeated measurements and 2. the uncertainty introduced by the test equipment. Among them, there are two main parameters for repeated measurement: 1. the impact energy of the drop weight; 2. the pit depth of the test piece.

Most researchers use Hertz's contact law [7] to model local deformation, according to Hertz's contact law, the impact force F of the drop weight tester is related to the α of indentation depth as follows:

$$F=k_c\alpha^{\frac{3}{2}} \tag{1}$$

The impact energy can be determined using the functional principle, and the work done by the impact force during the indentation from 0 to the maximum pit depth is:

$$Ec = mgh = \int_0^{\alpha_{max}} F d\alpha = \frac{2}{5}k_c\alpha^{\frac{5}{2}} \tag{2}$$

In Equation (2), E_c is the impact energy (J); m is the mass of the drop weight (kg); g is the acceleration due to gravity (m/s^2); h is the drop weight impact height (m); k_c is the contact stiffness, which is related to the material properties of the impactor and the impacted object, and α is the depth of the pit caused by the impact of the drop weight.

The contact stiffness k_c is as follows:

$$k_c = \frac{4}{3}E\sqrt{R} \tag{3}$$

R and E in Equation (3) are derived from the following equation:

$$\frac{1}{R} = \frac{1}{R_1} + \frac{1}{R_2} \tag{4}$$

$$\frac{1}{E} = \frac{1-\mu_1^2}{E_1} + \frac{1-\mu_2^2}{E_2} \tag{5}$$

R_1 and R_2 are the radius of curvature (mm) of the impact head and the impacted object, E_1 and E_2 are the elastic modulus (Gpa) of the impact head and the impacted object, and μ_1 and μ_2 are the Poisson's ratio between the impact head and the impacted object.

The impact energy and pit depth required to generate 1mm pits with different test piece configurations are shown in Table 2 below.

Table 2 Drop weight testing machine test data.

Ply Code	Impact Energy (J)	Pit Depth α (mm)
1201	25	1.005
	25	1.069
	24.5	0.867
	24.5	0.785
	25	0.979
	25	1.005
	25	0.891
	25	0.917
	25	1.001
	26	1.177
1202	25	0.784
	25	0.811
	29	1.186
	27	1.02
	27	0.915
	27	0.971
1203	25	0.974
	25	0.915
	26	1.145
	25	1.114
	25	0.981
	25	1.071

4. Uncertainty assessment

4.1. Measurement uncertainty component introduced by experimental repeatability

The measurement uncertainty component introduced by experimental repeatability was evaluated using class A standard [8-9]. Repeated observations of the same measured independently n times under repeatability conditions yielded n measured values, the arithmetic mean is calculated, and we use Bezier's formula to calculate the measured class A standard uncertainty u_A.

$$\overline{\alpha} = \frac{\sum_{i=1}^{n} \alpha_i}{n} \quad (6)$$

$$\overline{\alpha}' = \sqrt{\frac{\sum_{i=1}^{n} \left|\alpha_i - \overline{\alpha}\right|}{n(n-1)}} \quad (7)$$

$$u_{A1} = \frac{2}{5} \mathrm{k_c} \alpha^{\frac{5}{2}} = \frac{2}{5} \mathrm{k_c} \overline{\alpha}'^{\frac{5}{2}} \quad (8)$$

In Equations (6)-(8):

$\overline{\alpha}$ — the arithmetic mean of the pit depth.

$\overline{\alpha}'$— the deviation value of the pit depth.

n — Number of valid specimens.

$$\overline{E} = \frac{\sum_{i=1}^{n} E_i}{n} \quad (9)$$

$$u_{A2} = \sqrt{\frac{\sum_{i=1}^{n} \left|E_i - \overline{E}\right|}{n(n-1)}} \quad (10)$$

In Equations (9)-(10):

$\overline{E}$ — the arithmetic mean of the impact energy.

n — Number of valid specimens.

$$u_A = \sqrt{u_{A1}^2 + u_{A2}^2} \quad (11)$$

According to the test data in Table 2, we substitute Equations (6) ~ (11) to obtain Table 3.

Table 3 Repeated measurement of uncertain components.

Ply Code	Deviation of pit depth ($\overline{d}'$)	The uncertainty of the pit depth (u_{A1})	Impact energy ($\overline{E}$)	The uncertainty of impact energy (u_{A2})	The uncertainty of repeat measurement u_A
1201	0.096 mm	0.3 J	25 J	0.149 J	0.3349 J
1202	0.149 mm	0.4656 J	26.67 J	0.471 J	0.6622 J
1203	0.124 mm	0.3875 J	25.17 J	0.237 J	0.4542 J

4.2. The uncertainty component introduced by the test equipment

The uncertainty introduced by the metrological performance of the measuring instrument:

$$u_B = \frac{U_B}{k} \tag{12}$$

In Equation (12):

U_B — the extended uncertainty provided by the Measurement Equipment Calibration Certificate.

k — the coefficients provided by the equipment calibration certificate.

The measurement certificate of the third-party drop weight testing machine can be obtained, and the uncertainty is extended:

Weight m: $U_B = 0.208g, k = 2$.

Height h: $U_B = 0.06mm, k = 2$.

The above data can be substituted into the formula calculation, $u_{B1} = 0.104g; u_{B2} = 0.03mm$

The measurement uncertainty is introduced by the device u_B.

$u_B = u_{B1} \times u_{B2} \times g = 0.104 \times 0.03 \times 9.8 = 0.031\text{J}$

4.3. Synthetic standard uncertainty and extended uncertainty

The synthetic uncertainty for measuring the damage impedance of fiber-reinforced polymer matrix composites to drop weight impact events is:

$$u_c = \sqrt{u_A^2 + u_B^2} \tag{13}$$

In the equation:

u_c — Synthetic standard uncertainty

The extended uncertainty U is equal to the product of the synthetic standard uncertainty u_c and the inclusion factor k [10-11].

$$U = k \times u_c \tag{14}$$

The inclusion factor k is related to the measured σ distribution. In general, k = 2, substituting the data, we obtain the extended uncertainty values of different composite configurations with a 95% confidence probability, as shown in Table 4 below.

Table 4 Synthetic uncertainty and extended uncertainty.

Ply Code	The uncertainty of repeat measurement u_A	The uncertainty of the device u_B	Synthetic uncertainty u_c	Extended uncertainty U
1201	0.3349 J	0.031 J	0.3363 J	0.6726 J
1202	0.6622 J	0.031 J	0.6629 J	1.3258 J
1203	0.4542 J	0.031 J	0.4552 J	0.9105 J

5. Conclusion

It can be seen from the test data that in the impact test of composite materials, even composite materials of the same thickness, the same size, and different layers have different measurement uncertainties. Parameter repeatability measurement, drop weight and other test equipment have an impact on the results, repeated measurement caused by the error has the greatest impact on the uncertainty of the results, so the test process should be strictly in accordance with the standard test. Especially the part where people participate, need multiple measurements to improve the measurement accuracy.

Journal of Physics: Conference Series **2724** (2024) 012028 doi:10.1088/1742-6596/2724/1/012028

References

[1] Cui hai po,Wen wei dong Cui hai tao. Advances of study on Low Velocity Impact Damaged and Residual Strength of Composite Laminates [J]. Journal of Materials Science and Engineering, 2005, 23 (3): 466-472.

[2] Huang da fei. Research on Low-velocity Impact Damage of Stiffened Laminate and Parametric Modeling [D]. Nanjing University of Aeronautics and Astronautics. 2012. P2-P6.

[3] Zhang xiao juan, Zhang bo ping, Zhang jin kui, Wang lan- ming. Damage Analysis for Composite Material Subjected to Low-Speed Impact Based on Pit-Depth Analysis [J]. Journal of Experimental Mechanics.Jun.2010 Vol.25 No.3.P234-P238.

[4] Zhang xiao juan, Zhang bo ping. Damage Analysis of Composite Laminate under Low-velocity [J]. Aeronautical Computing Technique. 2010, 40(3): 68-74.

[5] Serge abrate.Impact Engineering of composite structures [M]. Italy:Springer Wien NewYork. 2011 by CISM,Udine. P83-P86.

[6] JJF 1059.1-2012 Measurement uncertainty assessment and representation [S].

[7] Serge Abrate. Impact on composite structures [M]. UK:Cambridge university press, 1998. P15-P24.

[8] Qiao tao. Tensile strength test uncertainty analysis of composite materials [J]. China computer& communication. 2016. No.4. P72-P84.

[9] Ran xuechen. Measurement Uncertainty Evaluation on Fatigue Testing Results [J]. Physical Testing and Chemical Analysis (Part A: Physical Testing).2017.No.53.P717-P719.

[10] WU Gang. Analysis on Uncertainty of Metallic Materials Tensile Test [J]. Journal of Test and Measurement Technology. Vol.28 No.5 2014.P405-P409.

[11] Yu li ping. Calibrating Method and Experimentation of Vertical Impact [J]. Metrology & Measurement Technology.2006, No5. P36-38.

Surface Contamination Measurement by Combining Detectors and Robots

Qiang Wang[1,2], Yifa Sheng[1]

[1]School of Electrical Engineering, University of South China, Hengyang 421001, China

[2]Corresponding author's e-mail: wangqiang_623@163.com

Abstract. Against the background of the rapid development of nuclear energy and nuclear technology, the number of places with radioactivity has gradually risen, which poses a significant safety hazard to human beings. Aiming at indoor radioactive places, this paper designs a detector that can be mounted on a robot, and experiments prove that this detector can accurately detect α-surface contamination and β-surface contamination. The surface contamination is measured using a combination of a remote-controlled robot and a detector.

1. Introduction

With the rapid development of nuclear energy and technology in today's world, there are more and more radioactive places, which are more common in our lives, such as the nuclear medicine department of medical science, isotope production workshops, and laboratories that need to do radioactive experiments [1]. These radioactive places are more prone to radioactive contamination. According to the national radiation safety management regulations, in places where radioactive contamination is likely to occur, the ground, equipment surfaces, etc., should be routinely monitored for contamination to prevent the spread of contamination [2].

In the routine monitoring of surface contamination of the ground, there are direct [3] and indirect [4] measurement methods. Direct measurement refers to the surface contamination measuring instrument to detect the level of contamination below the limit value of the regulations. The detector should be moved as close to the surface as possible during the measurement. Once the contamination is detected, the detector should be positioned in the area for a sufficient time to confirm the level of contamination. Indirect measurement refers to transferring contamination from a specific area of the object to be measured through wiping or decontamination, etc., to a fixed instrumentation in the laboratory for measurement. Both direct and indirect measurements avoid the risk of exposing personnel to radioactive scenarios and radiation hazards. On the other hand, the means of measurement using robots have been widely used in various industries, such as the inspection of 3D structures using robots [5], inspection of power systems [6], source localization of gas leaks [7], etc. Robots can also be applied to the measurement of surface contamination.

This paper designs a practical surface contamination detector and utilizes a robot to carry the detector out surface contamination measurements. This method not only avoids human contact with potentially radioactive hazardous locations, but also reduces the time that personnel are exposed to risky locations while dealing with surface contamination, and can reach inaccessible places to carry out measurements.

2. Surface Contamination Detector Design

The detectors used for surface contamination detection in the market are mainly hand-held, such as the CoMo 170 [8] surface contamination meter from Germany, which is unsuitable to be mounted on a robot. Therefore, in this paper, we design a detector suitable for mounting on a robot.

The surface contamination detector is mainly based on the STM32F103C8T6 as the core, the operating voltage is 24 volts, and the data detected on the surface of the detector is counted to characterize the severity of surface contamination. It mainly consists of STM32 and its peripheral circuits, multistage buck circuit, operational amplifier circuit, high voltage power supply module, and comparator circuit module [9].

2.1. Buck circuit

Since the detector needs to be mounted on the robot, the robot supplies 24 volts. The high-voltage power supply module operates at 24 volts, so the detector's operating voltage is also set to 24 volts. Since the STM32 operates at 3.3 volts, it also needs to supply 3.5 volts and 7 volts to the operator amplifier circuitry, so it is necessary to set up a buck module to enable the detector to operate normally.

The buck circuit to step down the detector operating voltage from 24 volts to 3.3 volts to make the STM32 work properly is shown in Figure 1.

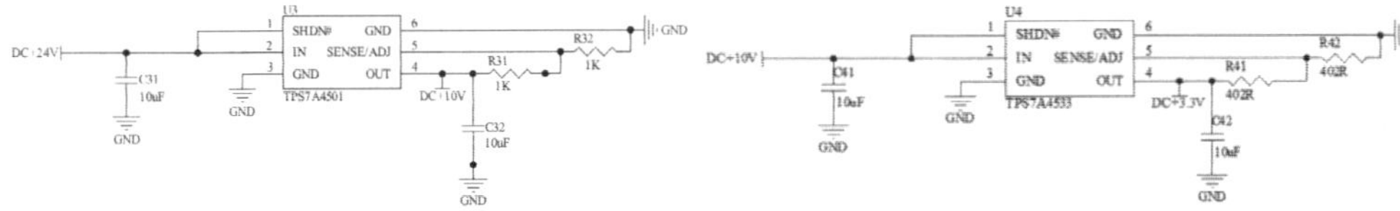

Figure 1. The circuit diagram of 24 volts stepped down twice to 3.3 volts.

The circuit diagram of the detector operating voltage of 24 volts stepped down to 7 volts versus 3.5 volts allowing the high voltage module, operational amplifier, and comparator circuits to operate properly is shown in Figure 2.

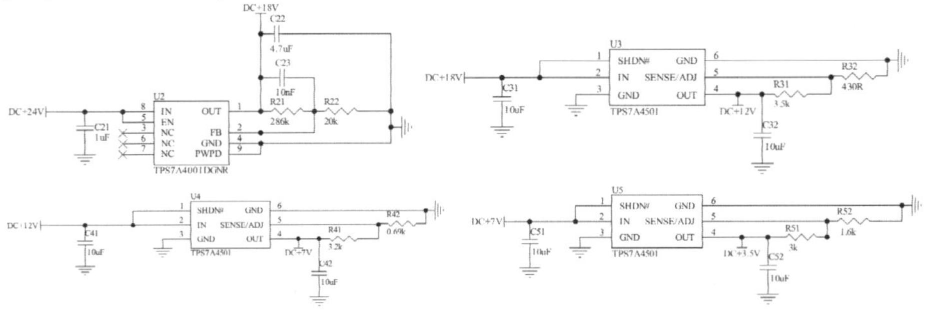

Figure 2. 24-volt to 7-volt with 3.3-volt circuit diagram.

The buck circuits all use multiple bucks during the bucking process, mainly to minimize the ripple effect on the circuit, improve the voltage quality, and ensure that the detector works better.

2.2. High voltage circuits

The primary function of the high-voltage module is to power the photomultiplier tube of the detector. When the rays emitted by surface contaminants such as α and β pass through the scintillator and emit photons, the photons pass through the photomultiplier tube, which converts the weak light signal into an electronic signal. The high-voltage module circuit is shown in Figure 3.

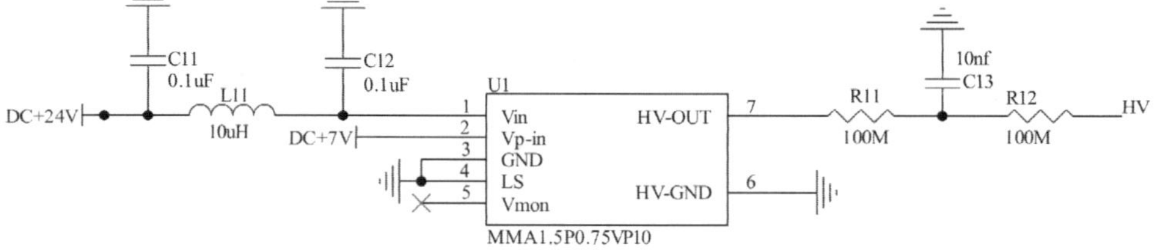

Figure 3. Circuit diagram of the high-voltage module.

2.3. Operational amplifier circuit

When the weak light signal through the photomultiplier tube produces electronic signals, the electronic signal through the ADA4817 voltage feedback amplifier will be converted from current signals to voltage signals. It can amplify the voltage signal to the level that can be processed by subsequent circuits. The circuit diagram is shown in Figure 4. Converted voltage signal after the AD8066ARZ chip output input differential voltage and amplifier gain product for comparator use [10].

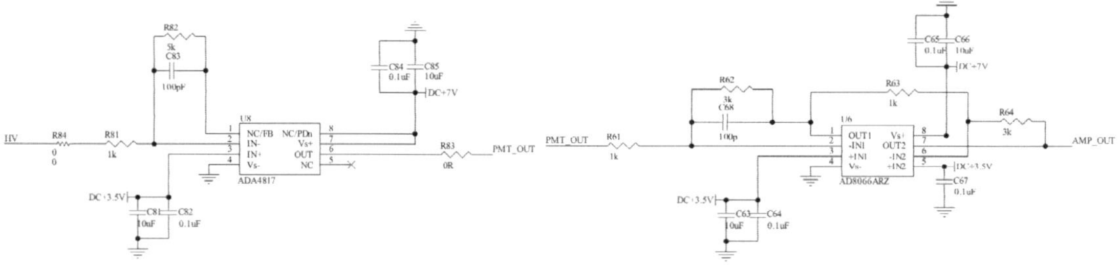

Figure 4. Operational amplifier module.

2.4. Comparator Circuit

This comparator compares the difference between the amplified signal and the preset threshold and controls the output voltage according to the comparison result. In this circuit, the output signal of the preamplifier is compared with the preset voltage to output a pulse signal, which is transmitted to the STM32 for counting. The circuit diagram is shown in Figure 5.

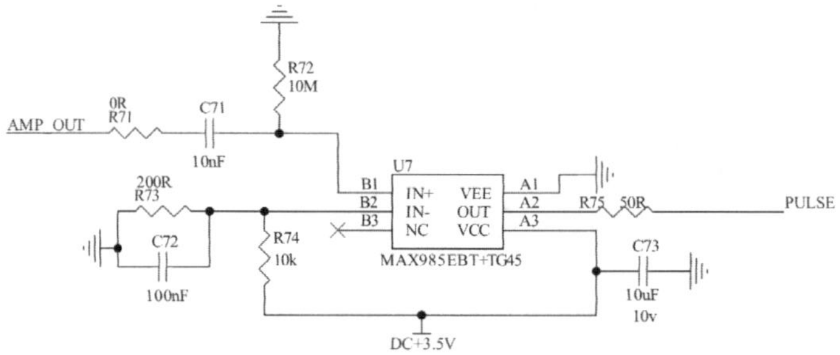

Figure 5. Comparison Amplifier Module.

3. Test and experimental results

The robot uses a mature robot chassis, the detector is mounted on the front of the robot, the detector probe faces the ground to ensure that it can detect surface contamination, and the robot is controlled by remote control. This combination of detector and robot can detect the degree of surface contamination in the forward direction of the robot. If there is contamination in the forward direction, you can use a remote control to avoid contamination of the robot's wheels, and at the same time to avoid surface contamination through the robot's wheels spread to the entire ground, expanding the scope of surface contamination.

The program of STM32 in this detector will count the number of pulses per second in cps, and the designed detector will be tested on α and β radioactive sources. The information on α and β radioactive sources is shown in Table 1 below. Since the nuclides have a half-life, ^{204}Tl has a half-life of 3.78 years and ^{239}Pu has a half-life of 24110 years, it is necessary to calculate the actual activity based on the half-life, the activity of ^{204}Tl varies a lot while ^{239}Pu remains unchanged, and the calculation of the number of particles emitted per minute is shown in Table 1.

Table 1 Information on α and β radioactive sources.

	^{204}Tl	^{239}Pu
Reference date	2015.8	2015.8
Cpm	9483	39300
Real cpm	2187	39300

In the same environment, the detector for radiation background, α and β, and other surface contamination of radioactive sources, the different radioactive sources will be under the detector for two minutes of counting. The results of the counts per second will be utilized in the moving average method and smoothed with a window of 30 data. The results of the experiment are shown in Figure 6.

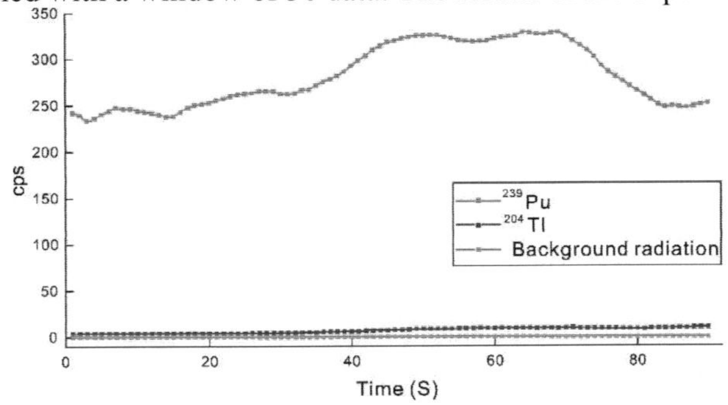

Figure 6. Detector counts per second.

Based on the experimental results, the detection efficiency of the detectors was calculated using the average counts per second of the detectors compared to the number of particles emitted per second from the sources. The radiation background was calculated to be 9%, the detection efficiency of the alpha source was 42.72%, and the detection efficiency of the beta source was 19.09%. The detector is mounted in front of the robot for detection. The data are transmitted to the robot through the serial port and then displayed on the computer using the remote desktop, which successfully realizes the detection of surface contamination by the robot. It has good detection efficiency.

4. Conclusion and outlook

The detector designed in this paper effectively detects the surface contamination of α and β radioactive sources. In remote-controlled robots using this detector, the detector is used for a fixed time for each position on the road to determine whether there is surface contamination in each position, which is proved to be feasible. However, there is the problem of a long time for detection, the robot is not easy to control, and the problem of detection efficiency is low. Next, work will be carried out to allow the robot to move autonomously to carry out the detection so the robot can work autonomously to complete the work.

References

[1] TANG, Y., WANG, X., WANG, Y., SHI, Y.J., ZHOU, Y. (2020) Research on Measurement Method of Small Area Surface Pollution. Nuclear Electronics & Detection Technology,40: 427-430.http://qikan.cqvip.com.http.yzn.proxy.chaoxing.com/Qikan/Search/Index?from=Qikan_Search_Index..

[2] LI, Y.Q., WEN, F.P., LU, Y. (2021) The study on field measurement technology of beta surface

contamination. Radiation Protection,41:20-25. http://qikan.cqvip.com.http.yzn.proxy.chaoxing.com/Qikan/Article/ReadIndex?id=7106048218&info=6JNLQ%2fTODDlXA8QIVAp9wnb3vVKWR0HTSc2MmkSjzI2CsfIYxvBUkg%3d%3d.

[3] LIU, Q.Y., XU, Z.J., MA, X.F., LIU, L. (2020) Study on Influencing Factors of Surface Contamination Measurement Accuracy. Nuclear Electronics & Detection Technology,40: 540-543.http://qikan.cqvip.com.http.yzn.proxy.chaoxing.com/Qikan/Article/ReadIndex?id=7104220816&info=jlbwTEX1BBwTayEjXgqe36blmD6GuShssgZXoBvlGkBqjp2dcBg%2bJA%3d%3d.

[4] LIU, C., ZHU, W.H., HU, Z.M., LUO, W.D. (2023) Rapid Measurement Method for Tritium Surface Contamination. SHANDONG CHEMICAL INDUSTRY,52:164-167. 10.19319/j.cnki.issn.1008-021x.2023.03.047.

[5] CHENG, X.M., LIU, Y.H., ZHAO, W.Z. (2023) Multi -robot coverage path planning for large 3D structure inspection. Computer Integrated Manufacturing Systems,29:246-253. 10.13196/j.cims.2023.01.021.

[6] HUANG, S., WU, Z.S., REN, Z.G., LIU, H.J., GUI, Y. (2020) ReView of electric power intelligent inspection robot. Electrical Measurement & Instrumentation,57:26-38. 10.19753/j.ISSN001-1390.2020.002.005.

[7] LI, S.X. (2017) Research on locating gas leakage source by robot detection. Experimental Technology and Management,34:46-48+54. 10.16791/j.cnki.sjg.2017.07.013.

[8] LI, Y.Q., WEN, Z. (2019) The Study on Measurement Factors of Beta Surface Contamination. Science & Technology Vision,88+92. 10.19694/j.cnki.issn2095-2457.2019.10.036.

[9] ZHANG, J., HOU, L., RAO, X.M., LIU, J.J., DU, X.Y., ZHANG, J., QIAO, M.J. (2020) The performance analysis of large area portable surface contamination survey meter.40:556-560.http://qikan.cqvip.com.http.yzn.proxy.chaoxing.com/Qikan/Article/ReadIndex?id=7103802255&info=JyCJtgxJZRYihryM9IrZxucYrYmdKiTlWg0NVJOJWoFmz%2f4gLMV9Jw%3d%3d.

[10] HU, T., SI, H.Y. (2010) Design and Research of Pre-amplifier Circuit from a Photoelectric Detector. ELECTRO – OPTIC TECHNOLOGY APPLICATION,25:52-55. http://qikan.cqvip.com.http.yzn.proxy.chaoxing.com/Qikan/Article/ReadIndex?id=33073875&info=GuaklyWva9WCwCw6%2fqa1D9NHX5wqHnK24kJBGeixkB0%3d.

A method of liquid refractive index measurement based on image correlation coefficient

Xinxin Xie [1], Junyi Chen [1], Mengna Zhou [1], Junjiang Xu [1], Zijian Zheng [1], Jiayao Yang [1], Jinming Luo [1, 2] and Yuancong Zhong [1]

[1]School of Physics and Electronic Engineering, Jiaying University, Meizhou, China

[2]Corresponding author's e-mail: zsuljm@163.com

Abstract. According to the difference of speckle image caused by the axial displacement due to the refraction of a transparent medium, the liquid refractive index is measured by using the image correlation coefficient. An optical measurement system based on laser speckle is built into the experiment. Firstly, the CCD camera is adjusted to move along the optical axis in air, and multiple speckle images at different positions are collected, which are correlated with the speckle image at zero position to obtain the image correlation coefficients, and then the exponential calibration relation between the correlation coefficient and the axial displacement can be obtained. Secondly, two speckle images are collected when the laser speckle spreads respectively through the liquid to be measured and water as the internal standard, and the correlation coefficient between them is calculated. Finally, according to the above calibration relation, the corresponding axial displacement between the liquid to be measured and water can be derived, and then the refractive index of the liquid to be measured is calculated using the axial displacement formula. The results show that this method can realize the measurement of liquid refractive index, and the optical structure is simple and easy to operate.

1. Introduction

The refractive index is a main parameter reflecting the characteristics of liquids, which has become more important in the fields of food hygiene and industrial production. Thus, simple and high-precision measurement methods of liquid refractive index are required for the liquid material analysis and also for its applications. At present, there are many optical methods for the measurement of liquid refractive index, including digital hologram method [1], Snell's law [2], grating diffraction method [3,4], Michelson interferometer method [5], and so on.

A laser speckle is a kind of irregularly distributed bright and dark image formed by the interference of scattered light when the laser irradiates the diffuser surface, which is widely used in surface roughness research, optical image processing, imaging quality evaluation, micro deformation measurement, and other fields [6-8]. For the measurement of liquid refractive index, one of the common methods using laser speckle is the internal standard method [9], which has been developed [10] recently. In this paper, laser speckle is used to measure the refractive index of liquid through image correlation coefficient. According to the difference in speckle image caused by the axial displacement of the image due to the refraction of the medium, an optical system for measuring the liquid refractive index based on the correlation coefficient of the speckle image was designed. Firstly, the relationship of the image correlation coefficient and axial displacement was calibrated, and then through measuring the image correlation coefficient between the water and the liquid to be measured, the axial displacement between them was obtained from the calibration relationship. Finally, the liquid refractive index can be calculated theoretically.

2. Principle

As shown in Figure 1, when a laser speckle irradiates a vertically placed transparent medium, refraction will occur. From geometrical optics, it can be seen that the speckle images before and after refraction will have an axial displacement in the optical axis direction, and the value of axial displacement z_0 relative to the air (refractive index of 1) can be expressed as

$$z_0 = D\left(1 - \frac{1}{n}\right) \tag{1}$$

where n is the refractive index of the transparent medium, and D is the thickness of the transparent medium.

In this paper, the transparent medium is a liquid, and water is used as the internal standard liquid. Then, the value of axial displacement z relative to the water can be deduced as

$$z = D\left(\frac{1}{n_w} - \frac{1}{n}\right) \tag{2}$$

where n_w is the refractive index of water and the reference value is 1.3330.

According to the axial displacement Formula (2), the difference in refractive index between the liquid to be measured and water will lead to different axial displacements, which cause the different speckle images collected by the CCD camera at the fixed position. Therefore, according to this image difference, the liquid refractive index can be measured by using the image correlation coefficient.

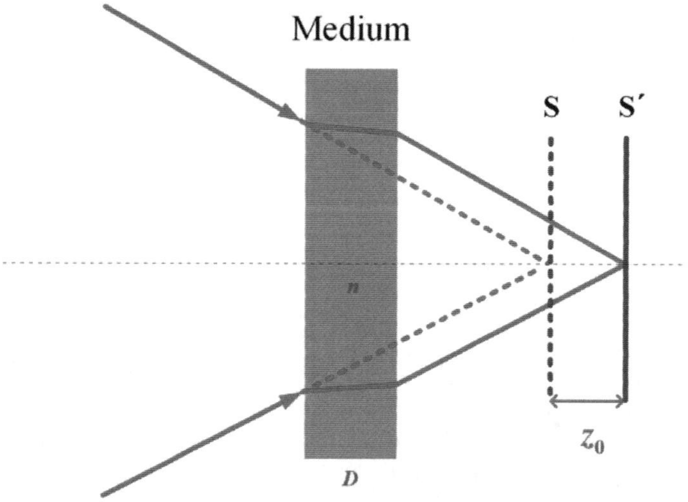

Figure 1. Refractive characteristic of a transparent medium.

3. Experimental

Figure 2 shows the experimental setup for the refractive index measurement of transparent liquid. The light beam emitted from a He-Ne laser source is expanded by an extender lens (focal length of 4.5 cm) and collimated by a collimator lens (focal length of 100 mm) to the plane wave and transmits through a ground glass diffuser to form a speckle image. Then, a rectangular parallelepiped glass cell (cross-section area is 50mm×50mm, two types of cell width is D=5 mm and 10 mm) containing liquid is vertically placed at a fixed position between the ground glass diffuser and a CCD camera (MER-500-14U3M-L, number of pixels is 2592×1944 and pixel size is 2.2 μm×2.2 μm). Finally, the speckle images (e.g. liquids using water and solution) are collected by the CCD camera and the image correlation coefficient between them can be calculated by MATLAB.

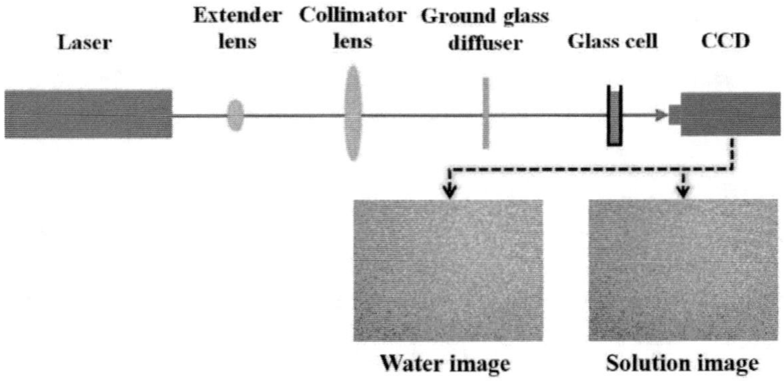

Figure 2. Schematic of the experimental setup of liquid refractive index measurement.

4. Results and Discussion

In this paper, the refractive indices of liquids are measured by the image correlation coefficient method, and the results are compared with those derived by the Abbe refractometer. Sodium chloride solution and sucrose solution with different mass percent concentrations C are used as the liquids to be measured.

4.1. Calibration relationship

According to measurement principle, the correlation coefficient of speckle images is related to the axial displacement. To obtain this relationship, the CCD camera was adjusted to move along the optical axis every 30 μm in air, and multiple speckle images at different positions were collected, which were correlated with the speckle image at zero position to obtain the image correlation coefficients. Then the calibration curve of the correlation coefficient r changing with the value of axial displacement z is obtained in Figure 3, which can be fitted with an exponential function for simplicity.

$$r = \exp(-0.00828z) \tag{3}$$

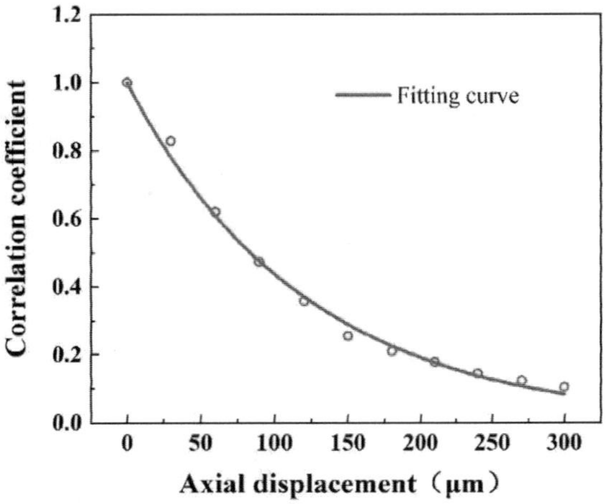

Figure 3. Calibration curve of image correlation coefficient and axial displacement.

As long as the correlation coefficient value of speckle images collected from water and the liquid to be measured is calculated, the corresponding axial displacement between them can be obtained through Formula (3), and then the liquid refractive index can be calculated according to the axial displacement Formula (2). However, when the axial displacement exceeds 300 μm, the correlation coefficient will be less than 0.1, that is, the two speckle images are almost uncorrelated, which can increase the calculation error. In addition, the calibration curve can be influenced by the structure parameters of the optical system. If the structure parameters are changed, the calibration curve should be measured again.

4.2. Refractive index measurement

4.2.1. Sodium chloride solution. Placing the glass cell vertically between the ground glass diffuser and the CCD camera, two speckle images are collected when the laser speckle spreads through sodium chloride solution with different concentrations and water respectively, and the correlation coefficient between them can be measured. Then, the corresponding axial displacement is obtained according to Formula (3), and the refractive index of the sodium chloride solution is calculated through Formula (2), which is compared with the measured value of the Abbe refractometer. To facilitate discussion, the refractive index and concentration of sodium chloride solution are linearly fitted shown in Figure 4.

From Table 1 and Figure 4, it can be seen that the refractive indices of sodium chloride solution with different concentrations measured by this method are close to the measured values of the Abbe refractometer, and the relative errors are all within 1%. With the increase of solution concentration, the refractive index of the liquid increases, and the corresponding axial displacement also becomes large, leading to the increasing difference of speckle image collected by the CCD camera, as a result, reducing the correlation coefficient. In addition, according to the measurement principle, the axial displacement is related to the cell width. For the solution with the same concentration, the larger the cell width is, the greater the axial displacement is, and the smaller the correlation coefficient is.

Table 1. Refractive indices of sodium chloride solution with different concentrations.

D (mm)	C (%)	Correlation coefficient	Refractive index		Relative error (%)
			Calculation	Abbe refractometer	
5	5	0.8951	1.3378	1.3415	0.28
	10	0.8295	1.3411	1.3505	0.7
	15	0.7142	1.3476	1.3595	0.88
	20	0.5177	1.3619	1.3685	0.48
	25	0.4633	1.3669	1.3767	0.71
10	5	0.536	1.3465	1.3415	0.37
	10	0.3611	1.3552	1.3505	0.35
	15	0.2842	1.3606	1.3595	0.08
	20	0.1959	1.3689	1.3685	0.03
	25	0.1357	1.3773	1.3767	0.04

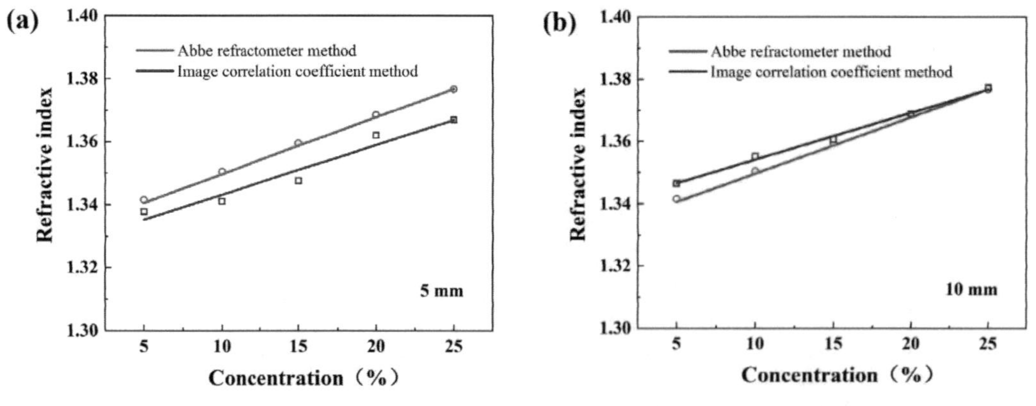

Figure 4. Fitting curves of refractive index and concentration of sodium chloride solution. Cell width: (a) 5mm; (b) 10mm.

 2724 (2024) 012030 doi:10.1088/1742-6596/2724/1/012030

4.2.2. Sucrose solution. Selecting sucrose solution with different concentrations, the liquid refractive index is also measured using the same method. The measurement data are listed in Table 2, and the linear fitting curves of the refractive index and concentration of sucrose solution are shown in Figure 5.

Similar to sodium chloride solution, the measurement results of the refractive index of sucrose solution are also close to the measurement values of the Abbe refractometer, and the relative errors are almost within 1% at all concentrations except that the concentrations exceed 40% for the cell width of 10 mm. The reason is that the axial displacements for these concentrations are too large, and the correlation coefficient values of sucrose solution are less than 0.1 accordingly, leading to a decrease in measurement accuracy. Therefore, an appropriate cell width for measurement can further reduce the error.

Table 2. Refractive indices of sucrose solution with different concentrations.

D (mm)	C (%)	Correlation coefficient	Refractive index		Relative error (%)
			Calculation	Abbe refractometer	
5	10	0.7236	1.347	1.35	0.22
	20	0.524	1.3614	1.3635	0.15
	30	0.2911	1.3882	1.3837	0.33
	40	0.2176	1.4018	1.403	0.09
	50	0.121	1.4303	1.4228	0.52
10	10	0.3384	1.3567	1.35	0.5
	20	0.2055	1.3678	1.3635	0.32
	30	0.1235	1.3794	1.3837	0.31
	40	0.0896	1.3869	1.403	1.15
	50	0.0553	1.3982	1.4228	1.73

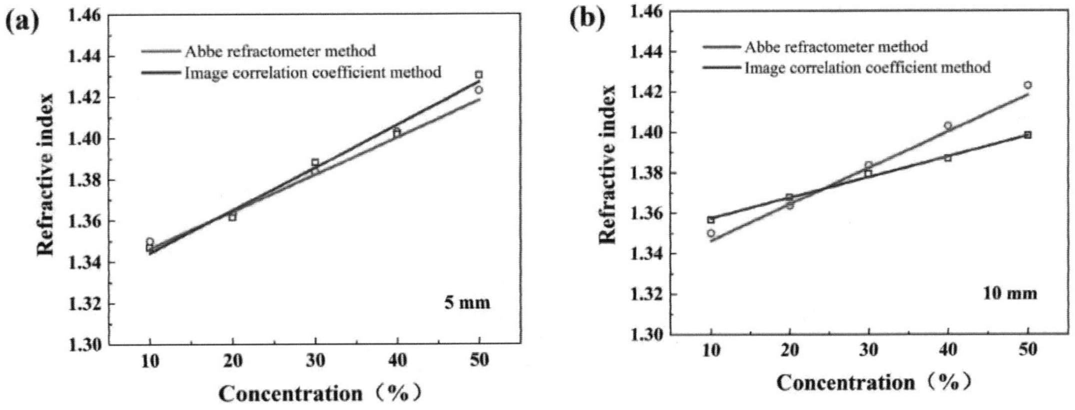

Figure 5. Fitting curves of refractive index and concentration of sucrose solution. Cell width: (a) 5mm; (b) 10mm.

5. Conclusion

In this paper, the refractive indices of sodium chloride and sucrose solution were measured by the image correlation coefficient method. First, the calibration relationship curve fitted with the exponential function between the correlation coefficient and the axial displacement is measured in air. Then, the speckle image correlation coefficient value of water and the liquid to be measured (sodium chloride and sucrose solution) is calculated, and the corresponding axial displacement is obtained according to the above calibration relationship. Finally, the refractive index of the liquid can be derived from the axial displacement formula, which is close to the measured value of the Abbe refractometer. Thus it can be seen that the image correlation coefficient method can be used to measure the refractive index of liquid, and the optical system is simple and easy to operate.

Acknowledgments
The authors would like to acknowledge the financial support from the Educational Science Planning Project of Guangdong Province (2023GXJK492), the Teaching Quality and Teaching Reform Engineering Projects of Jiaying University (JYJG202311O), and the Provincial Key Platform and Scientific Research Projects of Universities in Guangdong Province (2022KTSCX133).

References
[1] Huang M, Sun C, Qin H, Jiang Z 2023 Measurement of liquid refractive index by quantitative phase reconstruction of single frame dual-wavelength digital hologram[J] *Measurement* **206** 112325.
[2] Rivera-ortega U, Hernández-gómez C R, Vega-torres G, Lopez-Medina M E 2019 Simple apparatus to calculate the refractive index of liquids based on Snell's law[J] *Measurement* **134** 658-661.
[3] Zawisza R, Eftimov T, Mikulic P, Chinifooroshan Y, Celebańska A, Bock W J, Jaroszewicz L R 2018 Dual-resonance long-period grating in fiber loop mirror structure for liquid refractive index measurement[J] *Opto-Electron. Rev.* **26** 24-28.
[4] Jiang M, Wang Z M, Zhao Z Z, Li K, Yang F 2017 Long-period fiber grating cascaded to thin-core fiber for simultaneous measurement of liquid refractive-index and temperature[J] *Sensor Rev.* **38** 79-83.
[5] Abbas B, Alshikh Khalil M 2016 An experimental method for determination of the refractive index of liquid samples using michelson interferometer[J] *Acta Phys. Pol. A* **129** 59-63.
[6] Kim S, Kim E, Anguluan E and Kim J G 2022 Sample entropy analysis of laser speckle fluctuations to suppress motion artifact on blood flow monitoring[J] *Chin. Opt. Lett.* **20** 011702.
[7] Li Z, Zhong P, Chen Y, Tang X, Gao Y, Hu H 2021 Simultaneous measurement of three-dimensional deformation based on digital speckle pattern interferometry technology[J] *Opt. Commun.* **480** 126423.
[8] Zhao T, Zhang W, Gu S, Lv Y, Li Z 2020 Study on fracture mechanics of granite based on digital speckle correlation method[J] *Int. J. Solids Struct.* **193-194** 192-199.
[9] Docchio F, Corini S, Perini M, Gasana R S 1995 A simple and reliable system for measuring the refractive index of liquids using a position-sensitive detector[J] *IEEE T. Instrum. Meas.* **44** 68-70.
[10] Luo J, Li L, Yang G, Chen M, Weng L, Yang J, Zhong Y, Deng D 2023 Refractive index measurement of uniform liquid using an improved laser speckle method[J] *Opt. Eng.* **62** 044107.

Journal of Physics: Conference Series **2724** (2024) 012031 doi:10.1088/1742-6596/2724/1/012031

Temperature error compensation method of fiber optic gyroscope based on deep learning

Junjie Wu[1] and Lei Wang[1,2]

[1]School of Electronics and Information Engineering, Tongji University, China

[2]Corresponding author's e-mail: leiwang@tongji.edu.cn

Abstract. This work provides an error compensation strategy based on deep learning to address the temperature error of a fiber optic gyroscope (FOG). The Attention structure was used to improve the Long Short-Term Memory Network (LSTM), and the improved network was utilized to establish the prediction and compensation model of FOG error. From the two perspectives of the learning ability of the neural network and the improvement of FOG performance, this paper selects three indicators: prediction accuracy, mean square error, and FOG bias stability to comprehensively evaluate the performance of the compensation model. Through comparison, the compensation effect is significantly enhanced.

1. Introduction

A fiber optic gyroscope(FOG) is an inertial instrument without moving parts or worn components.[1] Due to its low weight, small size, quick start, high long-term stability, and other benefits, it is also widely used in the national military, commercial, civic, and other industries[2,3]. FOG has some common characteristics of traditional inertial instruments, as well as unique errors and characteristics. It is a kind of optical, mechanical, and electrical integration of the inertial instrument, the light signal from the light source, after hundreds of meters to thousands of meters of light path propagation after interference, detection by the detector, and then through the circuit modulation and demodulation, to get its sensitive angular rate signal. The phase difference detection accuracy of $10^{-3}(°)/h(1\delta)$-magnitude high-precision fiber optic gyroscope needs to reach more than 10^{-7} rad. Therefore, any factor at any position in the whole optical path has a weak influence on the nonreciprocal effect of two light waves propagating towards each other and the detection of circuit signals, and this influence will lead to errors in the fiber optic gyroscope.

There are two main technical difficulties in the error compensation technology of fiber optic gyroscope. First, it is difficult to establish the error estimation model accurately, that is, it is impossible to accurately determine the specific correlation quantities leading to errors. Second, the relation between error and correlation quantity cannot be accurately described. As a result, in order to address the issue of temperature drift error compensation in fiber optic gyroscopes, researchers at home and abroad mainly carried out in-depth research on the temperature drift error estimation model and its realization algorithm. The two categories of software compensation techniques now in use are classical fitting and machine learning prediction. Regression analysis and the least squares approach are traditionally used to create a linear model, and the error is then corrected using the difference method[4-5]. However, the conventional fitting techniques have a limited ability to generalize and have low application in the face of the complex changing environment, which limits the accuracy of

FOG after correction. By building intricate network architectures and utilizing suitable training and testing techniques, deep learning methods are able to fully learn the nonlinear properties of data, in contrast to typical fitting methods. For fiber optic gyroscope error compensation, Support Vector Machine (SVM)[6], back propagation (BP) neural network[7], and RBF (Radial Basis Function) neural network[8,9] have all been employed. These three models do, however, have certain limitations in terms of generality and compensation accuracy. This research trained the approach of merging the LSTM network with Attention and performed differential compensation in order to enhance the accuracy of the fiber optic gyro and the generalization ability of the model. The majority of the noise and error in the FOG output data were better corrected under the limited increase in the number of network layers.

2. Principle

The enhanced LSTM network is employed in this paper to compensate for the Shupe effect error of the fiber optic gyroscope. This section will introduce the temperature-affected fiber optic gyro principle and the algorithm principle in this work.

2.1. Shupe effect error principle

Under ideal circumstances, two coherent beams propagate counterclockwise in the optical fiber ring as the fiber optic gyroscope revolves around the normal line of the corresponding plane. When the two beams collide, the theory of relativity states that there is a phase difference. Sagnac came to the conclusion that its size is proportional to the device's rotational angular velocity with respect to inertial space.

The output interference light intensity and angular rate of FOG optical detector are:

$$I = \frac{I_0}{2}(1 + cos\Delta\varphi) \tag{1}$$

$$\Omega = \frac{\lambda c}{2\pi LD}\Delta\varphi \tag{2}$$

Where I_0 is the maximum light intensity, φ is the phase difference generated by the two beams of light propagating clockwise through the optical fiber ring, λ is the wavelength in vacuum, and c is the speed of light, D is the diameter of the optical fiber ring, and L indicates the total length of the optical fiber winding. The control circuit, optical fiber, detector, and mechanical mechanism of the fiber optic gyroscope are easily influenced by temperature, resulting in a change in the fiber optic gyroscope's performance index. The optical fiber is most affected by temperature. Because there are spatial temperature gradients in different directions of the optical fiber, the refractive index at different temperature gradients is different, and the time points of the two beams of light waves passing through are different, resulting in different phases of the light waves passing through. This effect is known as the Shupe effect [10]. Due to the Shupe effect, the phase difference between the two beams of light produces a non-reciprocity error,which satisfies:

$$\Delta\varphi(t) = \frac{2\pi}{\lambda}\int_0^L \frac{dn}{dT}\frac{dT}{dt}(z)(L - 2z)dz \tag{3}$$

Where n is the refractive index of the optical fiber, T is the real-time temperature of the optical fiber ring, and z is the distance from a certain position on the optical fiber ring to the beam splitter. The formula shows that the influence of temperature on the gyroscope's output is unaffected by the input angular rate, hence our model of zero bias stability can also account for the fiber optic gyroscope's correction for changes in angular speed.

2.2. Deep learning algorithm

Since its inception, the LSTM network has produced promising results in resolving the gradient explosion problem in RNN[11].The LSTM structure is shown in Figure 1.However, in this experiment, due to the slow temperature change, the temperature sensor collects nearly 20,000 pulse points for every one degree Celsius change. LSTM has a poor effect in solving such a long sequence. This article proposes an optimized network structure that uses Attention in parallel computing to solve the prediction problem of long sequences.

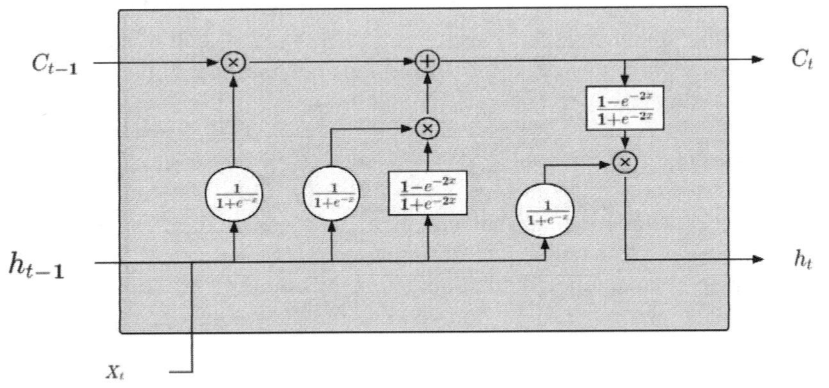

Figure 1. Structure of the LSTM neural network.

Figure 2 shows the method based on the combination of LSTM and Attention structure, which can obtain more nonlinear information in the model by encoding the similarity score of LSTM output. A residual module is added to the model to prevent gradient explosion and make the network easier to train.

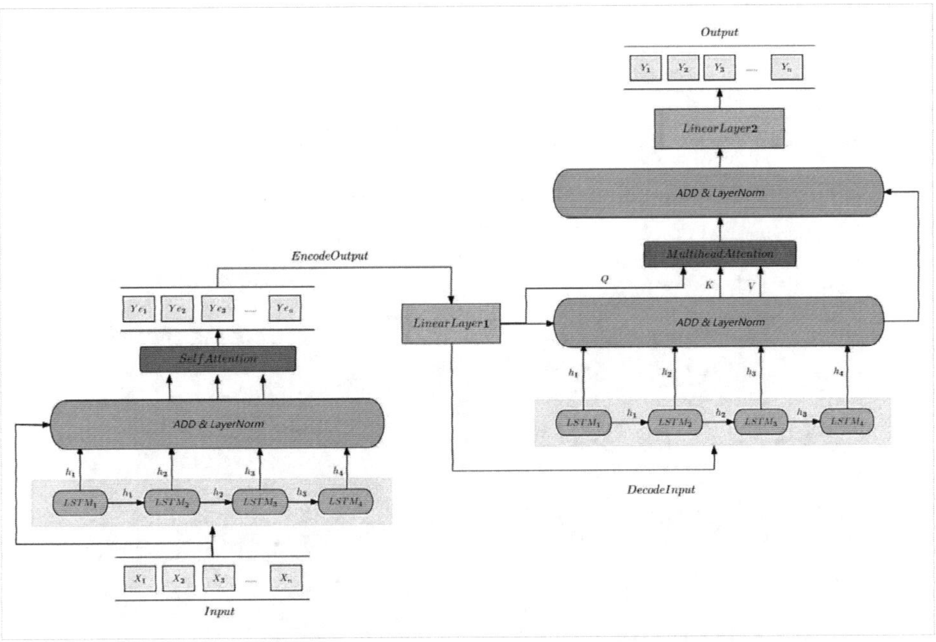

Figure 2. Deep learning network structure.

Attention is a method proposed by Dzmitry et al. [12] to determine the input-output relationship in translation projects. Attention calculates the similarity degree of the query tensor Q and the key tensor K matrix through softmax, and then adds the weighted sum with the value tensor V to obtain a new tensor instead of V to represent some relation between Q,K and V.

$$\text{Attention}(Q, K, V) = \text{softmax}(\frac{QK^T}{\sqrt{d_k}})V \quad (4)$$

The input $x_i(t)(i = 1,2...n)$ is a preprocessed FOG output and temperature sequence . The output $Y_j(j = 1,2...n)$from the decoder is subjected to differential compensation by subtracting it from the unprocessed fiber optic gyroscope data to obtain the final output. In the encoding process, we take the hidden states at each time step of the LSTM network, normalize them along with the input, and use them as input to the Self-Attention mechanism. In the decoding process, we use the decoder's input as the query tensor Q, normalize the LSTM outputs along with the encoder's outputs, and use the resulting values as the key (K) and value (V) tensors, which are then input into the Multihead Attention module.

In this paper, a 4-head attention mechanism was employed. To prevent overfitting, a linear learning rate decay was applied, starting with an initial learning rate of 0.005, which was reduced to 1/20 of its original value after 30 epochs. An optimizer was configured during the training process, and training was terminated when the Mean Squared Error (MSE) value remained unchanged for three consecutive epochs.

3. Data acquisition

3.1. Experimental equipment

The experimental setup for this study, as illustrated in Figure 3, consists of the following components: a Fiber Optic Gyroscope (FOG) with an embedded temperature sensor, a programmable thermostat, a data acquisition board, a computer, and a horizontal workbench. The specific optical components of the FOG include an super luminescent diode (SLD), a Panda-type polarization-maintaining fiber, and a PIN-FET, with the FET being an 8066 operational amplifier. The bandwidth of the fiber optic gyroscope is 796 kHz, and it has a scale factor of 1210 w.

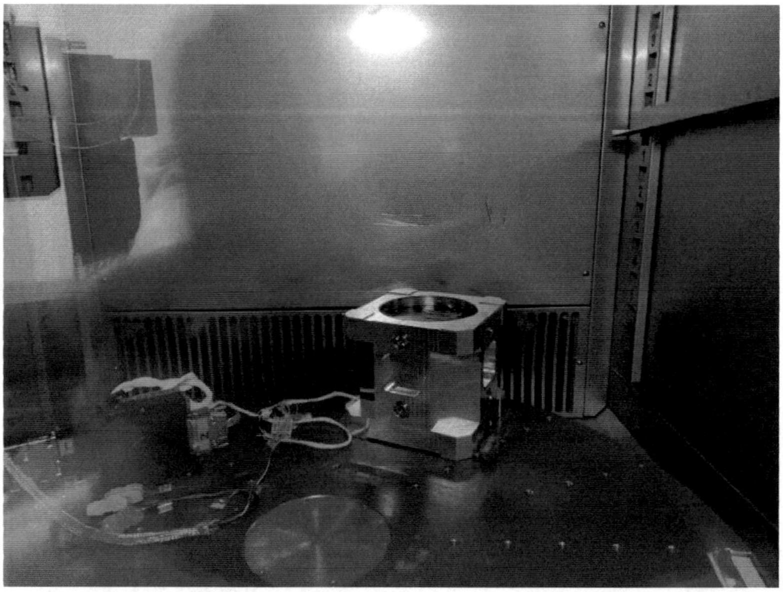

Figure 3. Experimental device.

The experimental temperature was controlled using a programmable thermostat, which involved a temperature profile consisting of a heating phase, a holding phase, a cooling phase, and another holding phase. The temperature inside the thermostat was varied within the range of -40°C to 60°C. In order to establish a detailed relationship between the output of the FOG and the temperature variations, three sets of temperature cycling experiments were conducted, totaling 28 hours of experimentation.

3.2. Data preprocessing

During the experimental process, data were collected at over 100,000 time points. Figure 4 displays the output data from the FOG alongside the temperature variation curve. The FOG output data contains numerous instances of noise and outliers. If left untreated, these issues can result in an overly complex neural network and poor generalization. Therefore, prior to training the network, data preprocessing is necessary.

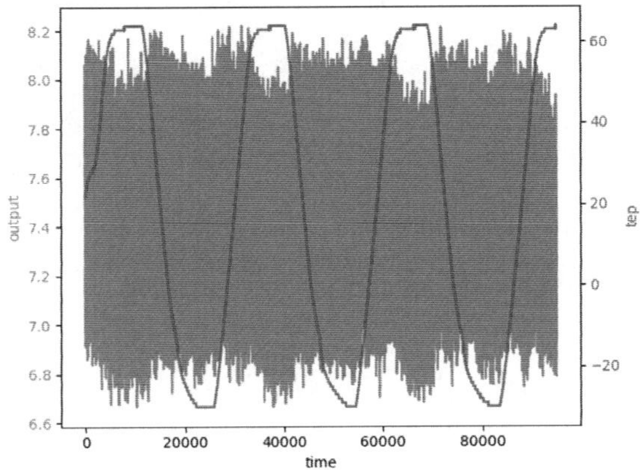

Figure 4. Experimental raw data.

The analysis of time series data necessitates outlier processing and the evaluation of data stationarity. For outlier handling, this study applied the widely utilized 3σ rule. Data stationarity was assessed through the Augmented Dickey-Fuller (ADF) test method, implemented using the "statsmodels. tsa.stattools. adfuller" function in Python. The ADF test as Figure 5 yielded a statistic of -7.3162, which is lower than the critical value of -3.4304 at the 1% significance level. Consequently, the null hypothesis can be confidently rejected, signifying the significant stationarity of the FOG output results.

ADF detection result: (-7.31616976743209, 1.2276793946907129e-10, 67, 94672, {'1%': -3.430419075094175, '5%': -2.8615705300904946, '10%': -2.5667862501021514}, -78139.61923363822)

Figure 5. The ADF test result.

In order to capture the long-term correlation between temperature and FOG outputs, this study treats $\{X_i, X_{i+1}, X_{i+2}, X_{i+3}, X_{i+4}\}$ as a sequence. It predicts the next output point X_{i+5} while incorporating temperature variations as features into the prediction model, as depicted in Figure 6. To meet the requirements of deep learning, the dataset was divided into an 80% training set and a 20% validation set. Finally, a new set of temperature cycling experiments was conducted on the FOG, serving as a test set to evaluate the model's generalization capacity.

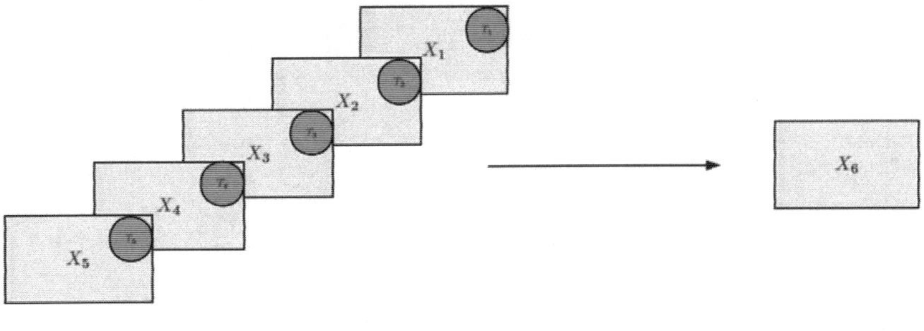

Figure 6. Prediction process.

4. Results discussion

4.1. Performance index

4.1.1. MSE. Mean square error (MSE) is a measure that reflects the degree of difference between an estimated value and the actual value being estimated. In the context of neural networks, a smaller MSE indicates a better predictive ability. Equation (4.1) represents the calculation method for MSE:

$$\mathrm{Mseloss} = \frac{1}{N}\sum_{i=0}^{N-1}(p_i - y_i)^2 \tag{5}$$

Where N represents the total number of data points, p represents the network's predicted results, and y represents the output results from the Fiber Optic Gyroscope.

4.1.2. Zero-bias stability. Zero bias stability is one of the crucial metrics for assessing the accuracy of the FOG. It quantifies the dispersion of FOG output values around their mean when the angular rate is zero. A lower zero-bias stability indicates higher FOG accuracy. The calculation method for zero bias stability is as follows:

$$\mathrm{BiasS} = \sqrt{\frac{1}{K}\sum_{i=0}^{K-1}(X_i - \overline{X})^2} \tag{6}$$

Where K represents the total number of data points after 100s smoothing, and X represents the FOG output smoothed over 100 seconds. After computation, the zero bias stability of the FOG before compensation, under 100s smoothing, is determined to be 0.057635315 °/h.

4.2. Analysis of experimental results

Figure 7 displays the results obtained by replicating the traditional LSTM method [13] under the same preprocessing steps for the same training set, test set, and validation set. This was conducted in five parallel experiments, and the average results were obtained to generate the predictive compensation results. Figure 8 illustrates the predictive compensation results obtained using the method proposed in this paper. In this context, the green line represents the original FOG output curve, the red line represents the model's prediction curve, and the blue line represents the FOG output curve after compensation.

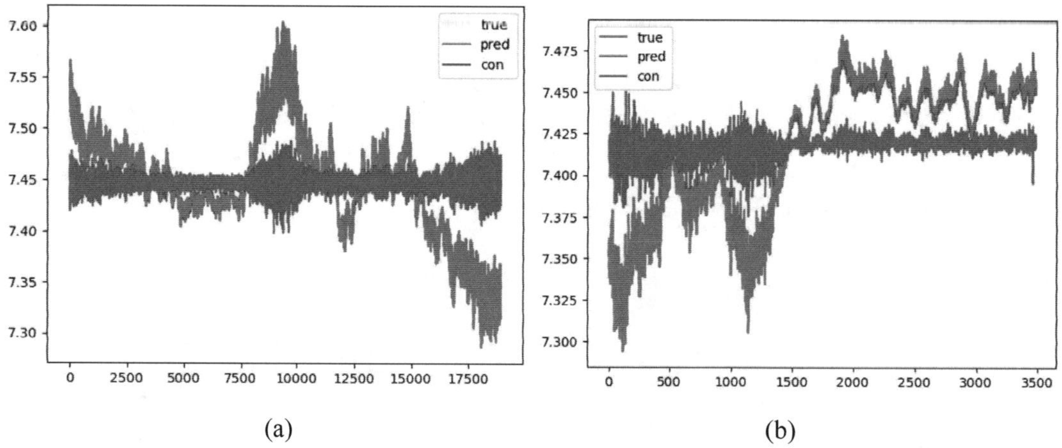

Figure 7. Traditional LSTM network experimental results. (a)Validation set; (b)Test set.

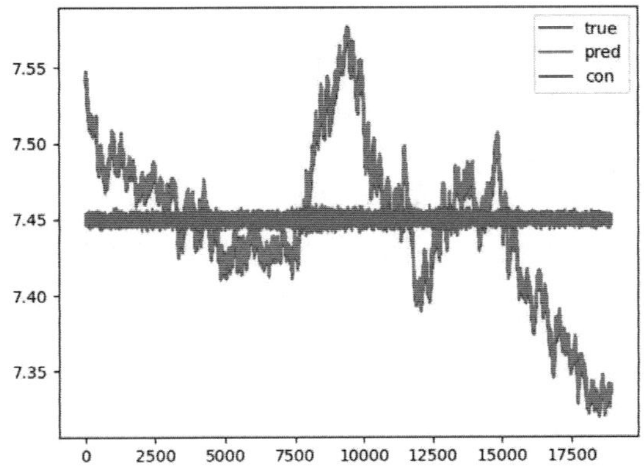

(a)

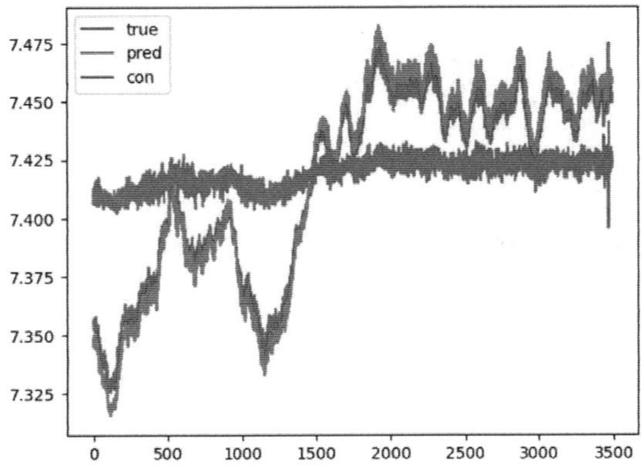

(b)

Figure 8. Att-LSTM network experimental results. (a)Validation set; (b)Test set.

As evident from the comparison in the figures, it is clear that the improved LSTM model's prediction results are more stable. In addition to the comparison with the traditional LSTM method, this paper also compares the results with the more commonly used BP neural network in fiber optic gyroscope compensation. The comparative results are as Table 1:

Table 1. Comparison of Results Across Various Neural Networks.

Network	MSE	Test set MSE	Bias stability (°/h)	Test set Bias stability (°/h)
BP	1.812e-4	8.744e-5	0.013	0.009
LSTM	5.778e-5	8.658e-5	0.007	0.008
Att-LSTM	4.733e-5	5.775e-5	0.002	0.006

From a model prediction perspective, models based on LSTM significantly outperform BP models, and the improved LSTM network exhibits stronger generalization capabilities on the test set compared to the traditional LSTM network.From the perspective of zero bias stability analysis, the zero bias stability of the FOG without compensation is 25 times higher than that after compensation using the method proposed in this paper. This substantial improvement greatly enhances the accuracy of the FOG.

5. Conclusion

This paper introduces a temperature compensation method for FOG based on an Attention-LSTM network. During the prediction process, temperature is incorporated as a feature to aid in predictions. Comparative analyses, when contrasted with existing temperature compensation methods, demonstrate the superiority of this approach in terms of prediction accuracy, compensation precision, and generalization capability. The method proposed in this paper underscores that by leveraging deep learning techniques, it is possible to significantly enhance the performance of mid- to low-precision fiber optic gyroscopes, indirectly reducing the cost of FOG in medium- to high-precision scenarios, and enabling their application in a wider range of fields. Furthermore, in stable environments where dynamic compensation for FOG errors is required, this method can serve as an important reference solution.

Acknowledgments
This work is supported by the Sino-German Intergovernmental Program for International Cooperation. The project number is 2018YFE0105000.

References

[1] Yang H, Qiao L, Yang Y, Huang W and Sun S, Thermally induced error analysis and suppression of optic fiber delay loop in the different variable rate of temperature[J], *Optik* 193, 162994 (2019).

[2] Liu H, Luo W and Lu J, High precision fiber optic gyroscope resolution test method based on low precision turntable[J], *IEEE Sensors Journal*, 20, 8656-8662 (2020).

[3] Dai M, Zhang C, Pan X, Yang Y and Li Z, A novel attitude measurement while drilling system based on single-axis fiber optic gyroscope[J], *IEEE Trans. Instrum. Meas.* 71, 9500311 (2022).

[4] Zhuo C, Du J, Tang H and Liu Q, Special thermal compensation experiment and algorithm design for inertial navigation system[C], In DGON Inertial Sensors and Systems (ISS) (2019),pp. 1-16.

[5] Li X, Zhang C, He Z and Zhong Z, Temperature errors of IFOG and its compensation in engineering application[C], In 9th International Conference on Electronic Measurement & Instruments (2009), pp. 2-230-2-234.

[6] Guo W, Geng L and Long X. Application of Least Squares-Support Vector Machine in system-level temperature compensation offing laser gyroscope[J]. *Measurement*, 2011, 44:1898-1903.

[7] Zhao D, Chen J, Han Y, Song C and Liu Z. Temperature compensation of FOG scale factor based on CPSO-BPNN[C]. In 2010 Chinese Control and Decision Conference,2010: 2898—2901.

[8] Han X and Fang J. Temperature compensation method of FOG-IMU based on RBF neural network[C]. In 2012 IEEE International Conference on Oxide Materials for Electronic Engineering (OMEE):671-674.

[9] Zhu R, Zhang Y and Bao Q. A novel intelligent strategy for improving measurement precision of FOG[J]. *IEEE Transactions On Instrumentation And Measurement,* 2000,12,49(6):1183-1188.

[10] D Smirnov, I Deyneka, A Kulikov, V Strigalev and I Meshkovsky, Methods for studying temperature characteristics of a FOG sensing coil[C], In 28th Saint Petersburg International Conference on Integrated Navigation Systems (ICINS) (2021).

[11] S Hochreiter, J Schmidhuber,Long short-term memory[J], *Neural Computation,*1997,9(8):1735-1780.

[12] D Bahdanau, K Cho and Y Bengio, Neural machine translation by jointly learning to align and translate[J], *Computer ence,* (2014.09), 10.48550/arXiv.1409.0473.

[13] Yin C *et al.* A method for temperature error compensation in fiber-optic gyroscope based on machine learning[J], *Optic*, 256, April 2022, 168765.

Geometric calibration of the linear-array whiskbroom optical satellites based on look-angle corrections

Yuanyuan Sun[1,2]

[1]School of Computer Science, Hubei University of Technology, Wuhan, China

[2]Corresponding author's e-mail: 102111120@hbut.edu.cn

Abstract. Linear-array whiskbroom optical satellites are equipped with scanning mirrors and compensation mirrors. They rotate and scan imaging with the satellite flight orbit as the rotation axis, and geometrically splice the scanned images along the orbit, achieving a balance between large width and high resolution of remote sensing images. The addition of scanning mirrors and compensation mirrors necessitates multiple reflections for the imaging ray to reach the focal plane. Hence, the calibration of the geometric aspects for linear-array whiskbroom optical satellites will be an intricate procedure. A geometric calibration model based on look-angle corrections is established according to the imaging characteristics of linear-array whiskbroom optical satellites. All errors are corrected using correction quantities, and cubic polynomial surfaces are used to fit correction quantities. The proposed geometric calibration method based on the look-angle corrections has a sensor positioning precision of better than 1.0 pixels for the calibrated image.

1. Introduction

Ground spatial resolution and imaging width are very important indicators in the design of modern optical remote sensing satellites, and linear-array push-broom optical satellites (LAPOSs) often cannot be taken into consideration. Linear-array whiskbroom optical satellites (LAWOSs) were proposed to increase the width of the image without reducing the spatial resolution. The images of LAWOSs consist of scanning mirrors, compensation mirrors, and multiple linear-array CCDs [1][2][7]. When the LAWOS is in orbit, the linear-array CCDs on the imager take the orbit as the rotation axis and rotate around the forward direction of the satellite to complete the imaging of the ground in the process of rotation [3].

Although the existence of errors has been taken into account in the design of the imager and possible errors are theoretically minimized, attitude and orbit measurements, instrument installation, and camera lens aberrations can affect the quality of the imaging. In particular, the installation of rotary scanning mirrors and compensation mirrors can also generate errors [4]. These errors affect the sensor positioning accuracy [5]. To improve the sensor positioning precision, the image can be geometrically calibrated.

Currently, there is a scarcity of research focused on the geometric calibration of LAWOSs. Li et al. [6] proposed a three-step estimation method using ground checkpoints to estimate the calibration parameters and used a fifth-degree polynomial to fit and correct the angular measurement error of the scanning mirror. Wang proposed a calibration method for China-YG-14 high-resolution thermal infrared images, which analyzed the errors introduced by the whiskbroom motion, and solved the model instability and low precision caused by the inconsistent pendulum sweeps and uneven angles.

Cao [7] proposed a geometric calibration method based on ELACCD for DaQi-1 and introduced a rotation matrix to describe the ELACCD-to-imager coordinate system conversion relation equivalent to the rotary scanning mirrors and compensation mirrors.

Section II describes the physical sensor model (PSM) of LAWOS, the calibration of geometric model look-angle corrections, and the calculation of parameters. Section III illustrates the analysis and validation of the feasibility and applicability of the proposed method by using five WSI image datasets. Section IV presents the findings.

2. Method

2.1. PSM based on LAWOSs

To geometrically calibrate LAWOS images, the first step is to establish a suitable PSM [5]. PSM precisely depicts the imaging process of LAWOS, thereby obtaining optical remote sensing satellite images with high geometric quality. Due to the complexity of the process of calculating the parameters of the scanning mirrors and the parameters of the compensation mirrors, a linear-array CCD (LACCD) rotating in the same manner as the scanning mirrors and compensation mirrors would be considered equivalent to a scanning mirror rotating [7]. The expression of the established PSM is as follows:

$$\begin{bmatrix} X \\ Y \\ Z \end{bmatrix} = \begin{bmatrix} X_S \\ Y_S \\ Z_S \end{bmatrix} + \lambda \cdot \boldsymbol{R}_{J2000}^{WGS84} \cdot \boldsymbol{R}_{Orbit}^{J2000} \cdot \boldsymbol{R}_{Body}^{Orbit} \cdot \boldsymbol{R}_{Imager}^{Body} \cdot \boldsymbol{R}_{Rotation}^{Imager} \begin{bmatrix} \tan\psi_x \\ \tan\psi_y \\ 1 \end{bmatrix} \tag{1}$$

where $[X\ Y\ Z]^T$ is the coordinates of the ground points; $[X_S\ Y_S\ Z_S]^T$ is the coordinates of the LAWOS; λ is the scale factor. $\boldsymbol{R}_{J2000}^{WGS84}$, $\boldsymbol{R}_{Orbit}^{J2000}$, $\boldsymbol{R}_{Body}^{Orbit}$, $\boldsymbol{R}_{Imager}^{Body}$, and $\boldsymbol{R}_{Rotation}^{Imager}$ are the rotation matrices for the transformation of the subscripted coordinate system to the superscripted one, respectively. $\boldsymbol{R}_{Body}^{Orbit}$ consists of the rotation angles $(\omega_B, \varphi_B, \kappa_B)$; $\boldsymbol{R}_{Imager}^{Body}$ consists of the mounting angles $(\omega_I, \varphi_I, \kappa_I)$; and $\boldsymbol{R}_{Rotation}^{Imager}$ consists of the rotation angles $(\omega_R, \varphi_R, \kappa_R)$. $\tan\psi_x$ and $\tan\psi_y$ are the coordinates of the image points.

The CCD detector look angles (ψ_x, ψ_y) and the LACCD rotation angles $(\omega_R, \varphi_R, \kappa_R)$ expressions [8][9][10] are as follows:

$$\begin{cases} \tan\psi_x = p_0 + p_1 r + p_2 r^2 + p_3 r^3 \\ \tan\psi_y = q_0 + q_1 r + q_2 r^2 + q_3 r^3 \end{cases} \tag{2}$$

$$\begin{cases} \omega_R = \alpha_0 + \alpha_1 c + \alpha_2 c^2 + \alpha_3 c^3 \\ \varphi_R = \beta_0 + \beta_1 c + \beta_2 c^2 + \beta_3 c^3 \\ \kappa_R = \theta_0 + \theta_1 c + \theta_2 c^2 + \theta_3 c^3 \end{cases} \tag{3}$$

where r and c are the row and column coordinates of the image point respectively.

2.2. Geometric Calibration Model and Parameters Solution

In the PSM based on LAWOSs, there is a strong correlation between the three types of parameters. Moreover, the step-by-step process of estimating the parameters is tedious and the computational process of geometric calibration will be very complicated. Therefore, this study proposes not to solve these three types of parameters in steps, but to use d_c and d_r directly for correction.

As shown in Figure 1, when the satellite is imaging, there is a ground point A. According to Equations (1) and (3), the coordinates $\tan\psi_x$ and $\tan\psi_y$ of the corresponding image point a can be calculated. In actual situations, there will be external orientation element errors and internal orientation element errors. The $\tan'\psi_x$ and $\tan'\psi_y$ are calculated based on Equation (2) to obtain the image point coordinates a'. The discrepancy between theoretical point a and actual point a' is evident in Figure 1. Therefore, this study proposes to directly use the correction quantities d_c and d_r to correct the image point coordinate errors.

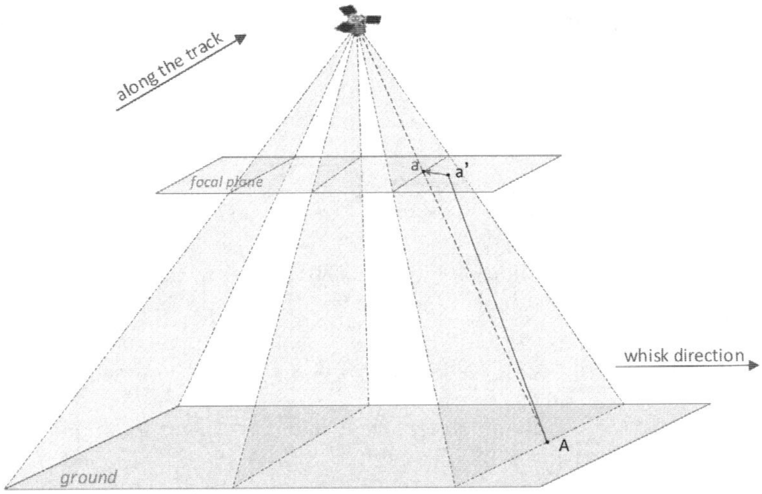

Figure 1. Imaging error map of LAWOSs.

The geometric calibration model based on look-angle corrections is established from Equations (1), (2), and (3) as follows:

$$\begin{bmatrix} p_0 + p_1 r + d_c \\ q_0 + q_1 r + d_r \\ 1 \end{bmatrix} = \frac{1}{\lambda} \cdot \boldsymbol{R}_{Imager}^{Rotation}(\alpha_0 + \alpha_1 c, 0,0) \cdot \boldsymbol{R}_{Body}^{Imager} \cdot \boldsymbol{R}_{Orbit}^{Body} \cdot \boldsymbol{R}_{J2000}^{Orbit} \cdot \boldsymbol{R}_{WGS84}^{J2000} \left(\begin{bmatrix} X \\ Y \\ Z \end{bmatrix} - \begin{bmatrix} X_S \\ Y_S \\ Z_S \end{bmatrix} \right) \quad (4)$$

where d_c and d_r are respectively correction quantities to column and row coordinates in image space. $\boldsymbol{R}_{Imager}^{Rotation}$, $\boldsymbol{R}_{Body}^{Imager}$, $\boldsymbol{R}_{Orbit}^{Body}$, $\boldsymbol{R}_{J2000}^{Orbit}$, and $\boldsymbol{R}_{WGS84}^{J2000}$ are the rotation matrices for the transformation of the superscripted coordinate system to the subscripted one, respectively.

The expressions of d_c and d_r are as follows:

$$\begin{cases} d_c = m_0 + m_1 c + m_2 r + m_3 c^2 + m_4 cr + m_5 r^2 + m_6 c^3 + m_7 c^2 r + m_8 cr^2 + m_9 r^3 \\ d_r = n_0 + n_1 c + n_2 r + n_3 c^2 + n_4 cr + n_5 r^2 + n_6 c^3 + n_7 c^2 r + n_8 cr^2 + n_9 r^3 \end{cases} \quad (5)$$

The parameter estimation process is as follows.

1) Some uniformly distributed ground checkpoints within the image are fetched from the digital orthophoto maps (DOM) and digital elevation models (DEM).

2) The CCD detector look angle parameters (p_0, p_1) and (q_0, q_1), the LACCD rotation angle parameters (α_0, α_1), and the camera mounting angles $(\omega_I, \varphi_I, \kappa_I)$ use the values calibrated in the laboratory as initial values The correction quantities d_c and d_r are set to 0. Based on Equations (1) and (3), $\tan\psi_x$ and $\tan\psi_y$ can be calculated.

3) The actual $\tan'\psi_x$ and $\tan'\psi_y$ are calculated based on Equation (2).

4) Calculate the difference by $d_c = \tan\psi_x - \tan'\psi_x$ and $d_r = \tan\psi_y - \tan'\psi_y$, and the difference is fitted by a cubic polynomial surface. The parameters$(m_0, \ldots, m_9)$ and $(n_0, \ldots, n_9)$ of the expression for the corrected quantity are found.

3. Experiment and Results

3.1. Geometric Calibration Analysis

The experimental datasets for this study are five sets of DaQi-1 images, each comprising 260 images. Each image measures 20480×120 pixels. The calibration image was selected from the images in dataset 1 with fewer cloud and water-covered areas. The distribution of 71 ground checkpoints manually fetched from the DOMs and DEMs is shown in Figure 2.

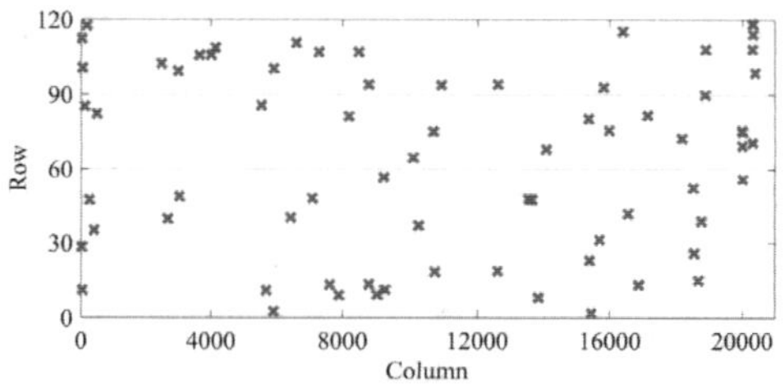

Figure 2. Ground checkpoint distributions in Image I1.

The following experiments were devised to validate the feasibility of the suggested calibration method.

1) Experiment A: No fitting of the correction quantities is performed, setting their parameters to the initial value of 0.

2) Experiment B: Fitting the correction quantities d_c and d_r with a primary polynomial surface, i.e., the parameters $(m_0, \ldots, m_2)$ and $(n_0, \ldots, n_2)$ are calibrated, and $(m_3, \ldots, m_9)$ and $(n_3, \ldots, n_9)$ are set to the initial value of 0.

3) Experiment C: Fitting the correction quantities d_c and d_r with a quadratic polynomial surface, i.e., calibrating the parameters $(m_0, \ldots, m_5)$ and $(n_0, \ldots, n_5)$ and setting $(m_6, \ldots, m_9)$ and $(n_6, \ldots, n_9)$ to the initial value of 0.

4) Experiment D: Estimate the three types of parameters separately according to the method in [7], i.e., calibrate the imager parameters $(\omega_I, \varphi_I, \kappa_I)$, $(\alpha_0, \ldots, \alpha_2)$, $(\beta_0, \ldots, \beta_3)$ in steps, and set the parameters $(\theta_0, \ldots, \theta_3)$, $(p_0, \ldots, p_3)$, and $(q_0, \ldots, q_3)$ to the initial value of 0.

5) Experiment E: Fitting of the correction quantities d_c and d_r with cubic polynomial surfaces, the parameters$(m_0, \ldots, m_9)$ and $(n_0, \ldots, n_9)$ are calibrated.

After each set of experimental calibrations, the position of the projection point of the ground coordinates of each ground checkpoint on the image was calculated based on the parameters of the obtained corrections, and the root-mean-square error (RMSE) of the residuals from the projection point to the image point was calculated as the sensor positioning precision. The precision of sensor positioning in the calibrated image is presented in Table 1. The fitting schematic of the correction quantities for the calibrated images is shown in Figure 3.

Table 1. Sensor positioning accuracies of Image I1.

Experiment	Maximum (pixels)		Mean (pixels)		RMSE (pixels)		
	c	r	c	r	c	r	planimetry
A	17.540	14.524	13.201	9.203	13.480	9.771	16.649
B	1.080	1.456	0.000	0.000	0.439	0.638	0.774
C	1.004	1.384	0.000	0.000	0.431	0.520	0.676
D	-1.036	-1.227	0.000	0.003	0.441	0.498	0.666
E	0.926	1.082	0.000	0.000	0.427	0.491	0.650

In Experiment A, the calibrated image exhibits a sensor positioning precision of just 16.649 pixels. As shown in Figure 3(a), sensor positioning error is exhibited as curved characteristics in three dimensions. This situation indicates that the jitter of the satellite during the launching process and the installation of the camera sensors always produce errors, which lead to the reduction of the sensor positioning precision.

In Experiment B, the fitting using the primary surface resulted in a sensor positioning precision of 0.774 pixels, which is a significant improvement over the uncalibrated results. This suggests that these errors have some regularity and can be predicted by fitting a function and then correcting the error using correction quantities.

In Experiment C, the precision of sensor positioning after quadratic surface fitting is improved by 0.1 pixels. This indicates that the use of quadratic surface function fitting can fit the correction quantities more accurately.

In Experiment D, the sensor positioning precision obtained by the method of step-by-step estimation of the parameters was 0.666 pixels.

In Experiment E, the sensor positioning precision is improved by 0.026 pixels after the cubic surface fitting compared to the quadratic surface fitting. This indicates that the cubic surface fitting can improve the fitting precision to some extent. Compared with the results of Experiment D, the sensor positioning precision was improved by 0.016 pixels, which indicates that the sensor positioning precision can be improved by using the surface fitting correction amount directly.

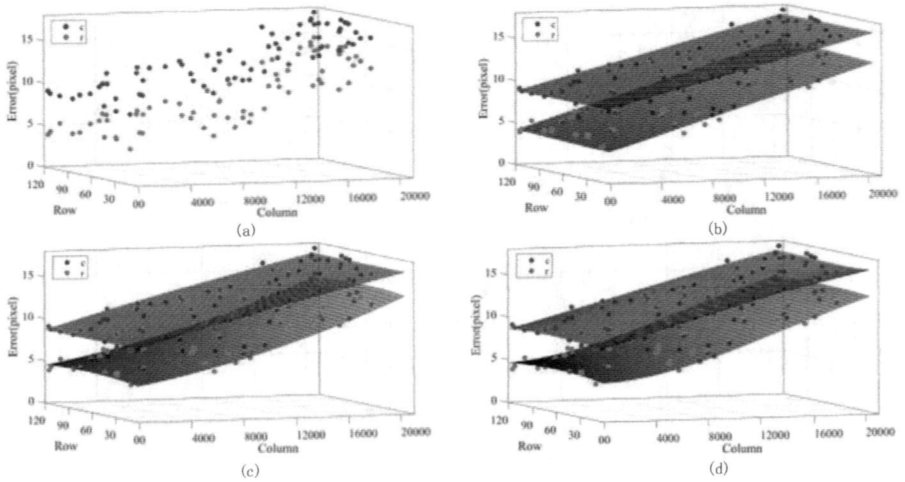

Figure 3. Correction quantities fitting surface in experiments: (a) A, (b) B, (c) C, and (d) E.

Derived from the outcomes of the aforementioned experiments, the method of calibrating images using polynomial surfaces fitted to correction quantities is feasible, and the effectiveness of the fit improves to a certain extent with the number of fitted functions.

3.2. Performance validation of calibration parameters

The parameters obtained by the geometric calibration of an image should apply to other images of the same satellite to improve the precision of the correspondence between image points and ground points. Therefore, four images I2-I5 were selected from datasets 2-5 respectively to evaluate the precision of the parameters obtained by image calibration of dataset 1. 52, 46, 53, and 36 checkpoints were fetched from the DOMs and DEMs of I2-I5, respectively, and Figure 4 shows the distribution of checkpoints in the four images.

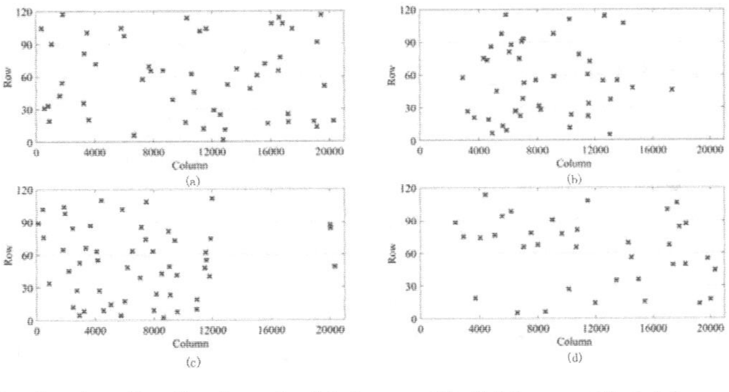

Figure 4. Ground checkpoint distributions in (a) image I2, (b) image I3, (c) image I4, and (d) image I5.

In the previous subsection, the scenarios of the five sets of experiments are noted as SA-SE, and Table 2 records the sensor positioning precision of I2-I5 under scenarios SA-SE, respectively. Figure 5 is the correction quantities fitting the surface of image I2.

In Figure 5(a), it can be seen that the errors in sensor positioning of the calibrated images exhibit curved surface features in the three-dimensional distribution, which is almost the same as the distribution in Figure 3(a).

Table 2. Sensor positioning accuracies of image I2-I5.

Image	Scenario	Maximum (pixels)		Mean (pixels)		RMSE (pixels)		
		c	r	c	r	c	r	planimetry
I2	SA	17.151	14.599	13.193	8.942	13.432	9.468	16.433
	SB	-1.122	-1.249	-0.127	0.074	0.507	0.629	0.808
	SC	-1.050	-1.188	-0.113	0.024	0.498	0.528	0.726
	SD	1.131	-1.384	0.131	-0.107	0.517	0.592	0.786
	SE	-1.004	1.183	-0.105	0.064	0.499	0.545	0.739
I3	SA	15.901	12.369	12.335	7.701	12.419	7.874	14.705
	SB	1.271	1.845	-0.123	0.282	0.579	0.684	0.896
	SC	1.344	1.570	-0.062	0.044	0.575	0.577	0.815
	SD	-1.263	-1.194	0.120	0.024	0.577	0.595	0.829
	SE	1.363	1.654	-0.076	-0.012	0.588	0.616	0.851
I4	SA	17.228	15.014	11.610	7.226	11.767	7.571	13.992
	SB	-1.396	-1.417	0.015	0.042	0.540	0.592	0.801
	SC	-1.305	1.152	0.032	-0.012	0.546	0.476	0.724
	SD	1.399	1.399	-0.021	0.060	0.535	0.571	0.783
	SE	-1.275	1.235	0.022	-0.107	0.568	0.516	0.768
I5	SA	17.562	14.032	13.654	9.400	13.820	9.850	16.971
	SB	-1.125	1.219	-0.159	0.140	0.597	0.621	0.861
	SC	-1.082	1.032	-0.133	0.063	0.574	0.521	0.775
	SD	1.105	-1.174	0.168	-0.082	0.596	0.540	0.804
	SE	-1.140	-1.201	-0.136	0.042	0.583	0.481	0.756

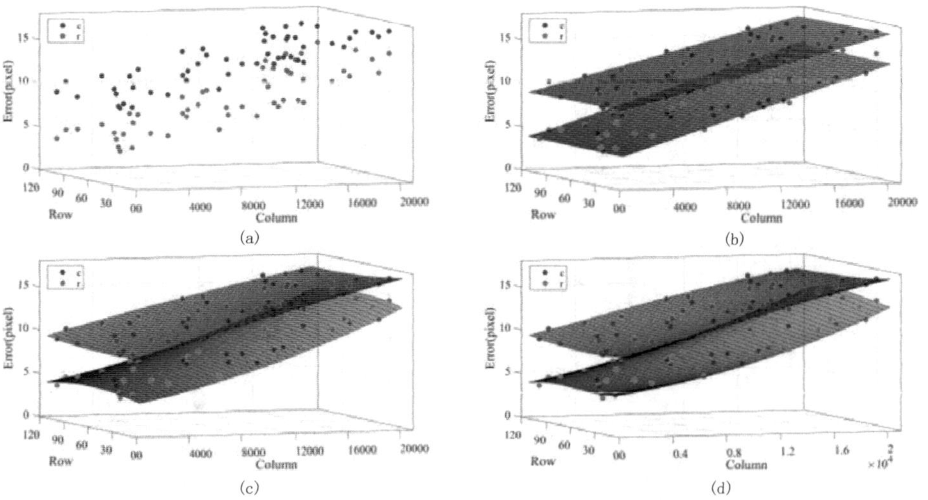

Figure 5. Correction quantities fitting the surface of image I2 in scenarios: (a) SA, (b) SB, (c) SC, and (d) SE.

In scenario SE, the precision of the proposed method in sensor positioning in this study has been improved to a certain extent compared with that of the method in [7]. This indicates that the calibration parameters obtained for the images in dataset 1 have precision and validity and can improve the sensor positioning precision for images I2-I5 in datasets 2-5.

From the data in Table 2, it can be seen that there is no improvement in the sensor positioning precision of images I2-I4 compared to the quadratic surface fitting when using cubic surface fitting corrections, which suggests that the results of the calibrations using the method of surface fitting have been stabilized on the experimental images.

4. Conclusion

A corresponding geometric calibration model is devised to address the imaging structure and process of LAWOSs, and a viable method for geometric calibration, utilizing look-angle corrections, is put forward in this study. The parameters to be estimated are directly corrected using the correction quantities d_c and d_r, which are obtained by fitting a cubic polynomial surface based on the errors of the calibration image.

From the experimental results of five sets of WSI images of DaQi-1, the correction quantities d_c and d_r are sufficient for correcting errors in DaQi-1 positioning, the method in this study also improves the sensor positioning precision compared with the method of solving the calibration parameters step by step, thus verifying the feasibility and effectiveness of the LAWOS in-orbit geometric calibration method based on look-angle corrections.

References

[1] Lee, D. S., Storey, J. C., Choate, M. J., and Hayes, R. W. (2004). Four years of Landsat-7 on-orbit geometric calibration and performance. *IEEE Transactions on Geoscience and Remote Sensing*, 42(12), 2786-2795. doi: 10.1109/TGRS.2004.836769.

[2] Wolfe, R. E., Lin, G., Nishihama, M., Tewari, K. P., Tilton, J. C., and Isaacman, A. R. (2013). Suomi NPP VIIRS is prelaunch and on-orbit geometric calibration and characterization. Journal of Geophysical Research: Atmospheres, 118(20), 11-508. doi: 10.1002/jgrd.50873.

[3] Jing, Z., Li, S., Hu, X., and Tang, F. (2021). Sub-pixel accuracy evaluation of FY-3D MERSI-2 geolocation based on OLI reference imagery. International Journal of Remote Sensing, 42(19), 7215-38. doi:10.1080/01431161.2021.1950941.

[4] Yang, B., Pi, Y., Li, X., and Yang, Y. (2020). Integrated geometric self-calibration of stereo cameras onboard the ZiYuan-3 satellite. ISPRS Journal of Photogrammetry and Remote Sensing, 162, 173-183. doi: 10.1016/j.isprsjprs.2020.02.015.

[5] Pan, H., Cui, Z., Hu, X., and Zhu, X. (2022). Systematic Geolocation Errors of FengYun-3D MERSI-II. IEEE Transactions on Geoscience and Remote Sensing, 60, 1-11. doi:10.1109/ TGRS.2022.3156999

[6] Li, X., Li, L., Zhao, L., Jiao, J., Jiang, L., Yang, L., and Sun, S. (2023). In-orbit Geometric Calibration for Long-linear-array and Wide-swath Whisk-broom TIS of SDGSAT-1. IEEE Transactions on Geoscience and Remote Sensing. doi: 10.1109/TGRS.2023.3268343.

[7] Cao, J. (2023). Geometric Calibration of the Whiskbroom DaQi-1 WSI Based on an Equivalent Linear-array CCD. IEEE Transactions on Geoscience and Remote Sensing. doi: 10.1109/T GRS.2023.3300013.

[8] Cao, J., Shang, H., Zhou, N., and Xu, S. (2022). In-orbit geometric calibration of multi-linear array optical remote sensing satellites with tie constraints. Optics Express, 30(15), 28091- 28111. doi:10.1364/OE.464112.

[9] Wang, M., Cheng, Y., Guo, B., and Jin, S. (2019). Parameters determination and sensor correction method based on virtual CMOS with distortion for the GaoFen6 WFV camera. ISPRS Journal of Photogrammetry and Remote Sensing, 156, 51-62. doi: 10.1016/j. isprsjprs.2019.08.001.

[10] Cheng, Y., Jin, S., Wang, M., Zhu, Y., and Dong, Z. (2017). A new image mosaicking approach for the multiple camera system of the optical remote sensing satellite GaoFen1. Remote Sensing Letters, 8(11), 1042-51. doi: 10.1080/2150704X.2017.1349960.

Noise reduction analysis of deformation data based on CEEMD-PE-SVD modeling

Qingda Duan[1], Hua Xia[1] and Ya Huang[2,3]

[1]College of Forestry, Xinyang Agriculture and Forestry University, Xinyang 464000, China
[2]Henan Academy of Geology, Zhengzhou, 450016, China

[3] Corresponding author's e-mail: kf1983@163.com

Abstract: In the field of high slope deformation monitoring, the deformation data obtained are often characterized by high volatility, strong nonlinearity, and noise content, due to the influence of factors such as the surrounding environment, human operation, and the complexity of the project. To overcome these problems, this paper proposes deformation data based on the coupling of complementary ensemble empirical modal decomposition (CEEMD), permutation entropy (PE), and singular value decomposition (SVD). Firstly, CEEMD decomposition is performed on the deformation data. Secondly, a randomness test is performed on the obtained modal components, and if the entropy value of the component arrangement entropy is greater than 0.5, SVD decomposition will be used for noise reduction. Finally, the components that meet the conditions are reconstructed to realize the noise reduction of the deformation data. The results show that the algorithm model has an excellent noise elimination effect, which can reflect the internal detail information of deformation data and provide a well-established method for future research on the processing of deformation monitoring data.

1. Introduction

The deformation data obtained from long-term observation of high slopes often contain noise (error) and the true value of the two parts of the external environment (temperature and humidity). The accuracy of the instrument and human operation and other factors resulting in the true value of the noise are often blended because of the presence of noise, so the deformation of the data fluctuation increased, increasing the difficulty of the later prediction of the deformation data. Therefore, it is very necessary to carry out noise reduction processing of deformation data [1]. Currently, the commonly used noise reduction methods mainly include wavelet analysis and complementary ensemble empirical mode decomposition (CEEMD) [2]. Wavelet analysis is usually suitable for processing low-frequency signals, and it is difficult to extract useful information from medium- and high-frequency signals and the noise cancellation effect varies greatly with different wavelet bases, thresholds, and decomposition scales [3]. CEEMD, on the other hand, is a new type of signal decomposition method without a priori knowledge, which can effectively separate low-frequency and high-frequency signals. At the same time, it can also improve the repeatability and robustness of signal decomposition, so the noise cancellation effect is better compared with wavelet analysis [4]. To further screen the hidden low-frequency useful information in the deformation data, this paper proposes a noise reduction method based on the coupling of the CEEMD algorithm, Permutation Entropy (PE), and Singular Value Decomposition (SVD), and the experiments show that the coupling algorithm has excellent effect on the deformation data of high slopes. The experiments show that the coupling algorithm has an excellent effect on noise reduction of high slope deformation data, which has great development prospects in practical application [5-6].

2. Introduction to the algorithm

2.1. Complementary ensemble empirical modal decomposition (CEEMD)

CEEMD decomposition is a new adaptive signal processing algorithm that greatly improves the efficiency of signal decomposition as well as reduces the reconstruction error of the signal during the decomposition of the signal [7]. The steps of CEEMD decomposition are as follows:

(1) We add a set of white noise $a_i(t)$ and $-a_i(t)$ of opposite sign to the original data sequence $S(t)$ to obtain new signals $B_i(t)$, $C_i(t)$.

$$\begin{bmatrix} B_i(t) \\ C_i(t) \end{bmatrix} = \begin{bmatrix} 1 & 1 \\ 1 & -1 \end{bmatrix} \begin{bmatrix} S_i(t) \\ a_i(t) \end{bmatrix} \tag{1}$$

(2) We perform EMD decomposition for $B_i(t)$ and $C_i(t)$, respectively.

$$\begin{cases} B_i(t) = \sum_j^J IMF_{ij}^+ \\ C_i(t) = \sum_j^J IMF_{ij}^- \end{cases} \tag{2}$$

where J is the number of components, IMF_{ij}^+ is the jth component of the signal after adding positive white noise reconstruction, and IMF_{ij}^- is the jth component of the signal after adding negative white noise reconstruction.

(3) We repeat Steps (1) and (2) P times respectively.

$$\begin{cases} \{\{IMF_{1j}^+, IMF_{2j}^+, \cdots, IMF_{pj}^+\}\} \\ \{\{IMF_{1j}^-, IMF_{2j}^-, \cdots, IMF_{pj}^-\}\} \end{cases} \tag{3}$$

(4) We perform integrated averaging of multicomponent quantities.

$$IMF_j = \frac{1}{2P} \sum_{i=1}^{P} (IMF_{ij}^+ + IMF_{ij}^-) \tag{4}$$

(5) The reconstructed signal $S(t)$ is:

$$S(t) = \sum_{j=1}^{P} IMF_j \tag{5}$$

2.2. Ranking entropy (PE)

Alignment entropy is a metric used to characterize the stochasticity of dynamic sequences with strong robustness, and the principle of the algorithm is described [8].

2.3. Singular value decomposition (SVD)

The SVD decomposition can effectively filter out the noise components in smooth and periodic signals, and the decomposition steps are as follows [9]:

(1) We construct a Hankel matrix for an arbitrary deformed signal $Y = \{y(1), y(2), ...y(N)\}$.

$$W=\begin{bmatrix} y(1) & y(2) & \cdots & y(n) \\ y(2) & y(3) & \cdots & y(n+1) \\ \cdots & \cdots & \cdots & \cdots \\ y(N-n+1) & y(N-n+2) & \cdots & y(N) \end{bmatrix} \tag{6}$$

(2) If the noise signal is T and the real signal is Z, the reconstructed matrix W is:

$$W = Z + T \tag{7}$$

(3) W decomposes into:

$$W=\begin{bmatrix} U_z & U_t \end{bmatrix}\begin{bmatrix} Z_z & 0 \\ 0 & Z_t \end{bmatrix}\begin{bmatrix} V_z & V_t \end{bmatrix}^{\mathrm{T}} \tag{8}$$

where the subscript t represents the noise signal matrix and the subscript z represents the true signal matrix.

2.4. Establishment of CEEMD-PE-SVD noise cancellation-based modeling

Aiming at the characteristics of monitoring data on high slope deformation with noise content and nonlinearity, this paper proposes a CEEMD-PE-SVD noise cancellation model based on the following steps:

(1) CEEMD decomposes the deformation monitoring data to obtain the sample vector y_i $(i=1,2,\cdots n)$;

(2) We create sample matrix Y, $Y=\left[y_1,y_2,\cdots y_n\right]$;

(3) Phase space reconstruction y_i and normalization process are conducted to get the arrangement entropy value of n [0, 1];

(4) We perform SVD decomposition for the components in (3) with entropy values greater than 0.5;

(5) We repeat Steps (3) and (4) until the entropy values of y_i are all less than 0.5;

(6) We reconstruct $\{y_1,y_2,\cdots,y_n\}$ to realize the noise reduction of deformation monitoring data.

3. Experimental analysis

In this paper, the deformation data of high slopes on both sides of the highway are used as the research object, and the cumulative horizontal displacement of monitoring point BD01 is selected as the research sample. The observation time is from January 23, 2020, to May 1, 2020, and the sampling interval is 1 d, with a total of 100 periods of data. The deformation monitoring data of monitoring point BD01 is shown in Figure 1.

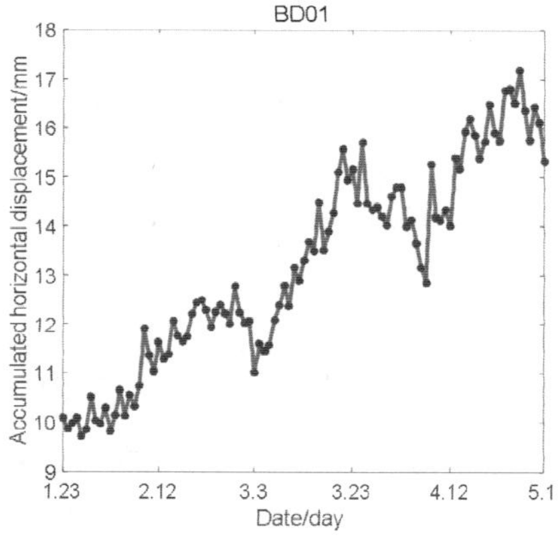

Figure 1. Deformation data for monitoring point BD01.

As can be seen from Figure 1, the influence of noise results in strong fluctuation of BD01 monitoring point deformation data, and there is no obvious regularity to follow. If the noise is not filtered out, it is difficult to get reliable results. Therefore, the deformation data for noise elimination is an essential part of the process.

3.1. CEEMD-PE-SVD noise canceling experiment

To observe the internal characteristics of the deformation data more clearly and reduce the volatility of the observed data, the deformation data are first decomposed by CEEMD.

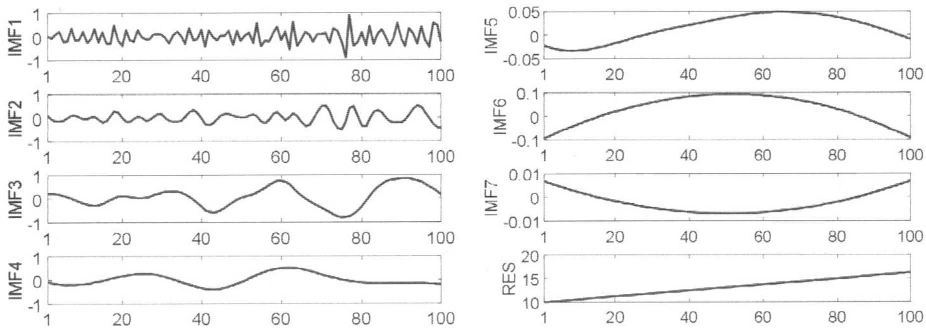

Figure 2. CEEMD decomposition of deformation data.

As shown in Figure 2, seven IMF components and one residual (RES) with frequencies from high to low are obtained by CEEMD decomposition. To extract the hidden detail information in the high-frequency IMF components, the entropy values of the aligned entropy of the seven IMF components and the RES are calculated respectively.

Table 1. Alignment entropy of each component.

Component	IMF1	IMF2	IMF3	IMF4	IMF5	IMF6	IMF7	RES
Entropy (physics)	0.6616	0.5671	0.3136	0.2702	0.1643	0.1363	0.1363	0

As can be seen from Table 1, the entropy value of IMF1 and IMF2 components is greater than 0.5, which is considered to contain more noise. It is necessary to carry out the SVD decomposition of these two components. The entropy value of the rest of the modal components is less than 0.5, which is considered to be more stable and is defaulted to be a pure signal. It does not need to be further processed for noise reduction.

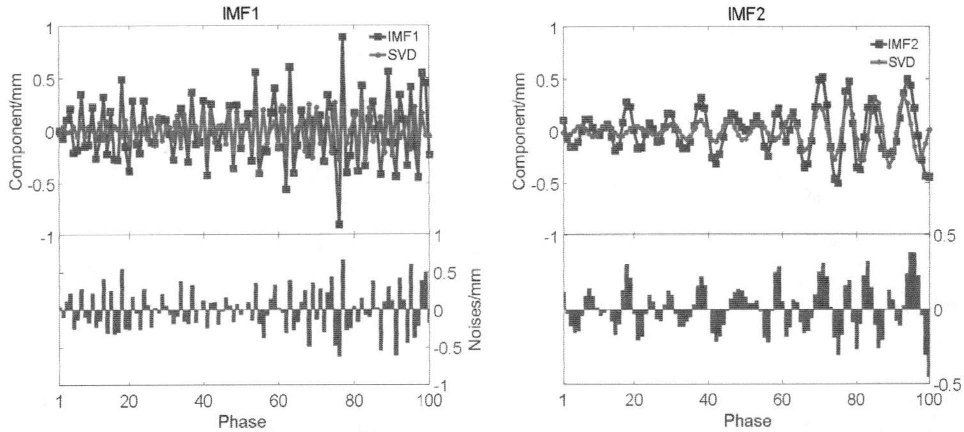

Figure 3. Comparison of the effect of IMF1 and IMF2 before and after noise cancellation.

As can be seen from Figure 3, after SVD decomposition, the volatility of IMF1 and IMF2 components are reduced, and the entropy value of the alignment entropy of these two components is less than 0.5 (entropy value after SVD decomposition of IMF1 component=0.4331 and entropy value after SVD decomposition of IMF2 component=0.4803). It reflects that the effect of the noise reduction processing by SVD decomposition is obvious and the noise reduction process using SVD decomposition is very effective, which can effectively filter out the noise in the high-frequency components (IMF1 and IMF2). From the comparison of the image effect before and after noise reduction, it can be found that the IMF1 component contains white noise and cycle error in the deformation data, and the IMF2 component contains white noise and the residual amount of cycle error in the deformation data, which makes the two components show large ups and downs and it difficult to capture the useful information hidden in the high-frequency signal. After SVD decomposition, the waveform undulations of both components are slowed down, and the curves of SVD decomposition at the inflection points and extreme points are smoother.

In general, the CEEMD-PE-SVD noise cancellation experiment mainly includes three steps. Firstly, CEEMD decomposition of the high slope deformation data is performed to obtain seven IMF components and one RES component from high to low frequency. Secondly, a randomness test is carried out for the multiple components obtained by using the entropy of the permutation. For the components larger than a set threshold (entropy value larger than 0.5), SVD decomposition is carried out, and the components after SVD decomposition are recalculated. If it meets the requirements, the next step will be carried out directly; If it does not meet the requirements, we continue to carry out the SVD decomposition, until all the components have an entropy value that is less than 0.5 of the arrangement entropy. Finally, the reconstruction of each IMF component that meets the conditions can be realized to the deformation of the high slope data of noise elimination processing.

3.2. Comparison experiment of noise canceling effect

To verify the effectiveness of the method proposed in this paper (based on the CEEMD-PE-SVD noise cancellation model), it is compared and analyzed with the more popular wavelet threshold noise cancellation method. After many experiments, when the deformation data of the BD01 monitoring point selects the db4 wavelet base and the number of decomposition layers is 8, the noise cancellation effect is optimal.

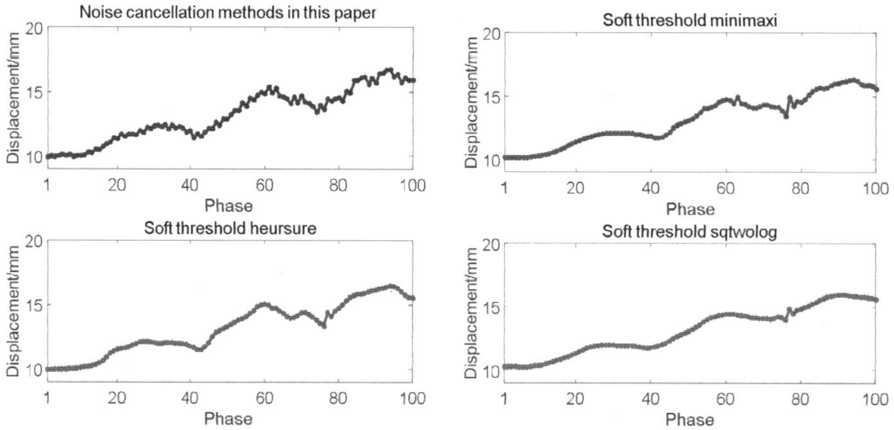

Figure 4. Comparison between the noise cancellation method in this paper and the wavelet thresholding noise cancellation method.

Figure 4 shows the comparison between this paper's noise cancellation method and wavelet threshold noise cancellation method. Compared with Figure 1 (deformation data of BD01 monitoring point), it is not difficult to see that these four noise cancellation methods have reduced the volatility of the deformation data, and the general trend of the curve is similar to that of the deformation data before noise cancellation. However, there are some shortcomings, for example, the curve after noise reduction in this paper is not smooth enough, and a large amount of deformation data details are lost after the soft threshold minimaxi, heursure, and sqtwolog noise reduction processes. To compare the advantages and disadvantages of several noise reduction methods more intuitively, the signal-to-noise ratio (SIGNAL-NOISE RATIO, SNR) and the root mean square error (RMSE) are introduced to judge which noise reduction method is superior [10-11].

$$SNR(dB) = 10\log_{10}(\frac{P_{signal}}{P_{noise}}) = 20\log_{10}(\frac{A_{signal}}{A_{noise}}) \tag{9}$$

$$RMSE = \sqrt{\frac{1}{n}\sum_{i=1}^{n}(f(t)-h(t))^2} \tag{10}$$

From Table 2, it can be found that compared with the wavelet thresholding noise cancellation method, the noise cancellation method in this paper is superior in SNR and RMSE. The curve after noise cancellation in this paper is not as smooth as the curve after wavelet thresholding noise cancellation, because the method in this paper not only filters out the high-frequency white noise in the deformation data but also preserves the detail information in the process of noise cancellation. On the other hand, the wavelet thresholding noise cancellation method has the problem of "over-filtering", in which the useful information is also filtered out in the process of noise cancellation, making it difficult to show the detailed information in the deformation data. The noise cancellation effect is not as good as the method proposed in this paper.

Table 2. Comparison of several noise cancellation methods.

Noise-canceling methods	Monitoring point BD01	
	SNR	RMSE
Noise cancellation methods in this paper	15.4666	0.3389
Soft threshold minimaxi	14.2850	0.3709
Soft threshold heursure	14.7007	0.3667
Soft threshold sqtwolog	11.3228	0.5006

4. Conclusion

This paper proposes a CEEMD-PE-SVD noise cancellation model based on the real high slope deformation data through the noise cancellation process. The results show that the method proposed in this paper, to a certain extent, overcomes the wavelet threshold noise cancellation method for the high and medium frequency signal noise cancellation effect. The reconstructed signal better reflects the deformation monitoring data, more clearly observes the deformation data of the cyclic changes, and is closer to the real deformation of high slope deformation law. The later study of the deformation analysis and deformation prediction provides a strong reference basis.

Acknowledgments

The paper is funded by the Project of Young Teachers' Research Fund of Xinyang College of Agriculture and Forestry (Project number: QN2021014).

References

[1] Gan R., Chen T., Zheng X., et al. Study on denoising deformation monitoring data by im proved wavelet threshold function [J]. Geodesy and Geodynamics, 2020, 40 (01): 17-22. DOI: 10.14075/j.jgg.2020.01.004.

[2] Andrews Athisayam, Manisekar Kondal. A multi-stage diagnosis method using CEEMD, A BC, and ANN for identifying compound gear-bearing faults[J]. Proceedings of the Ins

titution of Mechanical Engineers, Part C: Journal of Mechanical Engineering Science, 2023, 237 (8): 2030-2045. DOI: 10.1177/09544062221133757.
[3] Li C., Luo J., Shi H., et al. Recognition method of electric shock accident based on wave let analysis and probabilistic neural network [J]. Journal of Jiangsu University (Natura l Science Edition), 2023, 44 (01): 75-81+88. DOI: 10.3969/j.issn.1671-7775.2023.01.01 1.
[4] Zhang Z., Gao B., and Wei X. Weak fault diagnosis of rolling bearings based on stochasti c resonance and CEEMD [J]. Combined Machine Tools and Automated Machining Te chnology, 2023 (04): 46-49+53. DOI: 10.13462/j.cnki.mmtamt.2023.04.011.
[5] Smitha Bandi, V. B. S. Srilatha Indira Dutt. Single-channel fetal ECG extraction method b ased on extended Kalman filtering with singular value decomposition algorithm[J]. So ft Computing - A Fusion of Foundations, Methodologies and Applications, 2023, 27 (13): 9119-9129. DOI: 10.1007/s00500-023-08421-4.
[6] Zhou Y., Pan W., and Tan Q. Application of improved fine composite multiscale inversely weighted alignment entropy in gearbox fault diagnosis [J]. Combined Machine Tools and Automated Machining Technology, 2023 (05): 138-142+147. DOI: 10.13462/j.cnki. mmtamt.2023.05.032.
[7] Wu M., Xu J., and Wang Q. Research on CEEMD-Elman neural network prediction of A QI based on data decomposition [J]. China Environmental Science, 2019, 39 (11): 4, 580-4, 588. DOI: 10.19674/j.cnki.issn1000-6923.2019.0533.
[8] Zhou Q., Liang J., and Fu Y. Heterogeneous parallel acceleration of aligned entropy algori thms on Shenwei-TaihuLight [J]. Computer Engineering and Design, 2023, 44 (02): 4 00-406. DOI: 10.16208/j.issn1000-7024.2023.02.012.
[9] Dong X., Wu P., Ge C., et al. Joint estimation of lithium battery SOC based on paramete r online identification and SVD-UKF [J]. Control Engineering, 2022, 29 (09): 1, 713-1, 721. DOI: 10.14107/j.cnki.kzgc.20210155.
[10] Zhang H., Li C., Huang Z., et al. Conformal array beam synthesis method with maximu m SNR criterion [J/OL]. Telecommunications Technology: 1-7 [2023-10-11]. DOI: 10. 20079/j.issn.1001-893x.221019004.
[11] Ma J., Mu S., Su Y., et al. Particle swarm algorithm-based parameter extraction method f or combined wind speed model [J]. Laboratory Research and Exploration, 2022, 41 (08): 35-38+90. DOI: 10.19927/j.cnki.syyt.2022.08.009.

Multi-View consistency-based point cloud registration method with low overlap rate

Zhiqiang Cui[1], Zhaoyang Liao[1], Xubin Lin[1], Kezheng Sun[1,2], Taobo Cheng[1] and Xuefeng Zhou[1]

[1]Institute of Intelligent Manufacturing, Guangdong Academy of Sciences, Guangdong Key Laboratory of Modern Control Technology, Guangzhou, China

[2]Corresponding author's e-mail: kz.sun@giim.ac.cn

Abstract: Accurate and efficient workpiece measurement is crucial for workpiece processing and quality monitoring. Non-contact optical measurement methods have gained more attention due to their simplicity, efficiency, and flexibility compared to complicated and inefficient contact measurement methods. Multi-view registration of measurement data is a key issue in workpiece measurement, as it relies on the system's geometric accuracy and motion stability, presenting challenges such as the insufficient overlap of multi-viewpoint cloud data and cumulative error. To address these challenges, this paper proposes a multi-view planning and registration algorithm with a low overlap rate. The multi-view planning algorithm employs a greedy method to plan the scanning viewpoints of the workpiece to obtain complete point cloud data efficiently. The multi-view registration algorithm extracts features using a multi-scale geometric feature extraction network, matches the features based on the Hungarian algorithm, builds a graph, and optimizes workpiece positions based on the G2O algorithm for multi-view registration, effectively reducing cumulative error. Measurement experiments on blade workpieces confirm the feasibility of the proposed algorithms.

1. Introduction

Turbine blades, screws, and other complex peripheral components are typically manufactured by robots to achieve a high level of precision and accuracy in their three-dimensional morphology, including dimensions and angles. As a result, precise 3D topography measurement during machining plays a critical role in improving the accuracy of the finished parts.

Currently, contact-based probe measurements are widely used in the design, manufacturing, and quality inspection of casting parts for their high accuracy. However, these methods have low efficiency and limited online feedback. Optical-based surface measurements, on the other hand, are gaining attention for their simplicity, flexibility, and semi-automation. Line laser measurement is known for its high accuracy and efficiency, but it has limitations in acquiring complete point cloud data in one scan due to occluded reflections and field-of-view restrictions [1]. Therefore, it is necessary to obtain a complete and highly accurate model by measuring from multiple viewpoints and aligning multiple point clouds [2].

Aligning point cloud data from many coordinate systems into a single one is the goal of point cloud registration. Two categories of registration are two-view and multi-view, with the former aligning two-point cloud groups with overlapping areas. Identification of the source and target point clouds and their features are extracted. Next, using the features that were extracted, corresponding points are found. Lastly, the source point cloud is converted to the target point cloud's coordinate system and the rotational translation parameter is estimated.

Multi-viewpoint cloud registration is more complex and more important than two-viewpoint cloud registration due to the larger number of parameters to be solved and cumulative errors. In practical engineering applications, a common strategy is to first perform a coarse registration of the point cloud from multiple views to pre-align the point clouds of all views, and then a fine registration of the point cloud from multiple views to eliminate accumulated errors. Coarse registration accuracy depends on the accuracy of the calibration method used to obtain the initial positional relationship of the point cloud. If the initial position relationship cannot be obtained, the two-point clouds must be aligned to obtain the initial transformation. However, multi-view coarse registration algorithms based on spanning tree or shape generation are still in essence two-point cloud registration, which generates the cumulative error. Multi-viewpoint cloud detection involves aligning multiple point clouds of the same scene from different viewing angles. Coarse registration is the first step in this process, which involves obtaining an initial set of point cloud transformation matrices to be used as input for the fine registration stage. Fine registration aims to eliminate all transformation and cumulative errors, to achieve high accuracy in multi-view point cloud registration. In recent years, more attention has been paid to this problem, which has led to the development of various solutions for the fine registration of multiple viewpoint clouds. In general, the goal of multi-point point cloud registration is to obtain an accurate alignment between multiple point clouds, and both coarse and fine registration play an important role in achieving this goal.

Chen [3] proposes a direct extension of the two-by-two ICP algorithm to consider the multi-view point cloud fine registration as two-by-two point cloud ICP fine registration; through the two point clouds continuously using the strategy of ICP fine registration and merging the point cloud data, the data of two point clouds are sequentially aligned and merged until all the data of the point cloud are merged into a single point cloud, which is simple and feasible, but with the merging of the point cloud data and the ICP fine registration algorithm. The algorithm is simple and feasible, but as the point cloud data are merged, the search time of the ICP fine-registration algorithm will increase, and there is also the problem of the accumulation of registration errors. Bergevin [4] first proposed an algorithm for fine registration of multi-viewpoint clouds. The algorithm first establishes connections between all point clouds through a star network, and the center point cloud of the star network and the other point clouds search for point correspondences with each other; then the rigid transformation between point cloud data is estimated by the ICP algorithm; eventually, we use the minimum registration error as the objective function to optimize the hard transformation parameters for each point cloud. This algorithm considers all the point cloud data to optimize the registration parameters, which solves the problem of error accumulation to a certain extent. However, this algorithm is time-consuming due to the need to establish several point correspondences for the focal point cloud, and each focal point cloud must be aligned with all other point cloud data two by two. Han [5] proposed an improved RANSAC algorithm based on 3D domain covariance descriptors to improve alignment accuracy. Guo [6] proposed a practical method for coarse point cloud detection based on image feature points to reduce the number of iterations for fine-tuning. Zhang [7] utilizes an enhanced ISS key point extraction algorithm to address the issue of the point cloud's initial high position, and the low matching rate caused by the classical 3D Match's random sampling strategy that results in inconspicuous features. Toldo [8] proposes the use of a generalized Pluck analysis, which takes into account the simultaneous recording of all the point cloud data included in the ICP algorithm. Guo [9] embeds the Levenberg-Manelter (LM) algorithm into the ICP algorithm. The algorithm requires the creation of a minimum spanning tree for all point clouds along the shortest weighted path, globally aligned to the global reference coordinate system using the LM-ICP algorithm. Zhu [10] first estimates the overlap percentage of all pairs of point clouds in multi-view registration; then proposes a pruned scale iterative nearest point algorithm (TsICP) to compute the relative motions of two pairs of point clouds containing a high overlap rate, which can compute an accurate scale transformation; and finally the result of fine registration of multi-image point clouds is obtained using the MA algorithm. Gagnon [11] treats the multi-viewpoint cloud network as a whole and minimizes any view registration errors; the inter-point cloud transformations in the graph are computed by a series of matrix multiplications; if there are registration errors in the inter-point cloud transformation matrix, the matrix multiplication computation produces the cumulative errors in registration; it is crucial to obtain the network topology with minimum path lengths between the nodes; the algorithm proposes to improve a set of point cloud view registration transformations by a generic. The algorithm proposes a general method to improve the registration transformation of a set of point cloud views to minimize the cumulative error of registration among all views. Pulli [12] proposes to utilize the point cloud two-by-two registration results as constraints. Uniform elimination of two-by-two-point cloud registration errors in multi-view registration by successively estimating the stiff transformation of neighboring point clouds; spreading pairwise registration errors across multiple viewpoint clouds; combining neighboring views into a single point cloud when the registration error is small enough until all point clouds data are combined; and spreading registration errors uniformly across cyclic view pairs. Ding [13] proposes a deep mapping registration framework that uses a deep neural network to align multiple point clouds to a globally consistent coordinate system; two networks are used to accomplish global point cloud registration, one is used to estimate position, and the other is used to model scene structure by estimating the state of charge in global coordinates, transforming the registration problem into a binary occupancy classification problem, and uses gradient-based optimization to efficiently solve the fine registration of the point cloud from multiple viewpoints.

The scanning process requires the workpiece to be scanned from various angles to obtain complete point cloud data. However, to achieve higher scanning efficiency, it is practical to minimize the number of scanning angles. Additionally, multi-block point cloud registration can result in cumulative errors which can be reduced by minimizing the number of blocks to be aligned. Therefore, it is crucial to plan the scanning angle carefully. Our proposed algorithm based on the greedy idea optimizes scanning efficiency while reducing cumulative error by planning the viewpoints. The multi-view registration algorithm includes extracting features using a multi-scale geometric feature extraction network, performing feature matching based on the Hungarian algorithm, constructing graphs using workpiece positions as vertices and vertex connecting lines as edges, and optimizing workpiece positions based on the G2O algorithm for multi-view registration. This approach mitigates cumulative error caused by pairwise registration of multi-block point clouds.

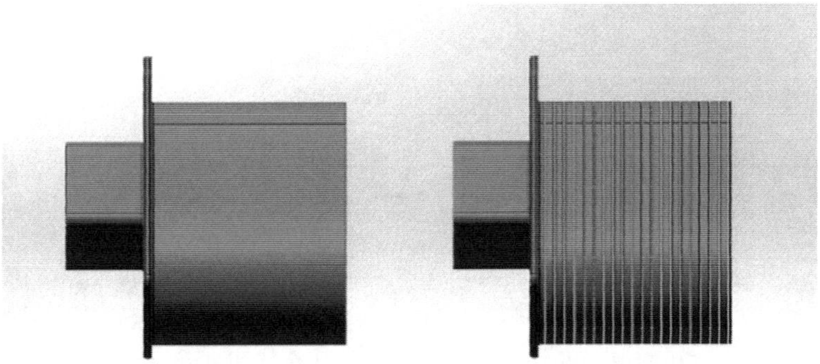

Figure 1. Sectional view of a workpiece.

2. Multi-Perspective Planning

To optimize the scanning efficiency and reduce the cumulative error of point cloud registration, it is important to carefully plan the scanning viewpoints. Our proposed algorithm uses a greedy approach to optimize scanning efficiency while minimizing the number of scanning angles. This is achieved through a multi-view registration process that incorporates feature extraction, feature matching, graph construction, and optimization using the G2O algorithm. As shown in Figure 1, by carefully planning the viewpoint for each cross-sectional profile curve, we can significantly reduce the number of scanning angles required for complete point cloud data acquisition. This approach helps to mitigate cumulative errors caused by pairwise registration of multi-block point clouds.

2.1 Construct the set of viewable points:

As shown in Figure 2, the rotation of the workpiece for scanning is transformed into scanning with multiple cameras placed around the workpiece axis in a 360-degree range. For the workpiece cross-section contour line C, here to define the concept of a visual point: for a point P_1 on the workpiece cross-section contour line C, if the point P_1 and the line scanning laser camera L_1 line of straight-line segment PL_1 and the workpiece cross-section contour line C in addition to the point P_1 no other point of intersection, then the angle of the line scanning laser camera L_1 is the point P_1 of a visual angle, the point P_1 is the camera L_1 of a visual point. The solution can be obtained by solving the set of all visible points PL_1 of the camera line scanning laser camera L_1 for S.

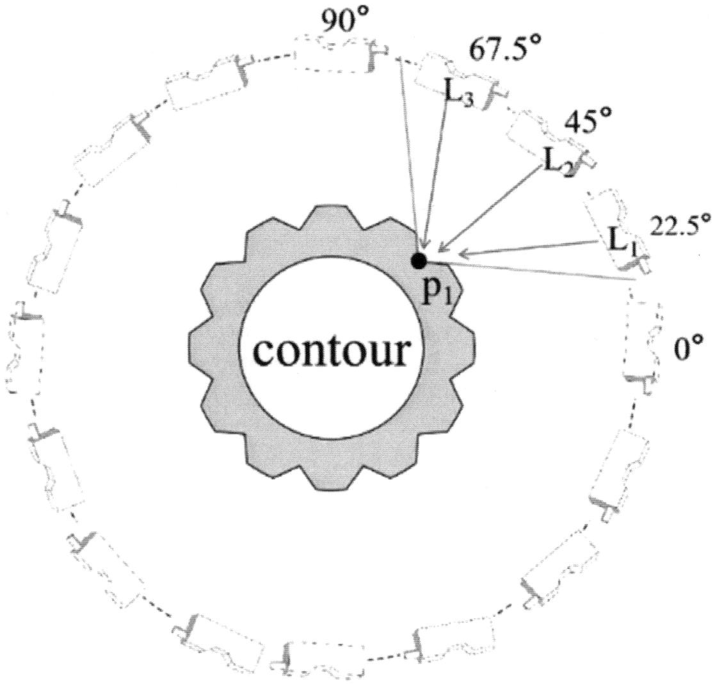

Figure 2. Schematic plan with visual angle.

2.2 Visual angle planning:

To optimize and streamline the scanning process, we propose the following steps:

(1) Firstly, count and sort the number of viewable points under each camera angle. Select the camera angle θ_0 with the highest number of viewable points as the first scanning angle.

(2) Remove the viewable points under camera angle θ_0 from the point set.

(3) Count the number of viewable points under each remaining camera angle. Select the camera angle θ_i with the highest number of viewable points.

(4) Remove the viewable points under camera angle θ_i from the point set.

(5) Repeat Steps 3 and 4 until there are no points left in the set of viewable points. The set of camera angles obtained at this time is the scanning angle planned for the slice cross-section C_1.

(6) Perform Steps 1 to 5 for each slice contour C_K to obtain the camera angle set θ_{C_K} for each slice contour.

(7) Concatenate all camera angle sets to obtain the final set of scanning angles.

By following this approach, we can efficiently determine the optimal camera angles for scanning each slice cross-section, minimizing the number of scanning angles required while ensuring comprehensive coverage of the workpiece.

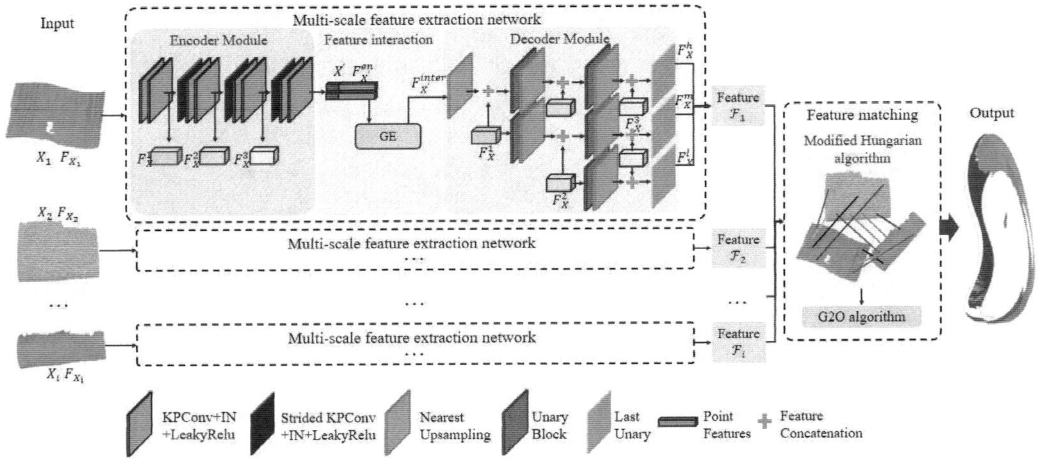

Figure 3. Multi-view point cloud feature extraction and registration method.

3. Multi-View Registration

In this paper, a multi-scale geometric feature extraction network is proposed for feature extraction, and at the same time the problem of registering a point cloud from multiple views is transformed into a graph optimization problem, where the vertices are defined as the initial positional attitude of each piece of point cloud output from the coarse registration of the multi-viewpoint cloud, and the vertices connecting the vertices are taken as the edges to build the graph, and the artifact position is nonlinearly optimized based on the G2O algorithm [14] to carry out the multi-view registration, and to alleviate the cumulative error caused by the two-by-two registration of the multi-viewpoint cloud.

3.1 Multi-scale geometric feature extraction network

The network initially uses a multidimensional coding structure to focus on semantically rich regions at different levels, followed by a geometric coding module (GE) to consider geometric features, which are decoded by a decoder to output features at three scales.

- Multiscale coding and decoding structure

The multiscale coding and decoding structure are divided into an encoder and a decoder, with the encoder being available for common use by all input point clouds. As shown in Figure 3, the encoder uses KPConvp [15] as the base unit. The neighborhood is expanded in a convolutional manner using 3-stride KPConv blocks. Record the feature mappings (denoted as F_X^1, F_X^2, F_X^3) before each stride KPConv block for the decoder to generate multi-scale features. This has the advantage of extending the neighborhood perception from F_X^1 to F_X^3 for each point feature. After the encoder has processed the input point cloud data, the output is the superimposed point $X'$and its feature $F_{X'}^{en} \in R^{N' \times Den}$.

The decoder takes as input the recorded feature mappings (F_X^1, F_X^2, F_X^3) as well as $F_{X'}^{inter}$ and output high, medium and low-level features (F_X^h, F_X^m, F_X^l) of the point cloud X.

Define the function φ（*,*,*） as follows:

$$\varphi(\mathrm{F}^1, \mathrm{F}^2, \mathrm{g}) = \mathrm{cat}[\mathrm{Up}(\mathrm{g}(\mathrm{F}^2)), \mathrm{F}^1] \tag{1}$$

where F^1 and F^2 is the input feature, g is a function of the MLP or Identity layer, cat [*, *] is the splicing operation, and Up (*) is up-sampling.

Then, F_X^h, F_X^m, and F_X^l are computed as follows:

$$F_X^l = MLP_2(\varphi(F_X^1, F_X^2, MLP_1)),$$

$$F_X^m = MLP_5(\varphi(F_X^1, \varphi(F_X^2, F_X^3, MLP_3), MLP_4)),$$

$$\mathrm{F}_{\mathrm{X}}^{\mathrm{C}} = \mathrm{MLP}_3(\varphi(\mathrm{F}_{\mathrm{X}}^1, \varphi(\mathrm{F}_{\mathrm{X}}^2, \varphi(\mathrm{F}_{\mathrm{X}}^3, \mathrm{F}_{\mathrm{X}'}^{\mathrm{inter}}, \mathrm{Identity}), \mathrm{MLP}_1), \mathrm{MLP}_2)). \quad (2)$$

The geometric coding module and the cross-attention module are also needed to connect the encoder and decoder.

- Geometric Encoding Module (GE)

As shown in Figure 4, the geometric encoding module (GE) takes as input the superposition point X' and its feature $F_{X'}^{en} \in R^{N' \times Den}$, and the output is a geometrically enhanced feature.

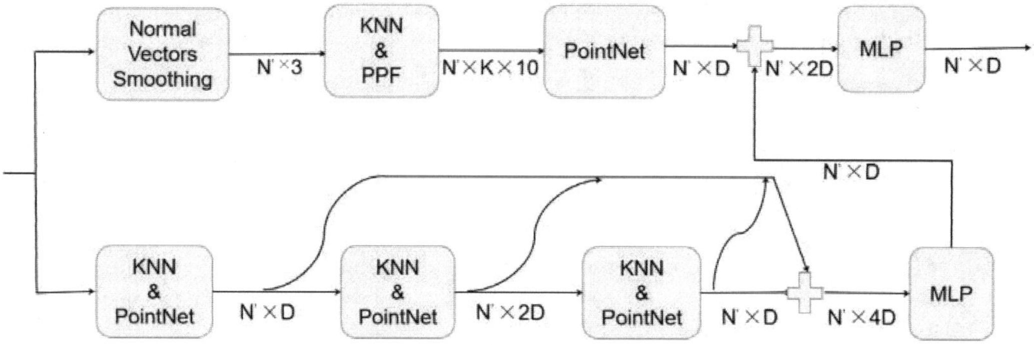

Figure 4. Geometric encoding module (GE).

The normal vector is a very important surface feature information in point cloud registration, which can help us to reduce the errors caused by noise or missing data, to more accurately describe and match the relationship between point clouds. As shown in Figure 4, to guarantee that normal vectors are represented geometrically correctly in the GE, normal vector smoothing techniques are applied. Specifically, first, the subsampled hyperpoints X'are mapped back to the original X, and then the normal N_X of X is computed using the classical API of Open 3D [16]. Finally, the normal vector of a point in X' is averaged over the normal vectors of the points surrounding it in X, for the point $x_i' \in X'$, the smooth normal vector $N_{x'}$ is computed as

$$N_{x_i'} = \frac{1}{|J_i^N|} \sum_{x_j \in J_i^N} N_{x_j} \quad (3)$$

where $J_i^N = \{x_j | \|x_j - x_i'\| < r^N\}, x_j \in X, r^N$ is the radius of the neighborhood.

The geometric characteristic $F_{x_i'}$ of point $x_i' \in X'$ is calculated by the following formula [21]:

$$\begin{aligned} \mathrm{PPF}(\mathbf{x}_i', \mathbf{x}_j') = & \left(\angle\left(\mathbf{x}_j' - \mathbf{x}_i', \mathbf{N}_{\mathbf{x}_i'}\right), \angle\left(\mathbf{x}_j' - \mathbf{x}_i', \mathbf{N}_{\mathbf{x}_j'}\right),\right. \\ & \left.\angle\left(\mathbf{N}_{\mathbf{x}_i'}, \mathbf{N}_{\mathbf{x}_j'}\right), \|\mathbf{x}_i' - \mathbf{x}_j'\|_2\right), \\ \mathrm{F}_{\mathbf{x}_j'} = & f_1\left(\mathbf{x}_i', \mathbf{x}_j' - \mathbf{x}_i', \mathrm{PPF}(\mathbf{x}_i', \mathbf{x}_j')\right), \\ \mathrm{F}_{\mathbf{x}_i'} = & \max\left\{\mathrm{F}_{\mathbf{x}_j'} \mid \mathbf{x}_j' \in J_i^{\mathbf{G}}\right\}. \end{aligned} \quad (4)$$

where $\angle(*,*) \in [0, \pi]$ represents the angle between two vectors, implemented by PointNet [17],

$$J_i^{\mathbf{G}} = \{\mathbf{x}_j' \mid \|\mathbf{x}_j' - \mathbf{x}_i'\| < r^{\mathbf{G}}\} \quad (5)$$

where $r^{\mathbf{G}}$ is the radius of the neighborhood of $\mathbf{x}_i'$, and $\max(*)$ denotes the maximum pooling.

The feature $\mathrm{F}_{\mathbf{X}'}^{\mathrm{inter}}$ is calculated as follows:

$$\mathrm{F}_{\mathbf{X}'}^{\mathrm{inter}} = \mathrm{GE}(\mathbf{X}', F_{X'}) \quad (6)$$

where GE (*,*) is the GE module.

3.2 Multi-view registration

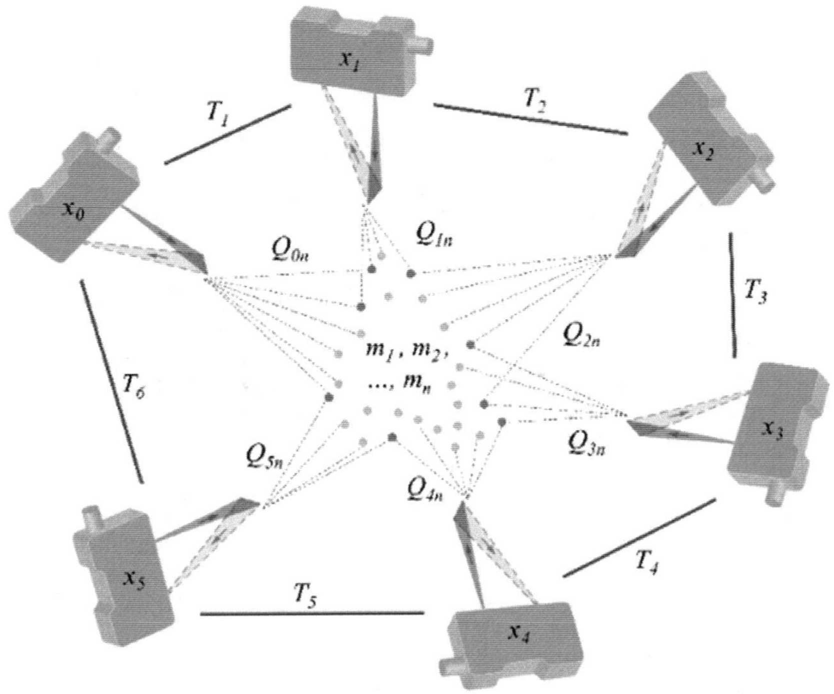

Figure 5. Multi-view registration.

The problem of point cloud registration can be described as follows: Given a source point cloud $X = \{x_i \in R^3\}_{i=1,2,...,N}$, and a target point cloud $Y = \{y_j \in R^3\}_{j=1,2,...,M}$, where N and M represent the number of points in X and Y respectively. The objective of point cloud registration is to determine the transformation $T \in SE(3)$ that achieves the optimal alignment between X and Y. This can be formulated as follows:

$$\underset{T}{argmin} \ \frac{1}{|\sigma|} \sum_{(i,j)\in\sigma} ||T(x_i) - y_i||_2 \tag{7}$$

where σ is the correspondence set and | . | is the base.

While the classical ICP [18] algorithm for finding the correct point correspondence between point clouds, after the initial coarse registration transformation, the algorithm calculates the distances from each point in the source point cloud to the target point cloud. It then identifies the point pair with the smallest distance as the corresponding point pair between the two-point clouds; ensures that there is a correspondence between the points in the source point cloud and those in the target point cloud, and at the same time constructs the residuals squared and the objective function; utilizing the least squares method, the algorithm minimizes the error function and performs iterative steps until the mean square error falls below the specified threshold.

The present work introduces an enhanced method for registering multi-view point clouds. It leverages the ICP algorithm and the feature extraction network to match corresponding features extracted from the point clouds. Additionally, as shown in Figure 5, it converts the registration problem into a problem of graph optimization, utilizing the Hungarian algorithm for feature matching. A graph is constructed with the workpiece positions as vertices and the connecting lines as edges. The position of the workpiece is then optimized using the G2O algorithm for non-linear optimization. This approach effectively addresses the cumulative error issue that arises from registering multiple blocks of point clouds in a two-by-two manner.

4. Experimentation

4.1System construction

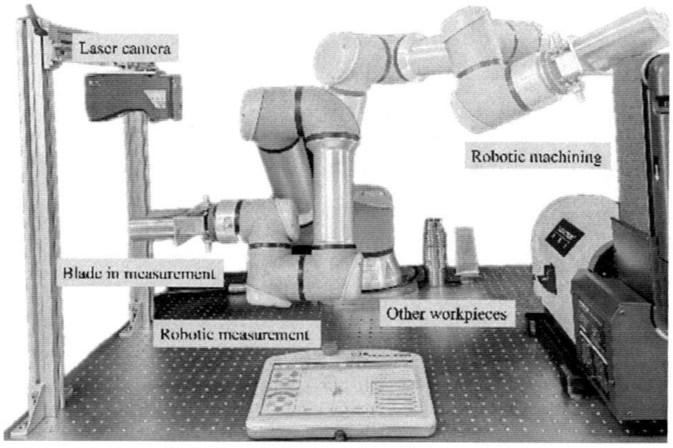

Figure 6. System diagram.

Figure 6 depicts the system configuration, comprising an OPT line laser camera and a UR16e robot. The robot is responsible for transporting the workpiece while continuously sending an external trigger pulse signal to initiate scans by the line laser camera. This enables capturing point cloud data from the workpiece's surface. The workpiece of interest is shown in Figure 7, and the measurement experiments primarily focus on its side.

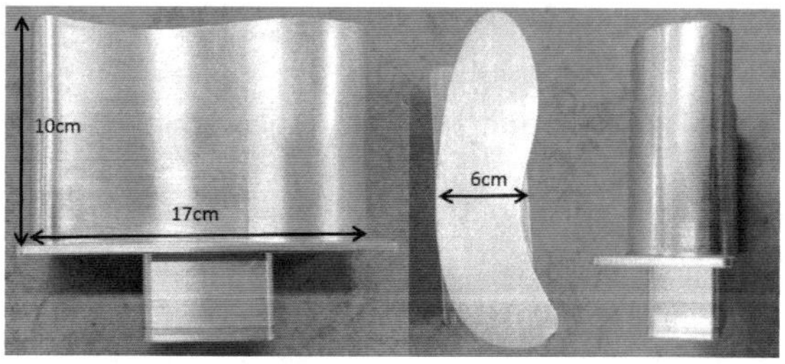

Figure 7. Blade to be measured.

4.2Perspective planning experiment

The view angle planning experiment is carried out on the side of the workpiece shown in Figure 7. Firstly, the view angle planning is carried out on a single cut surface, and then the planned angles of all cut surfaces are totaled to get the final planning angle. Following the mechanical position initial transformation of the workpiece cross-section shown in Figure 8 and Table 1, the manual estimation suggests that there are 6 angles involved in scanning the point cloud, the scanning time is 198s, while after the view angle planning algorithm calculates that only 4 scanning angles can cover the entire surface of the workpiece, the scanning time is 132 s, the scanning efficiency is improved by 33.33%.

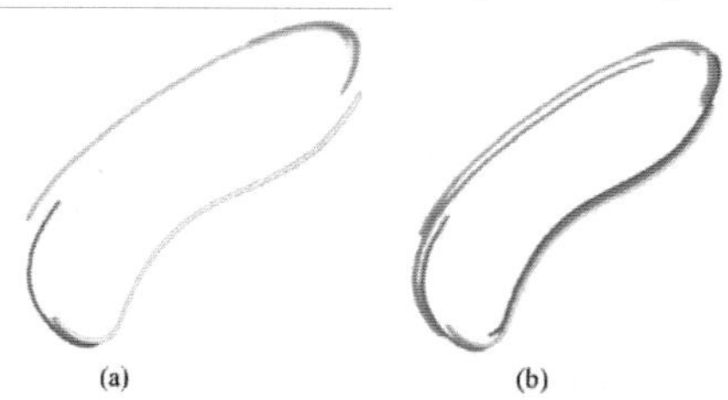

Figure 8. (a) cross-section in perspective plan case; (b) cross-section derived from manual experience.

Table 1. Perspective planning experimental results.

Method	View number	Scanning time
Perspective planning	4	132s
Manual instruction	6	198s

Table 2. Registration experimental results data.

Method	Point cloud matching rate
Multi-view registration	74.33%
Tree-based NDT+ICP	60.21%

4.3 Multi-view point cloud registration experiment

To enhance the registration speed, a voxel filter with a side length of 1.0f is initially employed to down-sample the multi-block point cloud data after collection, and then multi-view registration is performed. The method used in the comparison experiment is to first align two and two-point clouds using the NDT [20] algorithm based on the idea of a spanning tree [19], connect the successfully aligned point cloud pairs to build a spanning tree and then perform fine registration between two and two-point clouds according to ICP algorithm.

The experimental visualization results are shown in Figure 9. Figure 9. (a) is the cad cross-section of the workpiece during machining, Figure 9. (b) is the cross-section of the workpiece between two- and two-point clouds using the NDT algorithm for registration, and the point cloud pairs that have been successfully aligned are connected to build a spanning tree, then the workpiece cross-section between the two- and two-point clouds is fine-aligned according to the ICP algorithm, and Figure 9. (c) depicts the optimized cross-section of the workpiece achieved through the proposed multi-view point cloud registration algorithm in this paper. The intermediate result image reveals that the cumulative error has led to noticeable discrepancies, after one week of registration, a cross is formed between the red and white point clouds, while from the effect image in Figure 9. (c), it can be seen that the global optimization is optimized to alleviate the cumulative error generated by the two-by-two viewpoint registration. The 3D registration result is shown in Figure 10.

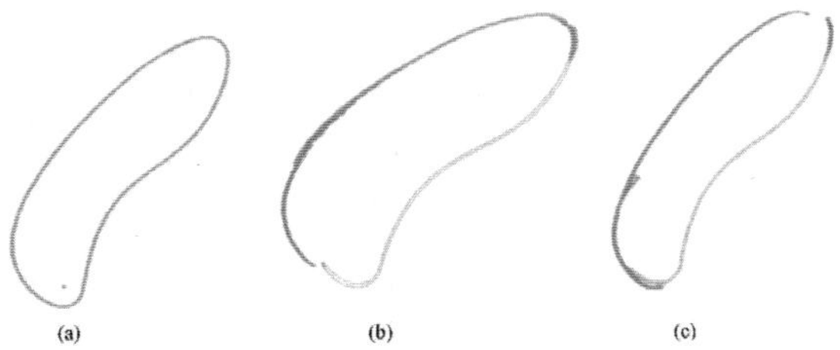

Figure 9. The cross-section of (a) the template point cloud; (b) the NDT+ICP method; and (c) the multi-view registration method.

Figure 10. The final point cloud registration result.

As shown in Table 2, the experimental evaluation in this paper utilizes the point cloud matching rate as the primary index. It involves the fusion of aligned point clouds into a complete point cloud, followed by down-sampling the fused point cloud to match the number of points in the CAD template. Subsequently, a match is established between the down-sampled point cloud and the CAD template by comparing the Euclidean distance between corresponding points. If the distance is below the threshold of 0.2, the point pairs are considered overlapping. The point cloud matching rate is then calculated as the ratio of the number of coincident points to the total number of points in the template point cloud.

5. Conclusion

This paper presents novel algorithms for multi-view planning and registration. The multi-view planning algorithm efficiently plans the scanning viewpoints for the workpiece by employing a greedy strategy. It aims to minimize the number of viewpoints while ensuring sufficient overlap between neighboring point clouds, thereby improving scanning efficiency and reducing the number of point cloud blocks for alignment.

The multi-view registration algorithm incorporates a multi-scale geometric feature extraction network to extract features from the point clouds. These features are then matched using the Hungarian algorithm. A graph is constructed, with workpiece positions as vertices and connecting lines as edges. The G2O algorithm is employed to perform non-linear optimization of the workpiece positions, enabling accurate multi-view registration and mitigating cumulative errors resulting from pairwise registration of multiple-point cloud blocks.

Experimental measurements conducted on blade workpieces provide evidence of the viability and effectiveness of these algorithms.

Acknowledgments

This work was co-supported by the National Key R&D Program of China [Grant No. 2022YFB4702500], and the National Natural Science Foundation of China [Grant No. U22A20176], GDAS' Project of Science and Technology Development [2022GDASZH-2022010108], the Guangdong HUST Industrial Technology Research Institute, Guangdong Provincial Key Laboratory of Manufacturing Equipment Digitization [Grant NO. 2020B1212060014], Guangdong Basic and Applied Basic Research Foundation [2022B1515120078].

Reference

[1] Haowen Deng, Tolga Birdal, and Slobodan Ilic. 3d local features for direct pairwise registration. In IEEE Conference on Computer Vision and Pattern Recognition (CVPR), 2019. 6, 7
[2] Li Ding and Chen Feng. Deep Mapping: Unsupervised map estimation from multiple point clouds. In IEEE Conference on Computer Vision and Pattern Recognition (CVPR), pages 8650–8659, 2019.
[3] Chen Y, Medioni G. Object modeling by registration of multiple range images[J]. Image and vision computing, 1992, 10(3): 145-155.
[4] Bergevin R, Soucy M, Gagnon H, et al. Towards a general multi-view registration technique[J]. IEEE Transactions on Pattern Analysis and Machine Intelligence, 1996, 18(5): 540-547.
[5] Han J, Wang F, Guo Y, et al. An improved RANSAC registration algorithm based on region covariance descriptor[C]//2015 Chinese Automation Congress (CAC). IEEE, 2015: 746-751.
[6] Guo Y, Sohel F, Bennamoun M, et al. TriSI: A distinctive local surface descriptor for 3D modeling and object recognition[C]//International Conference on Computer Graphics Theory and Applications. Scitepress, 2013, 2: 86-93.
[7] Z. Zhang et al., "Point cloud registration method based on 3DMatch network and improved ISS algorithm," 2022 International Symposium on Control Engineering and Robotics (ISCER), Changsha, China, 2022, pp. 161-16.

[8] Toldo R, Beinat A, Crosilla F. Global registration of multiple point clouds embedding the generalized Procrustes analysis into an ICP framework[C]//3DPVT 2010 Conference. 2010, 2: 5.

[9] Guo Y, Sohel F, Bennamoun M, et al. An accurate and robust range image registration algorithm for 3D object modeling[J]. IEEE transactions on multimedia, 2014, 16(5): 1377-1390.

[10] Zhu J, Zhu L, Jiang Z, et al. Scaling registration of multi-view range scans via motion averaging[J]. Journal of Electronic Imaging, 2016, 25(4): 043021-043021.

[11] Gagnon, Soucy, Bergevin, et al. Registration of multiple range views for automatic 3-D model building[C]//1994 Proceedings of IEEE Conference on Computer Vision and Pattern Recognition. IEEE, 1994: 581-586.

[12] Pulli K. Multiview registration for large data sets[C]//Second international conference on 3-D digital imaging and modeling (cat. no. pr00062). IEEE, 1999: 160-168.

[13] Ding L, Feng C. Deep Mapping: Unsupervised map estimation from multiple point clouds[C]//Proceedings of the IEEE/CVF conference on computer vision and pattern recognition. 2019: 8650-8659.

[14] Grisetti G, Kümmerle R, Strasdat H, et al. g2o: A general framework for (hyper) graph optimization[C]//Proceedings of the IEEE International Conference on Robotics and Automation (ICRA). 2011: 9-13.

[15] Thomas H, Qi C R, Deschaud J E, et al. Kpconv: Flexible and deformable convolution for point clouds[C]//Proceedings of the IEEE/CVF international conference on computer vision. 2019: 6411-6420.

[16] Zhou Q Y, Park J, Koltun V. Open3D: A modern library for 3D data processing[J]. arXiv preprint arXiv:1801.09847, 2018.

[17] Qi C R, Su H, Mo K, et al. Point net: Deep learning on point sets for 3d classification and segmentation[C]//Proceedings of the IEEE conference on computer vision and pattern recognition. 2017: 652-660.

[18] Besl P J, McKay N D. Method for registration of 3-D shapes[C]//Sensor fusion IV: control paradigms and data structures. Spie, 1992, 1611: 586-606.

[19] Zhu J, Zhu L, Jiang Z, et al. Local to global registration of multi-view range scans using spanning tree[J]. Computers & Electrical Engineering, 2017, 58: 477-488.

[20] Biber P, Straßer W. The normal distributions transform A new approach to laser scan matching[C]//Proceedings 2003 IEEE/RSJ International Conference on Intelligent Robots and Systems (IROS 2003) (Cat. No. 03CH37453). IEEE, 2003, 3: 2743-2748.

[21] Drost B, Ulrich M, Navab N, et al. Model globally, match locally: Efficient and robust 3D object recognition[C]//2010 IEEE Computer Society Conference on computer vision and pattern recognition. IEEE, 2010: 998-1005.

Journal of Physics: Conference Series **2724** (2024) 012035 doi:10.1088/1742-6596/2724/1/012035

Pipeline inertial measurement mileage correction method based on pipeline junction detection

Xiaopan Chen[1], Ningbo Wang[1], Zhipeng Chen[2], Yu Yin[2,3], Peng Jia[1], Chuanyong Yu[1], Likui Wang[1], Jianwei Yu[2] and Shiwang Lv[2,3,4]

[1]State Grid Xinyuan Henan Tianchi Pumped Storage Co., Ltd., China
[2]Guangdong Key Laboratory of Urban Informatics, Shenzhen University, China
[3]College of Civil and Transportation Engineering, Shenzhen University, China

[4]Corresponding author's e-mail: nplxdsy@163.com

Abstract. A pipeline measurement robot(PMR) is an important tool for pipeline shape measurement and disease detection. Flexible pipelines in the horizontal plane inside dams are used for internal deformation monitoring, which is characterized by small caliber, no ground marking assistance, and higher accuracy requirements, so the accuracy of the data fusion algorithm needs to be improved. In this paper, we propose a pipeline inertial measurement odometry correction method based on pipeline junction detection, which establishes a reference model of pipeline junction locations, detects pipeline junction gap locations along the measurement route using k-mean clustering, corrects odometer data based on the difference between the detected and reference locations, fuses IMU and odometer data and evaluates the internal compliance accuracy. The proposed method is validated using the measured data of the internal pipeline of Tianchi Dam, and the results show that the method proposed in this paper reduces the root-mean-square error (RMSE) by about 46% compared with the results without detection correction and direct fusion. Therefore, the method has good results and practical applications.

1. Introduction

Pipelines have an irreplaceable role in energy transportation, such as oil and gas transportation. The use of pipeline measurement robot (PMR) for trenchless and contact measurement of underground small diameter pipelines is an important pipeline shape measurement and disease detection technology [1,2], which is of great significance for the safe operation of pipelines. PMR is based on a strapdown inertial navigation system (SINS), which can complete the measurement of the whole pipeline without relying on other information and other energies, and is characterized by autonomy, continuity and high local relative accuracy. Therefore, more and more research for PMR has appeared in the field of pipeline inspection, which has become a popular research direction.

The measurement of the PMR is accomplished by the inertial measurement unit (IMU), which is used to measure the angular velocity and acceleration of the motion, thus calculating the attitude angle, velocity and position of the PMR by recursion. Since recursion leads to the accumulation of IMU errors, it is also necessary to integrate additional navigation sensors or set up external auxiliary localization information on the PMR, and then calculate the position and motion routes of the PMR by means of a data fusion algorithm based on Kalman filtering [3,4].

In recent years, some scholars have utilized flexible small-caliber pipelines to conduct research on deformation detection and safe operation and maintenance of large-scale infrastructures, such as the measurement of settlement deformation inside dams. Li et al. [5–7] buried flexible pipes in the horizontal plane inside the dam, and measured the shape change of the flexible pipes by PMR (Figure 1) with an integrated odometer, so as to obtain the vertical and horizontal deformation inside the dam. Unlike general urban underground pipelines and energy pipelines, where the measurement accuracy is about 0.2% to 0.03% of the length of the pipeline, the measurement accuracy of the pipeline for monitoring the internal deformation of the dam is much higher, and the accuracy of the pipeline for a length of 100 meters needs to reach the millimeter level, i.e., no more than 0.01% of the length of the pipeline. At the same time, limited by the internal diameter of the pipeline, the size of the PMR itself cannot be increased indefinitely, which makes it difficult to integrate IMUs with higher accuracy but larger size, and it is difficult to use auxiliary devices external to the pipeline to improve the positioning accuracy, such as above ground markers (AGMs). Therefore, it is necessary to dig deeper into the characteristics of the flexible pipeline inside the dam and the connection with the data obtained from the PMR to improve the accuracy of the data fusion algorithm, in order to provide a more reliable and accurate positioning method for the PMR.

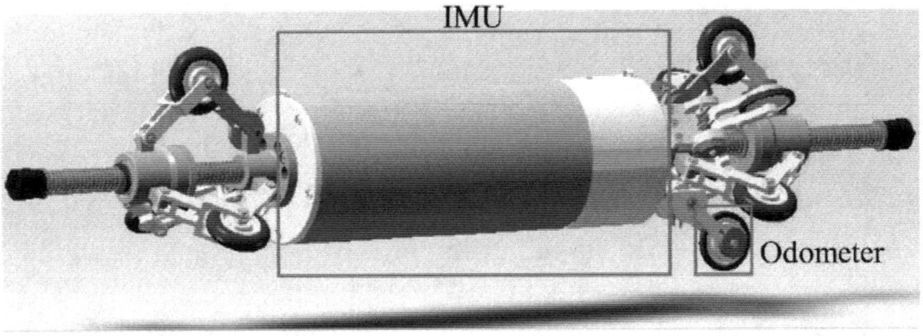

Figure 1. Pipeline measurement robot (PMR) with odometers.

There are many research results on methods to improve the accuracy of pipeline inertial measurement data fusion algorithms. The pipeline junction (PJ), such as weld, bend pipe, valve, flange, etc., is regarded as a class of motion constraints [8], which can be used as the position constraint information of PMR. At present, many researchers have explored how to obtain and utilize the position of pipeline junctions [9–12]. The specific methods include detecting the peak of IMU acceleration through wavelet decomposition to obtain the position of pipeline junctions, and adding the position of pipeline junctions as an observation to the Kalman filter, which ultimately improves the accuracy of the PMR localization. However, there are some shortcomings in this method. The peak detection needs to traverse all the IMU data, which leads to a complicated process and large data computation. Adding constraints at the pipeline connections in Kalman filtering will lead to a rise in the complexity of the algorithm, resulting in a reduction in the efficiency of data analysis. The selection of parameters for wavelet decomposition, such as the analytic formula of the mother wavelet and the number of decomposition layers, is also a problem that has not been solved quantitatively and depends on the experience of researchers.

In order to solve the above deficiencies and improve the accuracy and efficiency of inertial measurement of flexible pipelines in the horizontal plane inside dams, this paper proposes a mileage correction method for pipeline inertial measurement based on pipeline junction detection. First, based on the pipeline construction information and unit pipeline length, k-mean clustering of IMU acceleration is performed to detect the peak mileage location when PMR passes through the pipeline junction. Then, the odometer speeds between neighboring pipeline junctions are adjusted so as to constrain the same pipeline junctions in different measurements to the same mileage position. Finally, data fusion was performed using Kalman filtering to evaluate the internal accuracy of multiple measurements.

In Section 2 of this paper, we illustrate the information about the pipeline junctions. In Section 3, we explain the peak detection algorithm and the method of odometer speed adjustment. In Section 4, we demonstrate the effectiveness of the detection-correction method proposed in this paper by evaluating and validating it with measured data.

2. Information about pipeline junctions

The layout of the pipeline in the internal horizontal plane of the dam is shown in Figure 2, with (a) a three-dimensional view and (b) a vertical section view. After the dam is constructed to a certain elevation, the pipeline burial trench is excavated in the top horizontal plane, the flexible pipeline is put in, and the entrances and exits at both ends of the pipeline are led to the existing observation room. The slope of the pipeline is controlled to be no more than 1%, which means that the elevation of the whole pipeline is generally no more than one meter in elevation rise and fall. The connection method of the flexible pipe is shown in Figure 3. The whole pipe is connected by unit pipes (Figure 3(a)) of equal length, and the adjacent unit pipes are connected to become one by fixing (Figure 3(b)) and checking (Figure 3(c)). Their junction part is shown in the red dotted box in Figure 3(b) and Figure 3(c).

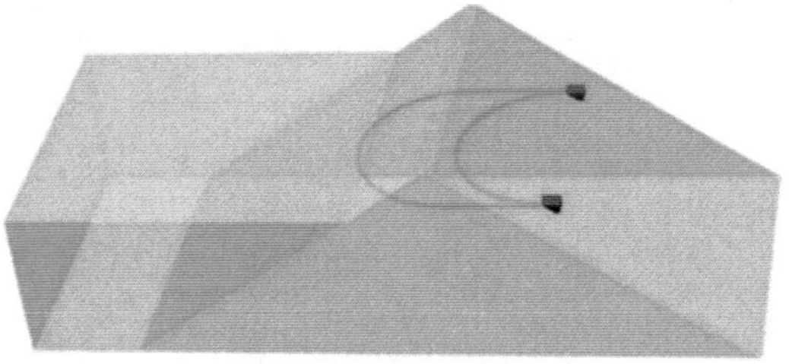

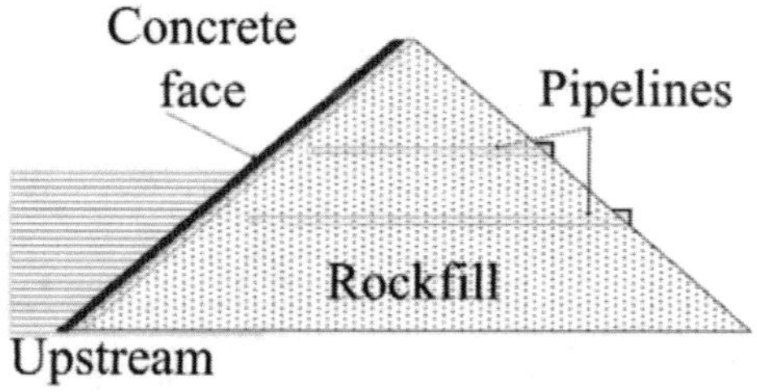

(a) (b)

Figure 2. Layout of flexible pipelines in the horizontal plane inside the dam.

(a) (b) (c)

Figure 3. Connection of flexible pipelines.

In practice, neighboring unit pipes do not connect perfectly, but there is always a gap, known as a pipe junction seam. The PMR will vibrate suddenly when passing through the pipe junction, and since the sensor of the IMU is sensitive enough, it will produce a sudden and obvious acceleration peak. For example, Figure 4 shows the accelerometer output data of PMR after moving through a horizontal pipe inside a dam, in which periodic and regular spikes appear in the graph and occur simultaneously in all three directions.

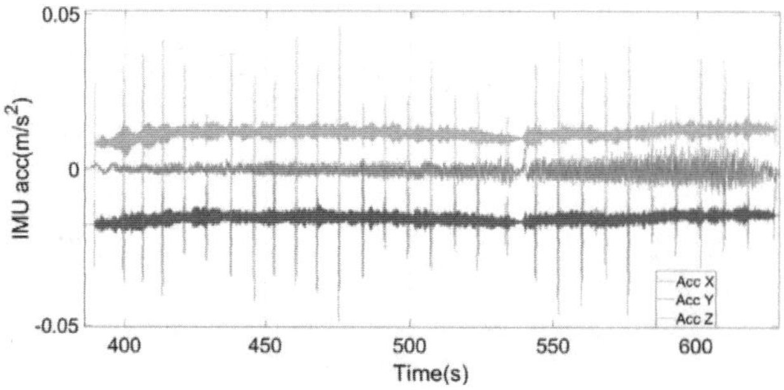

Figure 4. Accelerometer output.

When the pipeline is constructed, only the unit pipes are interconnected without cutting, so the mileage location where each spike occurs is known, i.e., $L_i = i * L_{\text{unit}}, i = 1,2, \dots, K-1$. Where L_i is the mileage location of the ith peak acceleration spike, L_{unit} is the length of the unit pipe, and K is the number of unit pipes contained in the overall pipe. After multiple traversals of the same pipe using PMR, the measured routes from each traverse should have peak acceleration spikes at the same mileage locations. This information is used as a constraint in terms of mileage.

3. Methodology

3.1. Pipeline junction detection method

The PMR performs one full movement from beginning to end or from end to beginning inside the pipeline and collects relevant data, which is called one complete measurement. When the PMR takes a complete measurement inside the pipeline, due to the small elevation fluctuation of the pipeline on the horizontal plane inside the dam, the overall fluctuation range of the acceleration data acquired by the IMU is not large, and basically fluctuates above and below the zero value after eliminating the zero bias error. However, the vibration generated when passing through the seam at the pipeline junction will produce significant acceleration peaks. When the overall acceleration is sorted from largest to smallest, these peaks occupy only a small portion of the front of the sorting table. On the other hand, compared to the overall length of the pipeline, the length of the seam at the pipeline junction is very small, and the resulting acceleration peaks have a very short duration and account for a very small percentage of the total acceleration data. As a result, the acceleration peaks to be detected show a strong tendency of clustering, in the direction of mileage they will be clustered to the location of the pipeline junction seam, while in the direction of acceleration, they will be clustered to the location with the largest value. Since the location of the seam at the pipeline junction is known, we can grasp the two clustering directions of the acceleration peaks to be detected.

Based on the above analysis, we propose to use the k-mean clustering [13] algorithm to detect the location of pipeline junctions. First, the measured route accelerations are sorted by numerical magnitude and the top 1% of the sorted table is intercepted. Then k-mean clustering is performed on this top 1% maximum acceleration data, where the number of clusters is equal to $K - 1$ and the initial center of mass of each cluster is L_i, so as to obtain the point clusters of the gaps at each pipe connection. Finally, the mileage corresponding to the point with the largest acceleration in each point clustering is selected as the position of the pipeline junction in the route of this measurement, thus completing the detection of the pipeline junction.

3.2. Mileage correction method

Because the mileage wheel can skid or slide as it moves, the distance increments measured by the odometer do not match the distance increments actually moved by the PMR. This can lead to an error in the mileage location of the peak acceleration of the seam at the pipeline junction in each measurement route, which does not appear at the same location on each trip. As shown in Figure 5(a), the red straight lines in the figure are the mileage axis and the reference spike mileage location L_i, the blue and green curves are the acceleration data obtained from the two measurements, and the circles are the detected pipeline junction locations.

In order to eliminate the mileage location error, the output data of the odometer needs to be corrected. Firstly, it is necessary to calculate the mileage difference between the acceleration peak locations and the corresponding reference spike locations L_i. Then the output data of the odometer between the acceleration peak at that location and the previous peak or starting point is adjusted. If the cumulative distance of the odometer is greater than the actual distance, the output data of the odometer is cut, and if the cumulative distance of the odometer is less than the actual distance, the output data of the odometer is compensated. The effect of the correction is shown in Figure 5(b), where the same pipeline junction seams are calibrated to the same locations in different measurement routes, thus improving the accuracy of the odometer output data.

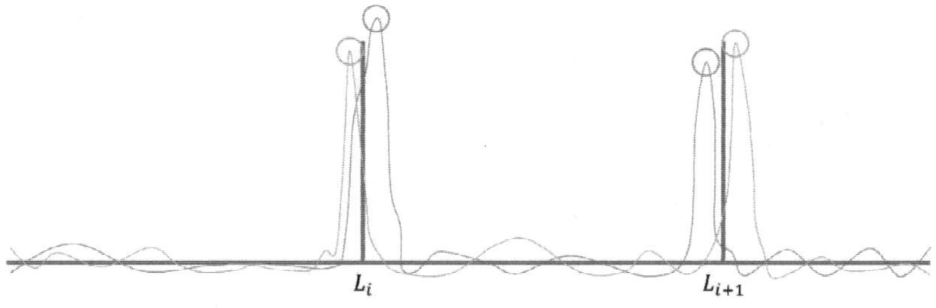

(a)

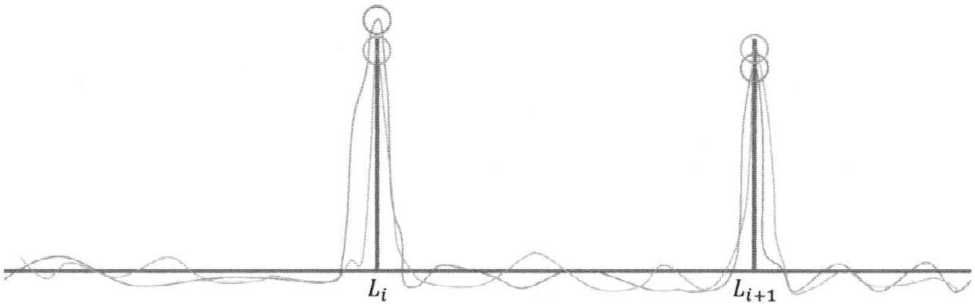

(b)

Figure 5. Seam locations of pipeline junctions for multiple routes before and after correction.

3.3. Accuracy evaluation method

Internal accuracy is of practical significance for assessing the reliability and stability of pipeline measurement robots, and is identified as the basis for evaluating the impact of measurement methodology on the accuracy of results [14,15]. After mileage correction, we use Kalman filtering to fuse the acceleration and angular velocity acquired by the IMU and the distance increment acquired by the odometer to obtain the shape of the flexible pipeline in the horizontal plane inside the dam, i.e., the mileage-elevation curve, which is described in reference [16–18].

Each complete measurement obtains a mileage-elevation curve $h_i = f(d_i)$ for the pipeline once, h_i and d_i are the elevation and mileage of the ith sampling point, $i = 1,2, \dots N$ respectively, the subscript i represents different locations of the same pipeline, and N is the total number of sampling points. We let one operation make M complete measurements of the monitoring pipeline, and get M mileage-elevation curves $h^j = f(d^j), j = 1,2, \dots, M$, with the superscript j representing different measurements of the same operation. The internal accuracy of this operation is expressed by the root mean square error (RMSE) σ. The mean value is used as the reference estimator, and σ is calculated as shown in Equation (1) and Equation (2). Obviously, the smaller the value of σ is, the higher the internal accuracy is.

$$\sigma = \sqrt{\frac{\sum_{i=1}^{N} \sum_{j=1}^{M} (h_i^j - \overline{h}_i)^2}{MN}} \tag{1}$$

$$\overline{h}_i = \frac{1}{M} \sum_{i=1}^{M} h_i^j \tag{2}$$

4. Results

The application experiment site is located in Tianchi Dam in Nanyang City, Henan Province, which is a panel rockfill dam with a maximum height of 118 meters. The monitoring pipeline layout of Tianchi Dam is shown in Figure 6. Three monitoring pipelines, named 1010 s (short), 1010 l (long) and 1037, with lengths of 261 m, 369 m and 261 m, respectively, are connected by unit pipes with lengths of 9 m on two horizontal surfaces with elevations of 1010 m (Figure 6(a)) and 1037 m (Figure 6(b)). The inspection results and internal accuracy of the pipe junction were analyzed using the 1010 s pipe as an example.

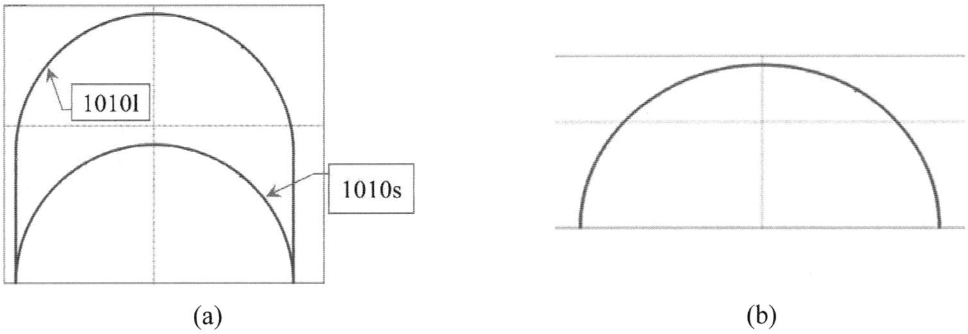

Figure 6. Map of Tianchi Dam monitoring pipelines at 1010 m and 1037 m height plans.

4.1. Results of pipeline junction detection

The 1010 s pipeline consists of 29 unit pipes connected with a length of 9 m, so the reference spike mileage location is $L_i = 9i, i = 1,2, \ldots ,28$. As an example, for the September 2021 data from the 1010 s pipeline, three measurements of the monitoring pipeline were conducted in this period, and the PMR made three round trips during each measurement, for a total of 18 complete measurements, which took a total of 105 minutes. The detection results are shown in Figure 7.

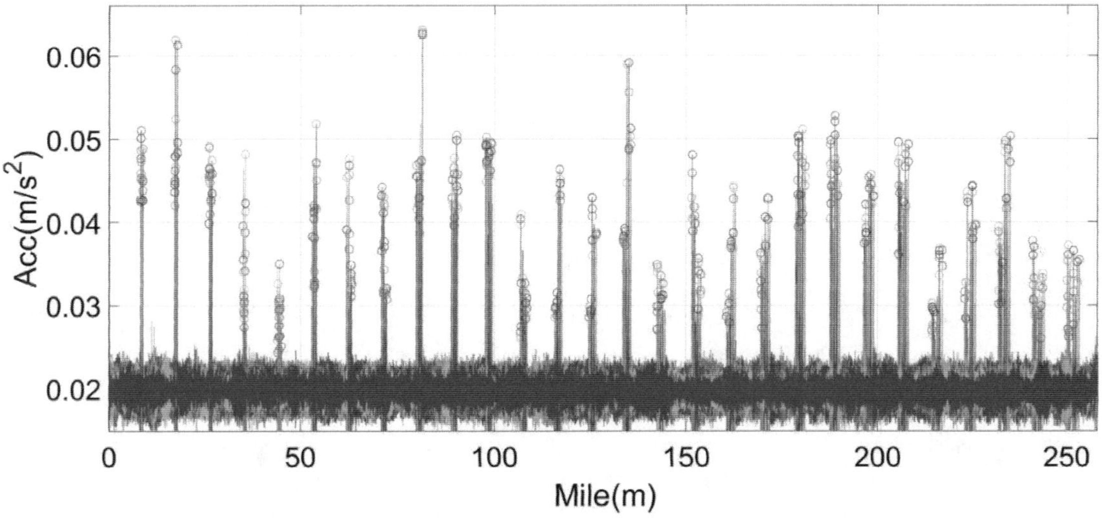

Figure 7. Detection results of the locations of 1010 s pipeline junctions.

As can be seen from the figure, the pipeline junction detection method proposed in this paper has detected a total of 504 acceleration peak points, which are consistent with the values of the reference spikes to be detected, and they are all located in the pipeline junction seams, which indicates that the pipeline junction detection method has a very good effect.

4.2. Results of internal accuracy analysis

Fusion calculations were performed on the September 2021 data for the 1010 s pipeline. Figure 8 and Figure 9 show the mile-relative elevation graph (a) and the mile-elevation residual graph (b) relative to the mean for each of the routes, where Figure 8 has not been mileage corrected and σ_{before} =7.02 mm, and Figure 9 has been mileage corrected and σ_{after}=3.75 mm. In Part (a) of both Figures, the images of the same area have been zoomed in so that it is easy to see the degree of clustering of multiple survey routes at that location.

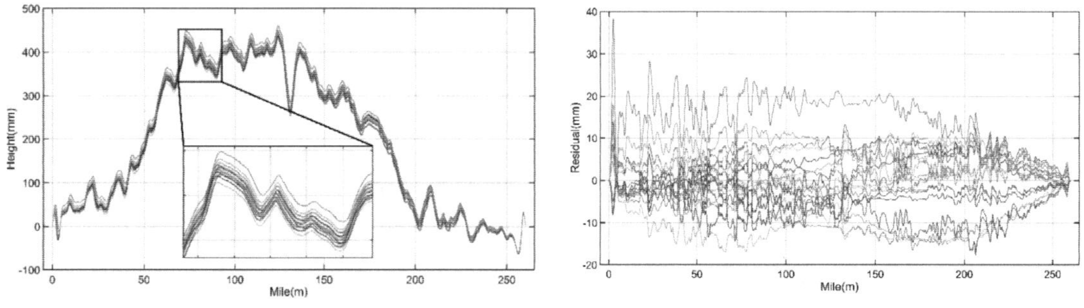

(a) (b)

Figure 8. Graph of pipeline's route before mileage correction.

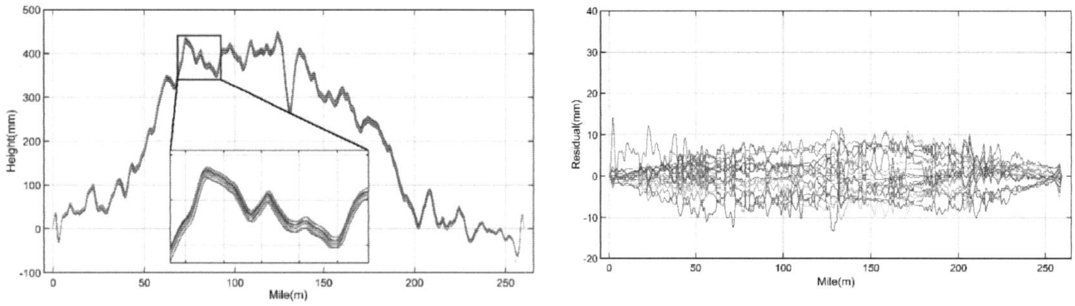

(a) (b)

Figure 9 Graph of pipeline's route after mileage correction.

As can be seen from the figure, after the mileage correction, the route of multiple measurements of the same pipeline is more centralized, the standard deviation of the whole route at almost every position is reduced, and the σ of this period is also reduced from 7.02 mm to 3.75 mm, which is reduced by 46.58%, thus indicating that the mileage correction is able to effectively improve the internal accuracy of PMR.

5. Conclusion

The main objective of the work carried out in this paper is to improve the accuracy of inertial measurements of flexible pipelines in the horizontal plane inside dams. We first build a reference model of the location of the pipeline junctions, use it as the center of mass for k-mean clustering of the IMU acceleration measurements, and detect the location of the pipeline junction seams in each route of the PMR measurement. Then the odometer measurements are corrected according to the bias of the locations. Finally, the IMU and odometer data are fused and the internal compliance accuracy is calculated. The newly developed detection-correction algorithm also reduces the amount of data analysis and the complexity of the data fusion algorithm, thus improving calculation efficiency. The algorithm of this paper was validated using the internal horizontal pipeline of Tianchi Dam as a data sample. The results show that the new method can improve the concentration of multiple measurement routes, reduce the RMSE by 46.58%, and improve the internal accuracy of pipeline inertial measurement with good practical results.

Acknowledgement

This work is supported by Science and technology project funding by State Grid Xin Yuan Company Limited (SGXYKJ-2020-097).

References

[1] Mohammed M N, Shini Nadarajah V, Mohd Lazim N F, Shazwany Zamani N, Al-Sanjary O I, et al. Design and Development of Pipeline Inspection Robot for Crack and Corrosion Detection [C/OL]//2018 IEEE Conference on Systems, Process and Control (ICSPC). 2018: 29-32 [2023-09-28]. https://ieeexplore.ieee.org/abstract/document/8704127.

[2] Ab Rashid M Z, Mohd Yakub M F, Zaki bin Shaikh Salim S A, Mamat N, Syed Mohd Putra S M, et al. Modeling of the in-pipe inspection robot: A comprehensive review [J/OL]. Ocean Engineering, 2020, 203: 107206. https://doi.org/10.1016/j.oceaneng.2020.107206.

[3] Sahli H. MEMS-based aided inertial navigation system for small diameter pipelines [D/OL]. AB, Canada: University of Calgary Calgary, 2016. https://prism.ucalgary.ca/server/api/core/bitstreams/48bcad0d-f061-4b4a-8a72-c52940c866cb/content.

[4] Al-Masri W M F, Abdel-Hafez M F, Jaradat M A. Inertial Navigation System of Pipeline Inspection Gauge [J/OL]. IEEE Transactions on Control Systems Technology, 2020, 28(2): 609-616. https://doi.org/10.1109/TCST.2018.2879628.

[5] LI Q, CHEN Z, YIN Y, CHENG X, Zhu J. Pipeline three-dimensional curve measuring robot and implementation method therefor: CN109780370A [P/OL]. 2019-05-21 [2022-09-07]. https://patents.google.com/patent/CN109780370A/zh.

[6] Chen Z, Yin Y, Yu J, Cheng X, Zhang D, et al. Internal deformation monitoring for earth-rockfill dam via high-precision flexible pipeline measurements [J/OL]. Automation in Construction, 2022, 136: 104177. https://doi.org/10.1016/j.autcon.2022.104177.

[7] Yin Y, Chen Z, Li Q, Zhang D, Liang A, et al. High-Precision Measuring System for Internal Deformation Monitoring Pipeline of Rockfill Dams: Design and Validation [J/OL]. IEEE Sensors Journal, 2023, 23(11): 12031-12040. https://doi.org/10.1109/JSEN.2023.3265023.

[8] Zhang S, Dubljevic S. Trajectory determination for pipelines using an inspection robot and pipeline features [J/OL]. Metrology and Measurement Systems, 2021, 28(3): 439-453. https://doi.org/10.24425/mms.2021.137134.

[9] Zhou Y, Ma J, Li F, Chen B, Xian T, et al. An Improved Algorithm for Peak Detection Based on Weighted Continuous Wavelet Transform [J/OL]. IEEE Access, 2022, 10: 118779-118788. https://doi.org/10.1109/ACCESS.2022.3220640.

[10] Sahli H, El-Sheimy N. A novel method to enhance pipeline trajectory determination using pipeline junctions [J/OL]. Sensors, 2016, 16(4): 567. https://doi.org/10.3390/s16040567.

[11] Guan L, Cong X, Sun Y, Gao Y, Iqbal U, et al. Enhanced MEMS SINS Aided Pipeline Surveying System by Pipeline Junction Detection in Small Diameter Pipeline [J/OL]. IFAC-PapersOnLine, 2017, 50(1): 3560-3565. https://doi.org/10.1016/j.ifacol.2017.08.962.

[12] Lv Z, Wang G, Wang Z, Zhao H, Gao W. Application of Weld Scar Recognition in Small-Diameter Transportation Pipeline Positioning System [J/OL]. Electronics, 2022, 11(7): 1100. https://doi.org/10.3390/electronics11071100.

[13] Liu X, Jia M, Ding H. Uplink resource allocation for multicarrier grouping cognitive internet of things based on K-means Learning [J/OL]. Ad Hoc Networks, 2020, 96: 102002. https://doi.org/10.1016/j.adhoc.2019.102002.

[14] XU D, LUO Y, JIA W. A study of the evaluation method for data quality of aeromagnetic repeated linear measurement [J/OL]. Geophysical and Geochemical Exploration, 2014, 38(04): 729-731. https://doi.org/10.11720/wtyht.2014.4.17.

[15] Kane J S. Accuracy and Its Determinants in Distributional Assessment [J/OL]. Human Performance, 2000, 13(1): 47-84. https://doi.org/10.1207/S15327043HUP1301_3.

[16] Zhang P, Hancock C M, Lau L, Roberts G W, de Ligt H. Low-cost IMU and odometer tightly coupled integration with Robust Kalman filter for underground 3-D pipeline mapping [J/OL]. Measurement, 2019, 137: 454-463. https://doi.org/10.1016/j.measurement.2019.01.068.

[17] Xu X, Sun Y, Gui J, Yao Y, Zhang T. A Fast Robust In-Motion Alignment Method for SINS With DVL Aided [J/OL]. IEEE Transactions on Vehicular Technology, 2020, 69(4): 3816-3827. https://doi.org/10.1109/TVT.2020.2974524.

[18] Lv S, Chen Z, Li Q, Luo T, Xu R, et al. A Measurement and Accuracy Evaluation Method for Internal Dam Settlement [C/OL]//2022 7th International Conference on Mechatronics System and Robots (ICMSR). 2022: 20-24. https://doi.org/10.1109/ICMSR2020.2022.00012.

Application of particle filter algorithm in satellite positioning and navigation system

Fangbin Ye[1,3], Yingjie Xiang [1], Chi Jiang [1], Jinrong Wang [1] and Taoyue Wang [2]

[1] State Grid Zhejiang Electric Power Co., Ltd., Hangzhou 310000, Zhejiang Province, China

[2] Hangzhou Tianchenwei Technology Co., Ltd., Hangzhou 310000, Zhejiang Province, China

[3] Corresponding author's e-mail: 18658188509@163.com

Abstract: To understand the application of satellite positioning and navigation system, the application of a particle filter algorithm in satellite positioning and navigation system is studied. In this paper, firstly, the state equation of the system is established according to the correlation between the parameters of GPS output, and the particle filter algorithm is applied to filter the GPS-received data. In the CPS navigation system, DSP is used to decode and filter the output signal of the CPS receiver. The results show that when the GPS signal is disturbed, the particle filter algorithm can deal with the abnormal observation samples well, which shows that it can reduce the computational complexity and improve the GPS positioning accuracy, and have a certain robustness.

1. Introduction

GPS is a hot and potential research topic at present. GPS is a new generation of satellite navigation and positioning system with positioning capability, which can carry out omnidirectional real-time two-dimensional and three-dimensional navigation in the sea, land, and air. The biggest problem in the application of GPS is the phenomenon that positioning deviates from the facts. However, with the cancellation of SA, the positioning accuracy (2D) of GPS is improved from 100 m to 15 m to 20 m, even within 10 m. Sometimes, the positioning error deviates from the real position by more than 50 m. Global Positioning System (GPS) can provide positioning and navigation information around the clock, quickly and accurately. But in urban high-rise areas, forest paths, culverts, and deep mountain canyons, the above functions of GPS often fail. Dead reckoning (DR) system is a commonly used positioning technology, but the direction error of its direction sensor is large and will accumulate with time, so the DR system cannot be used alone for a long time. By combining the DR system with GPS+6', it is possible to maintain normal navigation by using the autonomous positioning results of the DR system when the GPS signal is missing and the navigation system cannot work normally. The positioning error of the DR system accumulates with time. When the accumulated error value is large, it can be corrected and calibrated by using the accurate positioning results of GPS in normal operation. The error accumulation of the DR system is mainly caused by the random errors of various sensors (rate gyro, odometer, etc.). To effectively improve the positioning accuracy and reliability of the system, it is necessary to choose a good estimation method. At present, the commonly used estimation method is the Kalman filter (KF) algorithm, but in practical problems, few cases can meet the conditions of the Kalman filter. To solve the nonlinear and non-Gaussian problems, a commonly used particle filtering method is the extended Kalman filter (EKF) algorithm. Although the EKF algorithm is simple and easy to implement, its linearization process will inevitably produce model errors, which will lead to a decline in estimation accuracy [1-2].

2. Equation design of the navigation system

In practical application, the parameters of GPS transmission include longitude, latitude, speed, and speed direction. We calculate the distance traveled in this second by using the speed parameters transmitted each time and then decompose this distance into longitude and latitude components according to the speed direction angle. At this time, the distance traveled in the longitude and latitude directions in one second is obtained. These two distance parameters are also converted into angle parameters, and finally, the converted angle is added with the longitude and latitude values of the previous moment. Thus, the system state and output equations are obtained, as shown in Equations (1) and (2).

System state equation:

$$\left\{ \begin{matrix} x^{k+1} = x_k + 0.5*(u_{k+1} + U_k)*\cos(\theta_{k+1} * pi/180 \\ (lat*\cos(yk*pi/180)) + u \end{matrix} \right\} \tag{1}$$

$$z^1_{k+1} = x_{k+1}$$
$$z^2_{k+1} = y_{k+1} \tag{2}$$

where x is the longitude state variable; y is the latitude state variable; Speed 0 is speed direction; Lat is the distance converted by moving a part on the equator (unit: m/min); Lon is the distance converted by moving one part online (unit: m/min); u_1 and u_2 is system noise; z^1 is longitude; z^2 is latitude [3].

2.1. Particle filter algorithm

The particle filter algorithm has a large amount of calculation, which is not conducive to the application in high real-time systems. This paper aims at improving the operation speed and filtering accuracy of the particle filter algorithm and adopts the methods of reducing division operation and improving weight calculation. Normalization of weights increases the complexity of the algorithm. Because division operation costs a lot of hardware, reducing division operation as much as possible is the key to improving the operation speed of the particle filter algorithm. After repeated research on particle filter algorithms, it is found that the step of weight normalization can be combined into several steps such as weight calculation, result output, and resampling, thus reducing a lot of division operations and hardware overhead.

The specific improvement methods are as follows:

The output result of the particle filter algorithm is that each particle value is multiplied by the corresponding normalized particle weight. The output result of the improved particle filter algorithm is changed to that each particle value is multiplied by the corresponding non-normalized particle weight and then divided by the sum of the weights, that is, the output result is expressed by Equation (3).

$$x_k = \left(\sum_{m=1}^{M} x_k^{(m)} * w_k^{*(m)} \right) / S_k \tag{3}$$

The effective sample size of the particle degradation detection process in the particle filter algorithm is the reciprocal of the sum of squares of M normalized weights. The effective sample size of the particle degradation detection process in the improved particle filter algorithm is the reciprocal of the sum of squares of M non-normalized weights multiplied by the square of the sum of weights, that is, the effective sample size of particle degradation detection process in particle filter algorithm [4-5].

$$N_{eff} = 1/\sum_{m-1}^{M} W_k^m \tag{4}$$

In the particle filter algorithm, the uniformly distributed random number generated by the resampling process is in the interval (0-1/M), and in the improved particle filter algorithm, the uniformly distributed random number generated by the resampling process is in the interval (0-Sk/M).

3. System design of GPS navigation

The structural block diagram of the CPS navigation system is shown in Figure 1. The satellite signal is received by the antenna of the GPS receiver and transmitted to the GPS receiver through the wire. The GPS receiver analyzes and preprocesses the received satellite signal, and then transmits the preprocessed information such as longitude and latitude to the DSP through the serial port. The DSP decodes and filters the transmitted GPS signal, and finally transmits the filtered longitude and latitude values to MATLAB for storage [6].

Figure 1. System structure block diagram.

The navigation system uses u-blox LEA-5H as the GPS receiving module. This module has 50 channels of u-blox positioning engine and more than one million correlators and super sense indoor GPS (-160 dbm acquisition and tracking sensitivity), which has strong anti-interference ability. The output of the GPS receiving module is the NMEA -0183 code. This coding is mainly composed of frame header, frame trailer, and intra-frame data. According to different frame headers, the intra-frame data is different, mainly including "$ GPGGA", "$ CPGSA", "$ G-PCSV", and "$ GPRMC". These frame headers identify the structure of data in subsequent frames, and each frame uses carriage return and line feed as the end of the frame to mark the end of a frame. In practical application, the position is estimated only by the longitude, latitude, speed, and speed direction transmitted by GPS, so $ CPRMC is selected [7].

RS232 serial communication is adopted between the GPS receiving module and the DSP development board. The selected parameters of serial communication are baud rate =9, 600 bps, data bit =8 bits, start bit = 1 bit, stop bit = 1 bit, and there is no parity. The output statement selected by the GPS module is the $ GPRMC statement, and the output frequency is 1 Hz. If the GPS receiver is working, it will continuously transmit GPS navigation and positioning information to DSP through the serial port [8].

CETEK-5509A is the DSP development board selected for the navigation system, and the core processing chip on this development board is TMS320VC5509A DSP (hereinafter referred to as C5509A) of TI company. C5509A has the characteristics of fast processing speed and high efficiency of multiplication and division, exponential operation, trigonometric function, and random number generation function in the library function. C5509A is a fixed-point digital signal processor DSPl5 in TMS320 series DSP chips produced in America. Its structure is specially designed for real-time signal processing. It has the characteristics of flexible instructions, strong operability, high speed, and support for parallel operation and C language. It is a DSP with high-cost performance and has been widely used in portable devices. The C5509A on the development board communicates with the GPS module through the Max232 (serial voltage conversion chip) and 16C550 (asynchronous serial communication transceiver) on the board. The serial port receives the data and puts it in the cache. Before further processing, the cache is a long string of byte streams, and this information cannot be used until it is classified and extracted. Therefore, the information on the longitude, latitude, speed, and speed direction of each field is extracted from the buffer byte stream by the DSP program and transformed into practical positioning information data.

After many programming simulation tests, it is found that the performance of C5509A is relatively stable when the main frequency is 120 MHz. According to the requirements of being real-time, the frequency of serial port transmission data is 1 Hz, so the DSP processor should decode and filter the transmitted data within one second and the number of particles should not be too much. After repeated tests, it is found that the number of particles is 60 [9-10].

4. Experiment and result analysis

4.1. Experimental methods

This experiment adopts the method of MATLAB and CCS joint simulation and real-time data exchange (RTDX) programming. RTDX is an excellent real-time data transmission technology, which provides a brand-new method for software debugging of DSP systems. It uses the internal simulation logic of the DSP and JTAG interface to realize the data exchange between the host and target. Without occupying I/O resources such as DSP system bus and serial port, data transmission can run in the background of application programs. C5509 and its integrated development environment CCS 2.21 are connected through the CCSLink tool integrated in MATLAB 7.1. In the MATLAB environment, the operation of the CCS and DSP target board can be completed, including data exchange with the target board, detecting the state of the processor and controlling the operation of DSP programs.

In this experiment, the data transmitted by CPS includes longitude, latitude, speed, and speed direction angle. The data received by MATLAB includes longitude and latitude, and the distribution of particles is also sampled. The time used is 60 seconds and the driving distance is about 0.8 kilometers. The improved BPF algorithm is used to filter the signal output by GPS. The system state equation and system output equation are Equation (1) and Equation (2), respectively, and the number of particles is 60.

4.2. Analysis of experimental results

The improved BPF algorithm filters the longitude and latitude of the GPS output, and the results and the time taken for each filtering are shown in Table 1.

Table 1. The longitude and latitude filtering results.

Latitude	27.7	28	29	28
Comparison before and after GPS filtering	12	14	15	12

It can be seen from Table 1 that the filtering effect of the improved BPF algorithm is very good. The longitude and latitude signals output by GPS are very noisy, and the error range is about 2 meters to 30 meters. After filtering, the longitude and latitude are very close to the truth, and the error range is reduced from 2 meters to 30 meters before filtering to 1.5 meters to 8 meters, thus improving the accurate positioning ability of CPS. The specific statistical results are shown in Table 2.

Table 2. Summary of particle filter algorithm in filtering results.

The variance of longitude and latitude values and filtered values (points)	0.0019
The variance of longitude and latitude value and noise value (points)	0.0048
The variance of longitude and latitude values and filtered values (points)	0.0021
The variance of longitude and latitude value and noise value (points)	0.02
Average running time (s)	0.0072

It can be seen from Table 2 that the variance between the filtered value and the true value after filtered by the improved BPF algorithm is smaller than the variance between the noisy value and the true value of GPS output. The greater the variance is, the greater the external interference (that is, the greater introduced noise) of GPS signals is, and the lower the positioning accuracy is. It can be seen from Table 1 that the improved BPF algorithm can be applied to GPS navigation, which can improve the accuracy and anti-interference ability of GPS navigation, and the operation time is very short, which can meet the requirements of CPS filtering with a frequency of 1 Hz. In this experiment, the motion direction angle provided by GPS is also disturbed by external noise, which reduces the filtering accuracy. If a dynamic north finder can be added to correct the motion direction angle provided by GPS, the filtering effect of the particle filter will be more perfect.

5. Conclusion

In the application of GPS positioning, the deviation caused by multiple influences limits its practical application. The particle filter algorithm optimizes the internal structure of the algorithm by improving the steps of importance weight calculation, resampling, and result output, and plays a role in reducing the deviation when applied to the GPS positioning system. The optimization of the algorithm makes it easy to realize hardware programming and improve the calculation speed, which is expected to be used in high-precision GPS navigation.

References

[1] Liu J., Hao J., Yang Y., Xu Z., Liu W., & Wu R. (2022). Design optimization of low earth orbit constellation based on Beidou satellite navigation system precise point positioning. IET Radar, Sonar, & Navigation 12 (8), 16.

[2] Yang L., Yang J., Yao W., & Liu H. (2021). Design of autonomous navigation system for electric patrol unmanned aerial vehicle based on Beidou positioning and satellite-based enhancement. Journal of Physics: Conference Series, 2005 (1), 012093.

[3] Ge H., Li B., Jia S., Nie L., Wu T., Yang Z., et al. (2022). Leo enhanced global navigation satellite system (LEGNSS): progress, opportunities, and challenges. Geo-spatial Information Science, 25 (1), 1-13.

[4] Lu Huang, Bao Guoyu, Hong Shengli, Jun LI, Hao Nanjia, Jian Qiangcheng, et al. (2022). Pseudolite fingerprint positioning method under GNSS rejection environment. Acta Electronica Sinica, 50 (04), 811-822.

[5] Song J., Wu H., Guo X., Li S., & Li Y. (2021). Credible navigation algorithm for GNSS attack detection by using an auxiliary sensor system. Applied Sciences, 11 (14), 6, 321.

[6] Filjar R., Damas M. C., & Iliev T. B. (2021). Modern science, economy, and society empowered by resilient satellite navigation. IOP Conference Series: Materials Science and Engineering, 1, 032 (1), 012001 (9pp).

[7] Li X., Wang H., Li S., Feng S., Wang X., & Liao J. (2021). Gil: a tightly coupled GNSS ppp/ins/lidar method for precise vehicle navigation. Satellite Navigation, 2 (1), 1-17.

[8] Jun L. U., Chengeng S. U., Min H. U., Tang Z. P., Yuan H. B., Junyi X. U., et al. (2021). Analysis of the Beidou navigation satellite system: interoperability and performance. SCIENTIA SINICA Physica, Mechanica, & Astronomica, 51 (1), 019504.

[9] Chen J. P., Chao Y. U., Zhou J. H., Wang A., Zhang Y. Z., & Song Z. Y. (2021). Decimeter-level algorithm for satellite-based augmentation systems and performance analysis of bds-2/bds-3. SCIENTIA SINICA Physica, Mechanica, & Astronomica, 51 (1), 019507.

[10] Li Q., Li Z., Zhang J., Cao X., Ma Y., & Wan Q. (2021). A vector ionospheric delay correction method for GNSS positioning and navigation-wide area augmentation service. IOP Conference Series: Earth and Environmental Science, 668 (1), 012079 (5pp).

Experimental and numerical study on wet steam flow metering overreading characteristics in venturi tube

Fang Wu[1], Zhaokai Xing[2,3], Yuanhua Ma[1], Rongyuan Hu[2], Zhao Cheng[1], Hongjun Xie[2], Shanshan Wang[2], Jinfeng Yang[2], Shanliang Qu[2] and Cuilian Zhao[2]

[1] Shandong Nuclear Power Co., Ltd, Yantai 265116, China
[2] State Nuclear Electric Power Planning Design & Research Institute Co., Ltd, Beijing 100095, China

[3]Corresponding author's e-mail: xingzhaokai@snpdri.com

Abstract: The precise measurement of wet-steam flow is of great importance in various industrial systems, including nuclear reactors, oil and gas transportation, and cogeneration systems. However, the measurement accuracy of wet-steam flow can be affected by overreading (OR) when using venturi meters. This paper presents a systematic numerical simulation using the DPM model and an experimental study on the characteristics of overreading in wet-steam flow metering. The experiments are carried out on a two-phase flow liquid and steam platform. Firstly, the OR characteristics of a venturi structure with fixed dimensions (a throttling ratio of 0.6, a divergence angle of 15°, and a venturi diameter of 20 mm) are quantitatively analyzed at different flow rates (0.14-0.32 kg/s) and qualities (ranging from 0.14 to 0.90). It is found that the Lockhart-Martinelli parameter (X_{LM}) plays a more significant role than the gas Froude number (Fr_g) in determining the overreading (OR) characteristics of a venturi under a given temperature, pressure, and venturi structure. Secondly, a venturi overreading prediction model based on X_{LM} is proposed, and its ability to predict overreading for larger mass flow rates and venturi diameters is verified by numerical simulation data. As evaluated through numerical simulation data, the proposed model's maximum relative error and average deviation are below 15% and 10%, respectively. This study provides valuable insights into developing an overreading model for industrial applications under large flow and venturi diameter conditions.

1. Introduction

Accurately metering wet steam flow is critical in various industrial applications, including nuclear reactors and oil and gas transmission [1]. The venturi meter has been identified as a promising solution for wet steam flow measurement due to its simple design and high reliability [2]. However, overreading occurs during wet steam flow measurement when a venturi is used, leading to distortion of the measured value [3]. Therefore, developing a reasonable venturi overreading correction model is essential for achieving accurate wet steam flow metering.

The factors contributing to overreading include the Lockhart-Martinelli parameter X_{LM}, gas Froude number Fr_g, pressure P, and throttling ratio β [4]-[6]. Among the existing studies on venturi overreading models, theoretical models such as the Homogeneous Model [7], Separated Model [8], and Chisholm Model [9] are deemed less accurate due to their oversimplification, which makes them unable to reflect the actual situation of two-phase flow. Therefore, these models must be revised with experimental data before being applied to engineering wet steam metering. To this end, semi-empirical and semi-theoretical models represented by the De Leeuw Model have been proposed [10]-[14] based on the abovementioned concept. However, these empirical models are constrained by factors such as the working fluid, venturi structure, and experimental conditions, which limits their application scope. Current empirical model studies typically employ working fluids such as air, nitrogen, or petroleum, focusing mostly on normal temperature and pressure conditions. There is a lack of research on the overreading characteristics of wet steam under high temperature and pressure conditions.

On the other hand, numerical simulation technology has also been widely used in overreading research [15]-[18], mainly to verify the performance of empirical models under difficult-to-conduct experimental conditions.

This paper carries out the following work. Firstly, based on a high-temperature and high-pressure experimental platform, the overreading characteristics of a venturi meter with a small diameter are studied under high temperature, high pressure, and low mass flow rate conditions. The overreading characteristics are quantitatively analyzed in different mass flow rates and quality ranges under specified venturi structures and pressure conditions. Subsequently, an overreading venturi model is proposed based on the Lockhart-Martinelli parameter X_{LM} (hereafter referred to as 'this model'). Finally, the generality of this model for larger venturi diameters and mass flow rates is verified using the Ansys Fluent platform. This study will contribute to developing an overreading model suitable for use in industrial applications involving large mass flows and venturi diameters.

2. Definition of overreading coefficient and its affecting factors

2.1 Definition of overreading coefficient [19]

As a result of the obstruction effect of the liquid on the gas flow in wet steam, the venturi pressure drop in wet steam flow is higher than that of the gas singly flowing through at an equal mass flow rate. Generally, the degree of this overestimation is reflected by utilizing the overreading coefficient,

$$OR_{before} = \sqrt{\frac{\Delta P_{ws}}{\Delta P_g}} \tag{1}$$

2.2 The affecting factors of overreading [20]

Current studies have indicated that several factors, including the Lockhart-Martinelli parameter X_{LM}, gas Froude number Fr_g, pressure P, and throttling ratio β, influence the overreading of venturi.

2.2.1 Lockhart-Martinelli parameter X_{LM}. X_{LM} is closely related to the gas-liquid density ratio and quality, and it is a dimensionless parameter that characterizes the gas and liquid superficial inertia forces,

$$X_{LM} = \sqrt{\frac{\rho_g}{\rho_l}\frac{1-x}{x}} \tag{2}$$

As the X_{LM} parameter increases, the influence of the liquid phase on the gas phase becomes more pronounced, leading to an increase in overreading.

2.2.2 Gas froude number Fr_g. Fr_g is a dimensionless similarity criterion number characterizing the ratio of the gas inertial force to the gravitational force,

$$Fr_g = \frac{v_{sg}}{\sqrt{gD}}\sqrt{\frac{\rho_g}{\rho_l - \rho_g}} \tag{3}$$

Assuming the decrease in Fr_g is equivalent to the decrease in gas superficial velocity, which reduces friction between the gas and liquid. As a result, the overreading decreases.

2.2.3 Wet steam pressure P. As the pressure decreases, the gas density also decreases, while the liquid density remains relatively constant due to its low sensitivity to pressure changes. Consequently, the gas velocity increases, while the liquid velocity increases only slightly (compared to the gas velocity). It results in the sliding velocity's increase, and the liquid increasingly hinders the vapor. Finally, the overreading becomes greater.

2.2.4 Throttling ratio β. When the inlet diameter of the venturi remains constant, a smaller throttling ratio reduces the effective flow cross-sectional area of the throat, resulting in a more pronounced obstruction of the liquid to the gas. So, a smaller throttling ratio will lead to an increase in overreading.

3. Materials and methods

3.1 Control of experimental variables

This paper investigates the characteristics of wet steam overreading for a specified pressure and venturi structure. In Eqs. (2) and (3), for a given gas and liquid density and venturi structure, the X_{LM} parameter is solely dependent on the quality, while the Fr_g parameter is solely dependent on the gas superficial velocity v_{sg}, which is a variable coupled with the steam mass flow rate. The v_{sg} can be calculated using the following equation,

$$v_{sg} = \frac{4xG_{ws}}{\rho_g \pi D^2} \tag{4}$$

It can be observed from Eqs. (3) and (4) that the Fr_g parameter is also dependent solely on the quality and mass flow rate. The variables of quality and mass flow rate can be considered for studying the venturi overreading characteristics on the wet steam pressure and venturi structure conditions.

3.2 Experimental system

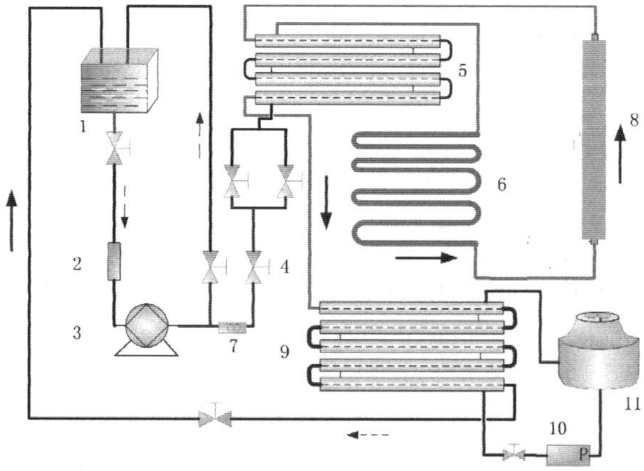

Figure 1. Layout of Experimental Apparatus.

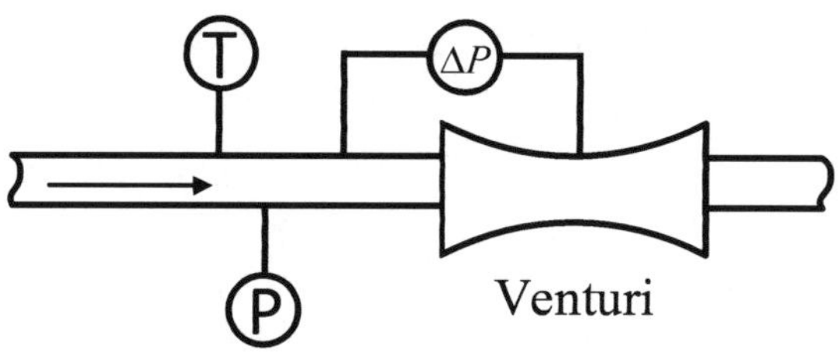

Figure 2. Test Section 8.

As depicted in figure 1, the experimental setup comprises several primary components, including Water Tank 1, Filter 2, High-pressure Piston Pump 3, Electric Valve 4, Heat Exchanger 5, Preheater 6, Mass Flow Meter 7, Test Section 8, Condenser 9, Water Tank Condensing Pump 10, Cooling Tower 11, and other auxiliary equipment.

The working fluid follows a specified process. Firstly, deionized water is pumped from Tank 1 by High-pressure Piston Pump 3 to reach the required pressure for the experiment. The water is then split into two paths, the bypass, and the main road. The bypass controls the mass flow rate, and the main road passes through the Mass Flow Meter 7 for flow rate measurement. The main road enters Heat Exchanger 5 to recover the heat of the working fluid at the outlet of the test section. The water then passes through Preheater 6 and Test Section 8. The resulting steam-water mixture from the test section travels through Heat Exchanger 5 and Condenser 9 before going through a series of pressure reduction devices to lower the pressure and ultimately returns to the water tank.

The experimental parameters vary within 1 MPa for pressure, 0.10-0.32 kg/s for mass flow rate, and 0.14-0.90 for quality. The venturi structure employed in the experiments is presented in table 1, figure 2 and figure 3.

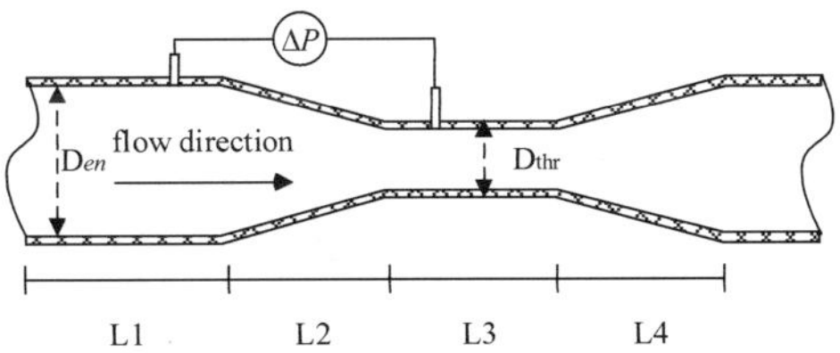

Figure 3. The Structure of Venturi.

The size of the venturi structure is as follows

Table 1. Size of Venturi.

β	D_{en} (mm)	$L1$ (mm)	$L2$ (mm)	$L3$ (mm)	$L4$ (mm)	wall thickness (mm)
0.6	20	21.6	21.6	12	30.4	2.5

3.3 Measurement techniques

Table 2. Device Uncertainty.

Parameters	Uncertainties
Pressure/Pressure Drop	±0.065%
Fluid Temperature	±1.5°C
Mass Flow Rate	±0.05%
Heating Power	±3.25%

The experiment involves measuring several parameters, including the temperature and pressure at the venturi inlet, venturi pressure drop, venturi mass flow rate, heating current, and voltage. The pressure and pressure drop are measured using Rosemount 3051 intelligent pressure or pressure drop transmitter, while the temperature is measured using NiCr-NiSi, K-type, and Φ3 mm stainless steel sheathed thermocouples. The approximate arrangement of temperature and pressure sensors is depicted in figure 2. The flow rate passing through the test section is measured using the RHEONIK-RHE08 Coriolis mass flow meter, as shown in figure 1 (Mass Flow Meter 7). The measurement uncertainties of these devices are detailed in table 2.

3.4 Experimental procedure
1) Adjust the pressure stabilization system to maintain the required pressure of the experimental system at the desired level.
2) Control the flow of the experimental section by adjusting the main and bypass flow regulating valves.
3) Adjust the power of the preheating section to achieve the desired level of quality in the working fluid.
4) Once the experimental system has stabilized, record the relevant experimental data.
5) Repeat the above steps to complete the data collection process for all experimental trials.

3.5 Experimental Data Processing
In previous studies, *OR*, a parameter used to assess the deviation of two-phase flow through the venturi, employs the pressure drop when the gas flowed singly through the venturi as a benchmark (all-gas-phase benchmark). However, a significant amount of energy is required to heat the water from the subcooled state to dry saturation due to the large latent heat of the vaporization. As a result, it is difficult to obtain an all-gas-phase benchmark. This paper redefines *OR* based on the pressure drop of liquid flowing alone through the venturi (all-liquid-phase benchmark),

$$OR_{this} = \sqrt{\frac{\Delta P_{ws}}{\Delta P_l}} \tag{5}$$

In addition, the gas Froude number Fr_g and Lockhart-Martinelli parameter X_{LM} can be calculated according to Eqs. (2) and (3).

4. Experimental results and analysis

4.1 Effect of XLM parameter on the OR
Figure 4 illustrates the relationship between the *OR* coefficient and X_{LM}. The figure shows a monotonically decreasing trend between X_{LM} and *OR*. This is because the higher the liquid content in the wet steam, the larger the obstruction effect of the liquid on the gas phase and the smaller the frictional pressure drop of the gas due to the acceleration of the liquid phase. As a result, the *OR* is lower. It is worth noting that this conclusion seemingly contradicts the one presented in Section 2.2.1, as the former definition of overreading is based on all-liquid-phase, while the latter is based on all-gas-phase. So, these two conclusions are consistent.

On the other hand, it can be observed that the trend of *OR* relative to X_{LM} is consistent under different mass flow rates, suggesting that the mass flow rate factor does not affect the establishment of the overreading model.

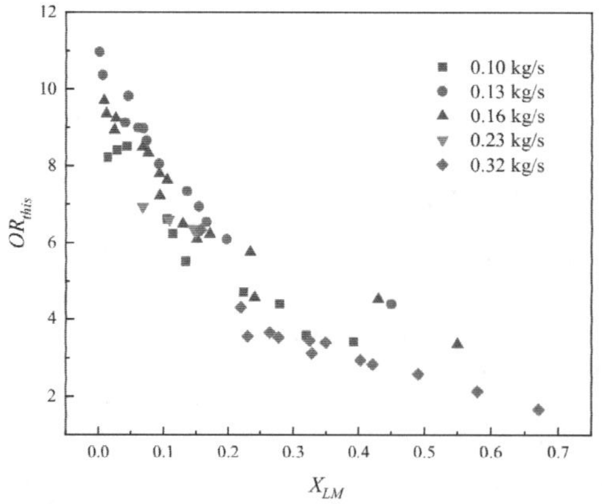

Figure 4. The Correlation Between the *Or* And X_{lm}.

4.2 Effect of Fr_g parameter on the OR

The relationship between Fr_g and *OR* is presented in figure 5. It is evident from the figure that Fr_g is not the primary factor affecting the *OR*, as there exist multiple Fr_g values corresponding to the same *OR* value. Thus, Fr_g does not significantly impact the reproduction of *OR*. Consequently, the construction of the overreading model can disregard the influence of Fr_g.

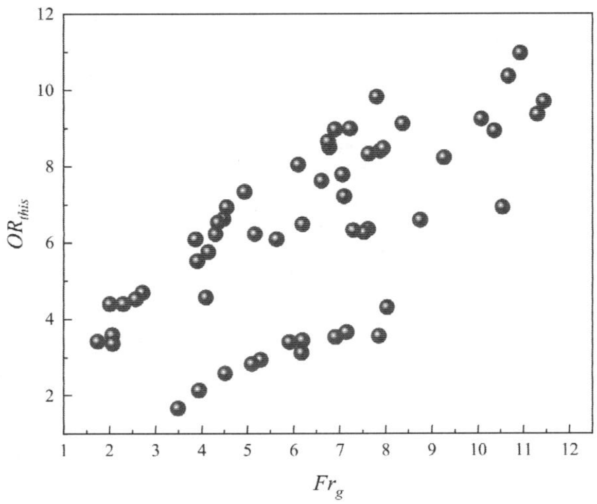

Figure 5. the Relationship Between the *or* and Fr_g.

4.3 Overreading model construction

The momentum equation for the steady-state wet steam flow is as follows,

$$-\frac{\partial P}{\partial z}=\frac{\tau c}{A}+\rho_{ws}g\sin\theta+\frac{1}{A}\frac{\partial}{\partial z}\left\{AM^2\left[\frac{(1-x)^2}{\rho_l(1-\alpha)}+\frac{x^2}{\rho_g\alpha}\right]\right\} \tag{6}$$

where z, c, A, τ, ρ_{ws}, θ, M, x, and α represent the length of the channel, wetted perimeter, cross-sectional area, friction force, the density of the gas-liquid mixture, pipe inclination angle, mass flux, quality, and void fraction, respectively.

The first term on the right-hand side of Eq. (6) represents the frictional pressure drop generated by the friction between the fluid and the pipe wall. The second term represents the gravity pressure drop caused by the change in gravitational potential energy. The third term represents the acceleration pressure drop caused by the density variation of the fluid and the change in flow area along the flow direction. When wet steam flow passes through a horizontally installed venturi tube, the change in gravitational potential energy is negligible, and the frictional losses in the venturi can also be neglected. Therefore, the main component of the venturi pressure drop is the acceleration pressure drop caused by the significant reduction in flow area. In summary, Eq. (6) can be simplified as follows,

$$-\frac{\partial P}{\partial z}=\frac{1}{A}\frac{\partial}{\partial z}\left\{AM^2\left[\frac{(1-x)^2}{\rho_l(1-\alpha)}+\frac{x^2}{\rho_g\alpha}\right]\right\} \tag{7}$$

Assuming that the density of the fluid along the pipe diameter direction remains constant and the void fraction of the fluid passing through the venturi does not change, the venturi pressure drop for wet steam flow can be obtained by integrating Eq. (7),

$$\Delta P_{ws} = M^2\left[\frac{(1-x)^2}{\rho_l(1-\alpha)}+\frac{x^2}{\rho_g\alpha}\right]\cdot\left[\left(A_{en}-A_{out}\right)+ln\frac{A_{en}}{A_{out}}\right] \tag{8}$$

Similarly, using Eq. (7), the pressure drops for the gas and liquid phases separately passing through the venturi tube can be obtained,

Liquid phase,

$$\Delta P_l = M^2\frac{(1-x)^2}{\rho_l}\left[\left(A_{en}-A_{out}\right)+ln\frac{A_{en}}{A_{out}}\right] \tag{9}$$

Gas phase,

$$\Delta P_g = M^2\frac{x^2}{\rho_g}\left[\left(A_{en}-A_{out}\right)+ln\frac{A_{en}}{A_{out}}\right] \tag{10}$$

Combining Eqs. (1), (8), and (9), the calculation method for *OR* can be obtained,

$$OR_{before} = \sqrt{\frac{(1-x)^2}{x^2}\cdot\frac{1}{1-\alpha}\cdot\frac{\rho_g}{\rho_l}+\frac{1}{\alpha}} \tag{11}$$

The continuity equation for the steady-state wet steam flow is as follows,

Liquid phase,

$$(1-x)MA = \rho_l v_l A_l \tag{12}$$

Gas phase,

$$xMA = \rho_g v_g A_g \tag{13}$$

Combining Eqs (11), (12), and (13), we can further obtain,

$$OR_{before} = \sqrt{1+C_{Ch}X_{LM}+X_{LM}^2} \tag{14}$$

where C_{ch} is the Chisholm coefficient, calculated according to the following equation,

$$C_{Ch} = \left(\frac{1}{s}\cdot\sqrt{\frac{\rho_l}{\rho_g}}+s\cdot\sqrt{\frac{\rho_g}{\rho_l}}\right) \tag{15}$$

where s represents the slip velocity ratio, v_g/v_l. Based on the Chisholm model, De Leeuw proposed that C_{ch} can be calculated by the following equation [10],

$$C_{Ch} = \left(\frac{\rho_l}{\rho_g}\right)^n+\left(\frac{\rho_g}{\rho_l}\right)^n \tag{16}$$

The n in Eq. (16) is the overreading index related to the X_{LM} and Fr_g.

Similarly, using the same method as deriving Eq. (14), the *OR* based on the all-liquid-phase benchmark in this paper can be obtained,

$$OR_{this} = \sqrt{1 + C_{Ch} X_{LM}^{-1} + X_{LM}^{-2}} \tag{17}$$

Following Eq. (17), this paper intends to construct an overreading model in the following form,

$$OR_{this}^2 = m log_{10}\left(X_{LM}^{-1}\right) + log_{10}\left(X_{LM}^{-2}\right) + n \tag{18}$$

After fitting the experimental data, the overreading correction model proposed in this paper can be obtained as follows,

$$OR_{this} = \sqrt{55.2 log_{10}\left(X_{LM}^{-1}\right) + log_{10}\left(X_{LM}^{-2}\right) - 11} \tag{19}$$

The fitting result is presented in figure 6 below,

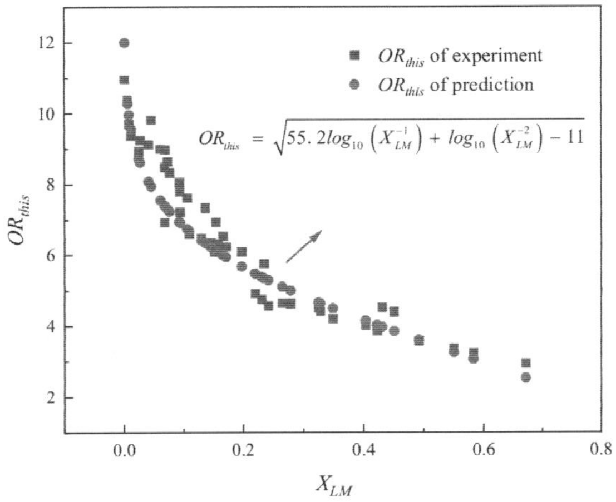

Figure 6. The Fitting Result of Overreading.

5. Cfd simulation

In industry, wet steam flow rate measurement often involves large flow rates and pipe diameters, making it challenging to conduct experiments. This section employs the CFD method to assess the model's applicability under large flow rate and pipe diameter conditions. The CFD geometric model is based on the same geometric model used in the experimental setup, which has been scaled up proportionally.

5.1 Computational conditions

The venturi diameter is scaled up from 20 mm to 930 mm, and the scaling ratio of the remaining portion is the same as that of the diameter. Eight different quality conditions are set at a mass flow rate of 190.3 kg/s, namely 0.14, 0.16, 0.19, 0.22, 0.23, 0.25, 0.27, and 0.28. The pressure conditions are consistent with the experimental setup in Section 3.

5.2 CFD model

When the wet steam qualities range from 0.14 to 0.32, the volume fraction of the liquid phase is sufficiently small, and the liquid phase exists in the form of dispersed mist within the gas phase. Consequently, the Discrete Phase Model (DPM) is deemed an appropriate simulation method. In the DPM, the gas phase is considered as a continuous medium and solved using the Euler method, and the liquid phase is treated as a discrete medium and solved using the Lagrange method. The two phases are solved alternately to couple their interaction.

5.3 CFD Data Processing

5.3.1 Theoretical Mass Flow Rate. Combining the single-phase flow calculation formula with the overreading prediction model, the theoretical mass flow rate can be obtained,

$$G_{th} = \frac{C\varepsilon}{OR_{this}\sqrt{1-\beta^4}} \times \frac{\pi}{4} D^2 \beta^2 \sqrt{2\rho_l \Delta P_{ws}} \tag{20}$$

The constants C, ε, β, and D represent the venturi discharge coefficient, expansibility factor, throttling ratio, and inlet diameter and are taken as 0.984, 1, 0.6, and 930 mm, respectively.

5.3.2 Predicted Mass Flow Rate Deviation. The relative error E of the predicted mass flow rate is defined as follows to evaluate the difference between the model proposed in this paper and the classic overreading prediction model,

$$E = \left| \frac{G_{th} - G_{exact}}{G_{exact}} \right| \times 100\% \tag{21}$$

where G_{exact} is the exact venturi mass flow rate, which is 190.3 kg/s.

5.4 Simulation results and analysis

To evaluate the applicability of this model constructed on the small flow rate and venturi diameter conditions to the large flow and venturi diameter conditions in the industry, the prediction results are compared with those of the Separated Model [8], Chisholm Model [9], and Graham Model [13]. More information on these three classic models can be found in their references. The difference in the *OR* prediction performance among the four models is presented in figure 7, and the comparison of predicted flow deviation is shown in figure 8. To quantitatively explain the performance differences between this model and the above three classic models, the maximum deviation and average deviation are listed in table 3.

Table 3. Predicted Average and Maximum Deviation.

	average deviation	maximum deviation
This Model	0.10	0.15
Separated Model	0.14	0.23
Chisholm Model	0.25	0.35
Graham Model	0.36	0.46

As depicted in figure 7, the predicted overreading exhibits an approximately linear relationship with the quality. Compared other three models, this model demonstrates the best prediction performance for large mass flow rate and venturi diameter conditions.

As shown in figure 8, the prediction deviation of mass flow for all models tends to decrease as quality increases. From table 3, it is evident that this model exhibits the best performance with a maximum deviation of 0.15 and an average deviation of 0.10 in the predicted mass flow rate among the four models compared.

Based on the numerical simulation results, this model demonstrates a high potential for industrial application in large venturi diameter and mass flow rate conditions.

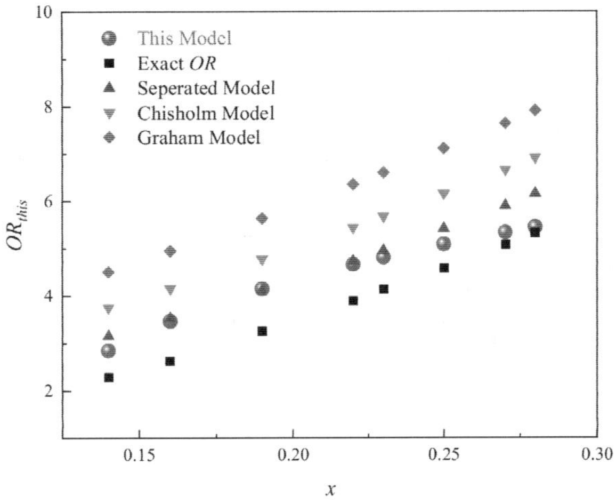

Figure 7. Overreading Prediction Results for Each Model.

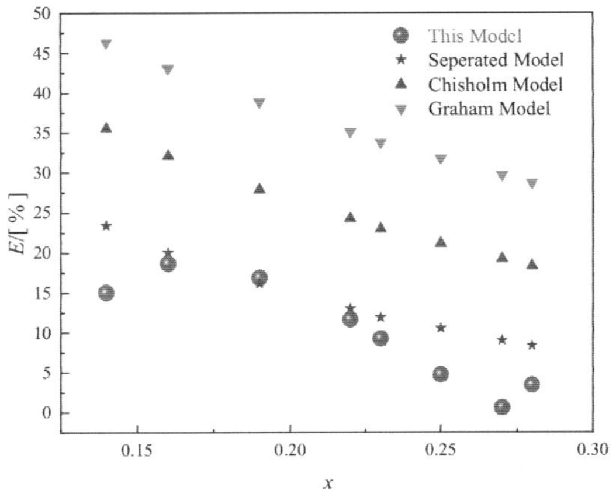

Figure 8. Predicted Mass Flow Rate Deviation for Each Model.

6. Conclusion

Firstly, this study conducts wet steam flow experiments to explore the impact of Lockhart-Martinelli parameter X_{LM} and gas Froude number Fr_g on the venturi overreading characteristics. The experiments are carried out at a small mass flow rate and venturi diameter conditions. Then, an overreading prediction model based on X_{LM} is established. Then, the applicability of this model on the large flow and venturi diameter conditions is evaluated through numerical simulations. Finally, the performance of this model is compared with that of three classical overreading models, represented by the Chisholm Model. Based on the results, several conclusions are drawn,

1) For a given pressure and venturi structure, the overreading characteristic of wet steam in venturi flow is mainly influenced by X_{LM}, while the mass flow rate and Fr_g have a relatively small impact on the overreading.

2) The overreading prediction model proposed in this study demonstrates better performance in industrial conditions with a large mass flow rate and venturi diameter. The average deviation and maximum deviation of the predicted wet steam mass flow rate are 0.10 and 0.15, respectively, which are superior to those of the classic Chisholm Model and the other two models.

3) This model is applicable for wet saturated steam with a pressure of 1 MPa, a quality range of 0.14-0.90, and a venturi throttling ratio of 0.60.

4) This study has proved that the overreading model, proposed for small mass flow rate and venturi diameter conditions, has the potential to be extrapolated to large flow rate and venturi diameter conditions. This finding offers an alternative method for flow measurement in industrial pipelines.

5) NOMENCLATURE

6) *A*	cross-sectional area, m^2
7) *C*	discharge coefficient, -
8) C_{ch}	Chisholm coefficient, -
9) *c*	wetted perimeter, m
10) *D*	diameter of the venturi, m
11) *E*	relative error of predicted mass flow rate, -
12) *Fr*	Froude number, -
13) *G*	mass flow rate of wet steam, kg/s
14) *g*	acceleration of gravity, kg/m^3
15) *L*1	straight part of venturi, mm
16) *L*2	convergence part of venturi, mm
17) *L*3	throat part of venturi, mm
18) *L*4	divergence part of venturi, mm
19) *M*	mass flux, $kg/(m^2 \cdot s)$
20) *OR*	overreading coefficient, -
21) *P*	pressure of wet steam, Pa
22) v_{sg}	superficial velocity of gas phase
23) X_{LM}	Lockhart-Martinelli parameter, -
24) *x*	quality of wet steam, -
25) *z*	length of the channel, m

26) Greek symbols

27) α	void fraction, -
28) β	throttling ratio, -
29) ε	expansibility factor, -
30) θ	pipe inclination angle, °
31) π	circle ratio, -
32) ρ	density, kg/m^3
33) τ	friction force, N/m

34) Subscripts

35) *before*	previous studies
36) *en*	entrance
37) *ex*	exact
38) *g*	vapor phase
39) *l*	liquid phase
40) *th*	theoretical
41) *thr*	throat
42) *this*	this paper
43) *ws*	wet steam

References

[1] Caliman R F, Ramos R. Wet gas metering by differential pressure meters: a literature review[J]. Multiphase Flow Dynamics, 2022, 87-99.

[2] He D, Bai B. Numerical investigation of wet gas flow in Venturi meter[J]. Flow Measurement and Instrumentation, 2012, 28: 1-6.

[3] Bizon E. Two-phase flow measurement with sharp-edged orifices and venturis[R]. Atomic Energy of Canada Ltd., Chalk River, 1965.

[4] Stewart D. The evaluation of wet gas metering technologies for offshore applications, part 1–differential pressure meters[R]. Flow Measurement Guidance Note, 2003.
[5] Skea A F, Hall A R W. Effects of two-phase flow on single-phase flowmeters[J]. Dti Flow Programme Guidance Notes, 1997 (3).
[6] Stewart D G. The evaluation of dry gas meters in wet gas conditions[J]. National Engineering Laboratory, London, 2002, 33: 58.
[7] Steven R N. Wet gas metering with a horizontally mounted venturi meter[J]. Flow Measurement and Instrumentation, 2002, 12(5-6): 361-372.
[8] Lin Z H, Wang S Z, Wang D, Two-phase flow and boiling heat transfer[M]. Xi'an Jiaotong University Press, 2003.
[9] Chisholm D. Two phase flow through sharp-edged orifices, research note[J]. J. Mech. Eng. Sci, 1977, 19: 127-129.
[10] De L R. Liquid correction of venturi meter readings in wet gas flow[J]. 1997.
[11] Steven R, Stobie G, Hall A. Further evaluation of the performance of horizontally installed Orifice plate and cone differential pressure meters with wet gas flows[C]//Proceedings of 7th International South East Asia Hydrocarbon Flow Meas Workshop, Kuala Lumpur. 2008.
[12] Reader H M, Nel T, Graham E. An improved model for Venturi-tube over-reading in wet gas[J]. Gas, 2009, 20: 23rd.
[13] Graham E, Reader H M, Hodges D, et al. Performance of a vertically installed venturi tube in wet-gas conditions[C]//Proc. North Sea Flow Measurement Workshop, St. Andrews, Scotland. 2014.
[14] Yuan C, Xu Y, Zhang T, et al. Experimental investigation of wet gas over reading in venturi[J]. Experimental Thermal and Fluid Science, 2015, 66: 63-71.
[15] Khayat O, Afarideh H. Numerical investigation of non-newtonian liquid-gas flow in venturi flow meter using computational fluid dynamics[J]. Iranian Journal of Science and Technology, Transactions of Mechanical Engineering, 2021, 45(2): 393-401.
[16] Jing J, Yuan Y, Du S, et al. A CFD study of wet gas metering over-reading model under high pressure[J]. Flow Measurement and Instrumentation, 2019, 69: 101608.
[17] Weise J, Baliño J L, Paladino E E. CFD study of the transient wet gas flow behavior through orifice plate flow meters[J]. Flow Measurement and Instrumentation, 2021, 82: 102077.
[18] Xu Y, Gao L, Zhao Y, et al. Wet gas overreading characteristics of a long-throat venturi at high pressure based on CFD[J]. Flow Measurement and Instrumentation, 2014, 40: 247-255.
[19] Xu Y, Cao L, Yuan C, et al. Study of the factors influencing the over-reading characteristics of the precession venturi[J]. Flow Measurement and Instrumentation, 2023, 89: 102272.
[20] Seraj H, Khaled M, Yusof R, et al. Review of wet gas flow measurement using venturi tubes and radio active materials[J]. International Journal on Smart Sensing and Intelligent Systems, 2010, 3(4): 672-689.

Accurate measurement method of contact angle within a small angle

Shuhui Wang[1], Xiaoting Liu [1,2], Tao Wang[1], Jing Luo[1] and Jian Yao[1]

[1]Army Academy of Armored Forces, Beijing, China

[2]Corresponding author's e-mail: greenleaf999@163.com

Abstract: The wetting phenomenon of liquid on a solid surface is a fundamental natural phenomenon in nature. It has a significant impact on people's production and life. This paper solves the problem that small-angle wetting cannot be accurately measured by changing the imaging direction and complete auxiliary methods.

1. Introduction

Wettability is one of the fundamental properties of solid surfaces. It plays a vital role in industrial and agricultural production and people's daily lives, encompassing oil extraction, pesticide spraying, fabric waterproofing, and washing. Surface wettability mainly hinges on surface roughness and surface free energy, which is usually measured by the contact angle between the droplet and the solid surface. In the equilibrium state, the tangent lines to the solid and liquid surfaces are drawn at the junction of the three phases of solid, liquid, and gas, respectively. The angle formed by two tangent lines inside the liquid is the contact angle. There are usually three methods for measuring the wetting angle [1-3]: Wilhelmy plate, session drop, and capacitive bubble. The larger the contact angle is, the better the wettability will be. The surface can be completely dripped to the ground, showing good wettability, indicating that the surface is completely covered by a substance. The smaller the contact angle is, the worse the wettability of the substance is, and the surface cannot be completely dropped by the liquid. Therefore, the contact angle is also widely used for wettability experimental research and surface control. The precise measurement of contact angle plays an essential role in the research of related fields.

Contact angle measuring instrument is a vital equipment for measuring contact angle. At the moment, the commonly used contact angle measurement is to image from the side and read the contact angle directly [4-5], as shown in Figure 1:

Figure 1. Framing direction settings for commonly used contact angle test.

The characteristic of the mentioned measurement method is that the frame is taken directly along the horizontal direction of the specimen surface. There is no theoretical error in the test method. It has become the primary method for contact angle measurement at present. However, when the contact angle is small and the droplet diameter is large, the depth of field at the edge of the droplet increases as the droplet radius increases (as shown in Figure 2). In addition, due to the limited height of the liquid, it is unable to provide enough light reflection, and the droplet edge imaging will generate a large error, resulting in a doubling of the final measurement error [6].

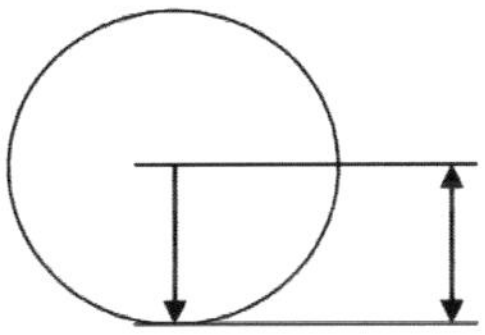

Figure 2. Variation of the detected depth of field in the contact angle with droplet radius.

Building upon the above reasons, the current contact angle test equipment will produce a relatively large error when the contact angle is below 10. The smaller the contact angle is, the greater the measurement error will be. When the accuracy requirements are not stringent and the contact angle is less than 5-7°C, researchers often assume that the liquid is thoroughly wetted [7], and further investigations are typically not pursued. In reality, many liquids, despite having small contact angles, still exhibit variations in their wetting capabilities. It is crucial to select the most suitable liquid among approximate candidates, differentiate the optimal wetting medium on specific surfaces, and research areas such as wetting on superhydrophilic surfaces. However, these endeavors require more advanced and high-precision measurement methods for support. Although current testing institutions have made unremitting efforts in the fields of light sources, testing methods, and platform stability, due to the limitations of testing methods, the accurate evaluation of small contact angles is still an insurmountable worldwide problem [8-10]. To alleviate the mentioned problems, an accurate evaluation of a small contact angle is the main motivation of this paper.

2. Top-down framing principle and related mechanism design

2.1. Framing direction settings

Given the above existing problems, the testing principle is improved and designed, commencing from side imaging to the top imaging of samples, to easily obtain a top-down view of the entire droplet after the droplet has spread. Then, the area of the droplet is used to solve the problem. This design solves the problem that small contact angles cannot accurately image. The specific framing direction is shown in Figure 3:

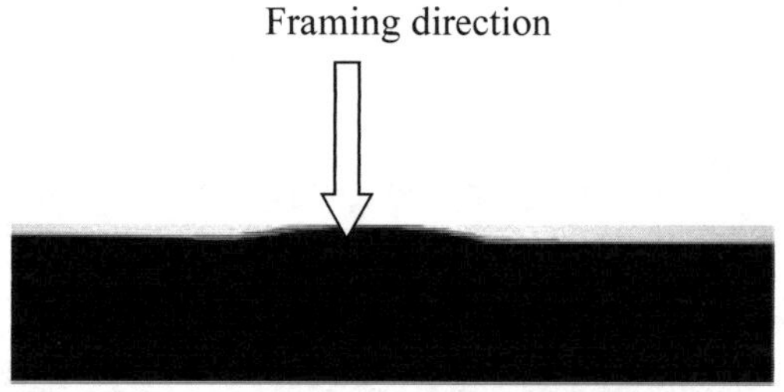

Figure 3. Framing direction of contact angle.

This testing method differs from conventional contact angle tests by imaging from above the specimen rather than from the side, eliminating the influence of depth of field on imaging. This modification facilitates obtaining the entire droplet shape after spreading, allowing for precise characterization of the wetting state of small-angle wetting droplets.

2.2. Principle and design of light source irradiation method

When the reagent drops on the sample, most of the droplets are transparent, similar in the color of the sample, and cannot be clearly imaged. During the experiment, a surface light source is provided to cover the entire sample. The droplets and the sample reflect light at different angles, resulting in a clear image in the camera. The schematic diagram is shown in Figure 4.

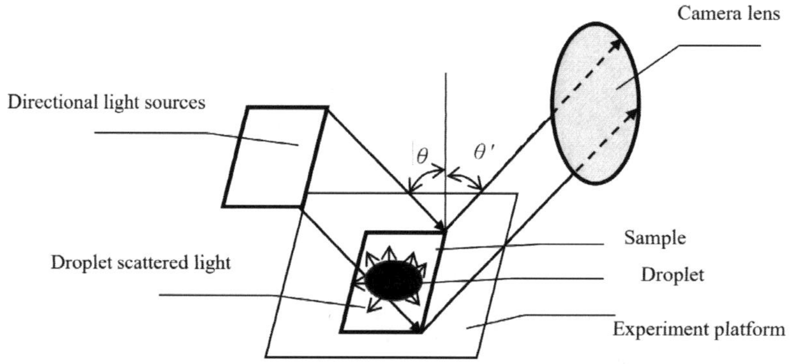

Figure 4. Droplet imaging schematic.

In Figure 4, θ and θ′ are the incident angle and reflection angle.

3. Wetting experiment operation method

A digital imaging device is placed at the camera lens of Figure 4 to obtain the wetting status of the liquid on the sample surface. Since this experiment requires continuous shooting, the camera equipment should have an automatic timing shooting function, and the pixels should meet the requirements for clear imaging (pixels greater than 2 million). With the development of digital equipment, there are many related equipment. For example, general-purpose single lens reflex (SLR) cameras and dedicated digital imaging equipment can achieve such functions. By focusing parallel light on the sample surface, the surface color of the wet sample is nearly consistent after being photographed (when the tolerance is less than 20 in Photoshop, the background color of all samples can be extracted), as shown in Figure 5.

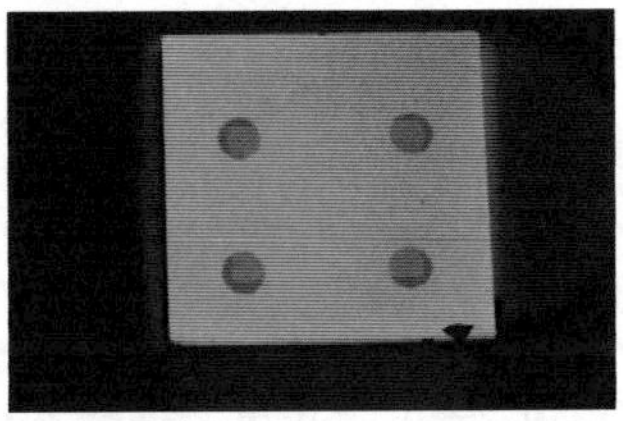

Figure 5. Imaging picture of wetting droplets.

If examining the wetting state at a specific moment, capturing a single photograph is sufficient. However, if investigating the dynamic wetting process of a droplet on the specimen surface, multiple photos need to be taken at regular intervals. When studying the dynamic wetting process, the observation time should exceed 1 minute, and the number of captured photos should be no less than 5.

4. Extraction of droplet wetted area

In Photoshop, the following method can be used to remove the area outside the droplets: (1) Areas other than the non-wetted sample are manually selected and removed; (2) All areas of the wetted sample are selected and removed using a magic wand tool with a tolerance of 20. The droplets' shape after differentiation is shown in Figure 6:

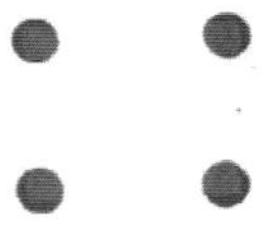

Figure 6. Droplet differentiation effect diagram.

5. Calculation method of contact angle

The calculation of the contact angle involves the following steps:

1) The pixel value of the length of the wetted sample in the picture in Figure 6 is measured using the ruler function in the PS software, and the total pixels of the wetted sample are calculated using length*width.

2) The actual length and width of the wetted sample are measured and the actual area of the sample is calculated.

3) The histogram module in the PS software is opened, droplets are selected, and the total pixel values of the four droplets in Figure 6 are read.

4) The area of four droplets is calculated through the pixel ratio of the droplet to the sample and the area of four droplets is divided by 4 to get the average area of each droplet.

5) the droplet radius r is derived according to the area equation of a circle: $A = \pi r^2$, and the actual radius value is calculated from the ratio of radius to piece length.

Substituting the radius r into the equation yields:

$$\frac{6V}{\pi r^3}=3\tan\left(\frac{\theta}{2}\right)+\tan^3\left(\frac{\theta}{2}\right) \quad (1)$$

The contact angle of the reagent on the sample surface can be obtained.

In Equation (1), v is the droplet volume, representing the average value of the volume of each droplet in previous experiments, r is the radius of the droplet, and θ is the contact angle.

Since Equation (1) is a linear equation with one unknown, an appropriate method is required to calculate the solution of the equation. For example, this can be achieved using the single-variable solution method in Excel or programming calculations.

6. Validation of the small contact angle testing

A contact angle wetting experiment is conducted using butanoic acid on the surface of silicon carbide ceramics due to its superior wetting effect.

6.1. Experimental environment

Since the experiment is not only oriented to a single experimental target, the surface tension and spread of organic chemical reagents on solid surfaces are different under different temperature conditions. In order to reduce the deviation caused by temperature on the measurement results of the wetting experiment and ensure that the experimental data is true and reliable, the test temperature must be constant. In the experiment, the experimental site and reagent temperatures are artificially set at 25°. Under the same temperature conditions, different reagents can show better compatibility. The imaging equipment utilized is the Canon 700D DSLR camera, assisted by the JGW-360r contact angle testing platform and a microinjection system. The lighting source employed is a three-stage adjustable LED light source.

6.2. Experimental process and results

The auxiliary information recorded through Photoshop is shown in Table 1:

Table 1. Auxiliary information for contact angle calculation.

Volume	4.3 microliter
Sample length in pixels	1350
Actual length of the specimen	49.8
The ratio of length in pixels to the actual length	27.10843

The wetting data and calculation results recorded through Photoshop are shown in Table 2:

Table 2. Contact angle calculation information table.

Sequence	Total pixels of four drops	Radius pixels	Actual radius	Contact angle
1	146652	108.06	3.99	4.95
2	169943	116.32	4.29	3.98
3	182359	120.49	4.44	3.59
4	187189	122.08	4.50	3.45
5	193274	124.05	4.58	3.30
6	197984	125.55	4.63	3.19
7	201962	126.81	4.68	3.08
8	203523	127.30	4.70	3.04
9	206322	128.17	4.73	2.99
10	207462	128.52	4.74	2.96
11	207432	128.51	4.74	2.96
12	209665	129.20	4.77	2.92
13	211241	129.69	4.78	2.90
14	210727	129.53	4.78	2.90
15	210939	129.59	4.78	2.90

As can be seen from Table 2, the wetting of n-butanol on the silicon carbide ceramic surface continues to decrease from the initial 4.95 degrees to 2.9 degrees and remains stable as time increases.

It should be emphasized that in this experiment, the final stable contact angle on the silicon carbide ceramic surface was 2.9 degrees, which cannot be measured with a traditional contact angle tester. Since this article uses a top-down view for measurement, the larger the wetted area is and the smaller the contact angle is, the higher the measurement accuracy will be, which makes up for the shortcomings of the traditional contact angle measurement method.

Moreover, theoretically, contact angle measurements should be conducted when the wetting liquid is stable. However, different wetting liquids may take varying amounts of time to reach a stable state. In this study, a timed measurement method was employed, allowing for the identification of a stable wetting state through the changing wetting process. This approach avoids prolonged waiting times and variations in contact angle due to liquid evaporation, thereby enhancing measurement accuracy.

7. Conclusion

1) Conventional wetting angle measurement devices often face challenges in accurately measuring small contact angles due to excessive depth of field, leading to difficulties or even impossibility in conducting precise measurement experiments. The top-down view measurement method proves effective in detecting the wetting conditions of small contact angles, offering higher detection accuracy.

2) The data corresponding to the top-down framing method in this article can be quantified by using the proportion of pixels in the wetted area to the entire area. Then, the wetting angle can be calculated through the equation $\frac{6V}{\pi r^3}=3\tan(\frac{\theta}{2})+\tan^3(\frac{\theta}{2})$.

3) The wetting angle test method for top-down framing is suitable for accurate measurement of small-angle wetting angles, especially for wetting angles below 7° that are difficult to measure with conventional wetting angle devices.

References

[1] Wang W Z H M Q Y H C L W H P G B. Wetting Angle Measurement of Liquid PbBi on RAFM Steel Surface Based on Ellipse Fitting [J]. Advanced Materials Research, 2013, 2730(820-820):102-105.

[2] Frédéric L. Revisiting the droplet simulation approach to derive force-field parameters for water on molybdenum disulfide from wetting angle measurements [J]. The Journal of Chemical Physics, 2016, 145(16):164705.

[3] Aleksei E S S. Effect of epoxy resin modifications with industrial fillers on wetting and water absorption [J]. Colloid and Polymer Science, 2023, 301(3):251-258.

[4] Wang Qijun, Yang Chao, Zhong Ye, et al. Development of a contact angle measuring instrument [J]. Modern Education Equipment in China, 2021(17):50-53.

[5] Xiong Jian, Tang Junfang, Zhou Xue, et al. Surface wettability of graphite modified with oxygen-containing functional groups and its impact on gas water distribution [J]. Geological Science and Technology Bulletin, 2023, 42(03):93-101.

[6] Liu Weilong, Liu Wenfang, Gao Xinyue et al. High precision contact angle measurement method based on spiral fitting [J]. Science, Technology and Engineering, 2023, 23(02):699-706.

[7] Youhua Y J J B L. Soft wetting: Modified Cassie-Baxter equation for soft superhydrophobic surfaces [J]. Colloids and Surfaces A: Physicochemical and Engineering Aspects, 2023, 677(PA).

[8] Chang Siyu, Yao Lihong, Wu Yueqi, et al. The effect of different silane coupling agents on the wettability of wood [J]. Forestry Industry, 2022, 59(07):14-18.

[9] Liu Zhen, Xu Xiongwen, Liu Jinping, et al. Study on the Wettability Changes of Liquid Droplets in a Circular Electric Field [J]. Chemical Engineering, 2021, 49(12):49-53.

[10] Yan Min, Yue Min, Lin Haifei, et al. Experimental study on the influence of functional groups in middle and low rank coal on coal wettability [J]. Coal Science and Technology, 2023, 51(05):103-113.

State of health estimation for lithium battery based on multi-source transfer learning

Tianhao Jiang[1,2] **and Youyu Wu**[1]

[1]Wuhan University of Technology, Wuhan, China

[2]Corresponding author's e-mail: wuyouyu1@whut.edu.cn

Abstract: Aiming at the problem that the state of health estimation accuracy decreases when the lithium battery in training and testing sets has different charging conditions, we proposed a solution based on the transfer learning with the multi-source method. This scheme uses the complete aging test data of 16 batteries under different charging conditions as the source domain and extracts aging characteristics respectively to train the basic model. The similarity coefficient was obtained by the similarity evaluation of the first 200 cycles of each source and target battery, and the learning weight of each source battery model was weighted by the similarity coefficient. The experimental results show that the MAE and RMSE of the transfer learning with a multi-source model decreased by 53.63% and 54.97% compared with the single-source estimation model under new charging conditions.

1. Introduction

Lithium battery (LiB) has the advantages of high energy density, long cycle life, and wide operating temperature range. However, with the widespread application of LiB in various devices, the problem of battery state of health (SOH) estimation has become increasingly prominent [1]. The SOH of a LiB is usually expressed by its capacity or internal resistance, and problems such as capacity degradation and internal resistance increase will occur during the cycle use [2]. Therefore, accurate estimation of LiB's SOH is helpful to ensure the safe use of batteries, extend battery life, and optimize battery management systems [3].

Machine learning methods have been widely used to estimate SOH during battery degradation [4]. Guo et al. [5], Su et al. [6], Lin et al. [7], Chen et al. [8], and Liu et al. [9] respectively studied the performance of SVM, GPR, BPNN, ELM, and CNN-GRU models in battery SOH estimation. The model can fit the relationship between different aging characteristics and SOH and can obtain high-precision results on SOH estimation.

However, the above premise for high-precision battery SOH estimation using machine learning models is that the training and testing data have the same distribution. However, due to differences in application conditions (such as differences in charge and discharge rate, charge and discharge depth, temperature, and humidity), there are obvious differences in the degradation modes of batteries in the same batch, resulting in a decline in the accuracy of SOH estimation [10]. The obvious solution is to run different complete aging tests for different conditions while collecting large amounts of data to train the estimation model, which is expensive. To solve this problem, transfer learning with a multi-source SOH estimation method was proposed. Complete test data under other charging conditions were used as the source domain and aging characteristics were extracted respectively to train the basic model. The similarity coefficient was obtained by the similarity evaluation of the first 200 cycles of each source-target battery. The transfer learning contribution of each source battery model of SOH estimation to the target battery model of SOH estimation was weighted by the similarity coefficient, to improve the SOH estimation accuracy of the target battery under new charging conditions.

2. Feature extraction and correlation analysis

In this section, the Health Index (HI) reflecting the aging characteristics of the battery was extracted, and correlation analysis was performed to verify the validity of the extracted features, providing feature input for the SOH estimation model in the next section.

2.1. Battery aging dataset

Stanford University, MIT, and the Toyota Research Institute [11] provided experimental data on the aging of batteries under an optimized fast charge protocol, consisting of 124 commercially available lithium batteries. Each battery in the dataset is named according to the test batch and the battery serial number in each batch. For example, the battery serial number 17 in the first batch of tests is named b1c17 (Target battery).. We extracted data from a total of 17 batteries with cycle lives between 800 and 900, and the information on all batteries used in the experiment is shown in Table 1.

Table 1. Battery information in the experiment

Charging conditions	Batteries	Charging conditions	Batteries
4.8C (80%)-4.8C	b1c07	6.0C (50%)-3.6C	b1c31
5.4C (60%)-3.0C	b1c14	5.6C (36%)-4.3C	b3c05, b3c14, b3c27
5.4C (60%)-3.6C	b1c17 (Target)	5.0C (67%)-4.0C	b3c08, b3c20, b3c24
6.0C (40%)-3.0C	b1c25	5.6C (19%)-4.6C	b3c11
6.0C (40%)-3.6C	b1c26, b1c27	5.9C (15%)-4.6C	b3c15, b3c29
6.0C (50%)-3.0C	b1c28	—	—

2.2. Aging feature extraction

The characteristic that reflects the SOH of the battery is called the HI, and the HI can be extracted from curves such as the constant current charge voltage curve and the Incremental Capacity Analysis (ICA) curve. The constant current charging voltage curve is the curve of the battery voltage changing with time during the constant current charging process, and the constant current charging voltage curve of different cycles will be different. The ICA curve is a battery cycle charge to voltage differential (dQ/dV) with the voltage change curve, and different cycles of the ICA curve will also be different.

Taking the b1c17 battery as an example, the change of the battery charging voltage curve with the number of cycles is shown in Figure 1. The change of the ICA curve with the number of cycles is shown in Figure 2.

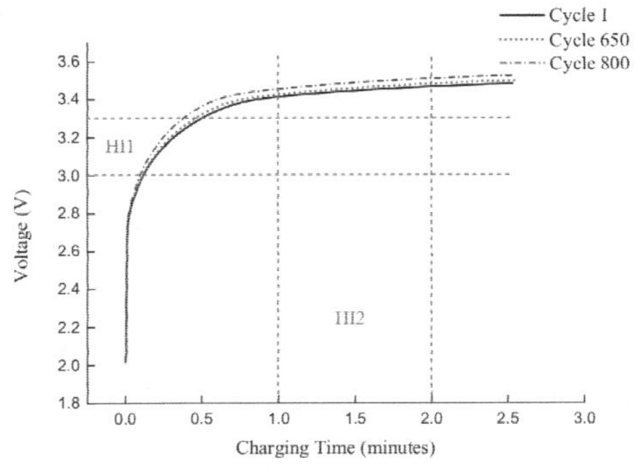

Figure 1. Charging voltage curve.

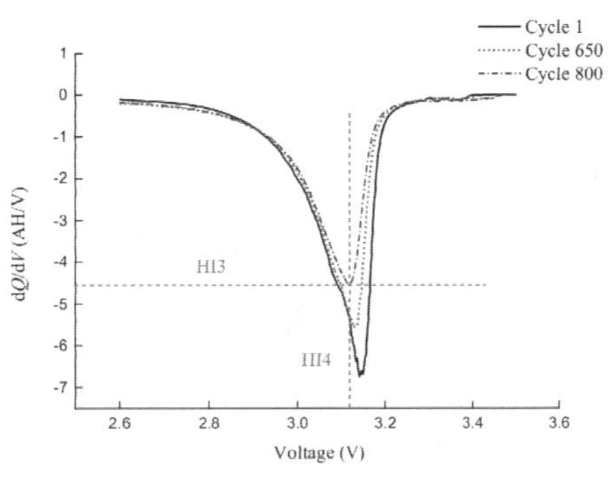

Figure 2. The ICA curve.

It can be observed from Figure 1 that in the voltage rise stage (C1: constant current charging stage), with the increase in the number of cycles, the time to reach the same voltage level will gradually shorten, and the voltage at the same time point will gradually increase. According to the variation characteristics of the voltage curve, two kinds of HI are extracted: charging time of fixed voltage segment (3.0-3.3 V) and average charging voltage of fixed period (1-2 minutes).

It can be observed from Figure 2 that the ICA curve has only one valley value. With the increase of cycles, the ICA curve valley value gradually increases, while the ICA curve valley value corresponding to the horizontal coordinate (voltage) gradually decreases. According to ICA curve valley value characteristics, two kinds of HI are extracted: ICA curve valley value and ICA curve valley value position.

2.3. Correlation analysis

To quantitatively analyze the rationality of HI selection, the correlation coefficient between HI and SOH was calculated. Commonly used correlation coefficients include Pearson coefficient and Spearman coefficient [12]. Since the Pearson coefficient requires variables to obey normal distribution, the Spearman coefficient was selected for correlation analysis. The calculation equation of Spearman correlation coefficient r_s is shown in Equation (1).

$$r_s = 1 - \frac{6}{N_1(N_1^2 - 1)} \sum_{i=1}^{N_1} d_i^2 \tag{1}$$

where d_i is the rank difference between HI_i and SOH_i, and N_1 is the total number of samples of HI and SOH.

3. Construction of the SOH estimation model

In this section, we proposed an SOH estimation method that combines Deep Neural Networks (DNN) with multi-source transfer learning methods. The similarity evaluation rules were established to evaluate the similarity between each source unit SOH estimation model and the target unit SOH estimation model. At the same time, the evaluation criteria were given to evaluate the model's accuracy.

3.1. Similarity evaluation

Batteries of the same batch will experience different degradation modes under different charging conditions. However, some training data with large differences may hurt the SOH estimation performance of the regression model in the target battery [13]. To measure the difference of such degradation modes, a similarity evaluation method was proposed to analyze the degradation modes of the source battery and target battery by using Euclidean distance. Firstly, the mean Euclidean distance δ_i of source battery i and target battery SOH samples was calculated, and the calculation equation is shown in Equation (2).

$$\delta_i = \frac{1}{N}\left(\sum_{j=1}^{N} (\mathrm{SOH}_{\mathrm{source},\, ij} - \mathrm{SOH}_{\mathrm{target},\, j})^2 \right)^{1/2} \tag{2}$$

where $\mathrm{SOH}_{\mathrm{source},\, ij}$ is the SOH value of source domain battery i at sample j, $\mathrm{SOH}_{\mathrm{target},\, j}$ is the SOH value of target battery at sample j, and N is the number of the first 200-cycle SOH samples of source and target battery. The smaller Euclidean distance means that the source battery is more similar to the target battery. Then, the similarity coefficient of the source battery and the target battery was obtained by using the reciprocal of δ_i and normalization. The calculation equation of the similarity coefficient w_i of the source battery i is shown in Equation (3).

$$w_i = \frac{1}{\delta_i}\left(\sum_{i=1}^{N_s}\frac{1}{\delta_i}\right)^{-1} \tag{3}$$

where i is the serial number of batteries in the source domain, N_s is the number of batteries in the source domain, and δ_i is the mean Euclidean distance of battery i in the source domain.

3.2. Model construction

In this subsection, a DNN estimation model was constructed for SOH estimation, and a transfer learning with a multi-source estimation model was constructed based on the DNN estimation model.

3.2.1. Single-source estimation model. For the construction of the basic battery model in a single source domain, DNN was used as the basic learner to fit the nonlinear relationship between the extracted features and SOH. The SOH estimation model based on DNN is shown in Figure 3.

The DNN model has an input layer, 2 hidden layers, and an output layer. The input layer takes the extracted four kinds of HI as the feature input, and the output layer gets the SOH estimate. In the training process, Adam optimizer is used to perform gradient descent to minimize the loss function. Adam's algorithm calculates the square value of the gradient and the deviation correction in each iteration to adjust the learning rate adaptively so that the network can converge to the optimal solution faster.

3.2.2. Multi-source transfer learning estimation model. After feature extraction and model training, the DNN-based SOH estimation model of each source battery was obtained, and then the similarity coefficient between each source region battery degradation curve and the target battery degradation curve was obtained by similarity evaluation. To apply the source domain cell estimation model to the target cell SOH estimation, a multi-source SOH estimation model weighted by similarity coefficient was proposed. The transfer learning with a multi-source estimation model is shown in Figure 4.

By extracting four kinds of HI of the target battery and inputting these four kinds of HI into the SOH estimation models of 16 source batteries respectively, the SOH output values of 16 single source estimation models (different working conditions) are obtained. Each SOH value was weighted and summed respectively, and the corresponding weight is the normalized similarity coefficient obtained after similarity evaluation.

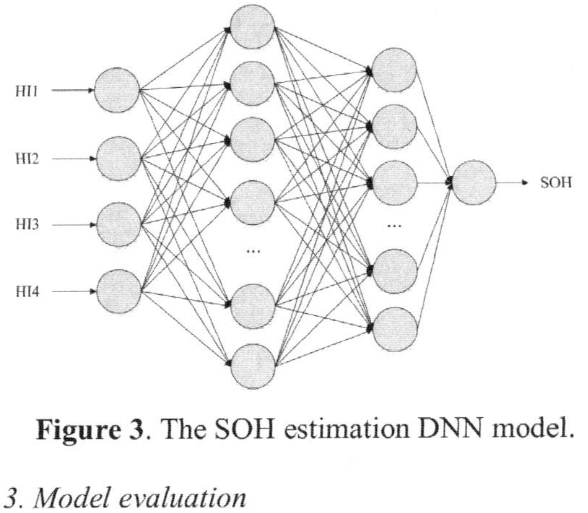

Figure 3. The SOH estimation DNN model.

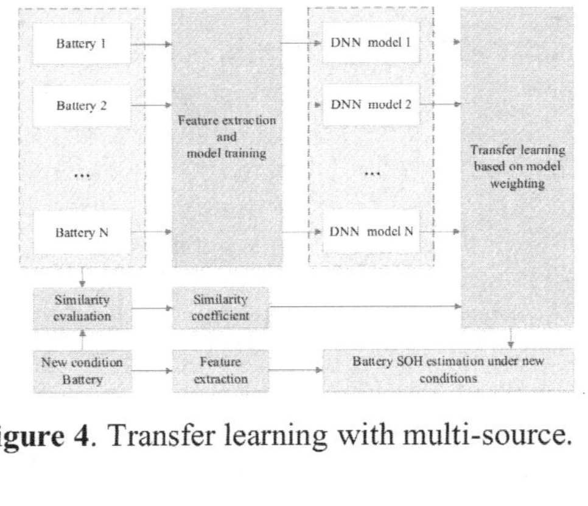

Figure 4. Transfer learning with multi-source.

3.3. Model evaluation

To evaluate the estimation accuracy of the transfer learning with a multi-source model for the SOH of the target battery, error indexes such as Mean Absolute Error (MAE) and Root Mean Square Error (RMSE) were introduced. The calculation equations are shown in Equation (4) and Equation (5) respectively.

$$\mathrm{MAE} = \frac{1}{N_1}\sum_{i=1}^{N_1}\left|\mathrm{SOH}_{\mathrm{actual},\,i} - \mathrm{SOH}_{\mathrm{predicted},\,i}\right| \tag{4}$$

$$\mathrm{RMSE} = \left(\frac{1}{N_1}\sum_{i=1}^{N_1}(\mathrm{SOH}_{\mathrm{actual},\,i} - \mathrm{SOH}_{\mathrm{predicted},\,i})^2\right)^{1/2} \tag{5}$$

where N_1 is the total number of samples of the actual and estimated SOH values, $\mathrm{SOH}_{\mathrm{actual},\,i}$ is the actual SOH value sample *i*, and $\mathrm{SOH}_{\mathrm{predictad},\,i}$ is the estimated SOH value sample *i*.

4. Analysis of experimental results

To compare the performance of the transfer learning with the multi-source estimation model proposed in this paper, three sets of comparison experiments were added. In the experiment of DNN (under the same working condition), the training set was the cyclic data of the b1c17 battery (5.4C-3.6C) and the test set was the cyclic data of the b1c17 battery (5.4C-3.6C). In the experiment of DNN (under different working conditions), the training set is the cycle data of the b3c11 battery (5.6C-4.6C) and the test set is the cycle data of the b1c17 battery. In the Long Short-term Memory Fully Connected Transfer Learning (LSTM-FC-TL) experiment [14], the basic model training set is the cycle data of the b3c11 battery. The LSTM layer was fixed during the model parameter fine-tuning phase and the fully connected layer parameters were fine-tuned by using the first 25% cycle data of the b1c17 battery. The SOH estimation error results of each algorithm model are shown in Table 2. All the above models adopt the same hyperparameters.

Table 2. SOH estimation error of each algorithm model.

Algorithm model	Parameters	MAE (%)	RMSE (%)
DNN (Same condition)	2, 201	0.42%	0.51%
DNN (Different condition)	2, 201	3.86%	4.53%
LSTM-FC-TL (Different condition)	27, 881	2.11%	2.78%
Ours (Different condition)	2, 201	1.79%	2.04%

As can be seen from Table 2, the DNN (Same condition) has the best fitting effect on the actual SOH curve of the target battery and obtains the highest accuracy among all models, with an MAE value of 0.42% and an RMSE value of 0.51%. However, complete test data under the same working conditions as the target battery are required. However, the DNN (Different condition) does not require the test data of the same working conditions as the target battery but obtains the worst fitting effect. Compared with the DNN (Same condition), the MAE value increases by 8.19 times, and the RMSE value increases by 7.88 times. When the training set and the test set have different charging conditions, the estimation accuracy of the SOH estimation will decrease significantly. Compared with the DNN (Different condition), the LSTM-FC-TL estimation model fits the SOH curve of the first 25% of the target battery better, and the MAE value decreases by 45.34% and RMSE value decreases by 38.63% in general. The Multi-source SOH estimation model proposed in this paper obtained an estimation accuracy second only to the DNN (Same condition) and had a good fitting effect for the actual SOH degradation curve. Compared with the DNN (Different condition), the MAE value decreased by 53.63% and the RMSE value decreased by 54.97%.

5. Conclusions

An SOH estimation method based on transfer learning with multi-source was proposed. Under different charging conditions of the training set battery and the test set battery, the SOH estimation accuracy of MAE=1.79% and RMSE=2.04% were obtained, and the aging test of the target battery under the new condition was reduced to 200 cycles.

Acknowledgments
We would like to express our thanks to the Department of Wuhan University of Technology for providing the experimental equipment.

References

[1] Dai H. F., Jiang B., Hu X. S., et al. (2021). *Renew & Sustain Energy Review*, **138**: 110, 480. https://doi.org/10.1016/j.rser.2020.110480.
[2] Che Y. H., Hu X. S., Lin X. K., et al. (2023). *Energy & Environ Science*, **16**: 338-71. https://doi.org/10.1039/D2EE03019E.
[3] Khalil A., Aljanaideh K. F., and Janaideh M. A. (2022). *IEEE Sensor J.*, **22**: 4, 164-74. https://doi.org/10.1109/JSEN.2022.3141932.
[4] Qin Y., Li W. T., Yuen C., et al. (2022). *IEEE Trans. on Industrial Infor.*, **18**: 4, 725-36. https://doi.org/10.1109/TII.2021.3075708.
[5] Guo Y. F., Huang K., Yu X. Y., et al. (2022). *Electrochimi. Acta.*, **428**: 140, 940. https://doi.org/10.1016/j.electacta.2022.140940.
[6] Su X. J., Sun B. X., Wang J. J., et al. (2022). *App. Energy*, **322**: 119, 516. https://doi.org/10.1016/j.apenergy.2022.119516.
[7] Lin H. Y., Kang L. Y., Xie D., et al. (2022). *Batterie Basel*, **8**: 19-29. https://doi.org/10.3390/batteries8040029.
[8] Chen L., Wang H. M., Liu B. H., et al. (2021). *Energy*, **215**: 119, 078. https://doi.org/10.1016/j.energy.2020.119078.
[9] Liu B. L., Xu J. L., and Xia W. (2023). *Interna. J. of Energy Research*, **2023**: 8, 569, 161. https://doi.org/10.1155/2023/8569161.
[10] Von Bülow F., Mentz J., and Meisen T. (2021). *J. of Energy Storage*, **44**: 103, 439. https://doi.org/10.1016/j.est.2021.103439.
[11] Severson K. A., Attia P. M., Jin N., et al. (2019). *Nature Energy*, **4**: 383–91. https://doi.org/10.1038/s41560-019-0356-8.
[12] Kalaiselvi B. and Thangamani M. (2020). *Measure: J of the Interna. Measure Confeder.*, **162**: 107, 885. https://doi.org/10.1016/j.measurement.2020.107885.
[13] Pan S. J. and Yang Q. A. (2010). *IEEE Trans. on Knowledge and Data Engineer*, **22**: 1, 345-59. https://doi.org/10.1109/TKDE.2009.191.
[14] Tan Y. D. and Zhao G. C. (2020). *IEEE Trans. on Industrial Electronic*, **67**: 8, 723-31. https://doi.org/10.1109/TIE.2019.2946551.

The effect of crystal phase noise on the proton precession magnetometer magnetic field measurement accuracy

Cheng Zhang[1,3], Tao Bo[1], Aimin Chen[2], Wang Luo[2] and Haobin Dong[2]

[1] Wuhan Digital Engineering Institute, Wuhan 430074, China
[2] School of Automation, China University of Geosciences, Wuhan 430074, China

[3]Corresponding author's e-mail: zhangc726@163.com

Abstract. The proton precession magnetometer is a crucial piece of equipment for measuring magnetic fields because it can determine the value of the magnetic field by measuring the frequency of the Free Induction Decay (FID) signal. The frequency source, one of the crucial parts of the frequency measurement system, has a direct impact on how well the system works as a whole. The communication quality and detection accuracy of the frequency measuring system are directly influenced by phase noise, which is a crucial indicator of frequency stability. By acquiring various output frequencies of crystal oscillators using the Phase Locked Loop (PLL) frequency doubling technique, this study investigates the extent to which crystal phase noise impacts the accuracy of magnetometer frequency measurement.

1. Introduction

The detection of weak magnetic signals is widely used in earth resource exploration, archaeology, and earthquake monitoring. Since the frequency of the FID signal generated by the proton magnetometer sensor is proportional to the magnitude of the measured magnetic field, the volume of the measured magnetic field can be calculated by measuring the frequency of the FID signal generated by the sensor. The measurement accuracy will directly determine the accuracy of the obtained magnetic field data [1]. Improving the frequency accuracy of the FID signal is the key to improving the measurement accuracy of the magnetometer.

High-precision time-frequency measurement and control technology play a crucial role in the development of measurement and control of various other physical quantities. Therefore, signals' generation, measurement, transformation, and control have been the focus of attention in countries worldwide [2]. Especially in recent years, high-precision frequency measurement has been playing an increasingly important role in many fields.

Currently, known instruments in China also use hardware frequency measurement [3][4]. Hardware frequency measurement is the primary measurement method, generally using direct frequency measurement, which generally uses three methods: the multi-period method, the frequency multiplication method, and the differential frequency method. Among them, the multi-period method of frequency measurement accuracy is the most effective in the entire frequency range required for better accuracy. This method, however, essentially only counts beyond zero and does not use all of the signal's amplitude information because the feed signal is an exponentially decaying sine wave that contains noise. The main disadvantage is that the auxiliary circuit noise spikes may lead to counting errors in the process of calculating the magnetic field. The error in one count during the calculation of the magnetic field will cause a substantial error in the overall measurement after the error accumulation.

The equal precision frequency measurement method and the direct frequency measurement method are used to achieve the measurement of the frequency of the signal to be measured by the method of gate counting. This measurement method, on the other hand, improves test accuracy and can achieve equal precision measurement in this frequency range. The measurement results of the equal precision frequency measurement method are not related to the size of the measured signal frequency but only to the size and stability of the reference frequency. For the equal precision measurement method, increasing the standard clock frequency can improve frequency measurement accuracy. If the gate time is 1 μs, the frequency measurement must be accurate to $\pm 1\times 10^{-3}$ orders of magnitude, which corresponds to a standard clock frequency of at least 1000 MHz. However, the current maximum counting frequency of the counter is about 500 MHz, which cannot meet the measurement requirements, and for close to 1000 MHz standard signal, it is pretty challenging to achieve.

In this study, we present a design method that multiplies the crystal output frequency using the ALTPLL IP core in Altera's Quartus II software, based on the crystal frequency phase-locked frequency multiplication from the crystal phase noise. The relationship between the crystal oscillator phase noise and the frequency measurement accuracy of the magnetometer is deduced from the experimental results after an experimental platform is constructed, circuit debugging and test experiments are carried out, and experimental results are obtained.

2. Crystal oscillator

Crystal oscillators are widely used as stable frequency reference sources in modern electronic systems requiring frequency control and management and in fields such as precision and accurate frequency metrology [5]. The application of crystal oscillators in various communication systems needs to focus on their short-term frequency stability, also known as phase noise. Good phase noise performance means a more desirable oscillation signal with shallow phase distortion, which can provide a more accurate signal.

2.1. Crystal performance evaluation

As crystal oscillators are susceptible to their own and external environment, resulting in their unstable working performance, in order to better use crystal oscillators, their working performance needs to be evaluated [6]. At present, there are two main parameters to evaluate the performance of a crystal oscillator: one is frequency accuracy, that is, whether its output vibration frequency is in line with its traditional value and whether it meets the factory standard; the other is frequency stability, that is, whether its output frequency signal sequence is smooth, whether there is frequency jump, and whether the output oscillation signal is within the safe range after long-term work. A study shows that the following formula can express the frequency accuracy of the crystal oscillator:

$$\mathrm{A}_f = \frac{f_t - f_0}{f_0} \tag{1}$$

In Equation (1), f_t is the actual operating output frequency of the crystal; f_0 is the expected frequency value; the unit to the left of the equal sign is ppm.

2.2. Phase noise of crystal oscillator

As the performance of radio communication systems improves, the performance of the signal source phase noise becomes a significant limiting factor in the overall system performance. Often referred to as phase noise, it refers to the short-term stability of the frequency, sometimes referred to as instantaneous stability. Short-term frequency stability is the relative frequency stability within one hour, usually the unwanted signal or noise and the interaction between the desired signal caused by the change in oscillator frequency, mainly caused by changes in circuit parameters, and does not include changes in frequency due to component aging or changes in the surrounding temperature. Frequency stability measures the oscillator's ability to produce the same frequency over the specified time range. If there is a transient change in signal frequency that cannot be kept constant. Then, there is instability in the signal source. Its cause is phase noise. Due to the presence of phase noise, it causes the expansion of the carrier spectrum, which can range from deviations from the carrier of less than 1 Hz to several megahertz. The oscillator's phase noise consists of various components, such as white noise and *1/f* noise [7].

In practice, all signal sources suffer from instabilities, i.e., there are useless signal amplitudes, frequencies, or phase undulations that never produce a pure sine signal. Since frequency is the derivative of phase concerning time, these instabilities can be equivalently viewed as useless frequency or phase undulations. The characterization of these phase undulations is often called phase noise. Phase noise is difficult to see in the time domain, and it exhibits zero crossover variation relative to a pure sinusoidal signal without phase noise. Because phase noise is generated by mixing low-frequency signals at much lower frequencies than the carrier frequency, the zero-crossing variation is not observed in the time domain until after many weeks of carrier waves.

Phase noise is an essential indicator of the stability of the frequency reference source, and in general, phase noise refers to the short-term stability of the signal frequency [8]. As shown in Figure 1, a represents a standard sine wave signal without any noise, and b represents a sine wave signal with phase noise. Relative to the standard sine wave signal, the phase of waveform b is arbitrarily disturbed at the set time grid, i.e., over or lagged.

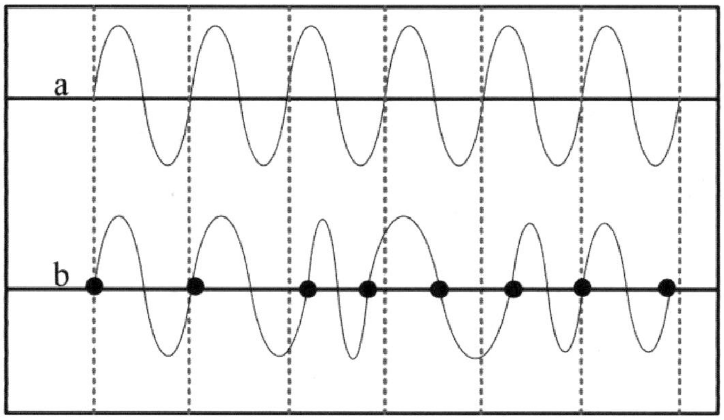

Figure 1. Noise-free vs. noise-containing sine wave.

From the frequency domain, an ideal crystal oscillator outputs a signal that behaves as a single spectrum. However, due to phase undulation or other factors, an actual crystal oscillator can never produce a single pure sine or cosine signal that behaves as an impure signal frequency. Figure 2 shows the spectrum of a pure sine signal and an impure sine signal.

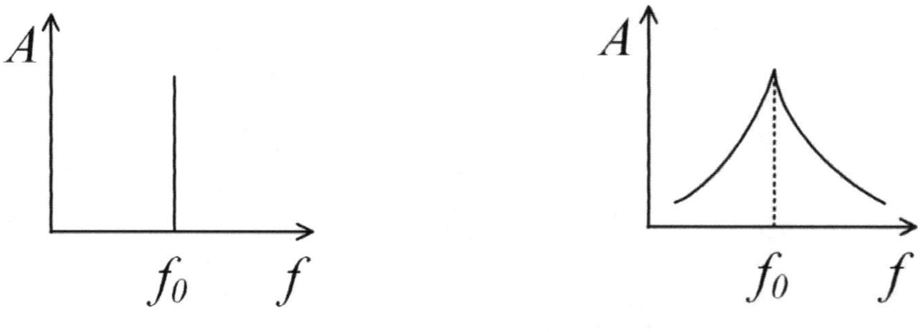

(a) Spectrum of a pure sine signal (b) Spectrum of an impure sine signal

Figure 2. Spectrum of pure and impure sine signals.

The output of an ideal oscillator is the ideal waveform, but the actual oscillator's output could be better. As shown in Figure 3, the ideal degree of the waveform can be measured from three perspectives: amplitude, frequency, and phase. For the oscillator output signal, its amplitude instability is shown in the fluctuation of peak value; its phase instability is shown in the fluctuation of voltage at zero point; its voltage frequency instability is shown in the period change. Due to the limiting characteristics of the oscillation circuit, the amplitude noise will be removed, and the non-idealities in the amplitude of the oscillation waveform will become less critical. However, the phase noise caused by the phase jitter of the crystal is introduced into the communication system referenced by it as the frequency reference source, significantly reducing the performance of the communication system.

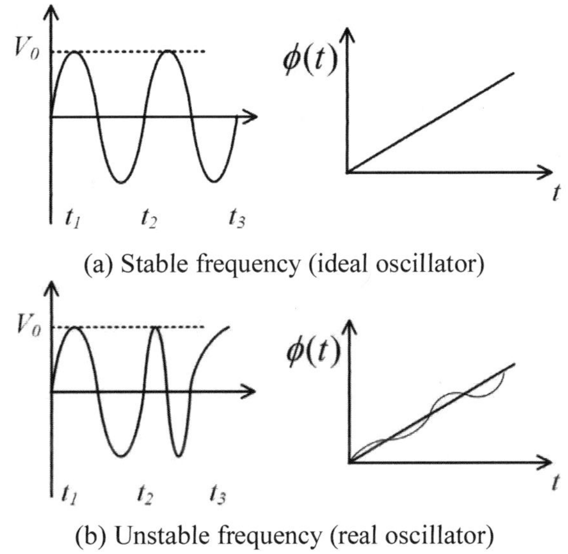

Figure 3. Ideal and non-ideal waveforms.

The phase noise at the low-frequency bias of the frequency synthesizer affects the modulated signal and reduces its signal-to-noise ratio; the phase noise at the high-frequency bias interferes with other signal channels. As shown in Figure 4, the frequency synthesizer's phase noise affects the transmitter performance process. Furthermore, the crystal oscillator is the primary source of low-frequency bias phase noise in the frequency synthesizer. Therefore, the low phase noise performance of the crystal oscillator is significant for the whole radio frequency (RF) transceiver system.

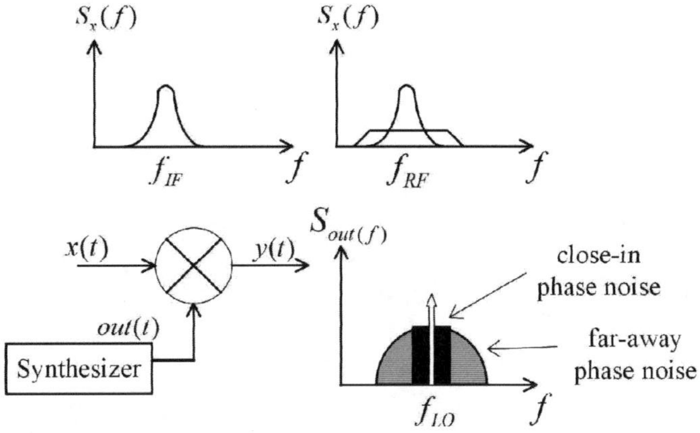

Figure 4. The process by which phase noise affects transmitter performance.

2.3. Frequency domain characterization of phase noise

The mathematical formula represents an ideal oscillator whose output is an ideal sinusoidal signal:

$$V(t) = A\cos(\omega_0 t + \varphi) \tag{2}$$

In Equation (2), $V(t)$ is the output waveform amplitude, A, ω_0, and φ are the vibration amplitude, angular frequency, and initial phase, respectively.

In practice, the crystal oscillator, due to the influence of other factors, its amplitude and phase are generally time-varying, which the mathematical formula can express:

$$V(t) = A(t)\cdot\cos(\omega_0 t + \varphi(t)) \tag{3}$$

The vibration amplitude $A(t)$ and the phase $\varphi(t)$ are functions that change with time. Generally, amplitude stabilization circuits are designed in high-precision crystal oscillators to suppress fluctuations in vibration amplitude $A(t)$, so the effect of phase $\varphi(t)$ variation (i.e., phase noise) is much greater than the effect of amplitude $A(t)$ fluctuations.

When using a spectrum analyzer to observe the signal noise sidebands, phase noise is symmetrically distributed on both sides of the center frequency. The actual engineering applications usually have single sideband phase noise (SSB). Single sideband phase noise is: at a particular frequency offset from the center corner frequency ω_0, the ratio of the power per unit bandwidth to the total power of the output signal, the unit is dBc/Hz.

2.4. Time domain characterization of phase noise

In addition to the frequency domain method to represent the phase noise, the time domain representation has also been proposed. With the ongoing in-depth study of the oscillator noise characteristics, the defining formula for the time domain representation of the phase noise has evolved from the classical variance form to the Arlen variance. In the case of a limited number of measurements, the best expression of the Arlen variance is given in [9] and [10]:

$$\sigma_y(t) = \sqrt{\frac{1}{2m}\sum_{k=1}^{m}\left[\overline{y_{2k}} - \overline{y_{2k-1}}\right]^2} \tag{4}$$

In Equation (4), t is the sampling time, and m is the total number of groups of measurements. $\overline{y_k}$ is the measured average frequency value, and every two measurements are taken as a group. It is important to ensure that there is no time interval between each two measurements.

Sometimes, to reduce the time required for the measurement, it is possible to use a sampling number of $m + 1$. This gives another expression for the Allan variance:

$$\sigma_y(t) = \sqrt{\frac{1}{2(m-1)}\sum_{k=1}^{m-1}\left[\overline{y_{k+1}} - \overline{y_k}\right]^2} \tag{5}$$

Allan variance applies to most of the current frequency sources and scales. It has gained the most widespread application due to its easy calculation and is an essential quantity for measuring frequency fluctuations. However, Allan variance also has some drawbacks and limitations. For example, the difference between adjacent frequency measurements in practice includes the random fluctuations caused by noise but also the frequency shift of the system itself.

2.5. Phase noise analysis based on the Leeson model

Leeson [11], [12] proposed the feedback oscillator model in the early 1960s. In the following years, Sauwvage [13] proved the feasibility of Leeson's model mathematically and theoretically. Figure 5 shows the schematic diagram of the feedback oscillator model proposed by Leeson.

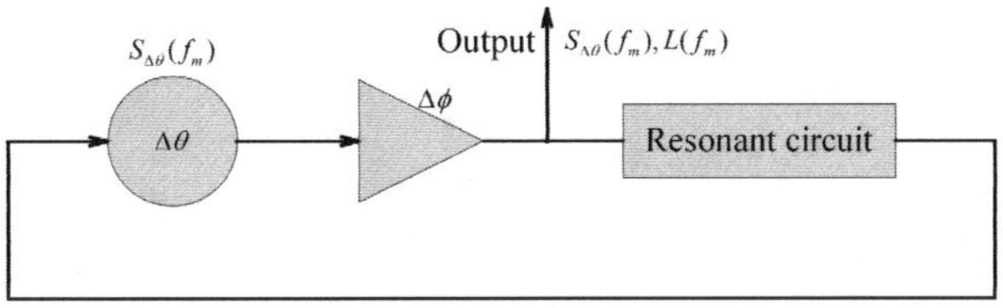

Figure 5. Feedback type oscillator Leeson model.

As seen in Figure 5, the Leeson model consists of two main parts, the amplifier and the filter resonant loop, and it is a phase-positive feedback system. Since the equivalent circuit model of the quartz crystal is a resistor-inductor-capacitor (RLC) circuit, the resonant loop in Figure 5 can be viewed as an RLC circuit, i.e., a resonant loop that acts as a bandpass filter. We relate the phase noise spectral density of both at the oscillator's output and inside the amplifier so that the total power spectral density of the oscillator can be found.

$$S_{\Delta\phi}\left(f_m\right)=S_{\Delta\theta}\left(f_m\right)\cdot\left\{1+\left(\frac{f_0}{2Q_L f_m}\right)^2\right\} \tag{6}$$

$$S_{\Delta\theta}\left(f_m\right)=\frac{FkT}{P}\left(1+\frac{f_c}{f_m}\right) \tag{7}$$

in the equation, $S_{\Delta\theta}(f_m)$ is the phase undulation spectral density of the oscillator. $S_{\Delta\theta}(f_m)$ is the phase undulation spectral density at the input of the amplifier; P and Q_L are the input excitation power and the on-load quality factor, respectively. F_c, f_m, f_0 represent the corner frequency, offset carrier frequency, and oscillator's carrier frequency, respectively. T, k, F represents the absolute temperature, Boltzmann's constant, and the feedback coefficient.

Equation (6) represents the phase undulation spectral density of the output of the Leeson model oscillator, i.e., the phase noise. It can be seen that due to the presence of positive phase feedback caused by the phase noise $S_{\Delta\theta}(f_m)$ at the input of the amplifier, the oscillator output phase noise increases by a factor of $1 + (f_0/2Q_Lf_m)^2$. We can derive the single-sideband phase noise of the oscillator from the relationship between the corner frequency fc of the flicker noise and the half bandwidth $f_0/2Q_L$ of the resonant loop and the relationship between the single-sideband phase noise and the output power spectral density as:

$$\iota\left(f_m\right)=\frac{1}{2}\left(1+\frac{f_c}{f_m}\right)\frac{FkT}{P},\frac{f_0}{2Q_L}<f_m<f_c \tag{8}$$

From Equation (8), we can know that the single sideband phase noise spectral density of the oscillator is related to many circuit factors, only a comprehensive weighing of the choice of various parameters of the oscillator circuit, especially the choice of parameters such as f_c, f_0, P, F, and Q_L. For example, for the carrier frequency f_0 of the oscillator, the higher the frequency is, the higher the output phase noise of the oscillator will be.

3. PLL phase-locked frequency doubling

The frequency multiplication circuit is the core of the atomic clock signal source. The input crystal signal can be multiplied by the frequency multiplication circuit to the required frequency of the atomic clock. The traditional frequency multiplication method uses analog frequency multiplication, and discrete devices make the circuit large in size. In order to meet the demand for the miniaturization of atomic clocks, digital phase-locked frequency multiplication is now used. Phase Locked Loop (PLL) is integrated, which reduces the size and has better performance in terms of phase noise and spurious. The loop filter is an essential part of the phase-locked loop and consists mainly of capacitors, resistors, or amplifiers. The loop filter filters out the high-frequency components of the discriminator output signal.

It dramatically impacts important loop parameters such as spurious rejection, phase noise, loop stability, and lock time of the phase-locked loop.

The PLL consists of four main modules: phase discriminator, loop filter, voltage-controlled oscillator, and frequency divider, as shown in Figure 6. The input signal introduced through the oscillator enters the phase discriminator. The output signal after the frequency divider is multiplied to discriminate the phase. The discriminator outputs a voltage signal related to its phase difference, which is processed by the loop filter and enters the voltage-controlled oscillator (VCO). The VCO voltage control terminal controls the VCO output signal. Finally, the feedback loop stabilizes the output frequency, and the VCO outputs the target frequency.

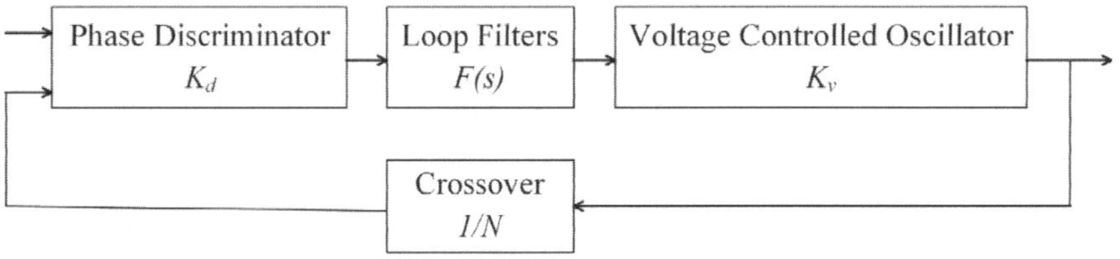

Figure 6. PLL building blocks.

4. Experiments and analysis of results

The experimental platform is shown in Figure 7. The experimental platform includes a computer, a signal generator, a magnetometer, and a sensor.

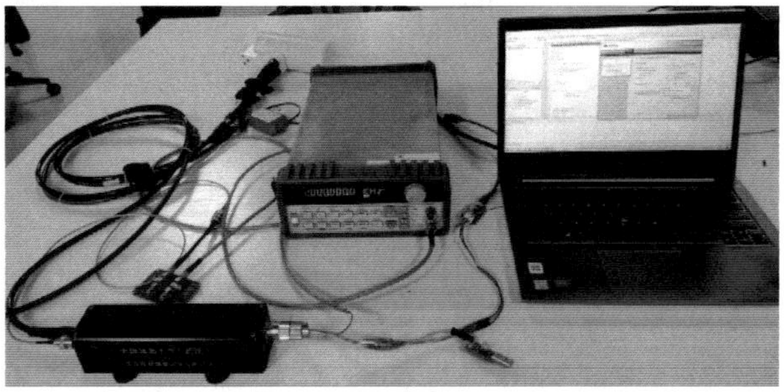

Figure 7. Test scenario map.

The crystal uses an SG-8018 series quartz programmable oscillator with a fixed output frequency of 50 MHz and the phase jitter (Root-Mean-Square, RMS), as shown in Figure 8.

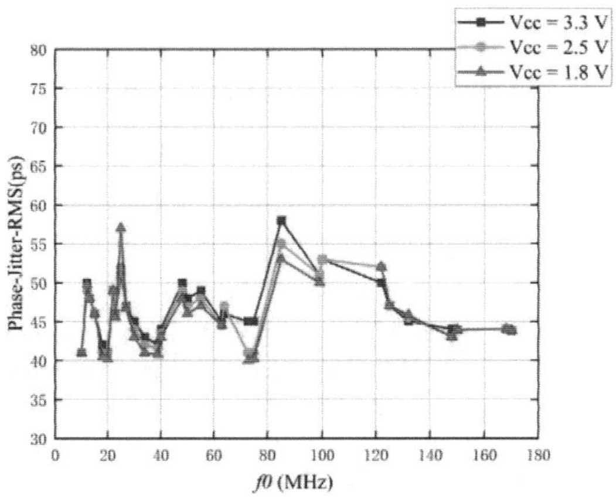

Figure 8. Phase jitter.

The frequency multiplication link mainly uses the IP core ALTPLL in Altera's Quartus II software to multiply the crystal frequency. The IP core ALTPLL in Altera's Quartus II software is mostly used by the frequency multiplication link to multiply the crystal frequency. This IP core can establish or alter the scale factor to produce the desired frequency value as soon as the procedure is finished, as indicated in Figure 9. The frequency multiplication to 100 MHz and 200 MHz is selected as the output in this experiment.

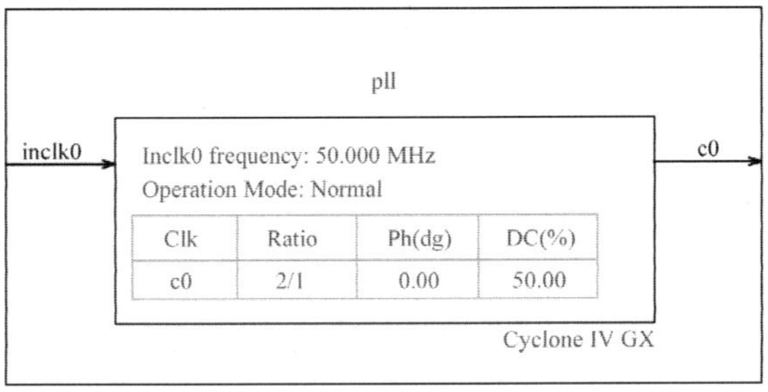

Figure 9. PLL frequency doubling.

The interface lets you directly set or adjust the scale factor to output the desired frequency value. For this experiment, the frequency multiplier to 100 MHz and 200 MHz is selected as the output.

The signal generator has a fixed output of 2 kHz and 200 mV frequency. The signal amplitude of 100,000 times is attenuated through the attenuator into the magnetometer. Finally, the measured frequency value of the magnetometer is obtained to study the effect of the crystal phase noise on the measurement accuracy after frequency doubling.

The test results are shown in Figure 10 and Figure 11.

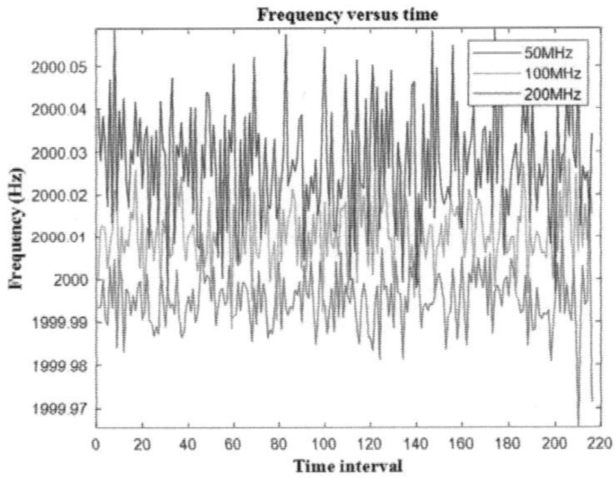

Figure 10. Test data waveforms.

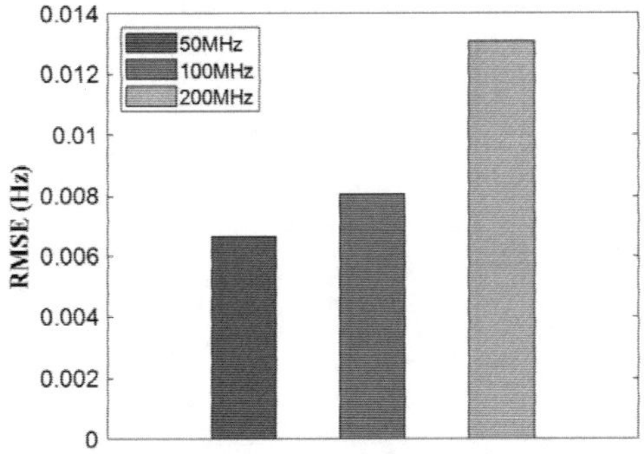

Figure 11. Root mean square error for different frequencies.

In summary, after multiplying the crystal frequency, it is known from the Leeson model that the oscillator output phase noise increases, resulting in a decrease in frequency measurement accuracy.

5. Conclusion

Phase noise is one of the main parameters of crystal oscillators, and it has an important impact on the performance of various electronic devices. This paper details the concept of phase noise and the characterization methods of phase noise, primary domain characterization method, and frequency domain characterization method. The phase noise model based on the Leeson model is analyzed, and the impact of the phase noise of the crystal oscillator on frequency measurement accuracy is analyzed theoretically. Finally, the frequency doubling experiment proves that increasing the crystal frequency will increase the phase noise and reduce the measurement accuracy. Since the signal frequency value is closely related to the magnetic field strength, the magnetic measurement accuracy of the proton precession magnetometer will also decrease when the measurement frequency accuracy decreases.

References

[1] Pan D, Li J, Jin C, et al. A new calibration method for triaxial fluxgate magnetometer based on magnetic shielding room. *IEEE Transactions on Industrial Electronics*, 2019, 67(5): 4183-4192.

[2] X. Wang, Research and design of ultra-high precision frequency meter. *Huazhong Normal University*, 2018.

[3] Luo W, Ge J, Dong H, et al. An intelligent full-wavenumber fitting algorithm for proton precession magnetometer frequency measurement. *IEEE Sensors Journal*, 2023.

[4] Li R, Baynes F N, Luiten A N, et al. Continuous high-sensitivity and high-bandwidth atomic magnetometer[J]. *Physical Review Applied*, 2020, 14(6): 064067.

[5] Robert J.Matthys, Crystal Oscillator Circuits. New York, 1983, pp.9-17.

[6] Y. Hou, A least-squares-based algorithm for crystal frequency drift prediction under alternating temperature. *Hebei University of Technology*, 2022.

[7] AN1026, Application Note, 1/f Noise Characteristics Influencing Phase Noise. *California Eastern Laboratories*, 1998.

[8] C. Liu, Analysis of crystal oscillator phase noise under vibration conditions and its compensation processing technology. *National University of Defense Technology*, 2014.

[9] Floyd M.Gardner, Phase Locked Loop Technology (Third Edition). *Peking*, 2007, pp.119-139.

[10] Vadim Manassewitsch. Frequency Synthesis Principles and Design (Third Edition). *Electronic Industry Press*, 2008, pp.66-92.

[11] A. Hajimiri, T. H. Lee, A general theory of phase noise in electrical oscillators, J. *IEEE J Solid-State circuit*, vol. 33, pp.179-194, February 1988.

[12] D. B. Leeson, A Simple of Frequency oscillator Noise Spectrum, J. *Proceedings of the IEEE*, vol. 54, pp.329-330, February 1966.

[13] Sauvage G, Phase noise in oscillators: A mathematical analysis of Leeson's model. *IEEE Transactions on Instrumentation and Measurement*, vol. 26(4), pp.408-410, December 1977.

Journal of Physics: Conference Series **2724** (2024) 012041 doi:10.1088/1742-6596/2724/1/012041

Temperature compensation for piezoresistive pressure sensor based on deep learning on graphs

Mingyao Xue[1], Ting Fang[1, 2], Chenghe Hu[1], Chong Dong[1] and Jiaming Han[1]

[1]School of Electrical and Information Engineering, Anhui University of Technology, China.

[2] Corresponding author's e-mail: Flyting_69@163.com

Abstract: With the escalating demand for precision in piezoresistive pressure sensors to cater to a wide spectrum of applications, a significant challenge emerges due to the material properties of these sensors, which induce a substantial temperature coefficient. This limitation restricts the operational temperature range and is further exacerbated by ambient temperature variations, resulting in pronounced nonlinearity in sensor responses. To address these challenges, A new method using Graph Neural Networks (GNNs) is introduced to address nonlinearity and temperature drift in piezoresistive pressure sensors. GNNs improve the sensors' accuracy and reliability across different temperatures, expanding their applicability. Test results show a notable accuracy enhancement, with a maximum full-scale error below 0.05% and even lower in specific ranges (under 0.04%). This precision makes the method ideal for industrial pressure sensor applications and production, offering significant benefits.

1. Introduction

Pressure measurement, a fundamental parameter in process control and automation systems, finds extensive application across diverse sectors, notably automotive, industrial, and healthcare [1][2]. This ubiquity stems from the straightforward nature of its quantification and its adaptability in various contexts. Crucially, pressure readings offer insights into other key process variables, including level, volume, flow, and density, enhancing their applicability in complex systems. Pressure sensors, integral to these applications, are subjected to a broad spectrum of operational conditions and environments. Despite these varying demands, they consistently conform to rigorous industry standards, ensuring unwavering performance and reliability. This alignment underscores their critical role in maintaining the precision and efficiency of contemporary automated systems.

The unveiling of the piezoresistance effect in silicon by Smith [3] spurred the development of the first piezoresistive semiconductor sensors in the early 1970s, initially designed for the automotive industry. This breakthrough heralded a new era, with silicon piezoresistive technology becoming a focal point of intensive research and extensive commercial application. These sensors are highly esteemed for their exceptional sensitivity, sturdy resistance, and compact form. Nonetheless, the manufacturing and packaging processes of these pressure sensors still present unresolved challenges, resulting in notable nonlinearities and temperature-dependent drift in their response characteristics [4]. To address these challenges, a range of approaches, encompassing a range of hardware and software solutions, has been suggested to incorporate thermal compensation into pressure sensor design [5]. Hardware compensation in pressure sensors, typically using analogy circuits for calibration, incorporates elements like thermistors and adjustable amplifiers. While simple, this approach has drawbacks in terms of reliability and precision [6]. In contrast, software temperature compensation relies on digital data processing with intelligent algorithms, particularly neural networks [7]. This software approach offers improved adaptability, accuracy, and the ability to handle complex data. The focus is increasingly on refining and enhancing intelligent optimization algorithms rather than just introducing new neural networks [8]-[10].

In this study, a novel method employing Graph Neural Networks (GNNs) is introduced [11]. Focused on addressing nonlinearity and temperature drift, this method utilizes the advanced computational power of GNNs, specifically two models: Graph Convolutional Networks (GCN) [12] and Graph Attention Networks (GAT) [13]. By leveraging these GNNs, sensor accuracy and reliability are significantly enhanced under varying temperature conditions, expanding the practical applications of piezoresistive pressure sensors. Through simulations and practical tests, the method has been shown to substantially improve accuracy. Notably, the maximum full-scale error across the sensor range is reduced to better than 0.05%, and in localized ranges, this error is further minimized to less than 0.04%. Comparative analysis revealed that while both GCN and GAT models contribute to performance improvements, the GAT model exhibits superior results, making it particularly effective. This level of precision establishes the method as highly suitable for the development and production of industrial pressure sensors, presenting multiple advantages.

2. Principles and Methods

2.1 Principle of piezoresistive pressure sensor

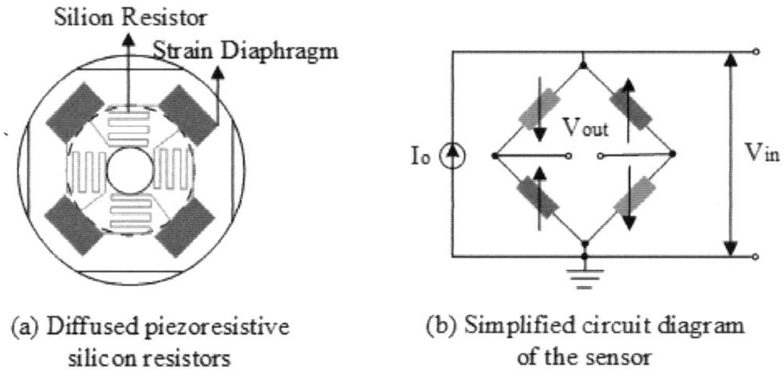

Figure 1. The main structure.

The silicon-based piezoresistive pressure sensor consists of a core that senses pressure through the piezoresistive effect and a module for signal processing. Illustrated in Figure 1(a), this sensor utilizes the piezoresistive properties of semiconductor materials within the core to detect pressure alterations. As shown in Figure 1(b), the sensor's sensitive diaphragm is equipped with four silicon strain resistors, forming a Wheatstone bridge. The unique positioning and orientation of these resistors result in resistance changes that are opposite on adjacent arms of the bridge but identical on the diagonal arms. In an ideal scenario, the bridge's output voltage, V_{out}, would be directly proportional to the pressure if the sensitivity of the core remains constant. Nonetheless, the sensitivity and accuracy of the sensor are impacted by nonlinearity and temperature drift caused by changes in the temperature coefficient and piezo-resistance coefficient. Therefore, implementing thorough compensation strategies is crucial to maintain accuracy.

This article details the design of a hardware solution for signal acquisition and processing, as shown in Figure 2. The core processor used is the STM32L1 microcontroller. The high-performance NSA2862X, a 24-bit Analog-to-Digital (A/D) converter chip, interfaces with the CPU through the Inter-Integrated Circuit (I^2C) communication protocol. The NSA2862X integrates a 24-bit primary signal measurement channel and a 24-bit auxiliary temperature measurement channel, allowing for simultaneous data acquisition from both pressure and temperature sensors. After processing through the signal acquisition circuit, the thermocouple PT1000 embedded in the pressure module achieves an accuracy of 0.1 μ°C, enabling more precise experimental temperature measurements.

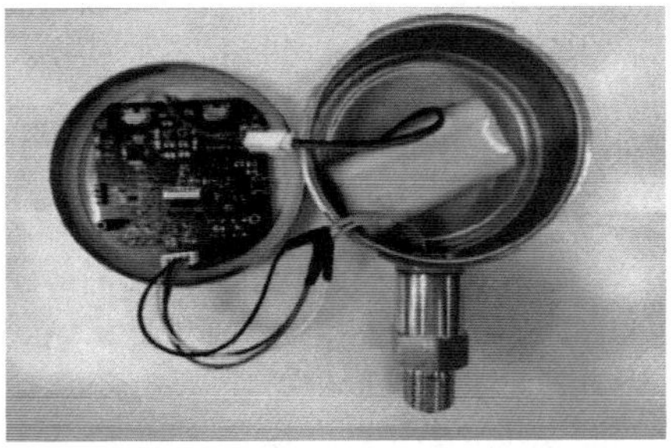

Figure 2. Sensor and signal acquisition and processing circuit.

2.2 Deep learning on graphs

GNNs are described as a class of deep learning models designed to handle graph-structured data. Unlike traditional neural networks, GNNs can capture the dependency of nodes in a graph, making them suitable for applications where data is inherently relational, such as social networks, molecular structures, and communication networks. GNNs operate by aggregating information from a node's neighbors through layers, thereby learning node representations that capture both their own features and the structure of the surrounding graph. This approach allows GNNs to effectively process and interpret complex and interconnected data.

2.2.1 Graph convolutional networks. The core function of a GCN is to extract features from data structured in graphs. GCN employs convolution operations on graphs, enabling features and messages to circulate and spread throughout the network. At its most basic level, this process involves state changes in each node that are proportional to the influence exerted by the surrounding graph structure, typically quantified using the graph's Laplacian operator. GCN has skillfully developed a technique to extract features from graph data. These extracted features can be used for node classification, graph classification, edge prediction, and representing the graph in an embedded form, making GCN widely applicable in various domains.

The structure of a GCN starts with a graph as input. In this graph, each node goes through a convolution process in one convolutional layer, where each node and its neighbors are convolved, updating the node based on this convolution. Subsequently, an activation function is applied. This sequence of convolution and activation is repeated until the network reaches its intended depth. In the end, an output function converts the states of the nodes into relevant output labels. In this model, each node continuously alters its state, influenced by its neighbors and more distant points in the graph, with closer neighbors having a more significant impact. The specific structure of the GCN is illustrated in Figure 3.

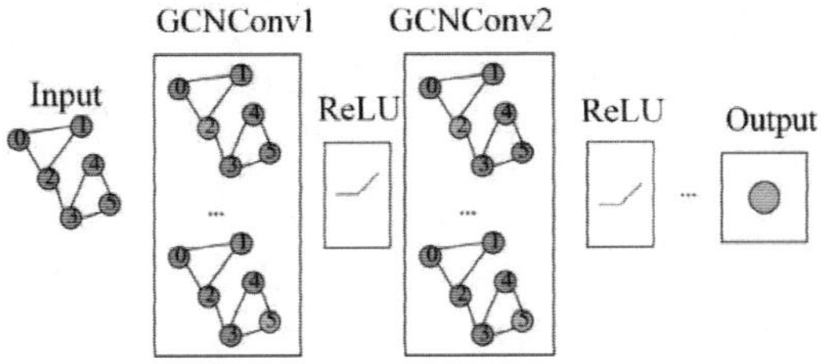

Figure 3. The structure of GCN.

GCNs represent a distinct layer within neural network architectures. It is assumed that we have a dataset comprising N nodes, with each node characterized by its unique attributes, constituting an N by D matrix X, with D signifying the attribute count for each node. Inter-node connections are encoded as another matrix of dimension $N \times N$, termed the adjacency matrix A, which, together with X serves as the input to our model. The method for propagating information across GCN layers is described by the following formula:

$$H^{(l+1)} = \sigma(\widetilde{D}^{-\frac{1}{2}}\tilde{A}\widetilde{D}^{-\frac{1}{2}}H^{(l)}W^{(l)}) \quad (1)$$

in this formula, $\tilde{A} = A + I$, where I is the identity matrix. $\widetilde{D}$ is the degree matrix of $\tilde{A}$, defined as $\widetilde{D}_{ij} = \sum_j \tilde{A}_{ij}$.$H$ is the feature matrix for each layer, with H being X for the input layer. $W^{(l)}$ is a randomly initialized weight matrix. σ represents a nonlinear activation function.

2.2.2 Graph attention networks. GAT, akin to GCN, is an advanced mechanism for extracting features from graph-structured data. What sets GAT apart is its integration of attention mechanisms, allowing for nuanced, adaptive feature aggregation from a node's neighbors, focusing on those most relevant to the task. In detail, GATs dynamically evaluate the importance of each node's neighbors using an attention mechanism, assigning scores that prioritize influential ones during feature aggregation. This process often employs a multi-head attention approach, enhancing stability and performance, and culminates in attention-weighted feature combinations, further refined through non-linear activation functions. This sophisticated approach enables GATs to effectively manage complex graph-structured tasks. The core formula of GAT can be expressed as:

$$Attention\left(H^{(l)}, a\right) = softmax\left(\frac{e_{ij}}{\sum_{k \in N_i} e_{ij}}\right) \quad (2)$$

Here, e_{ij} is the attention coefficient computed using a learnable weight vector a and the features of node i and its neighbor j.

$$H^{(l+1)} = \sigma(\textstyle\sum_{j \in N_i} Attention\left(H^{(l)}, a\right)_{ij} \cdot W^{(l)} \cdot H_j^{(l)}) \quad (3)$$

in this formula, $H^{(l)}$ represents the feature matrix of the nodes at layer l. N_i denotes the set of neighbors of node i. $W^{(l)}$ is the learnable weight matrix at layer l. σ is a non-linear activation function.

For regression tasks, while the attention mechanism (with $softmax$) remains effective for computing attention scores, the final activation function in the feature update may need to be adjusted to suit the continuous nature of the output, as regression involves predicting continuous values rather than discrete classes.

3. Experiment

3.1 Pressure data acquisition

The characterization experiment of the pressure sensor encompassed eleven distinct temperature points, with each temperature phase extending for a minimum of two hours. Maintaining this period was vital for the stability of the temperature's influence on the pressure sensor, which in turn helped to mitigate errors in measurement due to thermal inertia. Once the designated standard pressure was reached, each pressurization was maintained for at least 5 minutes to negate measurement inaccuracies arising from mechanical hysteresis. To reduce systemic errors, the mean of three measurements was calculated and used as the final experimental data.

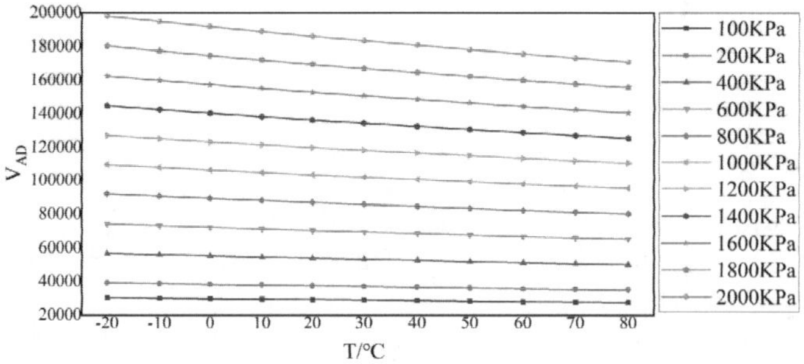

Figure 4. Output V_{AD} with different pressures and temperature.

As evident from Figure 4, the output values of the pressure sensor (VAD) under varying pressures exhibited significant changes with temperature fluctuations. At a fixed pressure, VAD demonstrated a linear decrease with an increase in temperature. However, within the high-pressure range, the influence of temperature on VAD became increasingly pronounced, indicating a marked non-linearity of VAD across the entire range of pressure. According to Zhao et al.'s work [10], the maximum full-scale error is the most critical parameter for evaluating the performance of the pressure sensor, and it is defined as follows:

$$\delta P[\%FS] = \left|\frac{\hat{P}_k - P_k}{P_{FS}}\right| \times 100\% \quad (4)$$

where P_k represents the actual standard pressure value produced by the automatic pressure calibrator, $\hat{P}_k$ is the pressure value determined through calculation, and P_{FS} denotes the complete pressure scale applicable to a specific sensor.

Figure 5 presents a set of full-scale error measurements plotted as a function of temperature across various pressures. These findings indicate that the pressure sensor's practical applicability is limited due to its diminished accuracy and uncorrected output, which are primarily attributed to temperature-induced nonlinearity and drift.

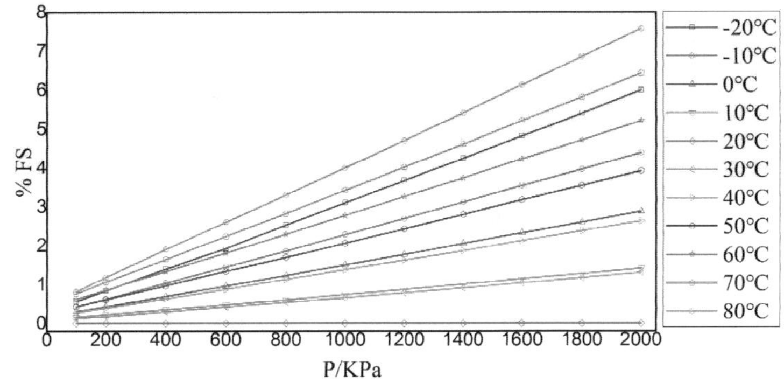

Figure 5. %FS at different temperatures and pressures.

3.2 Graph Data Generation
Temperature and V_{AD} are utilized as input features, while the actual pressure values serve as the target output. An edge index, representing the relationships between data points, is constructed based on the observed changes in temperature and V_{AD} values. The generated graph data is shown in Figure 6:

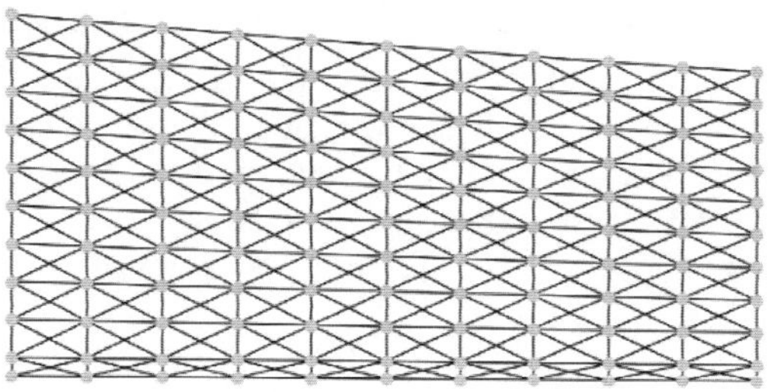

Figure 6. Graph data structure

3.3 Training Model
As depicted in Figure 7, the training model consists of two hidden layers, two LeakyReLU activation layers, and a single linear layer. The nature of the model—whether it is a GCN or a GAT—is determined

by the convolution operation in the hidden layers: GCNConv typifies a GCN model, whereas GATConv characterizes a GAT model. The first hidden layer transforms the input features into a higher-dimensional hidden representation, while the second layer further refines these representations. In the case of GATConv, it dynamically weights the importance of adjacent nodes within the graph, thereby delicately capturing the complex dependencies between nodes. A LeakyReLU activation function is applied following each hidden layer to introduce non-linearity, enabling the model to discern more intricate patterns within the graph data. The final output is procured via a linear layer that projects the hidden features onto the desired output dimensions.

Before training, the sample data is randomized to ensure unbiased learning, with a split into training and prediction groups at a ratio of 7:3. The batch size is set at 32, and the hidden layers are dimensioned at 128. The model is optimized using the Adam gradient descent algorithm, with a maximum of 200 iterations and an initial learning rate of 0.01.

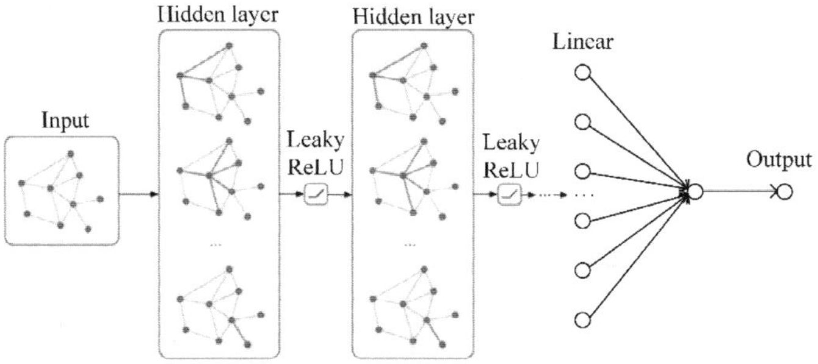

Figure 7. Training model.

4. Results

The data presented in Table 1 provides a succinct yet compelling comparative assessment of two distinct models: the GCN model and the GAT model. According to the table, the GAT model exhibits superior performance, with a maximum full-scale error (Max($\delta P[\%FS]$)) of only 0.0491%, a minimum full-scale error (Min($\delta P[\%FS]$)) of 0.0436%, and a mean full-scale error (Mean($\delta P[\%FS]$)) of 0.0460%. In contrast, the GCN model reports slightly higher errors across these metrics.

Table 1. Comparison of metrics of each model.

Model	Max ($\delta P[\%FS]$)	Min ($\delta P[\%FS]$)	Mean ($\delta P[\%FS]$)
GCN Model	0.0823%	0.0764%	0.0803%
GAT Model	0.0491%	0.0436%	0.0445%

These findings suggest that the GAT model's attention mechanism, which dynamically assigns importance to the nodes in the sensor's data graph, results in a more accurate representation of the sensor's response, reducing the full-scale error. Consequently, the GAT model's enhanced accuracy and lower error rates establish its potential for industrial applications where precise pressure measurements are crucial, affirming the significant benefits of utilizing GNNs in sensor technology.

5. Conclusion

This study focuses on enhancing the accuracy and reliability of piezoresistive pressure sensors under varying temperature conditions. By conducting a thorough analysis of the material properties of these sensors, particularly their nonlinear response and temperature drift when subjected to temperature changes, this research introduces a novel approach utilizing GNNs, specifically GCN and GAT, to address these challenges.

Through detailed experiments, we demonstrate the efficacy of GNNs in processing complex, interrelated data. Our findings reveal that the GAT model, employing an attention mechanism, exhibits higher precision and lower full-scale error compared to the traditional GCN when dealing with pressure sensor data. This discovery underscores the significant advantages of applying GNNs in sensor technology, especially in industrial applications where there is a growing demand for precise pressure measurement.

Acknowledgment
Anhui Province Industrial Internet Intelligent Application and Security Engineering Laboratory Open Fund (IASII21-05); Anhui University of Technology School Youth Fund (QZ202109); Anhui University of Technology School Youth Fund (QZ202109).

References

[1] Balavalad, K. B. A Review on Silicon Nanowires and their use in the Development of Nano Piezoresistive Pressure Sensors. Nanoscience and Nanotechnology - Asia, 13(5). 2023. https://doi.org/10.2174/2210681213666230619153413.

[2] Devi, R., Gill, S.S. Performance Investigation of Carbon Nanotube Based Temperature Compensated Piezoresistive Pressure Sensor. Silicon 14, 3931–3938 (2022). https://doi.org/10.1007/s12633-021-01153-w.

[3] C. S. Smith, "Piezoresistance effect in Germanium and silicon," Phys.Rev., vol. 94, no. 1, pp. 42–49, Apr. 1954. Doi:10.1103/PhysRev.94.42.

[4] Ruirui Wu, Huaijiang Li, Lei Gao; Research on temperature drift mechanism and compensation method of silicon piezoresistive pressure sensors. AIP Advances 1 March 2023; 13 (3): 035323. https://doi.org/10.1063/5.0135401.

[5] Tian, B., Shang, H. & Wang, W. Research on Temperature Zero Drift of SiC Piezoresistive Pressure Sensor Based on Asymmetric Wheatstone Bridge. Silicon 14, 5445–5451 (2022). https://doi.org/10.1007/s12633-021-01330-x.

[6] Wang H, Zeng Q, Zhang Z, Wang H. Research on Temperature Compensation of Multi-Channel Pressure Scanner Based on an Improved Cuckoo Search Optimizing a BP Neural Network. Micromachines. 2022; 13(8):1351. https://doi.org/10.3390/mi13081351.

[7] Y. Ruan, L. Yuan, W. Yuan, Y. He and L. Lu, "Temperature Compensation and Pressure Bias Estimation for Piezoresistive Pressure Sensor Based on Machine Learning Approach," in IEEE Transactions on Instrumentation and Measurement, vol. 70, pp. 1-10, 2021, Art no. 1008610, doi: 10.1109/TIM.2021.3089236.

[8] Cao, Yingying and Lyu, Guohui, A High Accuracy Temperature Compensation Method Based on Ga-Bp and Kalman Filter Algorithm for Optical Fiber F-P Pressure Sensor. 2022. http://dx.doi.org/10.2139/ssrn.4222844.

[9] Zou M, Xu Y, Jin J, Chu M, Huang W. Accurate Nonlinearity and TemperatureCompensation Method for Piezoresistive Pressure Sensors Based on Data Generation. Sensors. 2023; 23(13):6167. https://doi.org/10.3390/s23136167.

[10] Xueliang Zhao, Ying Chen, Guanghua Wei, LiLi Pang, Chongxuan Xu, A comprehensive compensation method for piezoresistive pressure sensor based on surface fitting and improved grey wolf algorithm, Measurement, Volume 207,2023,112387, ISSN 0263-2241, https://doi.org/10.1016/j.measurement.2022.112387.

[11] Petar Veličković, Everything is connected: Graph neural networks, Current Opinion in StructuralBiology, Volume79,2023, 102538, ISSN0959440X, https://doi.org/10.1016/j.sbi.2023.102538.

[12] Kipf, Thomas N.; Welling, Max. (2017) Semi-Supervised Classification with Graph Convolutional Networks. https://doi.org/10.48550/arXiv.1609.02907.

[13] Veličković, Petar; Cucurull, Guillem; Casanova, Arantxa; Romero, Adriana; Liò, Pietro; Bengio, Yoshua. (2018) Graph Attention Networks. https://doi.org/10.48550/arXiv.1710.10903.

Research on low power consumption of wireless MEMS methane sensor

Haiqing Zhang[1,2]

[1]Tiandi (Changzhou) Automation Co., Ltd., Changzhou 213015, China

[2] Corresponding author's e-mail: yyu0825@163.com

Abstract: Aiming at the problem that the high power consumption of traditional methane sensors does not realize the comprehensive detection of coal mines underground, based on MEMS methane detection technology, wireless methane sensing low-power research is proposed. First of all, based on the analysis of power correlation, we select the modularized design with low power, which includes power supply, sensing element driver, acquisition, and processing modules. We study the change of sensitive resistance, response time, and power consumption of the MEMS methane sensing element and sensor in different operating modes. Meanwhile, experimental tests are carried out. The experimental result shows that under the premise of ensuring the timeliness and stability of methane gas detection, the power consumption of the wireless MEMS methane sensor in one minute is only 2.4% of the traditional catalytic methane sensing element.

1. Introduction

As an important traditional energy industry in China, to enhance the level of coal mine intelligence and promote the high-quality development of China's coal industry, the National Energy Administration clarifies the goals of coal mine intelligence development and forms intelligent devices for comprehensive perception, real-time interconnection, dynamic prediction, and other collaborative controllers [1]. Comprehensive detection in coal mines cannot be separated from the data support of front-end sensors, and methane monitoring is the top priority of the environmental management of coal mines [2]. The commonly used methane sensors for detection include catalytic, thermal conductivity, infrared, and laser spectral absorption principles. The catalytic principle is used for low-concentration detection at a low price. Thermal conductivity is only used for high-concentration detection of over 4%, while infrared and laser can be used for full-range detection, but it is expensive [3]. The catalytic combustion methane sensor has the highest cost-effectiveness, the highest usage, and the highest power consumption. It is wired to transmit at a fixed electricity level, with fewer sensors on load, short transmission distance, and portable and wireless devices that require more battery energy. Its working time is short, making it impossible to achieve comprehensive underground detection in coal mines, and seriously restricting the intelligent development of the coal mining industry [4].

With the development of MEMS sensing technology, compared with traditional methods, MEMS sensing has the characteristics of low power consumption, easy integration, and intelligent implementation [5]. More and more scholars are investing in the research of MEMS sensing technology. Su et al. [6] are researching MEMS methane catalyst materials, Zhou et al. [7] are researching ultra-low power gas detection MEMS microheaters, and Dominik et al. [8] are researching MEMS hydrogen sensors for automotive detection. This article uses MEMS methane sensing elements for power consumption correlation analysis, conducts low-power research on wireless MEMS sensors, provides reliable technical solutions for product development, and promotes intelligent development.

2. Power consumption correlation analysis

The sensor power consumption means the loss of power during its normal operation. Sensor power consumption can be mainly divided into dynamic power consumption and static power consumption.

$$P_{\text{total}} = P_{dynamic} + P_{static} \tag{1}$$

P_{total} is the total power consumption, and P_{static} is the static power consumption when powered on but not working, which is related to V_{cc}. As V_{cc} increases, the static current of I_{static} increases, too [9].

$$P_{static} = V_{CC} * I_{static} \tag{2}$$

Dynamic power consumption ($P_{dynamic}$) is the power consumption caused by state changes during normal operation, consisting of the charging and discharging power consumption P_C of the load capacitor and the instantaneous conduction power consumption P_{TC}. The power consumption of the product during the state transition process of the P_{TC} signal is low. $P_{dynamic}$ mainly comes from P_C, accounting for over 80% of the total power consumption of the circuit [10].

$$P_{\text{dynamic}} = P_C + P_{TC} = \alpha * C * V_{CC}^2 * f + V_{CC} * I_{TC} \tag{3}$$

α is the activity factor which means the probability of changes in circuit nodes; C is the load capacitor; V_{cc} is the power supply for the circuit; f is the clock frequency at voltage V_{cc}; I_{TC} is the time average of the pulse current.

According to the above analysis, to reduce power consumption, the first step is to reduce dynamic power consumption. When the power supply voltage meets the application requirements, we try to reduce the type and voltage value of the power supply voltage as much as possible. When there are no special requirements for clock frequency, it can be reduced to the lowest value. Static power consumption mainly aims to reduce the voltage value of the power supply. At the same time, hardware modular design and power management can be adopted to turn off the power supply of nonworking modules to reduce the static current value. This article mainly conducts low-power research on wireless MEMS methane sensors through hardware modular design, reducing power supply voltage, working mode and frequency, and other methods.

3. Sensor design

3.1. Hardware design

The MEMS methane sensor adopts a modular design approach, mainly consisting of a power processing module, sensing elements and control driver module, signal processing module, microprocessor, and output module. The module diagram is shown in Figure 1.

The power processing module is used to supply power to each module, the power supply voltage should be minimized as much as possible, and the selection of components should minimize the inconsistency of power supply voltage among various modules. The power supply voltage of the MEMS methane sensing element needs to be 3.3 V, the power supply range of the microprocessor and output module is 1.7 V~5.5 V, and the power supply range of the signal processing module is 2.0 V~5.5 V. This article adopts a unified 3.3 V power supply.

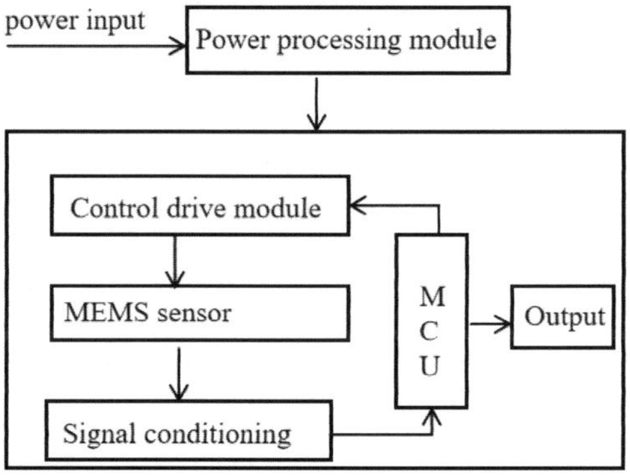

Figure 1. MEMS methane sensor module diagram.

The control driver module controls the sensing components and signal processing circuits to power on or off through a microprocessor. The MOS transistor driving switch circuit used in this article has low conduction resistance, low static power consumption, fast switching speed, and low driving current. To improve sampling accuracy and anti-interference ability, the MEMS methane sensing element is shaped and amplified by an operational amplifier and then passed through a low-pass filter. It is directly converted to a digital signal by a 16-bit analog-to-digital conversion chip, and the sensor supply voltage is collected to detect power supply stability, as shown in Figure 2.

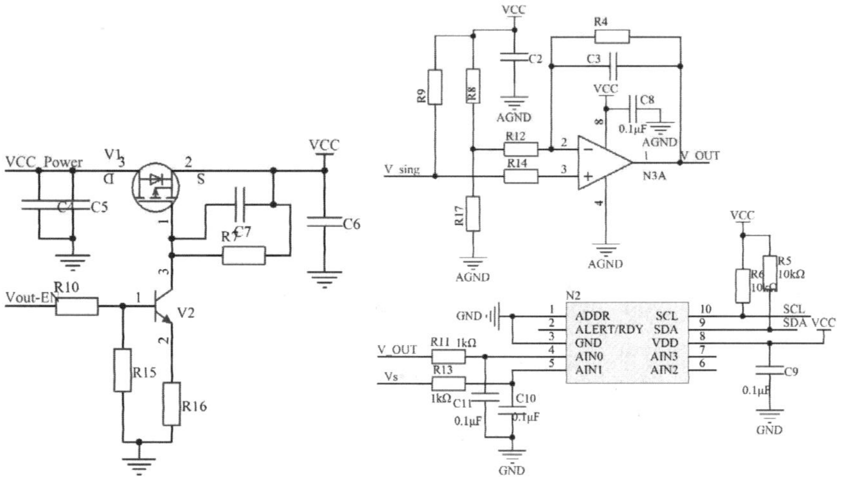

Figure 2. Diagram of drive control and signal processing.

The output and microprocessor units of this article use low-power Bluetooth with built-in microprocessors for data exchange. At the same time, the Bluetooth temperature detection function can be used to obtain the temperature of the tested environment, compensate for the temperature impact of the sensing components, and improve environmental adaptability. In application, only a simple peripheral circuit is needed.

3.2. Software design

The main function of MEMS methane sensor software is to control the power supply voltage of the sensing element through the microprocessor port. The microprocessor collects the sensing element signal and detects environmental temperature parameters. Then it compensates and processes them, converts them into concentration values, and outputs wirelessly. The entire software framework of the device is shown in Figure 3.

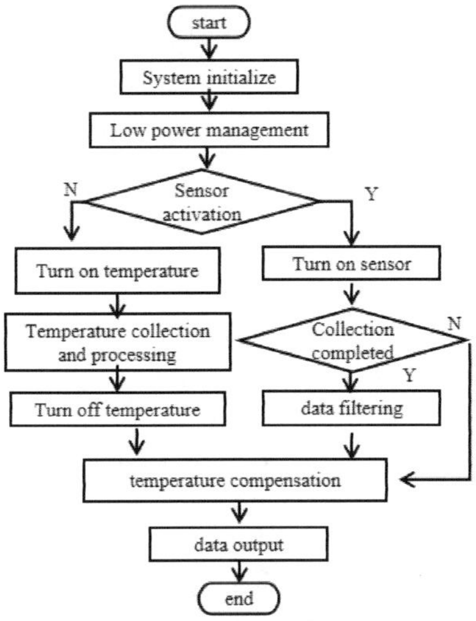

Figure 3. Software framework diagram.

The microprocessor needs to control the sensing element to turn on and cannot be completely dormant. The sensing element has switching losses during conduction and cut-off, which are usually much greater than conduction losses. Moreover, the faster the switching frequency is, the greater the loss is. We try to shorten the switching time to reduce the loss during each conduction and the switching frequency, thus reducing the switching frequency per unit of time.

The low-power design is achieved by adjusting key parameters through low-power management units, setting extended wireless connection intervals, broadcast intervals, and their durations, opening or closing corresponding peripheral resources, reducing the operating frequency of peripherals, and enabling the processor to operate in a low current state. Protocol optimization is used to avoid duplicate data transmission.

4. Experimental testing

When MEMS methane sensing elements react with the target gas, the heating resistor uses a direct current power supply to heat the sensitive material to a specific temperature to react with the target gas. The power consumption of the sensing element mainly comes from the heating resistor. Based on not affecting the performance of the sensing element, to reduce power consumption, the performance of the sensing element is studied by comparing the changes in component power consumption, response time, and sensitive resistance between continuous heating and pulse heating.

The experiment used four methane-sensing elements to test the power consumption under a continuous power supply. We applied a heating voltage of 3.30 V at both ends of the heating resistor. By recording the resistance values in air and a 0.50% methane environment, the sensitivity of the element is shown in Table 1.

Table 1. Continuous heating power consumption and sensitivity testing of methane elements.

Number	Voltage (V)	Current (mA)	Consumption (mW)	Air (Ω)	0.50% (Ω)	Sensitivity
No. 1	3.30	18.67	61.61	12, 175	1, 492	8.16
No. 2	3.30	18.81	62.07	10, 420	928	11.22
No. 3	3.30	18.38	60.65	10, 726	730	14.69
No. 4	3.30	18.19	60.02	10, 947	999	10.95

The response time test is under the condition of continuous power supply and heating, with the sensing element operating stably in a clean-air environment. 0.50% methane gas is introduced, and the change in resistance over time is recorded, as shown in Figure 4. From the results, it can be seen that under continuous heating, the response time of MEMS sensing elements is very fast, reaching a response value of 90% within about 3 seconds. To further reduce power consumption, the performance of methane sensing elements was tested in a low-power mode by using pulse power heating.

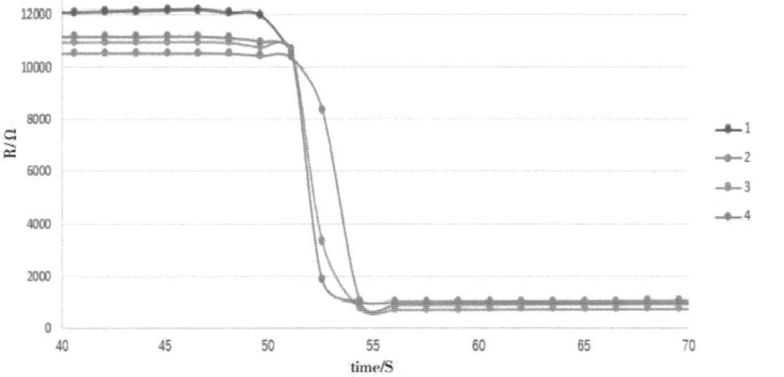

Figure 4. T90 test in the continuous heating state.

According to the "AQ6203-2006 Low Concentration Catalytic Methane Sensor for Mining", the response time shall not exceed 20 seconds. Based on not affecting the response time, its pulse working cycle shall be set as 20 seconds. When studying the pulse heating of sensing elements, the pulse is tested under different duty ratios of 15%, 25%, and 50%, and the impact on the performance of the sensing element is analyzed. According to the material-sensitive characteristics of the metal oxide methane sensing element, the sensing element is unheated and the sensitive resistance value is at the MΩ level. In the heating state, the resistance value of the sensitive resistor is below 20 kΩ. Prototype 1 was tested and the distribution of sensitive resistance at high and low levels during pulse heating in air is shown in Figure 5.

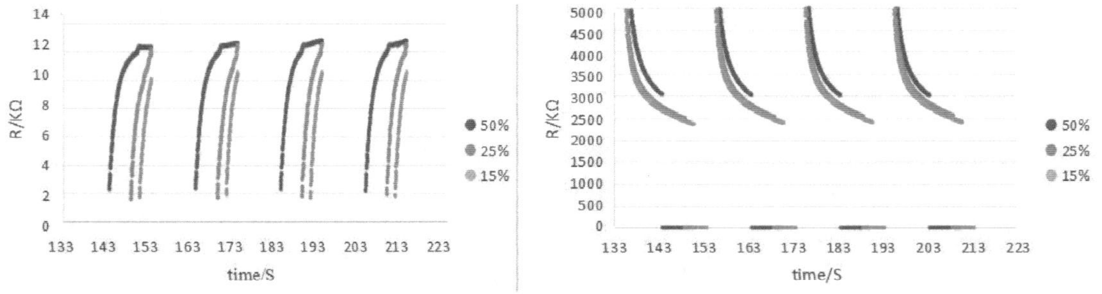

Figure 5. Resistance changes of sensing elements at high and low levels during pulse heating.

From Figure 5, it can be seen that as the pulse high-level heating time increases, the resistance value of the sensitive resistor continues to increase and then tends to stabilize. At a duty cycle of 25%, the sensitive resistance of the sensing element reaches a stable state of 50% duty cycle, and the average power consumption of pulse heating is reduced by 75% compared to continuous heating.

The power consumption of MEMS methane wireless sensors varies under different working modes. The power consumption of the sensor is tested under relative idle mode, continuous heating mode, and pulse heating mode. The average power consumption for testing is shown in Table 2.

Table 2. Average power consumption test for different working states.

Number	Idle (mW)	Pulse heating (mW)	Continuous heating (mW)
No. 1	2.98	24.45	85.33
No. 2	3.07	24.87	86.39
No. 3	2.95	23.91	85.25
No. 4	3.03	25.14	86.02

As the mode of the pulse cycle during 20 s is heating for 5 s, the average power consumption per minute is only 8.40 mW. The power consumption of traditional catalytic elements is 360 mW, which is only 2.4% of the traditional catalytic methane sensing elements. The practical application of MEMS methane sensors generally adopts a multi-point layout and wireless networking method. Low power design can change the connection interval, slave latency, and protocol optimization to reduce data retransmission based on actual application scenarios, further reducing power consumption.

5. Conclusion

With the rapid development of Internet of Things technology, the research and development of low-power methane wireless sensors is of great significance. This article is based on MEMS methane sensing technology for low-power device software and hardware design. While ensuring sensor performance, it reduces the power consumption of methane sensors by controlling the opening and closing of sensing elements and optimizing signal acquisition, processor sleep management, and data output. It also studies the changes in sensitive resistance, response time, and power consumption of MEMSmethane-sensing elements under different working modes. We conduct various performance comparison tests to reduce sensor power consumption in combination with practical applications. The paper provides reliable technical solutions for the development of wireless mine perception products and plays a huge promoting role in the development of underground IoT perception in coal mines.

Acknowledgments

The paper is funded by the International Science and Technology Cooperation Project (2021-2-GH003) of Tiandi Technology Co., Ltd.

Reference

[1] National Energy Administration 2020, Notice on Accelerating the Development of Intelligent Coal Mines. https://www.gov.cn/zhengce/zhengceku/2020-03/05/content_5487081.htm.

[2] Bu T. T., Qu S. J., Wu F. S., et al. (2021). Research on intelligent sensing method of gas disaster driven by edge computing. Safety in Coal Mines, 52: 110-116. Doi: 10.13347/j.cnki.mkaq.

[3] Zhang H. Q. (2018). Research on CH_4 Sensor in Gas Drainage System. Coal Mine Machinery, 39: 40-43. Doi: 10.13436/j.mkjx.201810013.

[4] Wang H. B. (2021). Research progress off low-power methane sensor. Industry and Mine Automation, 47: 16-23. Doi: 10.13272/j.issn.1671-251x.17754.

[5] Zhao X. H., Zhang K., Zhao Z. K. et al. (2021). Developing trends and key technologies of network technology of mine Internet of things. Industry and Mine Automation, 44 (4): 1-7. Doi: 10.13272/j.issn.1671-251x.17324.

[6] Su J. C., Cao L. H., Li L., et al. (2013). Highly sensitive methane catalytic combustion micro-sensor based on mesoporous structure and nano-catalyst, 5: 9, 720-9, 725. Doi: 10.1039/c3nr02916f.

[7] Zhou Q., Sussman C., Hang J ., et al. (2015). Fast response integrated MEMS microheaters for ultra-low power gas detection. Sensors and Actuators A: Physical, 223: 67—75. Doi: 10.1016 /j.sna.2014.12.005.

[8] Dominik B., Josef M., et al. (2020). MEMS-based thermal conductivity sensor for hydrogen gas detection in automotive applications. Sensors and Actuators: A. Physical. Volume 305, Issue prepublish. P 111, 670-111, 670. Doi: 10.1016/j.sna.2019.111670.

[9] Xu J. P., Geng S. J., Ma T. F., et al. (2003). Design principles for ultra-low power electronic circuit systems. Application of Electronic Technique, 02: 78-80. Doi: 10.16157/j.issn.0258-7998. 2003. 02. 025.

[10] Xie X. N. (2014). Research on energy efficient and reliable optimization algorithms of embedded real-time systems. https://x.cnki.net/kcms/detail/detail.aspx?dbcode=CDFD&dbname =CDF D&filename=1015712051.nh.

Study on calibration of surface temperature source

Li Feng[1], Wei Gong[1], Xianhong Li[1], Xuemei Yang[1,2], Qiuquan Zhang[1] and Tingting Ren[1]

[1]Chongqing Academy of Metrology and Quality Inspection, No. 1 Yangliu North Road, Yubei District, Chongqing 401123, China

[2]Corresponding author e-mail: 876536427@qq.com

Abstract. Surface temperature source is widely used to provide adjustable, stable and uniform temperature fields to calibrate surface thermometers. Parameters such as temperature deviation, temperature uniformity, and temperature fluctuation can directly decide the metrological performance of the instrument. At present, there is no relevant calibration specification for surface temperature sources in China. This article presents a novel calibration method for surface temperature sources by determining corresponding environmental conditions, calibration standards and calculation equations. The calibration results show that the calibration method is scientific and practical for the metrological performance evaluation of surface temperature sources.

1. Introduction

The surface temperature source mainly consists of a hot plate, a temperature control device, a measurement standard and other supporting measurement instruments [1-4]. The hot plate whose temperature is controlled by the heating device and measurement standard buried under the hot plate provides an adjustable, stable and uniform temperature field. Surface temperature source is widely used to provide an adjustable, stable and uniform temperature field to calibrate surface thermometers [5-8]. There are mainly two kinds of surface temperature sources according to the accuracy of the instruments, which are standard surface temperature source and surface temperature source for working. However, there is no calibration specification for the surface temperature source in China at present and its metrological performance cannot be ensured. At present, most surface temperature sources are calibrated based on JJF1409-2013 *Calibration Specification for the Surface Thermometers* [9], whose appendix D provides the measurement method of temperature stability and uniformity for surface temperature source, but the measurement method of temperature deviation is missing; HJ 1133-2020 *Ambient Air and waste gas - Determination of arsenic, selenium, bismuth, antimony in ambient particle matter - Atomic fluorescence spectrometry* [10] only stipulates the temperature control precision of surface temperature source, but the calibration method of temperature precision is not specified; JJF1101-2019 *Calibration Specification of Environment Testing Equipment for Temperature and Humidity Parameters* [11] is mainly applied to the temperature calibration of confined space, but with limitations when applied to surface temperature source, whose temperature field is open in the air. Therefore, to comprehensively improve the accuracy of metrological performance evaluation of surface temperature sources to ensure the reliable working performance of surface temperature sources, there is an urgent need to develop an effective calibration method for surface temperature sources. This paper studies the influential factors on essential technical parameters of surface temperature sources such as “temperature deviation”, “temperature uniformity” and “temperature fluctuation” to present a suitable and scientific calibration method for the temperature source, which meets the metering requirements of the equipment in practical use.

2. Calibration conditions

2.1. Environmental conditions
During the whole process of the measurement, the temperature should be (10-28)°C with fluctuation of no more than 0.5°C. Humidity should be no more than 80% RH. The power supply should be with good grounding. Its AC voltage should be (220 ± 22) V or (380 ± 38) V. Its frequency should be within the range of (50 ± 1) Hz. The surface temperature source should be firmly installed. There should be no electromagnetic interference or mechanical vibration around it to affect its normal operation. No explosive gases or corrosive liquids are allowed to be present.

2.2. Metrological standards
Multi-channel surface temperature measuring instrument equipped with the same type of surface temperature thermometer with a weight of 100 g is used to calibrate the surface temperature source. The number of temperature thermometers should meet the requirement of calibration point arrangement of the calibration method presented in this article. The measurement range of the instrument should be no narrower than (0-400)°C. The definition should be better than 0.01°C. The maximum permissive error is ±0.5°C. The whole set of the multi-channel surface temperature measuring instrument should be used with the corrected value participating in the calculation.

3. Calibration method
50°C, 100°C, 200°C, 300°C and 400°C are selected as the temperature calibration points. The shape of the hot plate of the surface temperature source is usually in two forms: square and circular, as shown in Figure 1. Measurement point O is the geometric center point of the working area of the surface temperature source. The other measurement points A, B, C, and D are evenly distributed within the effective working area of the surface temperature source at a distance of 1/10 from the boundary of the effective working area.

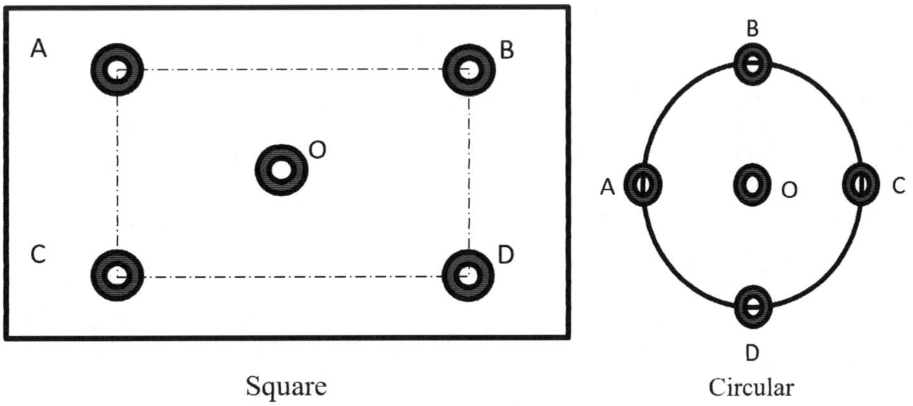

Figure 1. Plan sketch of the hot plate of the surface temperature source.

The temperature sensors of the multi-channel surface temperature are installed to measure the instrument. According to Figure 1, the surface temperature source is set to the calibration temperature, starting the operation. After the surface temperature source reaches a stable state, the indication value of each temperature sensor at a time interval of 1 minute is recorded for 10 minutes. 11 indication values of each sensor are recorded.

3.1. Check before calibration

The surface temperature source should have a nameplate indicating the name, model, serial number, manufacturer and production date of the instrument. The surface of the temperature source should be clean, flat without oil stains, obvious scratches and unevenness, etc. and free from surface oxidation which may affect the measurement accuracy of the instrument.

3.2. Temperature deviation calibration

The temperature deviation of the surface temperature source is calculated based on the measurement results of each temperature sensor of the multi-channel surface temperature measuring instrument. Temperature deviation includes upper deviation Δt_{max} and lower deviation Δt_{min}, which are calculated according to Equations (1) and (2):

$$\Delta t_{\max} = t_{\max} - t_{\mathrm{S}} \tag{1}$$

$$\Delta t_{\min} = t_{\min} - t_{\mathrm{S}} \tag{2}$$

where t_{max} is the maximum of measurement results of all temperature sensors, t_{min} is the minimum of measurement results of all temperature sensors and t_s is the set temperature of the surface temperature source.

3.3. Temperature uniformity calibration

The temperature uniformity of the surface temperature source is calculated based on the measurement results of each temperature sensor of the multi-channel surface temperature measuring instrument. Temperature uniformity Δt_u is the arithmetic mean of the 11 differences between the maximum and the minimum of the temperature measurement values of all temperature sensors recorded during one measurement every 1 minute when the surface temperature source is in a stable state, as demonstrated in Equation (3):

$$\Delta t_{\mathrm{u}} = \sum_{i=1}^{11} \left(t_{i\max} - t_{i\min} \right) / 11 \tag{3}$$

Where t_{imax} is the maximum of measurement results of all temperature sensors during measurement i (i = 1, 2,...,11), while t_{imin} is the minimum.

4. Experimental verification of calibration method

5 typical types of surface temperature sources are selected as the calibration subjects for the verification experiment. Type PLBMY-400 and type 3125 are standard surface temperature sources, while type RC-1AB, type TP300 and type DB-EFS are surface temperature sources for working. The calibration results are demonstrated in Table 1.

Table 1. Calibration results of 5 typical types of surface temperature source.

Type	Temperature range	Parameter	Result
PLBMY-400	Room temperature < $t \le 100$	Temperature deviation	- 0.9°C
		Temperature uniformity	0.5°C
		Temperature fluctuation	± 0.3°C
	100 < t ≤ 300	Temperature deviation	- 1.0°C
		Temperature uniformity	0.5°C
		Temperature fluctuation	± 0.3°C
	300 < t ≤ 400	Temperature deviation	- 1.0°C
		Temperature uniformity	0.4°C
		Temperature fluctuation	± 0.4°C
3125	Room temperature < $t \le 100$	Temperature deviation	- 0.6°C
		Temperature uniformity	0.5°C

		Temperature fluctuation	± 0.4°C
	100 < t ≤ 300	Temperature deviation	- 0.7°C
		Temperature uniformity	0.5°C
		Temperature fluctuation	± 0.5°C
	300 < t ≤ 400	Temperature deviation	- 0.8°C
		Temperature uniformity	0.5°C
		Temperature fluctuation	± 0.4°C
RC-1AB	Room temperature < *t* ≤ 100	Temperature deviation	- 1.9°C
		Temperature uniformity	1.6°C
		Temperature fluctuation	± 0.8°C
	100 < t ≤ 300	Temperature deviation	- 4.2°C
		Temperature uniformity	1.9°C
		Temperature fluctuation	± 1.1°C
	300 < t ≤ 400	Temperature deviation	- 4.4°C
		Temperature uniformity	1.9°C
		Temperature fluctuation	± 1.1°C
TP300	Room temperature < *t* ≤ 100	Temperature deviation	- 2.4°C
		Temperature uniformity	1.5°C
		Temperature fluctuation	± 0.7°C
	100 < t ≤ 300	Temperature deviation	- 3.9°C
		Temperature uniformity	1.8°C
		Temperature fluctuation	± 1.1°C
	300 < t ≤ 400	Temperature deviation	- 4.3°C
		Temperature uniformity	1.9°C
		Temperature fluctuation	± 1.5°C
DB-EFS	Room temperature < *t* ≤ 100	Temperature deviation	- 1.7°C
		Temperature uniformity	1.4°C
		Temperature fluctuation	± 0.7°C
	100 < t ≤ 300	Temperature deviation	- 2.9°C
		Temperature uniformity	1.9°C
		Temperature fluctuation	± 1.0°C
	300 < t ≤ 400	Temperature deviation	- 3.7°C
		Temperature uniformity	1.9°C
		Temperature fluctuation	± 1.2°C

As we can see from Table 1, the maximum temperature deviation of type PLBMY-400 and type 3125 standard surface temperature source is no more than ± 1.0°C, which can meet the requirement of JJF1409-2013 *Calibration Specification for the Surface Thermometers* of no more than ± 2.0°C. The maximum temperature uniformity of type PLBMY-400 and type 3125 standard surface temperature source is no more than 0.5°C, which can meet the requirement of JJF1409-2013 *Calibration Specification for the Surface Thermometers* of no more than 0.5°C within the temperature range of (room temperature -100)°C, no more than 1.0°C within the temperature range of (100-300)°C and no more than 1.5°C within the temperature range of (300-400)°C. The maximum temperature fluctuation of type PLBMY-400 and type 3125 standard surface temperature source within the temperature range of (room temperature-100)°C is no more than ± 0.4°C, which can meet the requirement of JJF1409-2013 *Calibration Specification for the Surface Thermometers* of no more than ± 0.4°C. The maximum temperature fluctuation of type PLBMY-400 and type 3125 standard surface temperature source within the temperature range of (100-300)°C is no more than ± 0.5°C, which can meet the requirement of JJF1409-2013 *Calibration Specification for the Surface Thermometers* of no more than ± 0.6°C. The maximum temperature fluctuation of type PLBMY-400 and type 3125 standard surface temperature source within the temperature range of (300-400)°C is no more than ± 0.4°C, which can meet the requirement of JJF1409-2013 *Calibration Specification for the Surface Thermometers* of no more than ± 1.0°C. The maximum temperature deviation of type RC-1AB, type TP300 and type DB-

EFS surface temperature source is no more than ± 4.3°C, which can meet the requirement of HJ 1133-2020 *Ambient Air and waste gas - Determination of arsenic, selenium, bismuth, antimony in ambient particle matter - Atomic fluorescence spectrometry* of no more than ± 5.0°C. The maximum temperature uniformity of type RC-1AB, type TP300 and type DB-EFS surface temperature source is no more than 1.9°C, which can meet the requirement of JJF1101-2019 *Calibration Specification of Environment Testing Equipment for Temperature and Humidity Parameters* of no more than 2.0°C. The maximum temperature fluctuation of type RC-1AB, type TP300 and type DB-EFS surface temperature source is no more than ± 1.5°C, which can meet the requirement of operation instructions of manufacturers. The feasibility and applicability of the calibration method have been proven.

5. Discussion and analysis

There are several factors that can influence the precision of the calibration result of the surface temperature source.

5.1. The weight of the surface temperature thermometer of the multi-channel surface temperature measuring instrument

The weight of the surface thermometer represents the pressure applied to the hot plate of the surface temperature source. The temperature measurement results of the same surface thermometer with different extra weights are presented in Table 2.

Table 2. The temperature measurement results of the same surface thermometer with different extra weights.

Extra weight	Parameter	Temperature/°C			
		50	100	200	300
0	Temperature measurement result	30.0	45.2	46.1	50.2
	Temperature uniformity	0.6	0.9	1.5	1.7
	Temperature fluctuation	± 0.9	± 1.1	± 2.5	± 2.7
20	Temperature measurement result	38.2	67.7	62.8	93.4
	Temperature uniformity	0.7	1.0	1.5	1.6
	Temperature fluctuation	± 0.8	± 1.0	± 2.3	± 2.6
40	Temperature measurement result	42.6	75.9	86.9	266.9
	Temperature uniformity	0.6	0.9	1.6	1.6
	Temperature fluctuation	± 0.8	± 0.9	± 2.2	± 2.5
60	Temperature measurement result	46.9	86.7	187.8	279.2
	Temperature uniformity	0.7	0.9	1.5	1.6
	Temperature fluctuation	± 0.7	± 0.9	± 2.2	± 2.3
80	Temperature measurement result	48.2	95.6	193.2	285.3
	Temperature uniformity	0.6	0.9	1.4	1.5
	Temperature fluctuation	± 0.6	± 0.8	± 1.7	± 2.0
100	Temperature measurement result	50.6	100.3	199.7	299.6
	Temperature uniformity	0.6	0.9	1.3	1.4
	Temperature fluctuation	± 0.4	± 0.5	± 1.5	± 1.8

As we can see from Table 2, the selection of a suitable weight for the surface thermometer is crucial for the calibration results of the surface temperature source.

5.2. Environmental condition

There should be no obvious air convection in the room where the surface temperature source is placed. If there is air conditioning in the room, it is necessary to add baffles around the work area of the surface temperature source to avoid the influence of air convection on the calibration accuracy of the surface temperature source.

5.3. The necessity of preheating

The surface temperature thermometers of the multi-channel surface temperature are placed to measure instrument at the edge of the surface temperature source to preheat the thermometers before calibration, which can reduce the influence of surface temperature thermometers on the already stabilized temperature field of the surface temperature source to improve the reliability of calibration results.

6. Conclusion

This article presents a novel method for surface temperature sources to calibrate temperature deviation, temperature fluctuation and temperature uniformity of the instrument by determining corresponding environmental conditions, calibration standards and calculation equations. The calibration results show that the calibration method is scientific and practical for the metrological performance evaluation of surface temperature sources. The follow-up study will develop a specific calibration device for surface temperature sources of suitable size, high accuracy and good portability.

Acknowledgments

Our work is supported by Special Project for Performance Incentive and Guidance of Scientific Research Institutions of Chongqing: Research on Traceability Method and Device for Surface Temperature Source (Grant No. cstc2021jxjl20016).

References

[1] Li, L., et al. (2019). Discussion on the Calibration Method of Surface Temperature Source Temperature Field. Industrial Metrology, 05: 13-14. doi: 10.13228/j.boyuan.issn1002-1183. 2019. 0139.

[2] Ji, Z. X., et al. (2020). Design of a Type of Surface Thermometer Calibration Instrument. Technology Wind, 20: 20. doi: 10.19392 /j. cnki. 1671-7341. 202020017.

[3] Wei, Q., et al. (2020). Design and Performance Analysis of a New Type of Surface Thermometer Calibrator. Quality and Technical Supervision Research, 02: 22-26, 46. doi:10.15902/j.cnki.zljsjdyj. 2019.02.006.

[4] An, N., et al. (2018). Optimization and Research on the Calibration Method of Surface Thermometers. Industrial Metrology, 03: 52-54, 57. doi: 10. 13228 /j. boyuan. issn1002 - 1183. 2018. 0016.

[5] Wang, X. L., et al. (2017). Development and Performance Check of the High Temperature Surface Thermometer Calibrator Development and Performance Verification Check. Metrology & Measurement Technology, 03:45-47. doi: 10. 11823 j. issn. 1674 - 5795. 2017. 03. 09.

[6] Meng, B. X., et al. (2022). Evaluation on Uncertainty of Measurement Results of Surface Thermometer. Modern Information Technology, 07: 29-31. doi:10.19850/j.cnki.2096-4706.2022.07.007.

[7] Qiao, Z. D., et al. (2017). Evaluation of Uncertainty in Indication Measurement Result on Surface Thermometer. Metrology & Measurement Technique, 11:85. doi:10. 15988 /j. cnki. 1004 - 6941. 2017. 11. 035.

[8] Zhang, Y. K., et al. (2019). Evaluation of Uncertainty in Calibration Results of Digital Surface Thermometers. Quality and Technical Supervision Research, 02: 2-5. doi:10.15902/j.cnki .zljsjdyj. 2019.02.006.

[9] Luo, T., et al. (2013). JJF1409 Calibration Specification for the Surface Thermometers. jjg. spc.org.cn/resmea/view/stdonline.

[10] Ministry of Ecology and Environment of the People's Republic of China. (2020). HJ 1133 Ambient Air and waste gas - Determination of arsenic, selenium, bismuth, antimony in ambient particle matter - Atomic fluorescence spectrometry. http://www.mee.gov.cn.

[11] Jin, Z. J., et al. (2019). JJF1101 Calibration Specification for Environmental Testing Equipment for Temperature and Humidity Parameters. jjg. spc.org.cn/resmea/view/stdonline.

Device for formaldehyde analysis in hydrogen energy

Feng Tang [1,3], Xiaojing Dang [1], Lei Zhu [2], Jin Zhu [2], Jianjian Wang [2], Chao Sun [2], Li Ren [2], Guang Yu [2] and Hua Fang [2]

[1]Shenzhen Power Supply Bureau Co., Ltd., Shenzhen 518002, China
[2]Shanghai Huaai Chromatography Analysis Co., Ltd., Shanghai 200940, China

[3]Corresponding author's e-mail: fhp755@163.com

Abstract: In this paper, a calibration device for the analysis of formaldehyde in hydrogen energy is designed, and the standard formaldehyde gas of different concentrations is obtained by using the principle of osmotic tube, and the standard formaldehyde gas obtained by gas chromatograph analysis is combined. The results of the experiments show that the concentration of formaldehyde standard gas obtained by this calibration device has high accuracy and excellent stability. This overcomes the difficulty of formulating low-concentration standard gas due to the adsorption of formaldehyde gas, replaces the traditional method of formaldehyde standard gas preparation, and provides a quantitative basis for the calibration of formaldehyde analysis in hydrogen energy.

1. Introduction

Under the objective of carbon neutrality, hydrogen energy has swiftly evolved as a nascent zero-carbon secondary energy source, presenting rare opportunities for advancing the power system. First and foremost, hydrogen is generated through renewable energy, thus promoting the absorption of renewable energy. Secondly, leveraging the energy storage characteristic of hydrogen enables large-scale, long-term, cross-seasonal electricity storage. Thirdly, the rapid response capability of hydrogen power stations provides flexible regulation methods for the new power system. Fourthly, it facilitates the interconnection of cross-domain, multi-type energy networks, expanding the comprehensive utilization pathways of electricity [1, 2]. As a flexible and efficient secondary energy source, hydrogen, through electrolyzers and fuel cells at energy consumption terminals, realizes the interconnection, complementarity, and coordinated optimization of power, heating, and fuel energy networks by converting electricity to hydrogen, thus propelling the development of distributed energy and enhancing terminal energy utilization efficiency [3, 4].

Moreover, given the significant potential of hydrogen fuel in emission reduction, hydrogen energy and fuel cell vehicle technologies have garnered widespread attention globally. In the forthcoming years, the industry of fuel cell vehicles is anticipated to witness explosive growth, with a concurrent rapid increase in hydrogen demand, thereby elevating the concern over the impact of hydrogen quality on fuel cell performance. Studies indicate that excessive formaldehyde in hydrogen may combust to produce CO and CO_2. While diluting hydrogen fuel, CO_2 forms CO through a reverse conversion reaction, leading to the poisoning of the anode platinum catalyst. CO adsorbs on the platinum catalyst, occupying the platinum catalytic active sites necessary for the oxidation reaction of hydrogen, consequently significantly diminishing fuel cell performance [5, 6]. As per the stipulation in China's national standard GB/T 37244-2018 "Proton Exchange Membrane Fuel Cell Vehicle Fuel Hydrogen", the maximum concentration of formaldehyde should not exceed 10 ppb. Therefore, accurately measuring the formaldehyde content in hydrogen energy is of paramount importance.

Usually, when utilizing analyzers to detect formaldehyde content, it is requisite first to prepare low-concentration formaldehyde standard gas for calibration. The preparation methods of formaldehyde standard gas generally fall into static and dynamic methods. Due to the unique adsorption properties of formaldehyde, when packed and stored in high-pressure cylinders, it easily adsorbs on the container body, valves, etc., making it challenging to obtain accurate and stable low-concentration formaldehyde standard gas, thereby increasing the difficulty of tracing the measurement value of formaldehyde gas detection instruments. Konopel'ko et al. [7] achieved a 2 mmol/mol level gas mixture of formaldehyde in nitrogen using a dynamic gravimetric method with trioxane vapor depolymerization. Liu et al. [8] prepared standard gas from formaldehyde aqueous solution standard substance through a static gas mixing device and measured the indication error of the formaldehyde gas detection instrument. Jiao et al. [9] prepared formaldehyde standard gas through heating catalytic decomposition of trioxane vapor. Liu et al. [10] placed formaldehyde solution in a D-10 diffusion tube, and with a standard gas generation device, a constant low-concentration formaldehyde gas could be generated.

Based on this, this paper designs a calibration device for formaldehyde analysis in hydrogen fuel using the principle of permeation tube, obtaining various concentrations of formaldehyde standard gas. Subsequently, the obtained formaldehyde standard gas, when tested with a gas chromatograph, showed high accuracy and excellent stability, providing a quantitative basis for calibrating formaldehyde analysis in hydrogen fuel.

2. Experimental Section

2.1. Instruments

Gas Chromatograph: GC 9560 Gas Chromatograph (equipped with a He purifier and PDHID detector), manufactured by Shanghai Huai Chromatography Analysis Co., Ltd.

2.2. Standard Gas

The details of standard gas are shown in Table 1.

Table 1. Standard gas.

Components	Concentration (ppm)
Formaldehyde	5.05
H_2	Balance

2.3. Operational Procedure for Formaldehyde Standard Gas Analysis and Calibration Device

The schematic diagram and actual image of the calibration device designed for formaldehyde analysis in hydrogen fuel in this paper are shown in Figure 1 and Figure 2 respectively. It mainly includes a purge gas, a mass flowmeter, a formaldehyde permeation tube assembly, a temperature regulation and circuit control assembly, and a gas chromatograph. First, open the purge gas and record the flow rate through the mass flowmeter. The flow rate adjustment range of the mass flowmeter is 0-2 L/min. The purge gas is 99.999% high-purity hydrogen. Then, set the temperature of the formaldehyde permeation tube. Different temperatures of the formaldehyde permeation tube result in different release rates and concentrations. The formaldehyde permeation tube assembly consists of an aluminum plate embedded with a heating rod, platinum resistance, and a formaldehyde permeation tube. The heating rod and platinum resistance are connected to the circuit board to control and display the temperature. The formaldehyde permeation tube includes a 316 L stainless steel outer shell and a 1/8 diameter SPK polyester tube filled with 5 ppm formaldehyde standard gas, sealed with stainless steel ferrules at both ends. The different concentrations of formaldehyde standard gas obtained are analyzed and tested using a helium ion gas chromatograph.

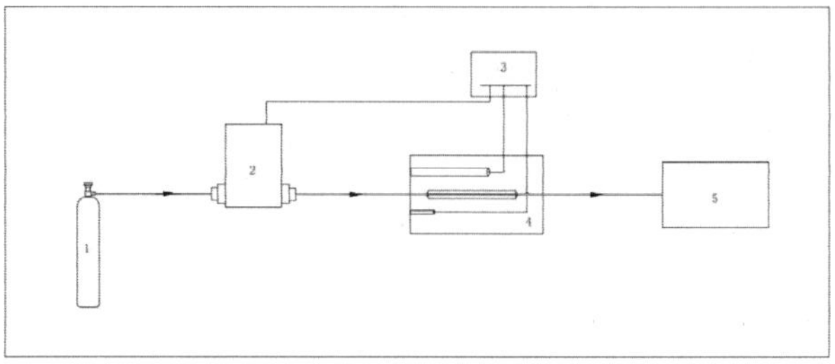

Figure 1. Schematic diagram of the calibration device.

The key components in the figure are labeled as follows:

1. Purge Gas.
2. Mass Flowmeter.
3. Formaldehyde Permeation Tube Assembly.
4. Temperature Regulation and Circuit Control Assembly.
5. Gas Chromatograph.

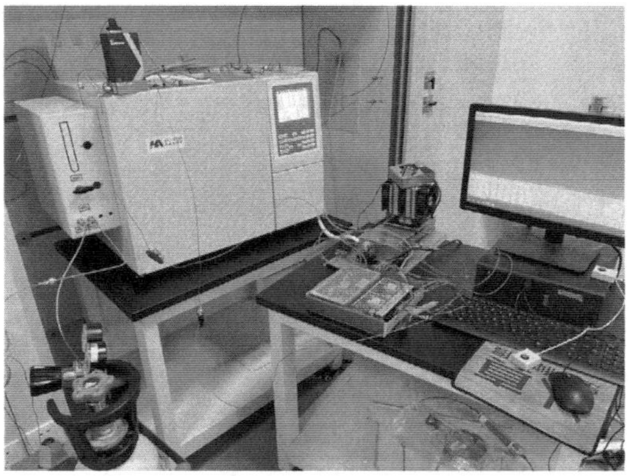

Figure 2. The actual image of the overall device.

2.4. Chromatographic Analysis System

The chromatographic analysis system primarily consists of a gas flow system comprised of a ten-way valve I, a formaldehyde analysis chromatographic column, and a helium ion detector. The analysis process is illustrated in Figure 3.

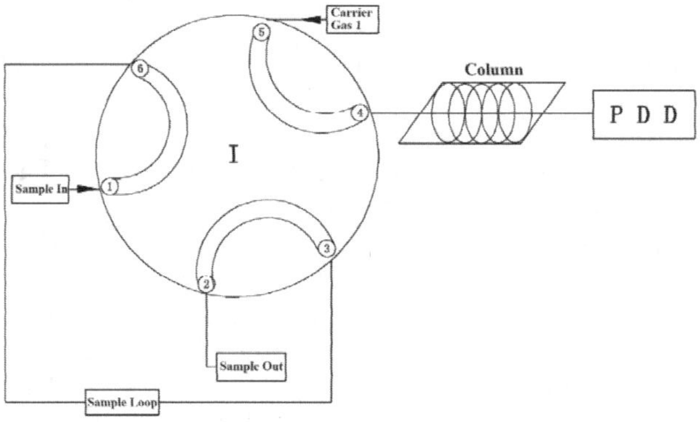

Figure 3. Chromatographic analysis process.

2.5. Chromatographic Conditions

- Quantitative tube volume: 0.5 mL.
- Injection pressure: 0.1 MPa.
- Temperature control of colume oven: 60℃.
- Carrier gas flow rate of carrier gas 1: 50 mL/min.

3. RESULTS AND DISCUSSION

3.1. Selection of Formaldehyde Permeation Tube Conditions

Principle of Permeation Tube: The permeation tube is placed in an environment with precisely controlled temperature. Volatile substances inside the permeation tube diffuse into the surrounding gas, precisely controlled regarding flow rate. This results in the production of the required gas mixture. Therefore, the concentration of formaldehyde standard gas obtained varies at different temperatures and purge gas flow rates.

According to experiments, it was found that under 60°C, the permeation tube releases formaldehyde at a rate of 165.82 ng/min, which is then diluted by hydrogen purge gas at a fixed flow rate. The results are shown in Table 2, and the concentration of the diluted sample is calculated using Equation (1).

$$C = \frac{V_f PM}{V_C PM} \tag{1}$$

where:

- C: Diluted Formaldehyde Concentration/ppm.
- V_f: Formaldehyde Release Rate from Permeation Tube/ng*min^{-1}.
- M: Formaldehyde Molar Mass/g*mol^{-1}.
- R: Gas Molar Constant/J*mol^{-1}*K^{-1}.
- T: Ambient Temperature/K.
- V_c: Carrier Gas Flow Rate/ml*min^{-1}.
- P: Ambient Pressure/kPa.

Table 2. Formaldehyde concentration at different purge air flow rates.

Component	M /g*mol^{-1}	V_{ms} /ng*min^{-1}	T /℃	V_c /ml*min^{-1}	C_s /ppm	P_s /kPa	A /μV*s
Formaldehyde	30.03	165.82	20	24.00	5.534	102.325	41134.0
				32.61	4.073	102.325	30057.0
				55.05	2.413	102.325	17724.4
				76.92	1.727	103.325	12282.5
				171.43	0.775	104.325	5540.0
				285.71	0.465	106.325	3424.0
				560.75	0.237	109.325	1861.8
				884.96	0.150	114.325	1482.8
				1125.70	0.118	118.325	1353.7

where:

- M: Formaldehyde Molar Mass.
- V_{ms}: Formaldehyde Release Rate from Permeation Tube.
- T: Ambient Temperature.
- V_c: Flow Rate of Helium Gas Passing Through the Permeation Tube.
- C_s: Calculated Theoretical Formaldehyde Injection Concentration.
- P_s: Injection Pressure.
- A: Peak Area.

According to the data in Table 2, the linear relationship between peak area and concentration is established by the standard curve method, the linear diagram and linear regression equation are shown in Figure 4. It can be seen that when the flow rate of helium gas passing through the permeation tube is high, the linearity error is relatively large. Therefore, divide the flow rate into two ranges, the high flow rate is >171 mL*min^{-1} and the low is 24~171 mL*min^{-1}, the linear diagram are shown in Figure 5 and Figure 6 respectively.

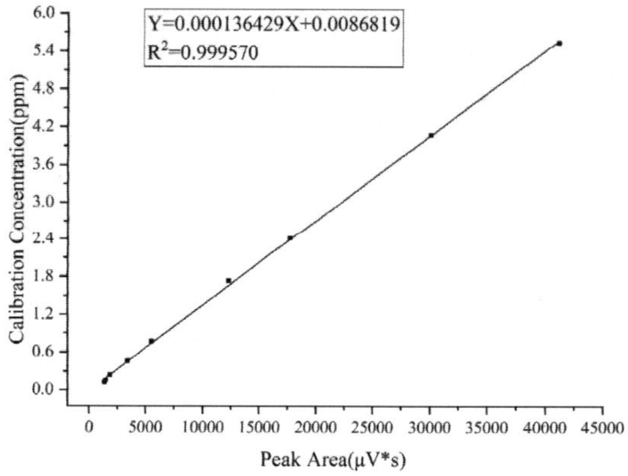

Figure 4. C_s-A linear diagram.

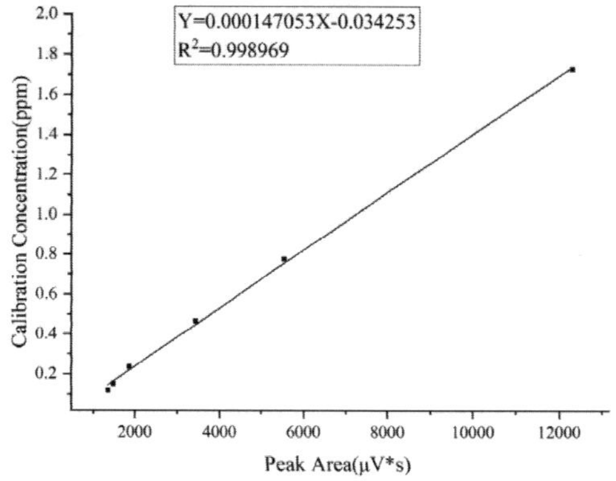

Figure 5. C_s-A linear diagram at high flow rate purge gas (>171 mL*min^{-1}).

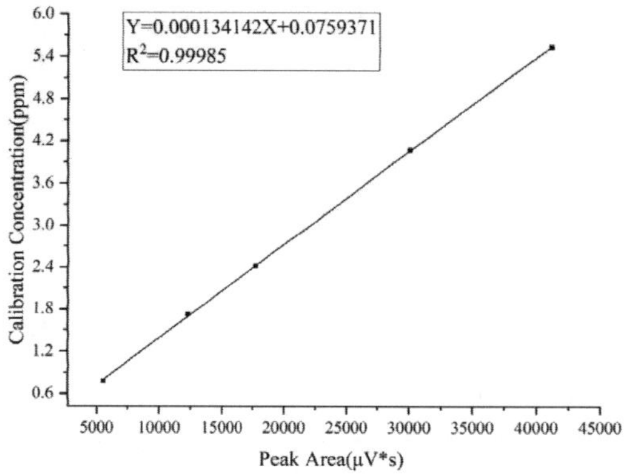

Figure 6. C_s-A linear diagram at high flow rate purge gas (24~171 mL*min^{-1}).

3.2. Chromatogram of Standard Formaldehyde Gas at the Outlet of the Calibration Device

The chromatogram of formaldehyde gas under the optimum condition is shown in Figure 7, and the formaldehyde chromatographic peak in the Figure 7 is preceded by the air peak.

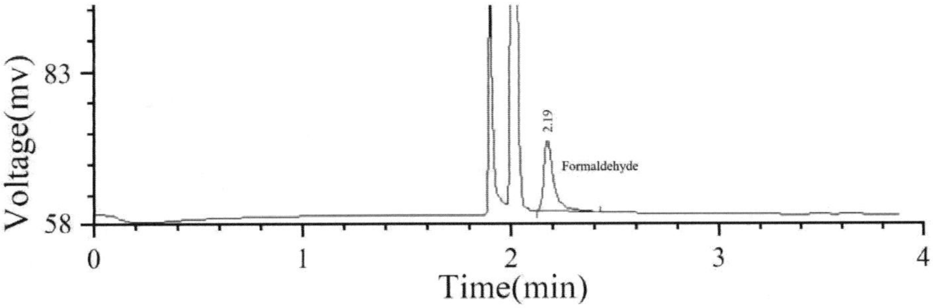

Figure 7. Detection of formaldehyde chromatogram by gas chromatography under preferred conditions.

3.3. Repetition Test

Under the optimal chromatographic analysis conditions, the analytical data results obtained from continuous injection of 1.7 ppm formaldehyde standard gas six times are presented in Table 3. The relative standard deviation (RSD) for each component is calculated. The results indicate that the RSD for peak areas of all components is less than 3%, indicating good stability of the formaldehyde standard gas obtained by this device.

Table 3. Repetition experimental results.

No.	1	2	3	4	5	6	Average	RSD/ %
Peak Area/μV*s Formaldehyde	11964. 8	12282. 5	12067. 6	12151. 9	12061. 3	12222. 5	12125.1	0.96

3.4. Discussion of Data Results

The experimental data presented above indicates that selecting the permeation tube at 60℃ results in the release of formaldehyde at a rate of 165.82 ng/min. Within the low flow rate range of the formaldehyde permeation tube (i.e., 24~171 mL*min^{-1}), R^2=0.99985, demonstrating good stability and high accuracy of the formaldehyde standard gas is obtained by this device. However, high purge gas flow rates and reduced permeation tube temperatures can affect linearity and result in lower accuracy of the formaldehyde standard gas.

The repetition test data show that the formaldehyde standard gas obtained using this device is accurate and stable.

4. Conclusion

This study presents the design of a calibration device for formaldehyde analysis in hydrogen energy applications. The device, grounded in the principle of permeation tubes, can prepare standard formaldehyde gases of varying concentrations. Compared to traditional methods, this approach boasts superior accuracy and stability. Experimental data indicate that, at lower flow rates of the permeation tube, the device yields highly accurate and stable concentrations of standard formaldehyde gas. However, at higher flow rates, a decrease in the temperature of the permeation tube can impact linearity, leading to reduced accuracy in the prepared standard formaldehyde gas. Additionally, repeatability tests further validate the accuracy and stability of the formaldehyde concentrations prepared by this device.

The invention of this device offers a novel and effective calibration method for formaldehyde analysis in hydrogen energy. Given the significance of formaldehyde in hydrogen energy and its potential impact on fuel cell performance, this research provides crucial technical support to ensure the quality and safety of hydrogen energy.

References

[1] Zuttel, A., Remhof, A., Borgschulte, A., Friedrichs, O. (2010) Hydrogen: the future energy carrier. Philosophical Transactions of the Royal Society A-Mathematical Physical and Engineering Sciences, 368: 3329-3342. http://doi.org/ 10.1098/rsta.2010.0113.

[2] Marchenko, O. V., Solomin, S. V. (2015) The future energy: Hydrogen versus electricity. International Journal of Hydrogen Energy, 40: 3801-3805. http://doi.org/ 10.1016/j.ijhydene. 2015.01.132.

[3] Schenk, Niels J., Schenk, Niels J., Schenk, Niels J., Benders, Rene M. J. (2007) Wind energy, electricity, and hydrogen in the Netherlands. Energy, 32: 1960-1971. http://doi.org/ 10.1016/j.energy.2007.02.002.

[4] Pan, G.S., Gu, W., Qiu, H.F., Lu, Y.P., Zhou, S.Y., Wu, Z. (2020) Bi-level mixed-integer planning for electricity-hydrogen integrated energy system considering levelized cost of hydrogen. Applied Energy, 270. http://doi.org/ 10.1016/j.apenergy.2020.115176.

[5] Zhang, F., Zhao, P.C., Niu, M., Maddy, J. (2016) The survey of key technologies in hydrogen energy storage. International Journal of Hydrogen Energy, 41: 14535-14552. http://doi.org/ 10.1016/j.ijhydene.2016.05.293.

[6] Pan, G.J., Bai, Y.P., Song, H.H., Qu, Y.B., Wang, Y., Wang, X.F. (2023) Hydrogen Fuel Cell Power System-Development Perspectives for Hybrid Topologies. Energies, 16: 6. http://doi.org/10.3390/en16062680.

[7] Konopel'ko, L.A., Efremova, O.V., Kadis, R.L., Klimov, A.Y., Kolobova, A.V., Mal'ginov, A.V., Chubchenko, Y.K. (2020) Reference Gas Mixtures of Formaldehyde in Nitrogen: Preparation by a Dynamic Gravimetric Method. MEASUREMENT TECHNIQUES, 62: 915-922. http://doi.org/10.1007/s11018-020-01713-3.

[8] Liu, B.Q., Liu, H.P., Yuan, Y.F., Luo, Y.L. (2021) Evaluation of the uncertainty of the error of the indicator value of formaldehyde gas detection by static gas distribution method (in Chinese). Measurement and testing technology, 48: 111-113. DOI: 10.15988/j.cnki.1004-6941.2021.6.037.

[9] Jiao, C.Y., Cui, J.S., Xing, B. (1994) Research on preparing formaldehyde standard gas by diffusion tube (in Chinese). Health research, S1: 100-112. DOI: 10.19813/j.cnki.wei shengyanjiu.1994.s1.015.

[10] Liu, G.Y., Yu, H.F., Hao, Y.J., Zhang, R. (2012) A new method for preparing low-concentration formaldehyde standard gas (in Chinese). Chinese Journal of Health Laboratory Technology, 22: 35-36. https://kns.cnki.net/kcms2/article/abstract?v=v0gKr Roz1UdYoKckgYmnVtPTuRS1z7Y6CuC-WhQNFwArqAWFitbNqAE9B58oApaLdgU AvpZ0Ku4r31dXNhFbBW4DcG0e6X6wL8PXW6zWEiAO6d1-pZ-jUJRaanyzD97u& uniplatform=NZKPT&flag=copy.

Thermal error modelling of gear measuring instrument based on principal component regression

Dongfeng He[1,2], Jianhua Wang[1] and Yanping Ren[1]

[1]Xi'an Technological University, Xi'an, Shaanxi 710021, China

[2]Corresponding author's e-mail: Dongfenghe@163.com

Abstract. As gear measuring instruments (GMI) are frequently used at the production site, the influence of temperature on their accuracy and stability becomes non-negligible. The existing standards and technologies only consider the error modelling and compensation problem when the GMI is in a uniform temperature field and is not suitable for the production site environment with a large temperature change rate and large temperature gradient. The paper delves into the problems of GMI thermal error modelling at the production site. First, the temperatures of the GMI and its surroundings were measured using multiple temperature sensors. The parallelism thermal errors between centres and the Z-axis (referred to as parallelism thermal errors) and the y-direction spatial thermal errors (referred to as spatial thermal errors) were also measured by the probe and laser interferometer synchronously. Then, the models of them were established based on principle component regression (PCR). As evidenced by the experimental results, the fitting determination coefficient of the models are both greater than 97%, and the prediction determination coefficient is greater than 90%, demonstrating that the models are precise and robust.

1. Introduction

The gear measuring instrument (GMI) is currently the most widely used equipment for gear measurement. In the process of gear measurement, temperature is an important factor affecting the measuring accuracy and stability of GMI. Existing specifications mainly reduce the impact of temperature on GMI by specifying the ambient temperature conditions [1]. As more and more GMIs are used at production sites, ensuring the required temperature conditions in specifications can be challenging, and the influence of temperature on GMI accuracy and stability becomes non-negligible. Some GMI compensates for only the thermal expansion of the workpiece and the scales by using the coefficient of temperature expansion (CTE) when the ambient temperature deviates from 20°C [2]. The compensation method only works well when there is a small temperature rise and uniform temperature. In the production site, where the temperature changes sharply and there is a large temperature gradient, the GMI will produce complex thermal deformation and compensating for the thermal expansion of the workpiece and the scales using the CTE are not enough.

Compared with the insufficient research on a thermal error of GMI, thermal error modelling has been widely researched in machine tools, and a series of modelling methods such as multiple linear regression [3], neural network [4], support vector machine [5] and autoregressive [6] have been proposed. By simultaneously measuring the temperatures of the primary heat sources and the thermal errors of the machine tool, these methods establish a relationship between them through data fitting. In these methods, because the temperature measured at different points affects each other, the collected temperature variables often show a very strong colinearity, which causes the deterioration of the prediction accuracy and robustness of the model. At present, the influence of the correlation between temperature variables is cut down by fuzzy clustering [7], grey correlation analysis [8], roughness theory [9] and so on. However, these methods also decrease the correlation between temperature variables and thermal errors while reducing the influence of the colinearities of temperature variables [10].

Measuring equipment, like GMI, usually has a small internal heat source, resulting in a strong correlation and a noticeable collinearity problem of temperature variables. In this paper, the thermal error models of GMI, including parallelism thermal errors between the centres and the Z-axis (referred to as parallelism thermal errors) and the y-direction spatial thermal errors (referred to as spatial thermal errors), are established using principal component regression (PCR). The method preserves the correlation between temperature variables and thermal errors as much as possible while reducing the collinearity of temperature variables. The experiment shows that the established models have a good fitting and prediction ability. They can be used to correct the thermal errors of the GMI at the production site.

2. Thermal error modelling method based on PCR

2.1. Modelling for parallelism thermal errors

The modelling steps of the parallel thermal error based on PCR [11] are as follows:

(1) Data normalization. The temperature data $\mathrm{T} = [\mathrm{T}_1, \mathrm{T}_2, \cdots, \mathrm{T}_\mathrm{m}]$ and parallelism thermal error data θ_y are normalized by the standard deviation normalization method, and the data normalized are $\tilde{\mathbf{T}} = \left[\tilde{\mathbf{T}}_1, \tilde{\mathbf{T}}_2, \cdots, \tilde{\mathbf{T}}_\mathbf{m}\right]$ and $\tilde{\theta}_\mathrm{y}$. Here, m is the number of temperature data measured.

(2) Calculate the covariance matrix of normalized temperature data by $\mathrm{COV_T} = \tilde{\mathrm{T}}^\mathrm{T}\tilde{\mathrm{T}}$.

(3) Calculate the eigenvalues $\lambda_1, \lambda_2, \cdots, \lambda_m$ and eigenvectors $\mathrm{V}_1, \mathrm{V}_2, \cdots, \mathrm{V}_\mathrm{m}$ of $\mathrm{COV_T}$.

(4) Construct the principal components.

First, the condition indices corresponding to each eigenvalue are calculated by $k_j = \left(\lambda_{\max}/\lambda_j\right)^{\frac{1}{2}}, \lambda_{\max} = \max(\lambda_j), j = 1, 2, \cdots, m$. Eigenvectors corresponding to k_j greater than the threshold k_{th} are selected to form a matrix $\mathbf{V} = \left[\mathbf{V}_1, \mathbf{V}_2, \cdots, \mathbf{V}_p\right]$, where $p \le m$. Then, we construct the principal components by $\tilde{\mathbf{C}} = \tilde{\mathbf{T}}\mathbf{V}$.

(5) Establish the PCR model. With principal components $\tilde{\mathrm{C}}$ as independent variables and $\tilde{\theta}_y$ as dependent variable, a least square regression model is established. Supposing that the coefficient matrix of the regression model is $\hat{\alpha} = \left[\hat{\alpha}_1 \quad \hat{\alpha}_2 \quad \cdots \quad \hat{\alpha}_p\right]^\mathbf{T}$, then the coefficient matrix corresponding to normalized temperature data $\tilde{\mathbf{T}}$ is $\hat{\boldsymbol{\beta}} = \mathbf{V}\hat{\boldsymbol{\alpha}}$.

(6) Do the reverse normalization. Let $k_j = \left(s_{\theta_y}/s_{T_j}\right)\beta_j, j = 1, 2, \cdots, m$ and $k_0 = \overline{\theta}_y - \sum_{j=1}^{m} k_j\overline{T}_j$, where $\overline{\theta}_y$ and s_{θ_y} are the mean and the standard deviation of θ_y, $\overline{T}_j$ and s_{T_j} are the mean and the standard deviation of the jth temperature measurement point data $\mathrm{T_j}$. Finally, the model of parallelism thermal error θ_y regarding temperatures $T_1, T_2, \cdots, T_m$ is:

$$\theta_y = k_0 + k_1T_1 + k_2T_2 + \cdots + k_mT_m \tag{1}$$

2.2. Point-by-point modeling for spatial thermal error

Because the spatial thermal error is not only related to temperatures but also to probe position, the spatial thermal error can be expressed as a mathematical formula:

$$\Delta y = f_y(\mathbf{T}, x, y, z) \tag{2}$$

where Δy indicates the y-direction spatial thermal error of the probe. x, y, z represents the probe position coordinates in the measuring space, which can be derived from the GMI grating. $\mathbf{T}$ is a matrix consisting of a temperature vector obtained from the different measurement points.

In the measurement of the spatial error, the number of position coordinates usually is much smaller than the number of temperature data for the consideration of measurement cost. Taking both temperature data and position coordinates as input for the spatial thermal error model will result in strong collinearity about position coordinates, reducing model convergence and robustness. Here, we use a point-by-point modelling approach to avoid the above problems. In point-by-point method, the entire measurement space is decomposed into multiple nodes, and then independent sub-models are established at each node, denoted as $\{f_y^1, f_y^2, \cdots, f_y^n\}$, n is the number of nodes. The sub-model of each node can be established using the PCR method similar to parallelism thermal error modelling:

$$\Delta y^i = f^i(\mathbf{T}^i) = k_0^i + \sum_{j=1}^{m} k_j^i T_j^i, i = 1, 2, \cdots, n \tag{3}$$

where $\mathbf{T}^i$ is the matrix composed of the temperature vector measured at the *i*th node, Δy^i and $k_j^i, j = 0, 1, \cdots, m$ are the predicted value of spatial thermal error and the PCR coefficient at the *i*th node.

The spatial thermal error model above can only predict the thermal error at nodes, and cannot give the thermal error at any point in space. Here, we use the inverse distance weighted interpolation method (IDWI) in Figure 1 to obtain the thermal error at any point in space.

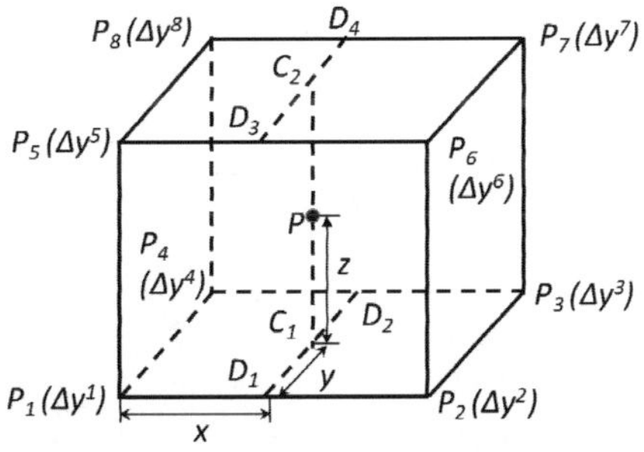

Figure 1. The inverse distance weighted interpolation (IDWI) principle.

Supposing the point P is a point anywhere in measurement space, $P_1, P_2, \cdots, P_8$ is the 8 nodes around the point, its coordinates are $(x_i, y_i, z_i), i = 1, 2, \cdots, 8$ and its thermal error values, Δy^1, Δy^2,..., Δy^8, have been predicted by the sub-models above. The spatial thermal error of the P point is obtained by (4).

$$\begin{aligned} E_p &= (1-x_w)(1-y_w)(1-z_w)\Delta y^1 + x_w(1-y_w)(1-z_w)\Delta y^2 + x_w y_w(1-z_w)\Delta y^3 + \\ &(1-x_w)y_w(1-z_w)\Delta y^4 + (1-x_w)(1-y_w)z_w\Delta y^5 + x_w(1-y_w)z_w\Delta y^6 + x_w y_w z_w \Delta y^7 + (1-x_w)y_w z_w \Delta y^8 \end{aligned} \tag{4}$$

where, $x_w = |x - x_1| / |x_2 - x_1|, y_w = |y - y_1| / |y_4 - y_1|, z_w = |z - z_1| / |z_5 - z_1|$ are the distance weight of the point P along the X, Y, and Z axes. x, y, z are the coordinates of the point P.

3. Experiment verification

3.1. Measurement of the temperatures

The temperatures of the GMI and its surroundings were acquired by 24 Pt100 thermal resistances and a 32-channel temperature detector. The allowable deviation of the thermal resistances is ± (0.15+0.002 | t|)°C, and the response time is <30 s. The acquisition accuracy of the temperature detector is 0.2%FS. The arrangement of thermal resistances and temperature measurement experiments are shown in Figure 2. Among them, T1 ~ T8 are used to collect the ambient temperature around the GMI and other thermal resistances are used to measure the temperature distribution in the GMI body.

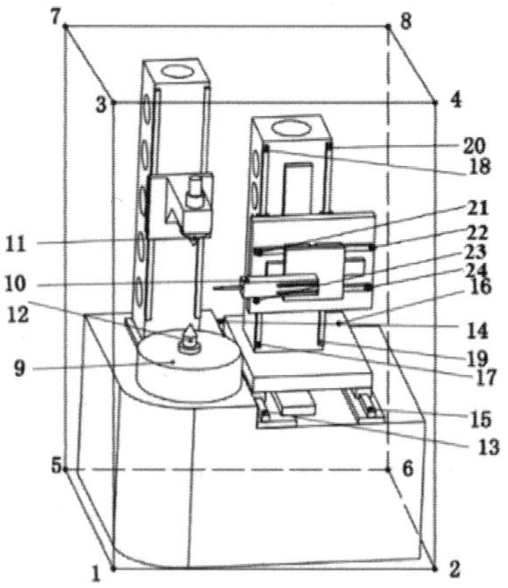

(a) Arrangement of thermal resistances.

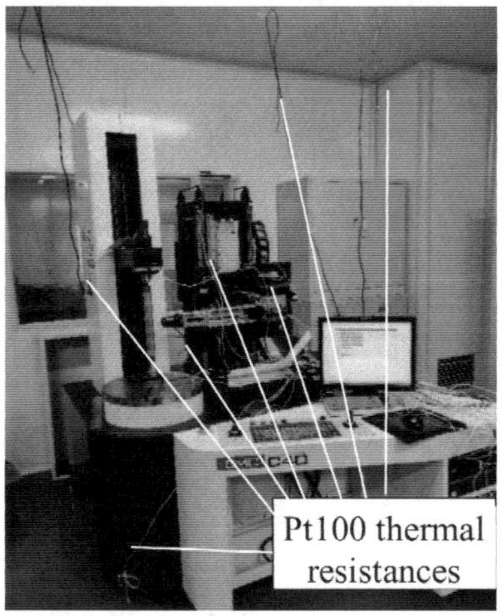

(b) Temperature measurement experiment.

Figure 2. Temperatures measurement.

3.2. Measurement of the thermal errors

The parallel thermal error and spatial thermal error were detected by referring to the inspection methods of parallelism error and spatial error of GMI in the ISO/TR 10064-5:2005 [1]. The detection devices are shown in Figure 3.

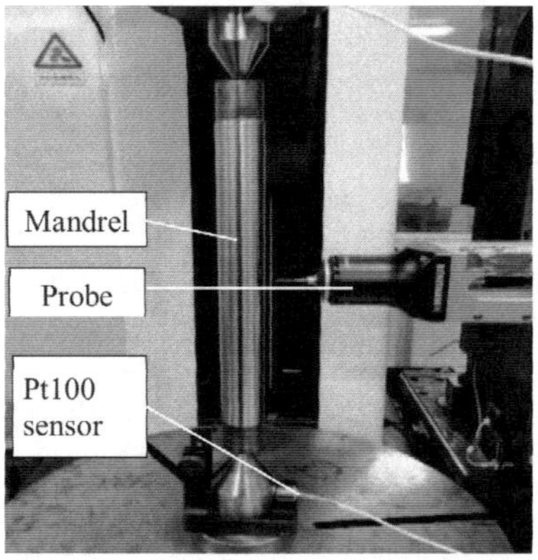

(a) Parallel thermal error measurement.

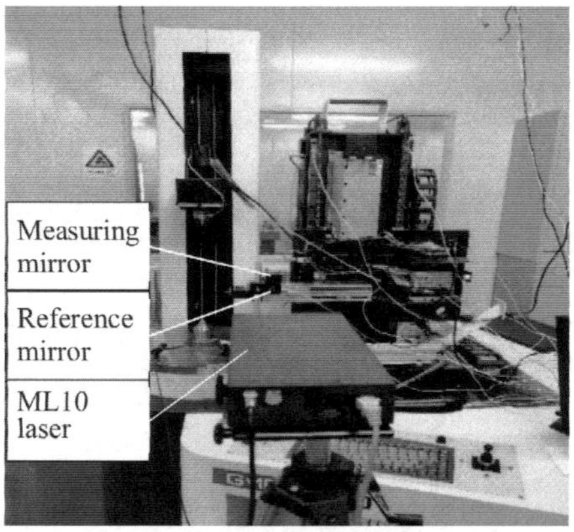

(b) Spatial thermal error measurement.

Figure 3. Thermal error measurement.

To measure parallelism thermal errors, a mandrel was installed between the top and bottom centres. The GMI probe was pressed against the mandrel in the Y direction and moved along the Z axis to obtain the coordinates of the multiple measuring points: $\{(z_i, y_i) i = 1, 2, \cdots, n\}$. Then, one line was fitted using the obtained coordinates, and the inclination of this line was taken as the parallelism error in the Y directions. The difference between the parallelism errors at each time and the starting time was the parallelism thermal errors at that moment, which was denoted as: θ_y.

The spatial thermal errors were measured with the Renishaw ML10 laser interferometer. In the measuring space of the GMI, 9 straight lines parallel to the Y-axis were selected, and 16 points were selected at equal intervals on each line. The probe y-direction positioning error at each point was measured by a laser interferometer. During measurement, the laser interferometer reflector was mounted on the probe mounting frame. By using the GMI grating indication in the Y direction as the reference and the laser interferometer indication as the actual position, the difference between them was the y-direction positioning error of the probe. The thermal error of each point was calculated by subtracting the measured positioning error at different times from the measured positioning error at the initial time.

3.3. Measurement experiment

The experiment started from the start-up state, without any preheating. The data acquisition program ran automatically every 6 minutes to collect the temperatures and two thermal errors above simultaneously. The probe returned to the reference position and waited for the next measurement interval after each measurement. The data collected underwent noise reduction using a Gaussian filter with a window size of 15.

3.4. Model verification for parallelism thermal error

The parallelism thermal error model was established using the PCR method described above. The threshold of the condition indices used for selecting the principal components was 30. The training result of the model is shown in Figure 4(a). Another set of newly measured data was used to verify the prediction ability of the models. The comparison of the predicted result with the measured result is shown in Figure 4(b).

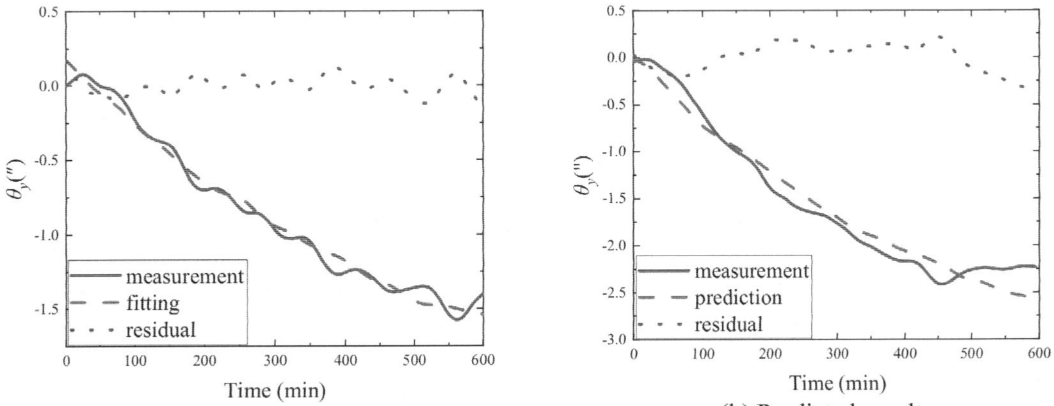

(a) Training result. (b) Predicted result.

Figure 4. Modelling results of the parallelism thermal error.

Figure 4(a) shows that the output value of the model after training aligns well with the measured value, suggesting a good fit between the model output and the parallelism thermal error data. In Figure 4(b), the model accurately predicts the trends in measurement results for the new measurement data, proving its strong prediction ability.

The goodness parameters of the parallelism thermal error model for fitting and prediction are shown in Table 1. In the table, RMSE is the residual's root-mean-square error, R2 is the model determination coefficient, and η is the model prediction ability.

Table 1. The goodness parameters of the parallelism thermal error model.

	Goodness parameters for fitting			Goodness parameters for prediction		
	RMSE (″)	R^2	η	RMSE (″)	R^2	η
θ_y	0.096	0.986	0.945	0.154	0.960	0.914

Table 1 illustrates that the fitting RMSE of the parallelism thermal error model is less than 0.1″, and the determination coefficient and prediction ability of the model are more than or close to 95%, indicating that the model has a good data fitting ability. The goodness parameters of the model for the prediction showed a slight decrease with the use of the new measurement data, but the decrease was not significant, suggesting that the model still maintains a good prediction ability and is robust.

3.5. Model verification for spatial thermal error

The spatial thermal error model was also established using the PCR method, in which the condition indices threshold was still set at 30. The fitting residual slice diagram of the model at different times is shown in Figure 5, and the maximum and minimum fitting residual of the model are shown in Table 2.

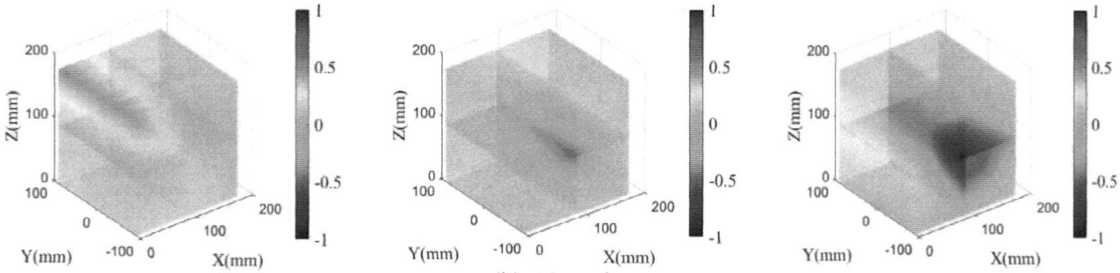

Figure 5. The fitting residual slice diagram of the spatial thermal error model at different times.

Table 2. Maximum and minimum fitting residua of the spatial thermal error model (Unit: μm).

	0 min		180 min		360 min	
	Minimum	Maximum	Minimum	Maximum	Minimum	Maximum
Δy	-0.180	0.692	-0.542	0.153	-1.050	0.365

Figure 5 and Table 2 show that the fitting residual of the spatial thermal error is less than 1 μm, and the residual distribution is relatively uniform from the initial time to 180 minutes, indicating that the established model fits the measured data well in these moments. Model fitting errors increase and the uniformity of error distribution decreases at 360 min. However, the maximum absolute residual is still less than 1.5 μm, and the model still has a good fitting ability. The fitting residual of the spatial thermal error is bigger than the fitting residual of the parallelism thermal error. This is mainly caused by the inverse distance weighted interpolation in section 2.2. With the increase in the number of nodes, the error caused by interpolation will gradually decrease.

For testing the fitting ability of the model at a node, a node (90 mm, 180 mm, 180 mm) was selected to observe the fitting result of thermal error change over time. The result is shown in Figure 6(a). For verifying the prediction ability of the model, a specific point (155 mm, -42 mm, 25 mm) was selected, thermal error over there and corresponding temperatures were measured, and then the thermal error at the point was predicted by the established model. The measured thermal error and the model predicted thermal error over time are shown in Figure 6 (b).

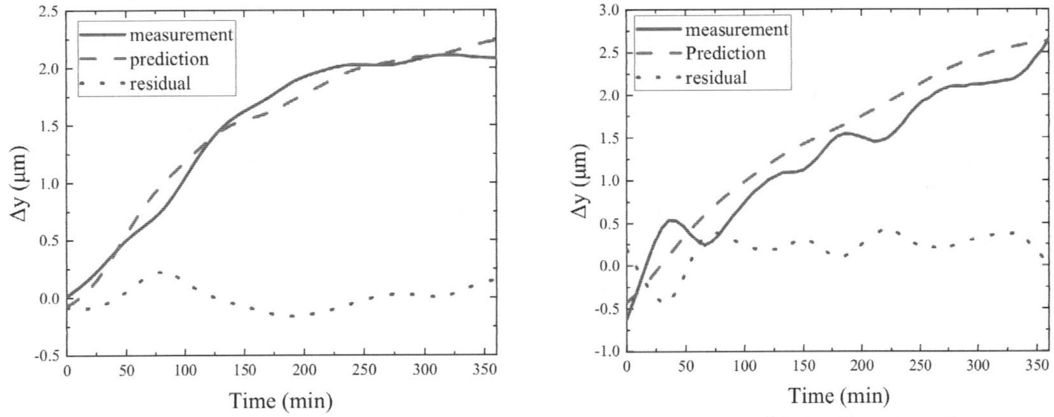

(a) Training result. (b) Predicted result.

Figure 6. Modelling results of the spatial thermal error.

Figure 6(a) shows that the measured curves and fitting curves of the thermal errors are almost the same, and the maximum fitting residual is within ±0.25 μm, indicating that the established spatial thermal error model fits the measured data well. Figure 6(b) illustrates that the curve tendency of the model predicted is consistent with the measured curve tendency over time, with residual within ±0.5 μm, demonstrating that there is a good prediction ability for the model.

The goodness parameters of the spatial thermal error model for fitting and prediction corresponding to Figure 6 are shown in Table 3.

Table 3. The goodness parameters of the spatial thermal error model.

	Goodness parameters for fitting			Goodness parameters for prediction		
	RMSE (μm)	R^2	η	RMSE (μm)	R^2	η
Δy	0.105	0.977	0.942	0.278	0.901	0.831

In Table 3, the RMSE for fitting is about 0.1 μm, and the determination coefficient for fitting reaches 97%, indicating that the model has a good fitting ability. Table 3 also shows that the RMSE for prediction are less than 0.3 μm, and the determination coefficient for prediction is more than 90%, demonstrating that the model has a good prediction ability.

4. Conclusion

Thermal error is a key factor which affects the measurement accuracy and stability of GMI. Our focus is on modelling of the thermal error resulting from temperature changes and complex temperature distribution at the production site in this paper. We measured the temperatures of the GMI body and its surroundings, the parallelism thermal error and the y-direction spatial thermal errors of the GMI, synchronously. The models of thermal errors were established based on the PCR method, and the fitting and prediction ability of the models was verified by experiments. The experimental results indicated that the fitting determination coefficients of the models are both greater than 97%, and the prediction determination coefficients are both greater than 90%, indicating that there is a good fitting and prediction ability for the models. The models are suitable for modelling and compensating for the GMI thermal error at the production site.

References

[1] ISO 2006 *ISO/TR 10064-5:2005* (Switzerland, ISO) pp 2-11.
[2] Klingelnberg 2004 P26 operating instruction (Hückeswagen: Klingelnberg GmbH) pp 703.01.
[3] Li G, Tang X, Li Z, Xu K and Li C 2022 *Precis. Eng.* 73 140-52.
[4] Li B, Tian X and Zhang M 2019 *Int. J. Adv. Manuf. Technol.* 105 1497-505.
[5] Tan F, Yin G, Zheng K and Wang X 2021 *Int. J. Adv. Manuf. Technol.* 116 99-114.
[6] Zimmermann N, Büchi T, Mayr J, Wegener K 2022 *J. Manuf. Syst.* 64 615-25.
[7] Zhang T, Ye W and Shan Y 2016 *Int. J. adv. Manuf. Technol.* 85 2761-71.
[8] Liu H, Miao E, Zhuang X and Wei X 2018 *Precis. Eng.* 51 169-75.
[9] Cheng Q, Qi Z, Zhang G, Zhao Y and Sun B 2016 *Int. J. Adv. Manuf. Technol.* 83 753-64.
[10] Liu H, Miao E, Wei X and Zhuang X 2017 *Int. J. Mach. Tool. Manu.* 113 35-48.
[11] Chatterjee S and Hadi A S 2013 *Regression analysis by example* (Hoboken: John Wiley & Sons, Inc.) chapter 10 pp 259–79.

Absolute attitude measurement based on multi-sensor fusion

Xuran He [1, 2], Kun Xiong[1], Jiabin Li[1] and Yingmin Zhou[1]

[1]Beijing Aerospace Institute for Metrology and Measurement, Beijing, China

[2]Corresponding author's e-mail: 15634328975@126.com

Abstract. In the field of positioning and navigation, the measurement and transmission of absolute attitude are often used in the initial calibration process of positioning and navigation equipment. This requires higher absolute attitude measurement accuracy and lower measurement complexity. In order to solve the problems of high complexity and time-consuming of traditional absolute attitude measurement, a multi-sensor fusion absolute attitude measurement method is proposed. This method uses an optical camera to calculate the angle offset, employs a biaxial inclination sensor to measure horizontal angles, and utilizes GNSS differential positioning to introduce true azimuth angle. The experimental results of the prototype show that under the condition of 60 m baseline length, the max uncertainty of yaw, pitch, and roll angle measurement is 14.13'', 4.41", and 3.53" respectively, and the average single measurement time consumption is 2.73 minutes.

1. Introduction

One of the main links in the existing absolute attitude measurement calibration process is the introduction of true azimuth and the measurement of horizontal angle. The commonly used absolute attitude measurement methods mainly include gyroscope measurement [1-3] and GNSS differential measurement [2,4-6].

Among them, the gyroscope uses the fixed axis and precession characteristic to measure the true north direction, and the gyroscope can sense the pitch angle and roll angle of its relative horizontal plane, and then calculate the three-axis attitude of the gyroscope itself. Furthermore, the absolute attitude of the measured object can be calculated by calibrating the relative relationship between the measured object and the gyroscope and converting the coordinate system. Limited by the level of manufacturing technology, the volume and weight of the current high-precision gyroscope are often large, and its portability is difficult to guarantee. Although the MEMS gyroscope is light in weight and small in size, its accuracy generally cannot meet the needs of high-precision measurement and calibration. In addition, the gyroscope inevitably has a drift error, which can only be suppressed but not eliminated [7,8] which also affects the measurement accuracy of the gyroscope to a certain extent. One of the methods in GNSS differential measurement is to measure the true azimuth angle of the connection line (baseline) of the two receivers by differentiating two GNSS receivers separated by a certain distance, and then transmit the true azimuth angle and the measured horizontal angle to the measured object through optical targeting equipment such as theodolite, and further calculate the absolute attitude of the measured object. This method needs to ensure that the perpendicular line between the phase center of the GNSS receiver antenna and the ground passes through the straight line where the optical axis of the optical aiming device is located, otherwise, the measurement error will be introduced. Therefore, the horizontal angle of the optical aiming equipment needs to be accurately adjusted to meet the above requirements. Another type of method is to install more than three GNSS receiver antennas on the measured object to form more than two intersecting baselines, and calculate the absolute attitude of the measured object according to the angle information of several baselines [9]. Since the azimuth measurement accuracy is directly related to the baseline length, this method requires a large volume of the measured object itself or a large volume of measurement tooling. The installation position of all antennas needs to be precisely adjusted or pre-calibrated, otherwise, the attitude measurement accuracy will be reduced.

In order to solve the above problems, this paper adopts the measurement method of multi-sensor fusion. The true azimuth information of the differential baseline is obtained by using the principle of GNSS differential positioning, the baseline vector is mapped to the reference hexahedral prism coordinate system by optical imaging [10]. Then, the biaxial inclination angle of the reference hexahedral prism coordinate system relative to the ground is calculated by using the biaxial inclination sensor. Finally, the absolute attitude of the reference hexahedral prism coordinate system relative to the ground is solved by the fusion algorithm [11], and the attitude is further transmitted to the measured object, so as to realize the high-precision measurement of the absolute attitude of the measured object.

2. Multi-sensor fusion attitude measurement method

2.1. Overall framework of the methodology

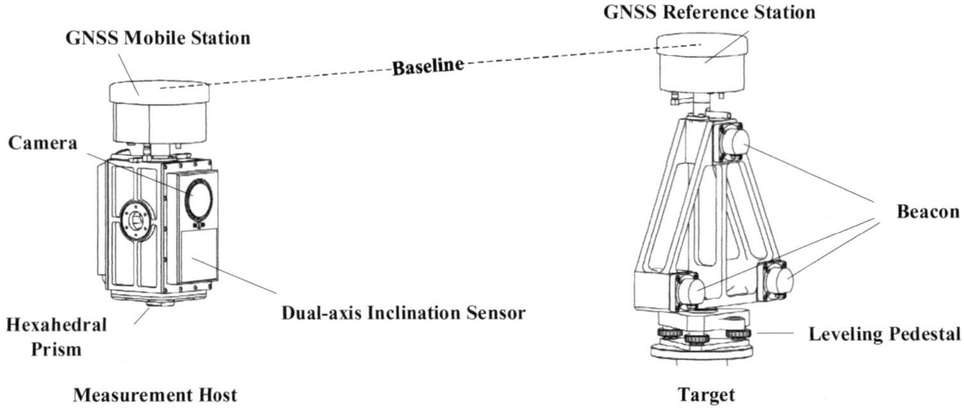

Figure 1. Multi-sensor fusion attitude measurement device.

As shown in Figure 1, the method measurement device in this paper is mainly composed of two parts: the measurement host and the target. Wherein, the measurement host comprises a horizontally mounted camera and a dual-axis inclination sensor module, the top of the measurement host is equipped with a GNSS mobile station, and the bottom of the measurement host is equipped with a hexahedral prism. The camera is used to image the optical beacon in the distance and calculate the amount of missed target, so as to obtain the mapping relationship between the differential baseline and the camera pointing. The biaxial inclination sensor can obtain the pitch and roll angles of the measurement host and participate in the subsequent fusion solution. The GNSS mobile station can calculate the accurate true azimuth information by differentiating it from the target GNSS reference station. The hexahedral prism is mainly used for calibration work and attitude transfer. The target is mainly composed of a GNSS reference station, an optical beacon, and a leveling pedestal. The GNSS reference station transmits the positioning information to the mobile station for calculation by differentiating it from the mobile station at the end of the measurement host. In order to facilitate the extraction and identification of the measuring host, the distance between the three optical beacons is precisely designed so that they form an isosceles triangle structure, and the extension line of the midline of the isosceles triangle passes through the phase center of the antenna of the reference station. The leveling pedestal is used to adjust the target to be level with the ground. The overall flow of the algorithm is shown in Figure 2.

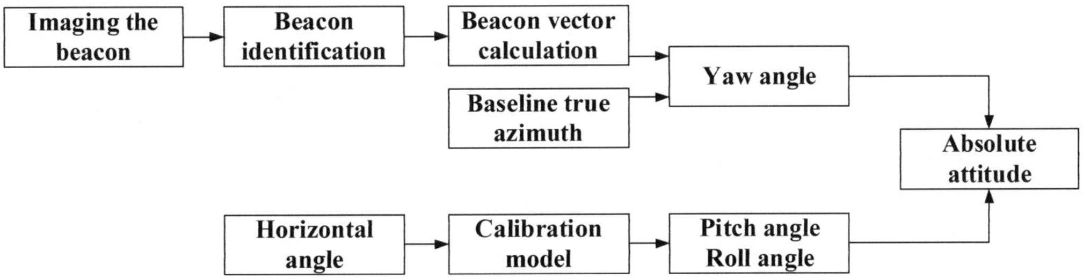

Figure 2. Algorithm flow chart.

2.2. Relationship between coordinate systems

In order to facilitate the calculation, four sets of coordinate systems are established, namely the camera coordinate system (X_c, Y_c, Z_c), the inclination sensor coordinate system (X_a, Y_a, Z_a), the hexahedral prism coordinate system (X_p, Y_p, Z_p), and the world coordinate system (X_w, Y_w, Z_w).

The camera coordinate system (X_c, Y_c, Z_c) takes the optical center O_c of the camera as the origin, and points from the image square focal plane to the object along the optical axis as the coordinate system Z_c axis. The image square focal plane is the $X_c O_c Y_c$ plane. The X_a axis and Z_a axis of the inclination sensor coordinate system (X_a, Y_a, Z_a) coincide with their sensitive roll and pitch axis, respectively. The three orthogonal normals of the hexahedral prism coordinate system (X_p, Y_p, Z_p) are X_p axis, Y_p axis, and Z_p axis. The world coordinate system (X_w, Y_w, Z_w) is often equated with the geocentric coordinate system and the ground coordinate system.

2.3. Optical offset measurement and calculation

In order to measure the amount of optical miss, it is necessary to distinguish between the beacon mark and the background environment in the captured image, and to identify the beacon mark below the differential reference station as the final target to participate in the calculation. Furthermore, the pre-calibrated camera model is used to convert the coordinates of the beacon points into the vector representation of the incident light in the camera coordinate system. The specific process is as follows:

- The Tophat operation is used to pre-process the captured images, which can filter a large area of the bright sky background and the complex environmental background with smooth brightness changes. This ensures that only the small connected domain of the highlight is left in the image. By calculating the theoretical spot size of the beacon under the current baseline length and adjusting the Tophat operation template based on this, the size of the bright connectivity domain in the image is not larger than the theoretical spot size of the beacon, which is convenient for subsequent identification and solving.
- The theoretical brightness value of the beacon spot under the current baseline length is calculated, and the pre-processed image is filtered globally with this brightness as a reference value to remove the interference of some false targets. The 4-connected domain method is used to extract the centroid positions of the remaining spots in the image and arrange them in descending order with brightness, which can effectively avoid the influence of isolated noise.
- According to the current baseline length and the known three-sided length of the beacon, and then according to the camera focal length and pixel size, the theoretical Euclidean distance between the three beacon imaging points can be calculated, and the Euclidean distance plus a certain tolerance can obtain four recognition thresholds.
- The extracted centroids are grouped according to the permutation selection, and all centroid groups that meet the recognition threshold are searched and recorded. The brightness sum of

each group of centroids that meet the requirements is calculated in groups, and the maximum judgment is the group of centroids corresponding to the beacon.

- In the identified centroid group, the Euclidean distance between the two centroids is calculated separately, and the centroid corresponding to the top light can be found by comparison.
- The representation of the top beacon vector in the camera coordinate system $\boldsymbol{v}_{c-meas}$ can be calculated by calibrating the completed camera model. Furthermore, the vector can be converted into the representation in the hexahedral prism coordinate system $\boldsymbol{v}_{p-meas}$ by using the structural calibration results.

2.4. Calculating Euler angles of the rotation matrix

It is assumed that the absolute attitude in the current hexahedral prism coordinate system is expanded according to the 3-1-2 rotation, and the triaxial Euler angles are $\theta_Z\theta_X\theta_Y$, respectively. Then the rotation matrix from the hexahedral prism coordinate system to the world coordinate system can be expressed as:

$$\boldsymbol{R}_{p2g} = \boldsymbol{R}_Y(\theta_Y)\boldsymbol{R}_X(\theta_X)\boldsymbol{R}_Z(\theta_Z) = \begin{bmatrix} \cos\theta_Y\cos\theta_Z - \sin\theta_X\sin\theta_Y\sin\theta_Z & \cos\theta_Y\sin\theta_Z + \sin\theta_X\sin\theta_Y\cos\theta_Z & -\cos\theta_X\sin\theta_Y \\ -\cos\theta_X\sin\theta_Z & \cos\theta_X\cos\theta_Z & \sin\theta_X \\ \sin\theta_Y\cos\theta_Z + \sin\theta_X\cos\theta_Y\sin\theta_Z & \cos\theta_Y\sin\theta_Z - \sin\theta_X\cos\theta_Y\cos\theta_Z & \cos\theta_X\cos\theta_Y \end{bmatrix} \tag{1}$$

At this time, the inclination angles of the inclination sensor in the roll axial direction and the pitch axis are θ_{X-a} and θ_{Z-a} respectively. The inclination angles of the roll axis θ_{X-p} and the pitch axis θ_{Z-p} in the hexahedral prism coordinate system can be obtained according to the conversion relationship between the inclination sensor and the hexahedral prism coordinate system. These two inclinations are the angles between the axis X_p and the axis Y_p and the geodetic horizon. Representation of the axis Z_p in the world coordinate system under the hexahedral prism coordinate system:

$$\boldsymbol{v}_{Z_p} = \boldsymbol{R}_{p2g}\begin{bmatrix}0\\0\\1\end{bmatrix} = \begin{bmatrix}-\cos\theta_X\sin\theta_Y\\ \sin\theta_X\\ \cos\theta_X\cos\theta_Y\end{bmatrix} \tag{2}$$

Equation (2) is compared with the $\boldsymbol{v}_{Z_p}$ representation in the polar coordinate system:

$$\theta_x = \theta_{Z-p} \tag{3}$$

Similarly, the representation of the axis X_p in the world coordinate system under the hexahedral prism coordinate system:

$$\boldsymbol{v}_{X_p} = \boldsymbol{R}_{p2g}\begin{bmatrix}1\\0\\0\end{bmatrix} = \begin{bmatrix}\cos\theta_Y\cos\theta_Z - \sin\theta_X\sin\theta_Y\sin\theta_Z\\ -\cos\theta_X\sin\theta_Z\\ \sin\theta_Y\cos\theta_Z + \sin\theta_X\cos\theta_Y\sin\theta_Z\end{bmatrix} \tag{4}$$

Equation (4) is compared with the $\boldsymbol{v}_{X_p}$ representation in the polar coordinate system:

$$\theta_Z = -\arcsin(\sin\theta_{X-p} / \cos\theta_X) \tag{5}$$

Therefore, according to the two-axis angle value of the biaxial inclination sensor, two Euler angles in the absolute attitude of the hexahedral prism coordinate system are obtained, and only the Euler angle θ_Y that constrains the yaw rotation remains.

The baseline vector $\boldsymbol{v}_{g-meas}$ is represented in the world coordinate system as:

$$\boldsymbol{v}_{g-meas} = \boldsymbol{R}_{p2g}\boldsymbol{v}_{p-meas} = \boldsymbol{R}_{p2g}\begin{bmatrix} \boldsymbol{v}_1 \\ \boldsymbol{v}_2 \\ \boldsymbol{v}_3 \end{bmatrix} \tag{6}$$

According to the matrix form expanded in Equation (1), the following known constants are simplified:

$$\begin{cases} a_1 = \cos\theta_Z \\ a_2 = \sin\theta_X \sin\theta_Z \\ a_3 = \sin\theta_Z \\ a_4 = \sin\theta_X \cos\theta_Z \\ a_5 = \cos\theta_X \end{cases} \tag{7}$$

Equation (6) can be rewritten as:

$$\boldsymbol{v}_{g-meas} = \begin{bmatrix} a_1\boldsymbol{v}_1\cos\theta_Y - a_2\boldsymbol{v}_1\sin\theta_Y + a_3\boldsymbol{v}_2\cos\theta_Y + a_4\boldsymbol{v}_2\sin\theta_Y - a_5\boldsymbol{v}_3\sin\theta_Y \\ \cdots \\ a_1\boldsymbol{v}_1\sin\theta_Y + a_2\boldsymbol{v}_1\cos\theta_Y + a_3\boldsymbol{v}_2\sin\theta_Y - a_4\boldsymbol{v}_2\cos\theta_Y + a_5\boldsymbol{v}_3\cos\theta_Y \end{bmatrix} \tag{8}$$

According to the definition of azimuth in the polar coordinate system and the true azimuth angle of the baseline in the world coordinate system θ_{dir} given from the differential by the GNSS, it can be further obtained:

$$\frac{-\left(a_1\boldsymbol{v}_1\cos\theta_Y - a_2\boldsymbol{v}_1\sin\theta_Y + a_3\boldsymbol{v}_2\cos\theta_Y + a_4\boldsymbol{v}_2\sin\theta_Y - a_5\boldsymbol{v}_3\sin\theta_Y\right)}{a_1\boldsymbol{v}_1\sin\theta_Y + a_2\boldsymbol{v}_1\cos\theta_Y + a_3\boldsymbol{v}_2\sin\theta_Y - a_4\boldsymbol{v}_2\cos\theta_Y + a_5\boldsymbol{v}_3\cos\theta_Y} = \tan\theta_{dir} \tag{9}$$

Finally, according to Equation (9), the Euler angle θ_Y in the rotation matrix can be calculated, and the absolute attitude of the measured object in the world coordinate system can be further obtained.

3. Experiments and analysis

We have made an embedded platform-based prototype to verify the feasibility of the proposed method, and some of the parameters of the prototype are as Table 1.

Table 1. Prototype parameters.

Hardware platform	XC7Z020
Image sensor specification	4 MP 12 bit gray
Camera field of view	18°
Inclination sensor data accuracy	3"
True azimuth measurement accuracy	10" (Baseline = 60 m)

We placed the target about 60 meters away from the measurement host and set it up according to the process. A complete measurement process of 840 times was performed at the same location. Finally, the gyro-theodolite is used to transmit the true azimuth angle and the two-axis inclination reference to the hexahedral prism system of the measuring host for comparison, and the measurement error of the Euler angle in the rotation matrix is shown in Figure 3.

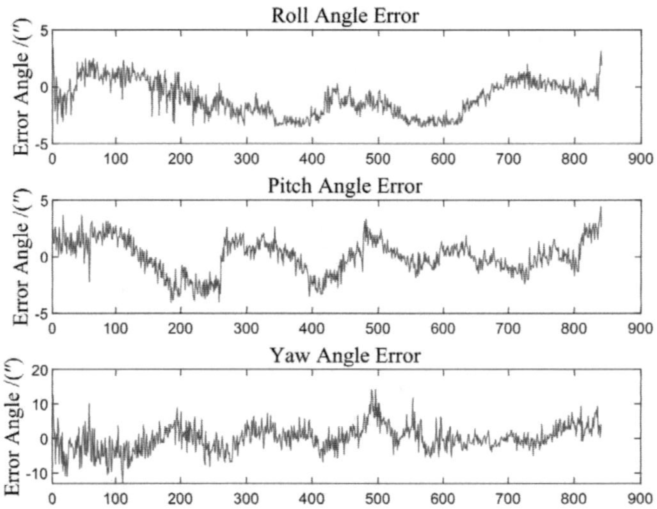

Figure 3. Euler angle measurement error.

Figure 3 shows that the calculation accuracy of roll angle and pitch angle is close to the output accuracy of the two-axis inclination sensor, and the solution accuracy of azimuth angle is close to that of satellite differential true azimuth measurement. This proves that the proposed method has a high absolute attitude solution accuracy.

In addition, in order to further verify the practicability of the method in this paper, the time consumption on each measurement was recorded during the measurement, as shown in Figure 4.

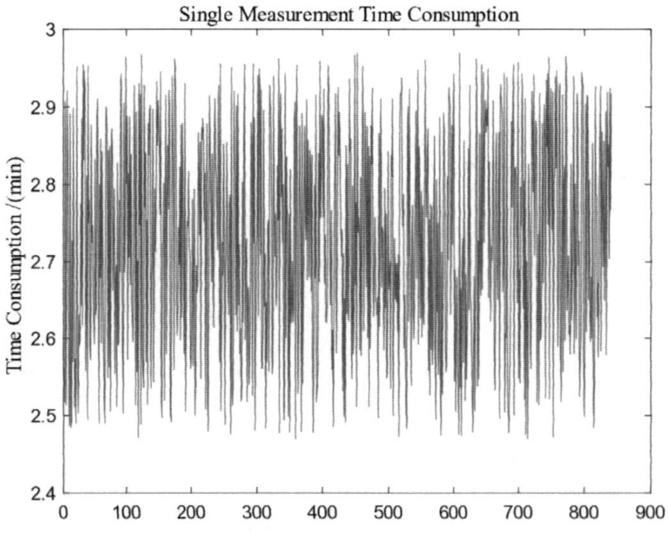

Figure 4. Measurement time consumption.

4. Conclusion

In this paper, we propose an absolute attitude measurement method based on multi-sensor fusion. It solves the missed target with an optical camera, measures the horizontal angle with a dual-axis inclination sensor, and introduces true azimuth from satellite differential positioning. Experiments show that the proposed method can ensure the accuracy of absolute attitude measurement while taking into account the measurement efficiency, and basically realize the automatic measurement of absolute attitude.

However, due to the limited research time, the current measurement method does not model the spatial relationship between the antenna phase center and the main point of the image sensor, and can only reduce the error as much as possible through structural design. In addition, the spatial relationship between the phase center of the antenna at the target end and the center of the top beacon needs to be modeled, and sensors may need to be added to transmit and measure this spatial relationship. Therefore, the follow-up research direction will mainly focus on these two aspects, so as to further improve the method to improve the accuracy of absolute attitude measurement.

References

[1] Zhang, Y., Guo, X.T., Shen, C., et al. (2019) Hybrid Multi-frequency Attitude Estimation Based on Vision/Inertial Integrated Measurement System. In: International Instrumentation and Measurement Technology Conference. Auckland. pp. 1-6. https://doi.org/10.1109/I2MTC.2019.8826925.

[2] Shen, X.R., Wang, Y.M., Dong, R.S. (2013) Implementation of micro-inertial measurement/GPS combinatorial attitude measurement system. In: Asian Control Conference. Istanbul. pp. 1-4. https://doi.org/10.1109/ASCC.2013.6606334.

[3] Wang, P., Wang, C.B., Zhang, B.S., et al. (2023) Adaptive Fusion Algorithm of Visual-Inertial Attitude Measurement in Moving Carrier Environment. Chinese Journal of Sensors and Actuators, 36(2): 218-224. DOI:10.3969/j.issn.1004-1699.2023.02.008.

[4] Xia, K.Q., Wang, B.H., Du, C.Y., et al. (2021) A Satellite Attitude Fusion Estimation Algorithm based on Antenna Measurement Information. In: Information Technology, Networking, Electronic and Automation Control Conference. Xi'an. pp. 780-783. https://doi.org/10.1109/ITNEC52019.2021.9587132.

[5] Xiang, J., Chao, Y., Wen, C., et al. (2017) GNSS 3D attitude measurement system based on dual-antenna receiver with common clock. In: Forum on Cooperative Positioning and Service. Harbin. pp. 223-227. https://doi.org/10.1109/CPGPS.2017.8075129.

[6] Xia, K.Q., Du, C.Y., Wang, B.H., et al. (2021) Research on Application of Antenna Measurement Information in Satellite Attitude Determination. In: Information Technology, Networking, Electronic and Automation Control Conference. Xi'an. pp. 788-792. https://doi.org/10.1109/ITNEC52019.2021.9587043.

[7] Hsu, Y.L., Chou, P.H., Kuo, Y.C. (2017) Drift modeling and compensation for MEMS-based gyroscope using a Wiener-type recurrent neural network. In: International Symposium on Inertial Sensors and Systems. Kauai. pp. 39-42. https://doi.org/10.1109/ISISS.2017.7935653.

[8] O-larnnithipong, N., Barreto. A. (2016) Gyroscope drift correction algorithm for inertial measurement unit used in hand motion tracking. In: IEEE SENSORS. Orlando. pp. 1-3. https://doi.org/10.1109/ICSENS.2016.7808525.

[9] Wei, Y.S., He K.F., Liu, D.X., et al. (2022) GNSS multi-antenna attitude measurement method based on baseline length weighting. Journal of Telemetry, Tracking and Command, 43(3): 18-23. http://dx.doi.org/10.12347/j.ycyk.20210915001.

[10] Li, Z.T., Liu, M.Y., Lu J.F. (2021) High Precision Autocollimation Measurement Technology Based on Image Recognition. In: International Conference on Computing and Data Science. Stanford. pp. 125-129. https://doi.org/10.1109/CDS52072.2021.00028.

[11] Liu, Y.L., Wu, M.P., Guo, Y., et al. (2022) A survey of algorithms for multi-vector attitude determination. Journal of Beijing University of Aeronautics and Astronautics, 11: 1-13. http://dx.doi.org/10.13700/j.bh.1001-5965.2022.0325.

Journal of Physics: Conference Series **2724** (2024) 012047 doi:10.1088/1742-6596/2724/1/012047

A windowed fourier transform-based method of numerically analyzing phase shifts of SPR sensors

Shikang Peng[1], Guofu Wang[1], Yansheng Liu[1,2], Xiaobo Jia[1] and Jin Zhou[1]

[1]College of Automation, Guangxi University of Science and Technology, Wenchang Road, Liuzhou 545001, Guangxi, China

[2]Corresponding author's e-mail: yansheng.liu@gxust.edu.cn

Abstract: This study presents a simple and reliable method for numerical evaluation of the sensitivity of SPR systems. A Windowed Fourier Transform (WFT) based method is applied to extract shifts of SPR interferogram. By applying the WFT, the frequencies consisting of SPR interferogram are extracted, and phase shifts of relevant frequencies are also obtained. The normalized amplitude of the frequencies larger than 0.35 is regarded as the most prominent frequency. The summation of the phase shifts of the most prominent frequencies is regarded as an index to numerically analyzing phase shifts of SPR sensors. Through this method, the small change in the sensing surface can be monitored. By using a homemade SPR system with the proposed calculation method, it can detect the NaCl solutions ranging from 0.00003% to 0.03% (wt%), indicating its great promise for applications in fields such as medicine, food safety, and biotechnology.

1. Introduction

With the advancement of the biomedical and chemical fields, Surface Plasmon Resonance (SPR) has found widespread application in the realms of biology and medicine, owing to its advantages of real-time detection, label-free capabilities, and high sensitivity[1-2]. SPR is based on the phenomenon of plasmon resonance on the surface of a metal film, allowing real-time monitoring of the adsorption and dissociation of biomolecules by tracking changes in the incident angle or wavelength [3]. The sensitivity of an SPR sensor is a crucial performance metric, determining the sensor's ability to effectively detect low concentrations of target molecules [4-5].

Lu et al. [6] reported a fiber-optic SPR system for detecting Hg ions, achieving a detection sensitivity as low as 94 nM. Huang et al. [7] achieved the detection of microplastics in water by using an ER/SPR system, with a sensitivity reaching as low as the particle size of 20 μm. Taylor et al. [8] described a multi-channel SPR system for the identification of four bacterial species, such as Escherichia coli. The limit of detection (LOD) for each of these four bacterial species in the tested samples ranged from 3.4×103 to 1.2×105 cfu/mL. In the realm of real-time and high-efficiency SPR biosensors, their extensive applications have led to a substantial surge in data analysis demands. Traditional methods involving direct comparisons with raw data often result in subtle variations, thereby limiting the full representation of the SPR detection limits. As a response to this challenge, novel approaches in data analysis have emerged in recent years, aiming to harness the full potential of real-time and efficient SPR biosensors. These advanced methods involve intricate signal processing, statistical techniques, and machine learning algorithms, which not only enhance the sensitivity and accuracy of detection but also enable the extraction of valuable insights from the wealth of generated data. In this study, we established a differential phase SPR system and assessed its sensitivity by using a WFT-based method. The sensitivity was initially evaluated by employing a NaCl concentration ranging from 0.00003% to 0.03% (wt%), and the LOD of the proposed SPR system is 0.00003% (wt%). The results indicate that the rapid assessment of SPR system sensitivity by using NaCl and Windowed Fourier Transform holds significant potential for practical applications.

2. Experimental part

2.1. Materials

NaCl (Spectral level) was procured from Shanghai Aladdin Biochemical Technology Co., Ltd. The gold target (99.99%) was obtained from Zhongnuo New Material Technology Co., Ltd. DI water has a resistivity of 18 MΩ·cm.

2.2 Apparatus

In the course of this research, we established a surface plasmon resonance (SPR) system incorporating differential phase measurements. The system primarily consists of a halogen light source, fiber collimators (I1 and I2), linear polarizers (P1 and P2), birefringent crystals (BC), focusing lens (FL), slit, SPR sensor, fiber spectrometer, injection pump, and PC, as shown in Figure 1. The SPR bio-chip was constructed using a standard attenuated total reflectance (ATR) structure, which included a microfluidic channel surface, a triangular prism, and a BK7 glass slide.

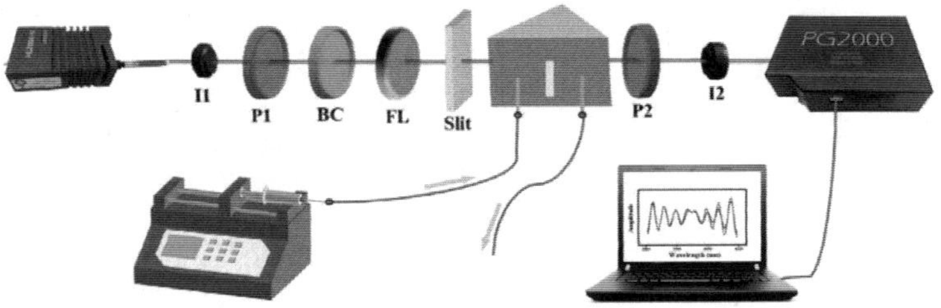

Figure 1. Schematic diagram of the differential phase SPR system.

2.3 Production of SPR chips and detection

The BK7 glass with a certain size has been cleaned by using an ultrasonic ethanol bath for 10 min. By employing thermal evaporation techniques, we successfully deposited a 50-nanometer-thick layer of gold onto the BK7 glass slide. Subsequently, utilizing a masking template, the gold film was shaped into a rectangular form with dimensions of 5 millimeters in length and 2 millimeters in width. This rectangular gold film was coupled with a microfluidic channel which was PMMA, and the total length and width are 6 mm and 2 mm, respectively. The deionized water was introduced onto the gold film surface through the microfluidic channel at a flow rate of 20 μL/min by using a syringe pump, and the spectral interferogram was collected. Subsequently, solutions of NaCl with concentrations of 0.00003%, 0.0003%, 0.003%, and 0.03% were individually introduced into the microfluidic channel at a flow rate of 20 μL/min, and the spectral interferograms were recorded.

In terms of sensor assembly, special attention is required. The coupling of the completed SPR chip with the prism necessitates the use of a custom fixture. This fixture securely holds the prism in place without obstructing the optical path. The SPR chip is then stably fixed at a specific position in the optical path through the clamping structure of the fixture. To ensure the sensor's normal operation, it is crucial to clean the prism's surface with a cleaning cloth before installing the fixture. Additionally, the gap between the well-coupled SPR chip and the prism must be filled with a matching medium of the same refractive index. It is imperative to avoid the formation of bubbles during the gap-filling process.

3. Results and discussion

Figure 2 (A1, A2, A3, and A4) presents the spectral interferogram of 0.03%~0.00003% NaCl solution detected by using the SPR sensor for calculating SPR sensing results. Figure 2 (B1, B2, B3, and B4) presents the calculated frequencies and phase shifts for each frequency. Frequencies with a normalized amplitude greater than 0.35 are selected, and the summation of phase shifts for these frequencies ($\boldsymbol{\Delta\Phi}_{\mathrm{Iops}}$) is used as a detection metric. The $\boldsymbol{\Delta\Phi}_{\mathrm{Iops}}$ greater than the machine response is considered a reliable phase shift in subsequent analysis. The summation of phase for $\boldsymbol{\Delta\Phi}_{\mathrm{Iops}}$ was calculated as follows:

$$\Delta\Phi_{Iops} = \sum \Delta\Phi \tag{1}$$

where $\Delta\Phi$ represents the phase shift in degrees obtained for each frequency greater than 0.35 by using the WFT. The SPR biosensor can be influenced by various factors such as light stability, spectrometer stability, and temperature fluctuations. To obtain reliable detection results, 10 measurements were conducted by using deionized water as a reference. The data was divided into two groups of 5 measurements each, and the error calculation is shown in Figure 2. In the first group, $\boldsymbol{\Delta\Phi}$Iops was 39°, and in the second group, $\boldsymbol{\Delta\Phi}$Iops was 28°. The maximum value of 39° was taken as the error.

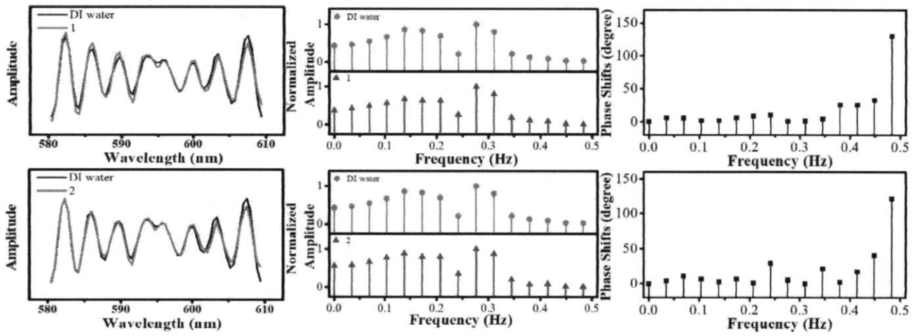

Figure 2. DI water as the detection target exhibiting spectral interference patterns along with the corresponding frequency domain plot and computed phase map.

The process of calculating the errors in the sensing system involves a systematic correction procedure. It is essential to note that, due to the significant impact of subtle vibrations on the optical path, the entire system needs to be installed on a stable optical platform. Simultaneously, precautions must be taken to prevent environmental light interference during experiments, and the entire optical path should be situated in a dark environment. During experimentation, it was observed that the movement of the optical fiber also had a substantial impact on the results. To address this, the optical fiber should be securely fixed using a ribbon, and throughout the experimental process, the sensor surface must remain bubble-free. Prior to each measurement, air bubbles should be purged.

In addition, the occurrence of bubble-induced vibrations during the detection process can significantly compromise the authenticity of experimental results. In such instances, it is imperative to clean the sensor with deionized (DI) water and initiate the experimental procedure anew. Only through these measures can reliable and accurate results be attained.

The total phase shift results for various concentrations of NaCl solutions and DI water are obtained by summing the phases of prominent frequencies in Figure 3. In the above context, (A1), (A2), (A3), and (A4) represent spectral interferograms captured at the SPR resonance wavelength for both deionized water and NaCl concentrations spanning from 0.00003% to 0.03%. The phase shift results were computed by analyzing the spectral interferograms. Among them, $\boldsymbol{\Delta\Phi}$Iops at 0.00003% is 108°, $\boldsymbol{\Delta\Phi}_{\mathrm{Iops}}$ at 0.0003% is 236°, $\boldsymbol{\Delta\Phi}_{\mathrm{Iops}}$ at 0.003% is 441°, and $\boldsymbol{\Delta\Phi}_{\mathrm{Iops}}$ at 0.03% is 772°. The results of our detection are presented in Figure 4. From the experimental results, it can be observed that within the margin of error, the minimum detection limit of this SPR system for NaCl is 0.00003%.

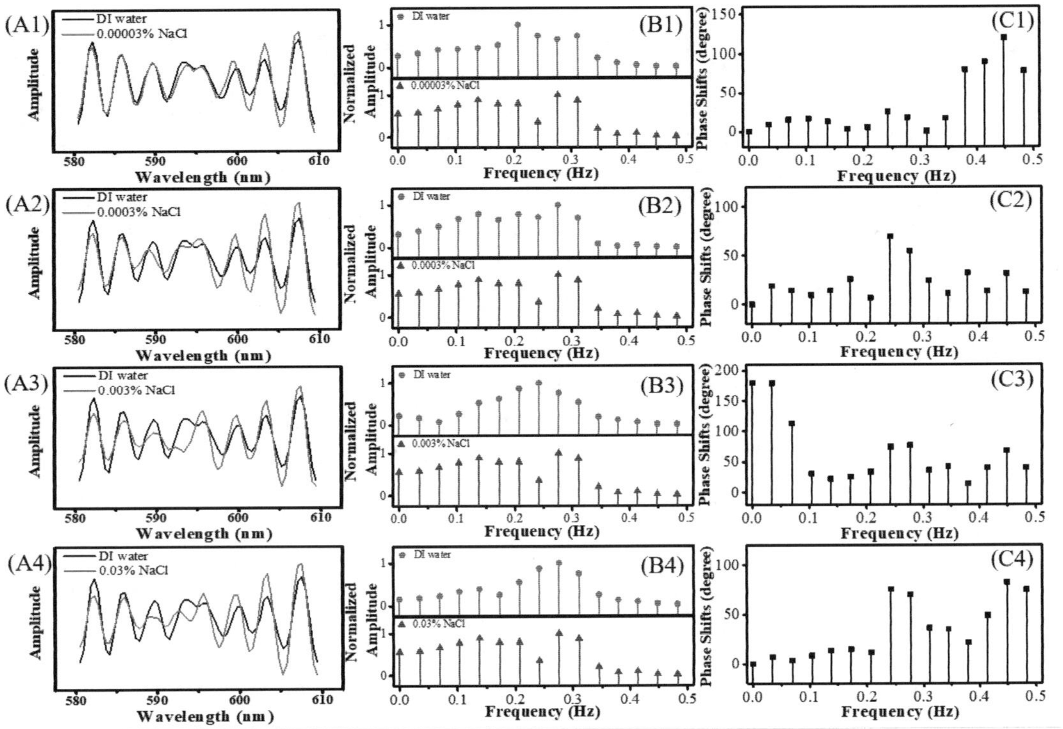

Figure 3. (A1), (A2), (A3), and (A4): Spectral interferograms of deionized water and 0.00003% to 0.03% NaCl captured at the SPR resonance wavelength; (B1), (B2), (B3), and (B4): The derived frequency domain resulted by the interferograms by using WFT; (C1), (C2), (C3), and (C4): The derived phase shifts ($\Delta\phi f$) of each frequency.

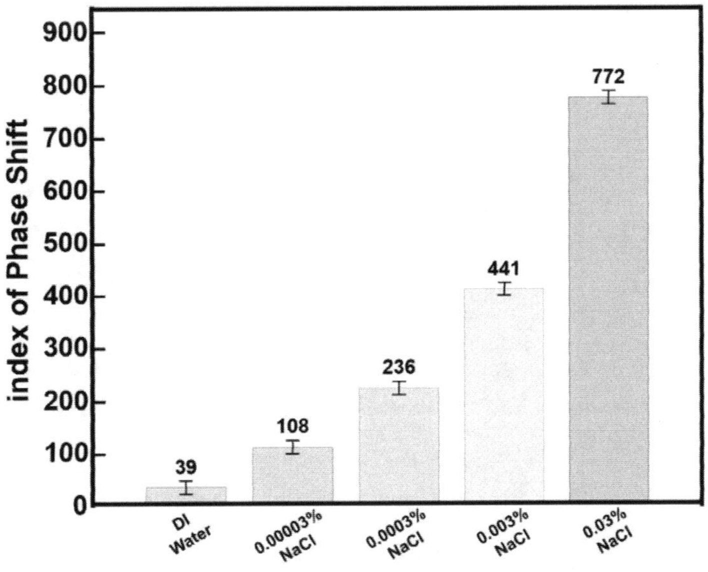

Figure 4. Detection of phase shift in NaCl solutions ranging from 0.00003% to 0.03% in DI water.

Journal of Physics: Conference Series **2724** (2024) 012047 doi:10.1088/1742-6596/2724/1/012047

4. Conclusion

In summary, we have successfully introduced and validated a method based on the WFT for analyzing the detection results of SPR systems. The application of the WFT method involves extracting frequency domain results from SPR spectral interferograms and determining phase shifts corresponding to each frequency. A numerical index, derived from the cumulative sum of phase shifts at prominent frequencies, is employed to portray and reflect the detected outcomes. Using NaCl as the analyte, the detection limit for NaCl solutions ranging from 0.00003% to 0.03% (wt%) was determined to be 0.00003% (wt%). The results of this study provide an effective computational method for discerning subtle shifts and hold practical application value in assessing the sensitivity of SPR systems.

Moreover, the versatility and robustness of the WFT-based analysis method make it a valuable tool for a wide range of applications. This approach not only enhances our understanding of SPR spectral interferograms but also offers insights into the detection limits for various analytes. Beyond its application with NaCl, this method can be extended to other compounds and complex mixtures, showcasing its potential for use in fields such as environmental monitoring, medical diagnostics, and biochemistry. As the demand for sensitive and accurate detection methods continues to grow, the WFT-based approach presented in this study holds promise as a valuable resource for researchers and practitioners in the realm of surface plasmon resonance technology.

References

[1] P. Englebienne, A. Van Hoonacker, and M. Verhas. *Spectroscopy,* Vol. 17, No. 2-3, pp. 255-273, 2003.
[2] J. Pollet et al. *Biosensors and Bioelectronics,* Vol. 25, No. 4, pp. 864-869, 2009.
[3] M. A. Syed, A. S. Bhatti, C. Li, and H. Bokhari. *Am J. Biomed Sci.,* Vol. 3, No. 4, pp. 253-257, 2011.
[4] S. Chen, Y. Liu, Q. Yu, and W. Peng. *Sensors, and Actuators B: Chemical,* Vol. 327, pp. 128, 935, 2021.
[5] M. F. Gouzy, M. Ke, and P. M. Krämer. *Biosensors and Bioelectronics,* Vol. 24, No. 6, pp. 1, 563-1, 568, 2009.
[6] M. Lu et al. *Journal of Lightwave Technology,* Vol. 39, No. 12, pp. 4, 034-4, 040, 2021.
[7] C. J. Huang, G. V. Narasimha, Y. C. Chen, J. K. Chen, and G. C. Dong. *Biosensors,* Vol. 11, No. 7, pp. 219, 2021.
[8] A. D. Taylor, J. Ladd, Q. Yu, S. Chen, J. Homola, and S. Jiang. *Biosensors and Bioelectronics,* Vol. 22, No. 5, pp. 752-758, 2006.

A multifunctional reference torque standard machine

Wei Yi[1], Dongli Qiu[1,2], Ying Liu[1], Guihong Yuan[1], Hua Dai[1] and Shiqian Li[1]

[1]National Institute of Measurement and Testing Technology, Chengdu 610021, China

[2] Corresponding author's e-mail: 1223214544@qq.com

Abstract. In response to the problem of using different single machines to achieve measurement and detection of existing torque instruments, this article introduces a multifunctional reference torque standard and proposes specific steps for torque measurement and detection. This standard machine can be used to detect various existing torque instruments and can solve the problem of using a single machine to achieve the detection of a certain type of torque equipment. After adopting this standard machine and its detection method, experimental data shows that the measurement of the torque transducer can meet the requirements of JJG995-2005 "Static Torque Measuring Machines" and the measurement of the reference torque wrench can meet the requirements of JJG1103-2014 "Reference torque wrench Verification Regulations".

1. Introduction

The measurement and testing of existing torque instruments such as torque transducers, reference torque wrenches, torque wrenches, and torque multipliers are usually achieved using different single machines [1]. The equipment investment is large, the land occupation is large, the universality is poor, the usage efficiency is low, and it is difficult to meet the measurement and testing needs of new torque instruments. Moreover, these single machines also face some problems when testing and measuring.

The measurement and detection of the torque transducer adopts a static weight torque standard machine [2], and its load at each level depends on the mass of the weights, while the mass of each weight is fixed and unchanged, so it cannot achieve torque loading beyond the number of weight combinations. At the same time, its loading efficiency depends on the lifting time of each level of weight, and the more weights are loaded, the lower its loading efficiency is. The metrological detection of reference torque wrenches is achieved by adding a reaction machine [3] to the traditional static weight torque standard machine. However, the existing static weight torque standard machine cannot maintain its lever and the handle of the tested reference torque wrench level, so it cannot offset the influence of gravity on the detection results. The detection results have large errors, which do not meet the technical requirements of the verification regulations, and the metrological risk is high [4]. At the same time, due to the force point being at the handle, rather than coaxial pure torsion, it will generate lateral thrust on the precision support blade and bearing on the lever of the static weight torque standard machine, which will cause structural damage to the static weight standard over time [5]. The internationally recommended method is to use the reference torque standard machine calibration method [6]. The measurement and testing of instruments such as torque wrenches and torque multipliers are carried out using torque wrench calibrators. However, the general range of torque wrench calibrators is ≤ 3 kNm, which cannot meet the testing requirements of torque multipliers ≥ 3 kNm. With technological progress, multifunctional reference torque standard machines have emerged [7-8], but the supports of the machines mostly adopt a vertical column layout. The upper and lower platforms of the columns need to be kept horizontal, and the columns also need to resist torsional forces. Due to the different lengths of the inspected equipment, the work plane also needs to slide up and down along the column to adjust the working space. This makes the entire reference torque standard machine structurally complex, inconvenient to use, and costly. In addition, due to the heavy weight of the large range torque transducer, some can weigh over 100 kilograms, making it difficult for manual handling. The reference torque standard machine with a vertical column layout is not conducive to installing or disassembling the large range torque transducer on the torque standard machine. At the same time, this machine adopts a single shaft output, and the torque output of the single shaft has a fixed optimal output range, which cannot accommodate a wide measurement range. The range of the torque transducer or torque multiplier being tested is often large, some can reach 10 kNm, while the range of the reference torque wrench being tested is generally ≤ 3 kNm. Single-axis output means that hardware characteristics such as transmission ratio and torque output resolution are fixed and unchanging. The torque output system with high power and high reduction ratio has no problem when detecting large torque, while the torque output will have a large jump when detecting small torque, which is difficult to stabilize and seriously affects the measurement results. On the contrary, a torque output system with low power and low reduction ratio makes it difficult to generate large torque output.

Therefore, in response to the current measurement and testing needs of torque instruments and the shortcomings of existing technology, a dual output multifunctional reference torque standard machine and method are proposed. On a servo control system and a machine, different standard torque instruments can be switched and configured to achieve measurement and testing of coaxial rotating equipment such as torque instruments and torque multipliers with a range of ≤ 10 kNm and an accuracy of ≤ 0.05%. It also achieves the detection of reference torque wrenches and torque wrenches with a range of ≤ 3 kNm and an accuracy of ≤ 0.1%, as well as the measurement and detection of non-coaxial rotating equipment such as other torque tools. It adopts three control modes: computer automatic control, electronic handwheel, and manual button, which can be flexibly selected according to usage habits and work needs. The three modes can be switched at any time, with compact structure, low manufacturing cost, flexible torque output, adjustable loading and unloading speed, and convenient operation. Thereby improving the automation level of torque equipment measurement and detection, improving work efficiency, and improving measurement accuracy through precise automatic control.

2. Research on reference torque standard machine

2.1. Introduction to machine structure design

The multifunctional reference torque standard machine is mainly composed of a servo motor, guide rail, coupling, support seat and support plate, support strip, measuring base slider, steering gear, reducer, steering gear, and computer control system. The basic technical principle is to control the rotation of the servo motor through three modes: computer automatic control, electronic hand wheel, and manual button control. The servo motor drives the rotation of the high-speed ratio planetary reducer. The planetary reducer is a left and right double shaft output structure, which produces two different application functions. The whole process realizes closed-loop control, with a high degree of automation, good system repeatability, and high data accuracy, and can also avoid Personal equations caused by manual operation. The three-dimensional view of the standard machine is shown in Figure 1, the front view is shown in Figure 2, the top view is shown in Figure 3, and the side view is shown in Figure 4.

The multifunctional reference torque standard machine utilizes a dual output shaft reducer to achieve two functional modules, and uses one machine to achieve measurement and detection of various torque equipment. For a better understanding, two typical examples of detecting torque transducers and reference torque wrenches are introduced in the article. Firstly, the state diagrams of the two examples are provided, as shown in Figure 5.

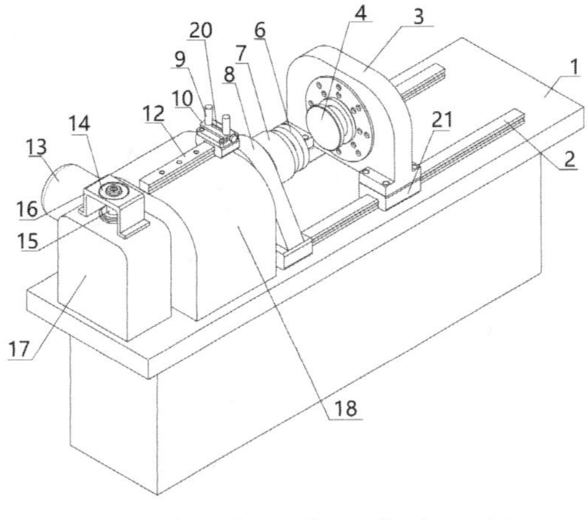

Figure 1. 3D view of standard machine.

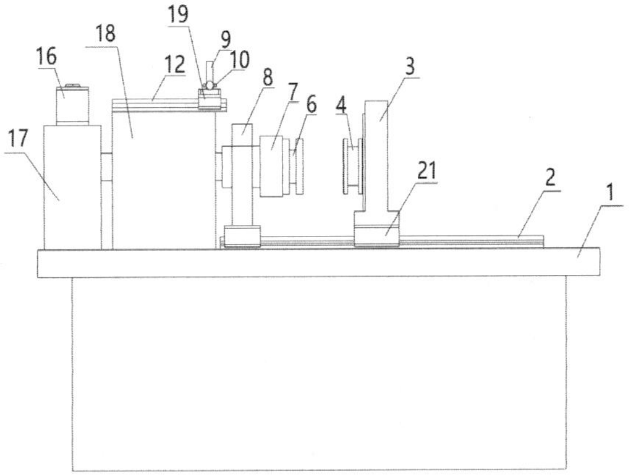

Figure 2. Front view of standard machine.

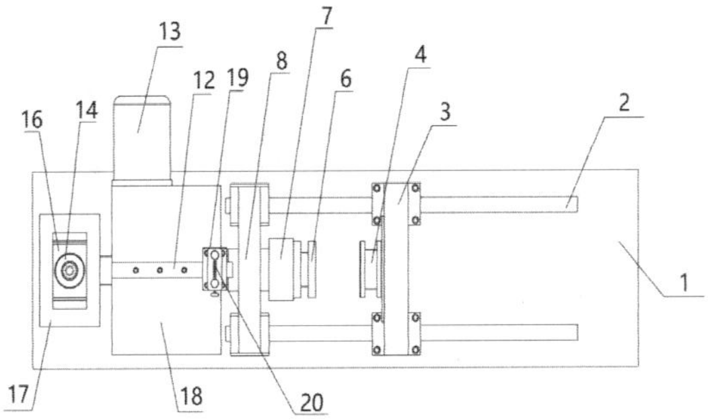

Figure 3. Top view of standard machine.

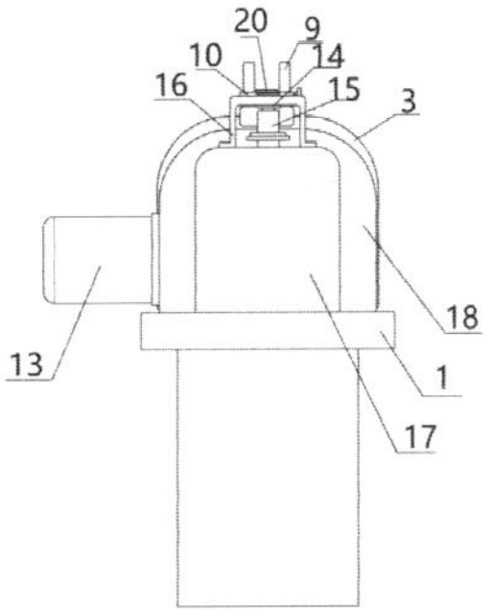

Figure 4. Side view of standard machine.

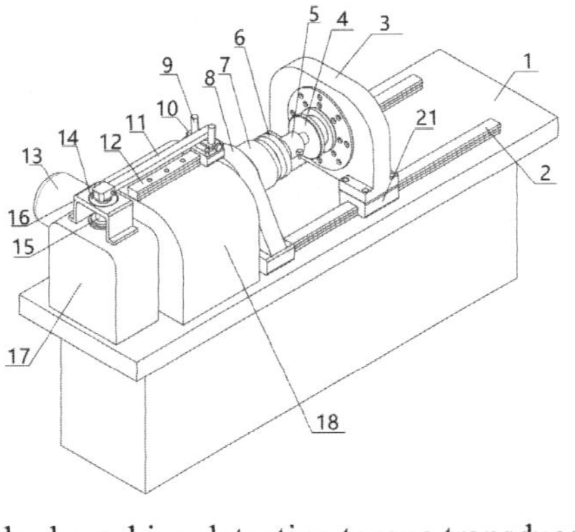

Figure 5. State diagram of standard machine detecting torque transducer and reference torque wrench.

The serial numbers in Figures 1 to 5 are 1-Machine, 2-First linear guide rail, 3-First support seat, 4-Second coupling, 5-Tested torque transducer, 6-First standard torque transducer, 7-First coupling, 8-Support plate, 9-Reaction baffle, 10 Support plate, 11-Tested reference torque wrench, 12-Second linear guide rail, 13-Servo motor, 14-First bearing, 15-Second standard torque transducer, 16-Reaction bracket, 17-Right angle commutator, 18-Dual output shaft reducer, 19-Second support seat, 20-Support strip, 21-Base slider.

2.2. Research on machine function

The multifunctional reference torque standard machine utilizes a dual output shaft reducer to achieve two functional modules, including the following functions:

A single machine has been used to achieve measurement and detection of various torque equipment, with low equipment cost, high integration, and small footprint. The first output shaft of the dual output shaft reducer is connected to the first standard torque transducer through a first coupling. The first standard torque transducer rotates coaxially with the first output shaft, and a second coupling is fixed on the first support seat. The second coupling remains coaxial with the first output shaft, and the first support seat can move horizontally along the axial direction of the first output shaft. This functional module is used for torque transducers. The measurement and detection of equipment that can rotate coaxially, such as torque multiplier, adopts a horizontal layout for this functional module, which has higher torsional strength and simple structure. Moreover, compared to the vertical column layout, the work plane needs to slide up and down along the column to adjust the detection space. The horizontal layout adjusts the detection space through the horizontal movement of the first support seat, making the operation easier and more convenient. In addition, the horizontal layout is also conducive to using the driving system to install or remove the large range torque transducer from the torque standard machine. The second output shaft of the dual output shaft reducer is connected to a right angle commutator, which is used to change the output direction to a vertical direction. The output end of the right angle commutator is connected to the second standard torque transducer, and the output end of the second standard torque transducer and the right angle commutator rotate on the same shaft. The end of the second standard torque transducer far from the right angle commutator is sleeved with a first bearing. The first bearing is used to withstand the bending moment generated by the tested equipment during the testing process. The second support seat is set on the top of the dual output shaft reducer. The second support seat is used to fix the handle of the tested reference torque wrench, so that the tested reference torque wrench remains stationary during the testing process. This functional module is used for measuring and testing non-coaxial rotating equipment such as reference torque wrenches and torque wrenches. The two functional modules of this machine are independent of each other. This machine can accurately measure a large number of torque transducers or torque multipliers, as well as small range reference torque wrenches or torque wrenches. This machine can take into account a wide measurement range.

In the structure of the detection machine, the first support seat is installed on the machine table through the first linear guide rail. The centerline of the first linear guide rail is parallel to the axis of the first output shaft, and the first support seat can move horizontally along the first linear guide rail. The first linear guide rail can achieve the axial horizontal movement of the first support seat along the first output shaft, which is conducive to adjusting the detection space through the horizontal movement of the first support seat for detecting different sizes of tested equipment.

In the structure of the detection machine, the support plate is installed on the machine table through a first linear guide rail, and a second bearing is fixed on the support plate. The second bearing is used to withstand the reaction force, and the support plate is used to support and fix the second bearing.

In the structure of the detection machine, the first output shaft is connected to the first coupling through the second bearing. The second bearing is used to withstand the reaction force of the first standard torque transducer and protect the first output shaft.

When measuring and testing the torque multiplier, the side of the first standard torque transducer that is not connected in series with the first output shaft is first connected in series with the support plate, and then connected to the torque multiplier. The support plate bears the reaction force of the torque multiplier, protecting the first standard torque transducer.

The second support seat is installed on the top of the dual output shaft reducer through the second linear guide rail, and can move along the second linear guide rail. The second linear guide rail can move the second support seat along it, which is beneficial for adjusting the position of the second support seat according to the length of the handle of the tested standard torque plate and torque wrench, and for keeping the tested reference torque wrench stationary during the testing process. The second support seat includes a base, a reaction baffle, and a support plate. The base is located on the top of the dual output shaft reducer, and the base is kept horizontal. The reaction baffle is in vertical contact with the base and fixed. The support plate is provided with a through hole, and the reaction baffle penetrates through the through hole. The support plate can move up and down along the reaction baffle. The reaction baffle is used to limit the horizontal movement of the handles of the tested reference torque wrench and torque wrench, so that they remain stationary during the testing process. The support plate is used to support the handles of the tested reference torque wrench and torque wrench, so that they remain horizontal. The support plate can move up and down along the reaction baffle, which is conducive to adjusting the height of the support plate according to the size of different reference torque wrenches and torque wrenches to keep it horizontal. A support strip is set at the contact point between the support plate and the handle of the tested reference torque wrench. A support strip is installed at the contact point between the handle of the support plate and the tested reference torque wrench, reducing the contact area between the handle and the support plate, thereby reducing frictional resistance, and making the output torque transmission of the second standard torque transducer better. At the same time, it can make the tested reference torque wrench and torque wrench easier to return to zero, improving measurement accuracy.

A reaction support bracket was also designed, which is fixed on the top of the right angle commutator. The reaction support bracket is provided with a through hole, and the first bearing is fixed in the through hole. The reaction bracket is used to support and fix the first bearing.

In the structure of the detection machine, the dual output shaft reducer is a dual output shaft planetary reducer. The planetary reducer has a large transmission ratio, small back clearance, low noise, low vibration, and compact structure, and is suitable for occasions that require precision transmission.

2.3. Research on machine detection methods

The multifunctional reference torque standard machine also provides a method for measuring and testing torque instruments, including the following steps:

For measuring and testing equipment that can rotate coaxially: The tested torque transducer is connected to the second coupling, making sure that the tested torque transducer cannot rotate and remains coaxial with the first output shaft. The first support seat moves horizontally along the axis of the first output shaft on the machine, and the space is adjusted according to the size of the tested torque transducer. After the spatial adjustment is completed, the detected torque transducer is connected to the first standard torque transducer, and the servo motor is turned on. The detected torque transducer and the first standard torque transducer each display data under the torsion force of the servo motor. The data of the detected torque transducer is compared with the data of the first standard torque transducer to obtain the measurement parameters of the detected torque transducer.

For measurement and testing of non-coaxial rotating equipment: The tested reference torque wrench is connected to the second standard torque transducer, the handle of the tested reference torque wrench is placed on the second support seat, and the second support seat is moved according to the size of the tested reference torque wrench to adjust the space. The handle of the tested reference torque wrench is kept horizontal, and the servo motor is turned on. The tested reference torque wrench and the second standard torque transducer each display data under the action of servo motor torque force. The data of the tested reference torque wrench is compared with the data of the second standard torque transducer to obtain the measurement parameters of the tested reference torque wrench.

3. Experimental results and analysis

A TB2-10 kNm torque transducer produced by HBM company with an accuracy of 0.03% was selected as the first standard sensor in the standard machine, and a TB2-3 kNm torque transducer with an accuracy of 0.03% was selected as the second standard sensor, both of which have been calibrated by German PTB and meet the requirement of 0.03%. The data acquisition instrument matched with the standard sensor is DMP41 produced by HBM company, with an accuracy of 0.0005 level. A T40B-10 kNm (accuracy 0.05%) torque transducer and a TTS-1 kNm (accuracy 0.1%) reference torque wrench were selected for the experiment. The torque transducer and reference torque wrench were measured using a multifunctional reference torque standard machine. Firstly, the torque transducer was installed as shown in Figure 5, and then the indication error of the torque transducer was measured according to the verification regulation of JJG 995-2005 "Static Torque Measuring Machines" [9]. The experimental results after correct measurement and data processing are shown in Table 1 in the following text. Then, the reference torque wrench was installed as shown in Figure 5, and the indication error of the reference torque wrench was measured according to the verification regulation of JJG1103-2014 "Reference Torque Wrench Verification Regulations" [10]. After correct measurement and data processing, the experimental results are shown in Table 2 below.

Table 1. Measurement data of torque transducer.

DIR	Torque /Nm	$\overline{X}$ /Nm	δ /%	R /%	R_{ot} /%
CW	2000	2000.5	0.02	0.02	0.01
	4000	4000.8	0.02	0.01	0.01
	6000	6001.0	0.02	0.01	0.01
	8000	8001.7	0.02	0.01	0.01
	10000	10002.5	0.02	0.01	0.01
CCW	2000	1999.5	-0.02	0.02	0.01
	4000	3999.2	-0.02	0.01	0.01
	6000	5998.5	-0.02	0.01	0.01
	8000	7998.2	-0.02	0.01	0.01
	10000	9997.8	-0.02	0.01	0.01

where $\overline{X}$ is mean value, δ is relative deviation, R is repeatability, and R_{ot} is relative orientation error.

Table 2. Measurement data of reference torque wrench.

DIR	Torque /Nm	$\overline{X}$ /Nm	δ /%	b' /%	b /%
CW	200	200.12	0.06	0.05	0.02
	400	400.25	0.06	0.03	0.02
	600	600.33	0.05	0.02	0.01
	800	800.43	0.05	0.02	0.01
	1000	1000.40	0.04	0.02	0.01
CCW	200	200.13	0.07	0.08	0.04
	400	400.25	0.06	0.05	0.02
	600	599.64	-0.06	0.06	0.02
	800	799.62	-0.05	0.06	0.02
	1000	999.51	-0.05	0.04	0.02

where $\overline{X}$ is mean value, δ is relative deviation, b' is repeatability under maximum lever length, and b is reproducibility under maximum lever length.

Data analysis and conclusion: From the experimental data results in Table 1, it can be analyzed that the indication error, repeatability error, and azimuth error of the torque transducer are all $\leq 0.02\%$, meeting the metrological performance requirement of 0.05 level specified in the JJG 995-2005 "Static Torque Measuring Machines" verification regulation. From the experimental data results in Table 2, it can be analyzed that the indication error, repeatability error, and reproducibility error of the reference torque wrench are all $\leq 0.1\%$, meeting the metrological performance requirements of level 0.1 specified in the JJG 1103-2014 "Reference Torque Wrench Verification Regulations" verification regulation. Therefore, after adopting the machine and method mentioned in the article, the measurement of the torque transducer and reference torque wrench can meet the requirements.

4. Conclusion

This article proposes a multifunctional reference torque standard machine and provides a detailed introduction to its structure, functions, and detection methods. This standard machine can be used to detect various existing torque instruments, and can solve the problem of using a single machine to achieve the detection of a certain type of torque equipment. At the same time, experiments were conducted on two typical types of torque equipment, torque gauges and reference torque wrenches. Based on experimental data analysis, the machine can meet the requirements of the verification regulations of JJG 995-2005 "Static Torque Measuring Machines" and JJG 1103-2014 "Reference Torque Wrench Verification Regulations", respectively. However, it should also be noted that such reference torque standard machines have very high requirements for supporting standard sensors and data acquisition instruments, and their accuracy and long-term stability should preferably be 5 times higher than that of the double measuring object. This has become an important factor restricting the development of such machines towards higher accuracy. This solution integrates multiple detection functions, reduces manufacturing costs, and is suitable for use as a standard machine of level 0.05 or 0.1. It can meet the detection needs of the vast majority of torque measuring instruments, and has good application prospects and promotion value.

References

[1] Feng Meng, Jing Pinjiao, Zhimin Zhang, Dianlong Zhang. Various Torque Tools Calibration at NIM [J]. Measurement: Sensors, Volume 18, December 2021, 100177
[2] Li Tao, Xu Guang, Ji Honglei, Ni Jinquan, Hu Fangfang. Research on Design Technology of Torque Standard Machine [J]. China Metrology Technology. 2019(10).
[3] H. Cheng, J.Z. Yao, W.B. Wen, B.H. Ma, W.J. Huang. Study on Reference Torque Wrench Tested by Deadweight Torque Standard Machine [J]. IMEKO 24th TC3, 14th TC5, 6th TC16 and 5th TC22 International Conference 11 – 13 October 2022, Cavtat-Dubrovnik, Croatia
[4] Mao Qinwei, He Jianxin. Research on the Quantity Traceability of the Reference torque wrench for the Deadweight Torque Standard Machine [J]. Metrology Science and Technology. 2019(04).
[5] Xu Tao. Reference torque wrench Calibration Technology [J]. Ship Engineering. 2 2019, 41(S2).
[6] DKD-R 3-7, "Statische Kalibrierung von anzeigenden Drehmomentschlüsseln".
[7] Wu Chengyong, Lin Jiejun, QI Hongli, Ma Xianglong. Development of 5 kN·m Horizontal Reference Torque Standard Machine [J]. Ship Engineering. 2020, 42(S1).
[8] F. Meng, Z.M. Zhang, Y. Zhang. Calibration of Reference Torque Wrench by New Reference Torque Standard Machine at Nim [J]. IMEKO 2018-September 3-6, 2018, Belfast United Kingdom.
[9] JJG 1103-2014, "Reference Torque Wrench Verification Regulations".
[10] JJG 995-2005, "Static Torque Measuring Machines".

Testability evaluation of radar equipment with improved matter-element extension method

Jiajun Chen [1,4], Bing Hu [1], Duanyang Shi [2] and Zhuo Min [3]

[1] Air Force Early Warning Academy, Wuhan, China
[2] Unit 95174 of PLA, Wuhan, China
[3] Unit 95867 of PLA, Bejing, China

[4]Corresponding author's e-mail: 517937473@qq.com

Abstract. In response to the problems of many influencing factors and different indexes, a radar equipment testability assessment index system is constructed based on the characteristics of equipment and testability connotation requirements. The Criteria Importance Though the Intercrieria Correlation (CRITIC) method is improved by using the variance coefficient to replace standard deviation and conflict coefficient to calculate the absolute value, and the objective weights of evaluation indexes are determined. An improved matter-element extension model is established, and the testability evaluation of different types of radar equipment is obtained by calculating the asymmetric close degree of the extension set. The effectiveness of the model is verified by an example analysis.

1. Introduction

As the battlefield becomes more complex, the integration of radars is increasing, and their performance and structure are becoming more complex, creating greater challenges for radar fault detection and isolation [1]. As far as radar equipment is concerned, good testability is of great significance in reducing the consumption of maintenance manpower and support resources, reducing life cycle costs, and improving the combat readiness and mission success of equipment [2]. Scientifically evaluating the testability level of radar equipment can check whether its testability design meets the specified testability requirements, identify the defects of equipment testability, and accurately grasp its performance status, which provides important decision support for the improvement of radar equipment testability.

At present, a large number of scholars have adopted a variety of methods to evaluate the testability of related equipment from their respective fields, and have achieved certain results. Liu [3] constructed six general quality characteristic index libraries of complex products, including testability, and carried out a comprehensive evaluation based on the VIKOR method. Su et al. [4] took fault detection rate as an example to study the testability evaluation method of aircraft prediction and health management (PHM) system based on outfield data. Li et al. [5] established a missile state assessment index system based on test data and proposed a missile state assessment and decision method with extended TOPSIS-grey correlation.

According to the analysis, the testability evaluation objects mainly focus on missiles, aviation, and other equipment, while the testability evaluation studies on radar equipment are relatively few, and most of them are based on the classification of test stages to establish evaluation indexes or rely on a few quantitative indexes such as fault detection rate and fault isolation rate to evaluate the testability of equipment systems. There are many types and complex systems of modern radar equipment, so it is of practical significance to establish a testability evaluation index system and model suitable for most radar equipment.

Based on the characteristics of radar equipment and the requirement of testability connotation, in this paper, the evaluation index system of radar equipment testability is constructed, and the extension evaluation model of radar equipment testability is put forward based on the weight of the improved CRITIC method. CRITIC method is an objective weighting method proposed by Diakoulaki et al [6]. This method not only considers the influence of index variation degree on weight but also takes into account the conflict among indexes. In this paper, the coefficient of variation is used to replace standard deviation to measure the contrast intensity between indexes, the correlation coefficient is taken as an absolute value in the calculation of conflict coefficient, and the CRITIC method is optimized and improved to make the weight results more scientific and accurate. By normalizing index assignment and replacing the grade criterion of maximum membership degree, the shortcomings of the traditional matter-element extension model are effectively improved, and a more accurate grade of testability evaluation is obtained, which provides a decision reference for the improvement of testability evaluation and testability design level of radar equipment.

2. Construction of Testability Evaluation Index System of Radar Equipment

The testability requirements of radar equipment specify the condition monitoring capabilities, fault diagnosis capabilities, and constraints that radar equipment should have, which are mainly divided into qualitative requirements and quantitative requirements [7]. Combining with the characteristics of radar equipment itself, analyzing the connotation of testability and the specific content of testability requirements, considering from the perspective of condition monitoring, fault diagnosis, equipment design, and resource guarantee, a radar equipment testability evaluation index system is constructed.

Condition performance monitoring parameters (A_1) are parameters used to observe and measure the condition performance of radar equipment to determine whether the condition performance of the equipment meets the specified requirements and to be able to perform fault diagnosis and trend analysis. Its indexes include parameter monitoring rate (B_{11}), condition monitoring rate (B_{12}), and station-level average manual parameter detection time (B_{13}).

Fault diagnostics parameters (A_2) describe the process of detecting and isolating faults in radar equipment and are the centralized expression of equipment testability. The index is described by the fault detection rate (B_{21}), fault isolation rate (B_{22}), false alarm rate (B_{23}), fault detection time (B_{24}), fault isolation time (B_{25}), fault isolation ambiguity group size (B_{26}), mistake dismantle rate (B_{27}), and BITE coverage rate (B_{28}).

A good testability design can effectively improve the efficiency of condition monitoring and fault diagnosis while saving costs, thus improving the operational readiness and mission reliability of radar equipment. The indexes of testability design parameters (A_3) are test controllability (B_{31}), test observability (B_{32}), UUT compatibility with external test equipment (B_{33}), structure design and function division rationality (B_{34}), software testability (B_{35}), and test security (B_{36}).

Testability resource parameters (A_4) refer to the resources planned and designed to ensure the testing and use of radar equipment. Scientific configuration of testability resources can improve the mission success and fault detection isolation of radar systems and effectively reduce the cost and time consumption of testing and diagnosis. The indexes are the quantity and quality of testers (B_{41}), technical data integration rate (B_{42}), random test equipment common rate (B_{43}), random test equipment integration rate (B_{44}), and fault information integrity (B_{45}).

3. Radar Equipment Testability Extension Evaluation Model Based on Improved CRITIC Weighting

3.1. Improve the CRITIC method to determine index weight

The basic steps of using the improved CRITIC method to determine the testability index weight of radar equipment are as follows.

1) Construct the original data matrix and normalize it.

Assuming that there are m objects to be evaluated and n evaluation indexes, then y_{ij} ($i = 1,2,\cdots,n; j = 1,2,\cdots,m$) is the observed value of index i in the jth evaluation object, from which the original data matrix $\mathbf{Y} = (y_{ij})_{n\times m}$ is established. To obtain the normalized matrix $\mathbf{X} = (x_{ij})_{n\times m}$, different types of index data need to be normalized.

$$\begin{cases} x_{ij} = \frac{y_{ij}-\min(y_{ij})}{\max(y_{ij})-\min(y_{ij})} & Benefit\ index \\ x_{ij} = \frac{\max(y_{ij})-y_{ij}}{\max(y_{ij})-\min(y_{ij})} & Cost\ index \end{cases} \tag{1}$$

2) Calculate the coefficient of variation and conflict coefficient of the index.

The coefficient of variation s of the ith evaluation index is:

$$S_i = \frac{\sigma_i}{\bar{x}_i} = \frac{\sqrt{\frac{1}{m}\sum_{j=1}^{m}(x_{ij}-\bar{x}_i)^2}}{\bar{x}_i} \tag{2}$$

where $\bar{x}_i$ and σ_i are the mean and standard deviation of the ith index respectively.

The correlation coefficient r_{ik} between the indexes is:

$$r_{ik} = \frac{\mathrm{cov}(i,k)}{\sigma_i\sigma_k} = \frac{\sum_{j=1}^{m}(x_{ij}-\bar{x}_i)(x_{kj}-\bar{x}_k)}{\sqrt{\sum_{j=1}^{m}(x_{ij}-\bar{x}_i)^2}\sqrt{\sum_{j=1}^{m}(x_{kj}-\bar{x}_k)^2}} \tag{3}$$

where $\mathrm{cov}(i,k)$ is the covariance of index i and k.

Therefore, the conflict coefficient R of index i is further obtained.

$$R_i = \sum_{k=1}^{n}(1-|r_{ik}|), i = 1,2,\cdots,n \tag{4}$$

3) Determine the weight of each index.

$$w_i = \frac{S_iR_i}{\sum_{i=1}^{n}S_iR_i} \tag{5}$$

3.2. Improved Matter-Element Extension Evaluation Model

Extension is a new discipline founded by Professor CAI in 1983. Based on matter-element theory, extension studies the extensibility and conjugacy of things with formal models and is used to solve contradictory problems [8]. There are some limitations in the application of the matter-element extension evaluation model: first, when there are differences in the type and dimension of the evaluation indicators, and the measured data value of the indicators exceeds the section range, it is impossible to calculate the value of the associated evaluation levels. Second, the traditional Topological model is based on the maximum affiliation as the discriminant criterion, and the assessment levels are determined by the approximate processing [9], which sometimes results in the loss of some information about the object to be assessed and reduces the validity of the assessment results. The validity of the evaluation results is reduced. Therefore, this paper addresses these two shortcomings: (1) normalizing the classical domain, nodal domain, and the actual value of indexes in the assessment model; (2) replacing the maximum affiliation with the asymmetric closeness, which can effectively avoid the loss of information of index data that leads to the failure of the discriminatory principle.

The concrete steps to improve the matter-element extension evaluation model are as follows.

1) Determine the classical domain, section domain, and matter element to be evaluated.

The testability of the target radar is divided into m levels, then the jth evaluation level is represented as N_j, where $j = 1, 2, \cdots, m$; There are n testability evaluation indexes c_i, that is $c_1, c_2, \cdots, c_n$; v_{ji} is the range of values of evaluation index c_i in the jth evaluation level, represented by $\langle a_{j1}, b_{j1}\rangle$; R_j is the jth level of the radar equipment testability element model.

$$R_j = (N_j, C_i, V_{ji}) = \begin{bmatrix} N_j & c_1 & v_{j1} \\ & c_2 & v_{j2} \\ & \vdots & \vdots \\ & c_n & v_{jn} \end{bmatrix} = \begin{bmatrix} N_j & c_1 & \langle a_{j1}, b_{j1} \rangle \\ & c_2 & \langle a_{j2}, b_{j2} \rangle \\ & \vdots & \vdots \\ & c_n & \langle a_{jn}, b_{jn} \rangle \end{bmatrix} \tag{6}$$

Section domain represents the total range of quantity values of index c_i in all evaluation levels, and the matter-element section domain matrix R_p is:

$$R_p = (N_p, C_i, V_{pi}) = \begin{bmatrix} N_p & c_1 & v_{p1} \\ & c_2 & v_{p2} \\ & \vdots & \vdots \\ & c_n & v_{pn} \end{bmatrix} = \begin{bmatrix} N_j & c_1 & \langle a_{p1}, b_{p1} \rangle \\ & c_2 & \langle a_{p2}, b_{p2} \rangle \\ & \vdots & \vdots \\ & c_n & \langle a_{pn}, b_{pn} \rangle \end{bmatrix}. \tag{7}$$

The testability matter element matrix R_o of radar equipment to be evaluated is expressed as:

$$R_o = (N_o, C_i, V_i) = \begin{bmatrix} N_o & c_1 & v_1 \\ & c_2 & v_2 \\ & \vdots & \vdots \\ & c_n & v_n \end{bmatrix}. \tag{8}$$

2) Normalized processing.

The evaluation of each index of the classical domain, section domain, and matter element to be evaluated is processed without dimensionality, then the standardized formulas of the benefit index and cost index are as follows:

$$v'_i = \begin{cases} 1 & v_i \geq b_{pi} \\ \frac{v_i - a_{pi}}{b_{pi} - a_{pi}} & a_{pi} < v_i < b_{pi}, \\ 0 & v_i \leq a_{pi} \end{cases} \quad v'_i = \begin{cases} 1 & v_i \leq a_{pi} \\ \frac{b_{pi} - v_i}{b_{pi} - a_{pi}} & a_{pi} < v_i < b_{pi} \\ 0 & v_i \geq b_{pi} \end{cases} \tag{9}$$

where v_i is the magnitude value of each index c_i of the classical domain, segment domain, and subject element to be evaluated; a_{pi} and b_{pi} are respectively the lower limit and upper limit of the section range. The new classical domain matter element R'_j, node domain matter element R'_p and subject matter element R'_o can be obtained after standardization.

3) Calculate the closeness.

The closeness degree can effectively measure the closeness degree between fuzzy sets, so as to avoid the limitation of applying the maximum membership degree to comprehensive evaluation. After sorting out the asymmetric proximity formula ($p = 1$) [10], the proximity degree of testability matter element R'_o of radar equipment is to be evaluated with respect to each grade j can be obtained.

$$K_j(N_0) = 1 - \frac{1}{n(n+1)} \sum_{i=1}^{n} D_j\,(v'_i) w_i \tag{10}$$

$$D_j(v'_i) = \left| v'_i - \frac{a'_{ji} + b'_{ji}}{2} \right| - \frac{1}{2}(b'_{ji} - a'_{ji}) \tag{11}$$

where $D_j(v'_i)$ is the distance between the matter element R'_o to be evaluated and the canonical classical domain; w_i is the weight of the evaluation index.

4. Instance Analysis

Taking three types of radar equipment as an example, the testability evaluation index of radar equipment is weighted by the improved CRITIC method, and the testability evaluation is performed by the improved matter-element extension model. Based on the existing national and military standards for the testability of radar equipment [7], [11][12], and through a large number of surveys and expert consultation, the evaluation indexes were divided into four grades: **N**={N_1, N_2, N_3, N_4}= {excellent, good, medium, poor}, and the evaluation index grade standards were assigned values as shown in Table 1. Among them, the indexes B_{13}, B_{23}, B_{24}, B_{25}, B_{26}, and B_{27} are cost indexes, and the other indexes are benefit indexes.

Table 1. Radar equipment testability evaluation index grade standard.

Index	Excellent N_1	Good N_2	Medium N_3	poor N_4
B_{11} /%	[95,100]	[90,95)	[80,90)	[0,80)
B_{12} /%	[95,100]	[90,95)	[80,90)	[0,80)
B_{13} /min	(0,5]	(5,10]	(10,15]	(15,25]
B_{21} /%	[95,100]	[85,95)	[70,85)	[0,70)
B_{22} /%	[95,100]	[85,95)	[70,85)	[0,70)
B_{23} /%	[0,1]	(1,3]	(3,5]	(5,10]
B_{24} /min	(0,0.5]	(0.5,1]	(1,3	(3,5]
B_{25} /min	(0,1]	(1,3]	(3,5]	(5,10]
B_{26}	(0,1]	(1,3]	(3,5]	(5,8]
B_{27} /%	[0,1]	(1,3]	(3,5]	(5,10]
B_{28} /%	[95,100]	[85,95)	[70,85)	[0,70)
B_{31}	[9,10]	[8,9)	[6,8)	[0,6)
B_{32}	[9,10]	[8,9)	[6,8)	[0,6)
B_{33}	[9,10]	[8,9)	[6,8)	[0,6)
B_{34}	[9,10]	[8,9)	[6,8)	[0,6)
B_{35}	[9,10]	[8,9)	[6,8)	[0,6)
B_{36}	[9,10]	[8,9)	[6,8)	[0,6)
B_{41}	[9,10]	[8,9)	[6,8)	[0,6)
B_{42} /%	[95,100]	[80,95)	[60,80)	[0,60)
B_{43} /%	[95,100]	[80,95)	[60,80)	[0,60)
B_{44} /%	[95,100]	[85,95)	[70,85)	[0,70)
B_{45}	[9,10]	[8,9)	[6,8)	[0,6)

Through the testability test of the physical prototypes of three types of radar equipment, the testability evaluation Indicator values of the radar equipment of each scheme are obtained, as shown in Table 2. Among them, indexes B_{31}, B_{32}, B_{33}, B_{34}, B_{35}, B_{36}, B_{41}, and B_{45} are qualitative indexes, and their index values are the average value of the scores of 10 experts from equipment development units, manufacturers and radar units, etc. The scoring standards are shown in Table 1.

Table 2. Testability evaluation index values of certain 3 types of radar equipment.

	B_{11}	B_{12}	B_{13}	B_{21}	B_{22}	B_{23}	B_{24}	B_{25}	B_{26}	B_{27}	B_{28}	B_{31}	B_{32}	B_{33}	B_{34}	B_{35}	B_{36}	B_{41}	B_{42}	B_{43}	B_{44}	B_{45}
1#	97	97	10	95	96	2	0.8	2.5	3	3.5	89	8.8	9	9.1	9.5	8.6	8.6	8.8	87	86	86	8.9
2#	93	92	15	90	89	3.2	1.5	3.5	4	5	84	7.8	7.6	8.2	8.9	7.6	7.6	6.8	74	80	75	7.5
3#	94	93	12	88	88	3	1	3	3	4.5	90	8	7.9	9	9.2	8.8	8	8.5	80	85	88	8

4.1. Improve CRITIC Method to Determine Weight of Evaluation Index

According to the calculation steps of the improved CRITIC method, the weights of each index for radar equipment testability assessment were determined as **W**= (0.0412, 0.0490, 0.0412, 0.0490, 0.0218, 0.1193, 0.1052, 0.0551, 0.0252, 0.0222, 0.0447, 0.0315, 0.0632, 0.0490, 0.0466, 0.0357, 0.0222, 0.0632, 0.0263, 0.0329, 0.0232, 0.0318, 0.0614, 0.0294).

4.2. Improved Matter-element Extension Evaluation

The standard assignments of the radar equipment testability assessment levels and the actual measured values of the indexes in Table 1 and Table 2 are normalized according to Equation (9) to obtain the new classical domain R'_j, section domain element R'_p and the object to be evaluated element R'_o.

Taking the radar equipment numbered 2 as an example, the distance between the object to be evaluated and the normalized classical domain is calculated according to Equation (11), and then substituted into Equation (10) together with the weight values of each index to derive the closeness of this radar equipment to be evaluated testability to each grade as $K_j(N_0)$= (0.999 716, 0.999 945, 1.000 021, 0.999 694). From the formula $\max\{K_j(N_0)\} = K_3(N_0)$=1.000 021, it can be seen that the radar equipment testability evaluation grade of number 2 is medium; the same calculation process as above, it can be concluded that the radar equipment testability grades of numbers 1 and 3 are good. The measured data of each index of type 2 radar are mostly lower than the evaluation standard of good grade, so it is necessary to improve the testability design of this type of radar equipment to effectively improve the testability level, which is also consistent with the actual investigation results. The testability grades of the three types of radars obtained by the traditional matter-element extension method [13] are all good. The evaluation process ignores the fuzziness of the testability of the radar equipment to be evaluated, and some data information is lost, leading to inaccurate evaluation results. The evaluation results were obtained by using the improved matter-element extension model, and the reasonableness and applicability of the evaluation model were verified.

5. Conclusion

Based on the definition and requirements of radar equipment testability, in this paper, a combined qualitative and quantitative radar equipment testability index system is constructed. By introducing the coefficient of variation and calculating the conflict coefficient by taking the absolute value of the correlation coefficient, the traditional objective weighting method of CRITIC is improved to make the weighting result more accurate and scientific. Then the improved matter-element extension model is used to evaluate the testability of different types of radar equipment and distinguish the advantages and disadvantages. By standardizing the matter-element domain values, the method makes the evaluation model grade criteria more flexible and practical. The closeness is also used to effectively compensate for the shortcomings of the maximum affiliation to discriminate the assessment level, which makes the assessment results more accurate and reasonable. The application of the method can expose the deficiencies related to the testability of radar equipment, and provide a certain basis for equipment development units, manufacturers, and user units to optimize the testability design of radar equipment in a targeted manner and enhance its fault diagnosis capability and maintenance test efficiency.

References

[1] Du, X.S., Hu, B., Qin, J. (2021) Testability Analysis Method of Radar Equipment Based on Dependency Model. Journal of Physics: Conference Series, 2093 (1): 1-7.

[2] Wang, H.X., Ye, X.H, Chen, B., et al. (2018) The Technology on Testability Design Analysis and Verification of Equipment. Publishing House of Electronics Industry, Beijing.

[3] LIU, X.Y. (2022) Research on Six Indexes Design and Evaluation of Complex Products. Value Engineering, 41 (34): 32-34.

[4] Su, M., Zhang, B. (2017) Testability Evaluation Method for Aircraft PHM Systems Based on Field Data. Electronic Technology & Software Engineering, 114 (16): 186-187.

[5] Li, H.J., Xu T.X., Ying, X.Y. (2021) Missile Condition Assessment Decision Based on Test Data and Extended TOPSIS-Grey Correlation. Aero Weaponry, 28 (06): 88-94.

[6] DIAKOULAKI, D., MAVROTAS, G., PAPAYANNAKIS, L. (1995) Determining objective weights in multiple criteria problems: the critic method. Computers and Operations Research, 22 (7): 763-770.

[7] (2012) GJB 2547A-2012 General requirement for materiel testability program. Army Standards Press of General Armament Department, Beijing.

[8] Yang, C.Y. (2017) Extension Innovation Methods. Science Press, Beijing.
[9] LIU, Y., LIU, X.Y., LI, Y.F. (2022) Evaluation of Campus Emergency Management Capability Based on Improved Matter-Element Extension Model. Chinese Journal of Management Science, 30 (11): 299-308.
[10] Zhang, X.P. (2004) The definition of product about fuzzy comprehensive evaluation methods based on closeness. Journal of Shandong University, (02): 25-29.
[11] (2000) GJB 3970-2000 Testability requirements for military ground radar. Standards Press of China, Beijing.
[12] (2001) GJB 4260-2001 General requirements for testability of reconnaissance radar sets. Standards Press of China, Beijing.
[13] Liu, J.K., Huang, J., Wang, C.X. (2023) Evaluation of emergency management capability of chemical industry park based on improved combination weighted extension model. Journal of Safety and Environment, 23 (05): 1423-1430.

MCAI-2023 IOP Publishing
Journal of Physics: Conference Series 2724 (2024) 012050 doi:10.1088/1742-6596/2724/1/012050

Error state prediction method for electronic voltage transformer based on empirical mode decomposition and auto-regressive moving average

Binbin Li [1,2], Chen Liu [2,4] and Jie Huang [3]

[1]Electric Power Research Institute of State Grid Anhui Electric Power Co., Ltd., Hefei 230601, Anhui, China
[2]Hefei University of Technology, Hefei 230009, Anhui, China
[3]State Grid Anhui Electric Power Co., Ltd., Hefei 230022, Anhui, China

[4]Corresponding author's e-mail: 2832006747@qq.com

Abstract: Electronic voltage transformers have been used in the power system in conjunction with the development of power grid intelligence due to their good insulation structure, economy, volume, and other advantages. However, in its long-term operation, the phenomena of error overshooting and unstable operation will inevitably occur, and the existing periodic off-line testing methods require a high calibration cycle, which is difficult to achieve. To solve these problems, an error state prediction method based on empirical mode decomposition and autoregressive moving average model is proposed. Firstly, according to the physical correlation between the secondary output signal and the error, the output signal is quantified as a statistic Q by the method of principal component analysis, and then the statistic is used as the prediction object to establish a prediction model for predicting the operating state of the electronic voltage transformer. The simulation results show that the model can accurately predict the trend of the transformer error, which meets the prediction requirements.

1. Introduction

For the power system measurement, protection, and other secondary equipment, Electronic Voltage Transformers (EVT) can convert the primary side of the high voltage into a certain proportion of low-voltage and easily measurable instrumentation of small signals, to provide accurate voltage signals [1-2]. However, under the complex operating environment of smart substations, the long-term stability of the metering error of EVTs is poor and the phenomenon of error overshooting will occur, which will have an impact on power metering [3-4]. Therefore, to ensure the safe and stable operation of intelligent substations, it is necessary to research the metering error state detection and prediction technology of EVT.

The detection of measurement errors in electronic transformers is now generally performed by comparison with the standard [5-6]. Compared with the traditional electromagnetic transformer, EVT has a shorter calibration period, and it is difficult to achieve frequent non-faulty outages, resulting in the EVT exceeding the calibration period, which is a potential safety hazard [7].

Existing online detection methods are mainly classified into three categories: error detection methods based on signal processing, model analysis, and knowledge diagnosis, respectively. A method based on a wavelet singular entropy algorithm is used for online monitoring of the operating state of the transformer [8]. An online assessment of the error state is proposed based on factor analysis, realizing the abnormal positioning of the three-phase electronic voltage transformer through the contribution rate of the statistics [9]. However, this method can only be used for short-term error state detection, and cannot adapt to the changing environment in the long-term operation of the power grid. A method to establish a π-type equivalent circuit is proposed, but due to the complex operating environment of the power system, the generality of the mathematical model is not strong [10]. Corresponding assessment methods for different operating states are proposed based on knowledge diagnosis, but this method is more subjective and the assessment reliability is not enough [11]. Therefore, it is necessary to study a more accurate online self-testing method of EVT to ensure the normal operation of the power system.

In the actual operation of EVT, it is not only necessary to make a fast diagnosis when its metering error is abnormal, but furthermore, it is necessary to make a timely prediction of the deterioration trend of EVT metering error to reflect the long-term operation status of the transformer. Aiming at the shortcomings of existing methods, this paper proposes a method for predicting the error state of electronic voltage transformers based on principal component analysis and EMD-ARMA. By using the principal component analysis method, the characteristic statistic Q is extracted from the secondary output signals, and the prediction model is established with the statistic Q as the prediction object. The EMD method is used to decompose and reconstruct the nonlinear and non-stationary statistic Q time series, and then the ARMA prediction model is established to realize the prediction of the error state of EVT.

2. Selection of prediction object

Due to the dynamic equilibrium of the grid operation state, the three-phase voltage vector sum is zero, and there is a time-invariant physical relationship in the power system. The secondary output signal of the three-phase transformer will be bound by the physical relationship of the primary voltage and can be analyzed by the method of statistical analysis.

2.1. Construction of the statistic Q

Principal Component Analysis (PCA) is a commonly used multivariate statistical analysis method. When the metrological error state of the EVT is abnormal, the projection of the three-phase measurement data in the residual subspace will be shifted. The degree of deviation of the error state of the error in the EVT secondary measurement data can be determined by comparing the Q statistic with the statistical control limit Q in the residual space.

2.2. Empirical mode decomposition

Empirical Mode Decomposition (EMD) is widely used in the field of signal decomposition and processing. EMD does not require any pre-set basis function and can decompose the original signal into multiple Intrinsic Mode Function (IMF) components and residual terms according to its time scale. IMF has the characteristics of localization and adaptivity, which can effectively decompose non-linear and Unstable signals.

The decomposition process of EMD is as follows:

(1) For the input signal, we find the extreme value point and the minimal value point;

(2) We construct the upper and lower envelopes by interpolating the cubic spline function to the extreme value point and the minimal value point, and calculate the mean value function of the upper and lower envelopes;

(3) We examine whether the condition of the IMF is satisfied. If so, we will go to the next step, and otherwise, we will carry out the first two steps until the condition of IMF is satisfied. Then, first IMF is obtained;

(4) We repeat the above until the signal is monotonic or only one pole exists, and the decomposition is complete.

The statistical *Q* data is decomposed by EMD to obtain a set of IMF components $c_i\ (t)$ (i=1, 2,......., s) and a residual residue *r*(*t*), and each IMF component represents the vibration mode of the signal on different time scales. The high-frequency IMF components in the front will appear in the noise sequence, through the correlation coefficient between the IMF components and the original data to determine whether the decomposition of the IMF components is valid or not. The correlation coefficient for the first time to obtain the smallest value of the corresponding component is the bounding IMF function, and then we reconstruct the low-frequency IMF components and residuals of the bounding to get the smoother statistical quantities of the sequence.

3. Prediction methodology

3.1. ARMA modelling

ARMA is an important method for studying the time series, and it is a prediction model based on the autoregressive moving average (AMAR) model.

ARMA prediction needs to go through the steps of model type identification, model ordering, parameter estimation, and residual series autocorrelation tests. The model type can be determined based on the nature of its autocorrelation and partial correlation coefficients, where *p* is the order of the auto-regressive model, and *q* is the order of the moving average model. The model order is often determined by the Akaike information criterion (AIC). The smaller the value of the information criterion is, the better the ARMA model is for the corresponding (p, q)-order. The model parameters are estimated based on the historical state data of the time series, which can be directly estimated by using the least-squares method. Finally, the obtained model is subjected to the DW (Durbin-Watson) statistic to detect the autocorrelation of the residual series and to verify the reasonableness of the model.

3.2. EMD-ARMA prediction model

The ARMA model is based on a smooth data series, so the Q-statistics need to be processed before prediction. After EMD decomposition, the complexity of the series can be reduced and the prediction accuracy can be improved. The EMD-ARMA prediction model proposed in this paper includes the following five steps:

(1) We collect data on the normal operation of the EVT, establish a PCA model, and quantify it as the statistic Q;

(2) We build an EMD model to decompose the statistics and test the smoothness of each disintegrated IMF component and residual term. If the sample series is a non-smooth time series, the series shall be processed until the data series becomes smooth;

(3) We select the effective IMF components and reconstruct them to obtain a smooth new signal;

(4) We determine the model order of the time series and each model parameter;

(5) We establish the ARMA model for the reconstructed statistic signals and make predictions to obtain the predicted value of statistic Q.

3.3. Predictive evaluation indicators

Root Mean Squared Error (RMSE) and Mean Absolute Error (MAE) were chosen as the basis for discriminating the prediction accuracy, and the RMSE and MAE were calculated as follows:

$$RMSE=\sqrt{\frac{1}{n}\sum_{i=1}^{n}(y_{pre}-y_{real})^2} \qquad MAE=\frac{1}{n}\sum_{i=1}^{n}|y_{pre}-y_{real}| \tag{1}$$

where n is the number of predicted statistics, y_{pre} is the predicted value of the statistic, and y_{real} is the true value of the statistic.

4. Simulation analysis

4.1. PCA-based EVT error state assessment

To truly reflect the fluctuation information of the primary voltage, the 110 kV three-phase EVT secondary output data is collected to obtain a data set of 3, 000×3. The PCA model is built by using MATLAB to analyze the secondary output information of the EVT and obtain the Q statistics under normal operating conditions, as shown in Figure 1 (a). We set the A-phase deviation in the EVT as 0.2% to simulate the change of the error state of the EVT and evaluate the error state, and the simulation results are shown in Figure 1 (b). The total number of abnormal points exceeding the statistic threshold in the normal operation state of three-phase EVT is 23, and the percentage of abnormal data is 2.3%. The control limit corresponding to the statistic Q is 0.1513. The total number of abnormal points exceeding the statistic threshold in the error state of the three-phase EVT is 1, 965, and the proportion of abnormal data is 98.25%. The proportion of abnormal data increases from 2.3% to 98.25% in the case of the error of 0.2% for the EVT. The simulation results illustrate that when the operating state of the three-phase EVT is changed, the secondary output data analyzed by the PCA model can reflect the change of the error state of the EVT for effective state assessment.

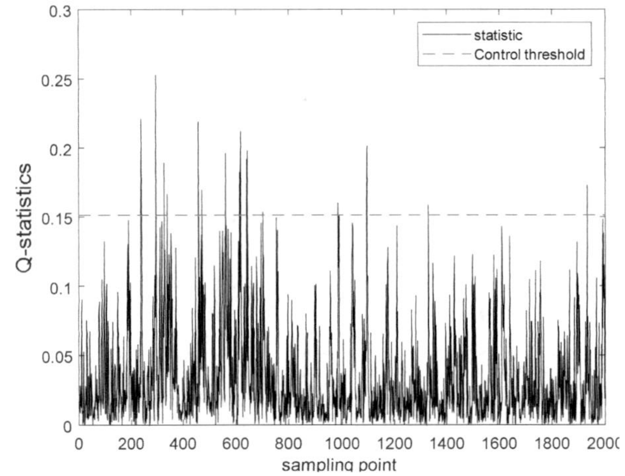

(a)

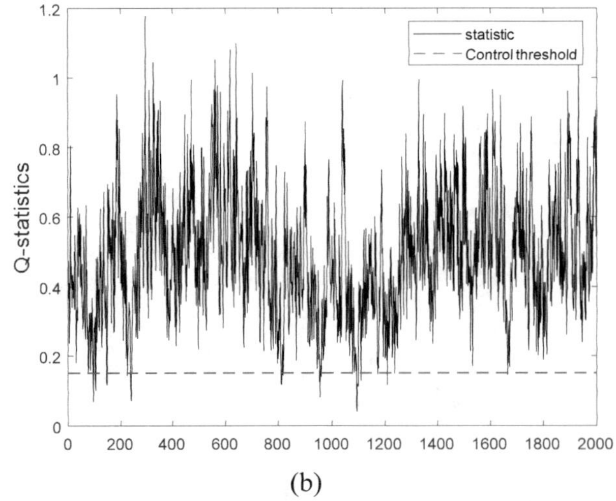

(b)

Figure 1. Statistic monitoring graph in normal state and error state.

4.2. EMD-ARMA prediction results

4.2.1. EMD-ARMA model prediction results under normal operating conditions. The establishment of a time series prediction model needs to accurately grasp its internal relationship and set appropriate parameters. Due to the non-stationarity and non-linearity of the Q-statistic series, it is not suitable to build the model directly for prediction. Therefore, the EMD model is used to decompose the Q-statistics data, remove the invalid components to reconstruct them, and then build an ARMA prediction model for the reconstructed signals to make predictions.

The first 1, 000 data sets of the data were selected as the training group for model training, and the last 1, 000 data sets were predicted. The autoregressive and sliding average model orders were selected based on the AIC values at different orders, as shown in Table 1. According to the autocorrelation and partial autocorrelation coefficients and the AIC values at each order, the autoregressive model order p of the ARMA model was selected as 4, the moving average model order q was selected as 3, and the model DW value was 2.05.

Table 1. Model parameter setting.

Model order		MA model					
		Phase 1	Phase 2	Phase 3	Phase 4	Phase 5	Phase 6
AR model		NaN	-15, 284.8	-17, 406.5	-18, 344.6	-18, 508.9	-18, 511.4
	Phase 1	-15, 424.6	-17, 673.9	-19, 082.7	-19, 690.7	-19, 869.2	-19, 575.8
	Phase 2	-19, 149.9	-18, 997.6	-20, 448.2	-19, 998.0	-20, 457.3	-20, 452.7
	Phase 3	-20, 152.1	-20, 460.3	-20, 410.3	-20, 432.7	-20, 448.0	-20, 556.3
	Phase 4	-20, 467.7	-20, 432.9	-20, 421.3	-20, 571.5	-20, 472.4	-20, 491.2
	Phase5	-20, 462.2	-20, 454.6	-20, 557.8	-20, 506.3	-20, 436.0	-20, 482.4

The final obtained statistic prediction map is shown in Figure 2. As can be seen from Figure 2 (a), the EMD-ARMA model can predict the Q-statistics data series of EVT under normal operation conditions with more accuracy. The prediction evaluation index RMSE is 0.13% and MAE is 0.08%.

The predicted values of the Q-statistics data series are compared with the statistic thresholds to determine the operational status of the EVT. The monitoring graph of the predicted value of the statistic in the normal operation state is shown in Figure 2 (b). The total number of abnormal points exceeding the statistic threshold is 0, and the EVT is in the normal operation state, which is consistent with the actual operation state.

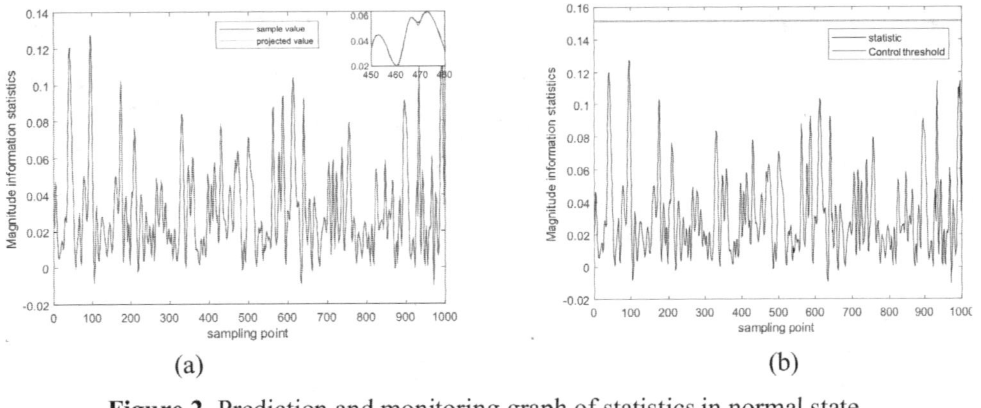

(a) (b)

Figure 2. Prediction and monitoring graph of statistics in normal state.

4.2.2. EMD-ARMA model prediction results under error states. Similarly, the Q-statistic data in the 0.2% error state is decomposed by using the EMD model.

After the EMD decomposition was completed, each IMF component and residual residue term obtained from the decomposition was reconstructed to build an ARMA prediction model for prediction. According to the AIC values at different orders, the autoregressive model order p of the ARMA model is selected as 5, the moving average model order q is 5, and the model DW value is 2.0084. The first 1, 000 data sets of the data are selected as the training group for training, and the prediction is carried out on the last 1, 000 data sets to obtain the predicted value of the Q-statistic under the error state of 0.2%. The Q statistic data series prediction graph obtained through simulation experiments is shown in Figure 3. As can be seen from Figure 3 (a), the EMD-ARMA model can predict the trend of the statistic more accurately. The prediction evaluation index RMSE is 0.48% and MAE is 0.26%.

Comparing the predicted value of statistics in the error state with the threshold value, the monitoring graph is shown in Figure 3 (b). The total number of abnormal points exceeding the statistics threshold is 1, 000, and the percentage of abnormal points is as high as 100%, so it can be judged that the EVT is in an abnormal operation state, which is consistent with the actual operation state.

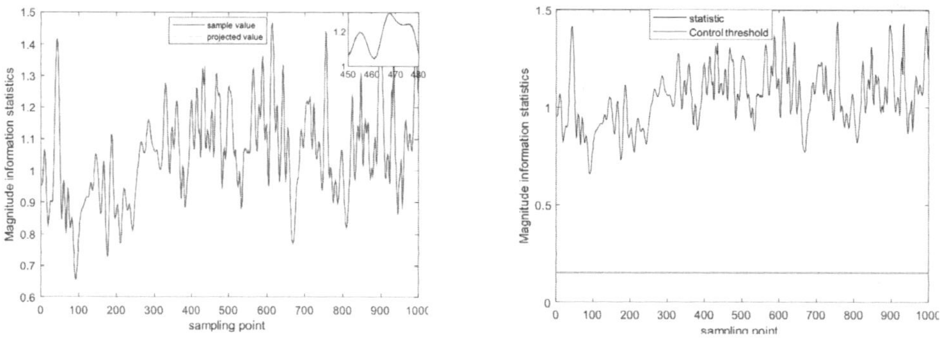

(a) (b)

Figure 3. Prediction and monitoring graph of statistics in the error state of 0.2%.

5. Conclusion

In this paper, a study is conducted on the online detection of metering error and prediction of the operating state of electronic voltage transformers (EVTs). By analyzing the secondary output data of three-phase EVTs, the signals characterizing the metering error are separated and quantified as Q statistics by the method of principal component analysis. To predict the trend of EVT metering error, an EMD-ARMA-based EVT error state prediction model was established with the Q statistic as the prediction object, and the ARMA model predicted the data after determining the smoothness of the data series. Because the Q statistic data is a nonlinear and non-smooth complex time series, it is difficult to guarantee prediction accuracy by direct prediction, so the EMD-ARMA model is proposed. Firstly, we decompose the data and then select the appropriate components to reconstruct, and finally, the reconstructed signals are used as the new statistical data to make the prediction. The simulation and comparison results show that the EMD-ARMA model can more accurately predict the trend of the electronic voltage transformer error state, and have a positive impact on the smooth operation of intelligent transformer stations.

Acknowledgments

This work was supported in part by the State Grid Science and Technology Foundation of China under Grant (52120523000K).

References

[1] Pei Y. Y. (2023). Application of electronic voltage transformer in intelligent substation. J. Electronic Components and Information Technology, 7 (02): 85-88. DOI: 10.19772/j. cnki.2096-4455. 2023.2.020.

[2] Li Q. Y., Yin D., Liu X. G., et al. (2022). Modeling and simulation analysis of electronic voltage transformer. J. Shandong Power Technology., 49 (10): 44-52. https://kns.cnki.net/kcms/deta il/detail.aspx? FileName=SDDJ202210007&DbName=CJFQ2022.

[3] Zhang L. H., Xu Z. H., Zhang G. X., et al. (2012). Design and implementation of electronic transformer calibration device based on virtual instrument technology. J. Electrical Measurement and Instrumentation, 49 (4): 31-33. https://kns.cnki.net/kcms/detail/detail. aspx?FileName=DCYQ201204010&DbName=CJFQ2012.

[4] Zhang M. Z., Li K. C., and Li Z. X. (2012). Key technologies affecting the design accuracy of electronic transformer calibration system. J. Electrical Measurement and Instrumentation, 49 (12): 41-45. https://kns.cnki.net/kcms/detail/detail.aspx?FileName=DCYQ201212011& DbName=CJFQ2012.

[5] Ma Y. Y. and Huang M. (2010). Development of digital output electronic transformer calibration system. J. Power System Protection and Control, 38 (01): 83-86. https://kns.cnki.net/kcms/ detail/detail.aspx?FileName=JDQW201001024&DbName=CJFQ2010.

[6] Yu C. Y., Ye G. X., Wang X. Q., et al. (2004). Calibration methods and techniques for electronic transformers. J. High Voltage Technology, (04): 20-24. DOI: 10.13336/j.1003-652 0.hve.2004.04.009.

[7] Zhang Z., Chen Q. Hu C., et al. (2018). Evaluation of the metering error of electronic transformers online based on VN-MWPCA. J. Measurement, 130: 1-7. https://kns.cnki.net/kcms/detail/detail.aspx?FileName=SJESD17B7DA6DF111E5E3CAEC1 22171CDC67&DbName=GARJ2018.

[8] Luo C., Liu Y. H., Xiao W., et al. (2015). Research on wavelet singular entropy algorithm for diagnosing electronic transformer fault. J. Value Engineering, 34 (32): 204-207. DOI: 10.14018/j.cnki.cn13-1085/n.2015.32.079.

[9] Shi Y. J., Yang M. Z., Yun Z. X., et al. (2023). Online assessment of electronic voltage transformer error state based on factor analysis. J. Modern Electronic Technology, 46 (06): 123-128. DOI: 10.16652/j.issn.1004-373x.2023.06.022.

[10] Anamitra P., Paroma C., James S. T., et al. (2016). Online Calibration of Voltage Transformers Using Synchrophasor Measurements. J. IEEE Transactions on Power Delivery, 31 (1): 370-80. https://kns.cnki.net/kcms/detail/detail.aspx?FileName=NSTL4DD5D99F0A79D17AA CFD712E6DDD78D5&DbName=GARJ2016.

[11] Cheng Y. Y., Wu H., Yang H. X., et al. (2012). Research on fuzzy comprehensive assessment and inspection strategy of electric energy metering device. J. Electrical Measurement and Instrumentation, 49 (564): 1-6. https://kns.cnki.net/kcms/detail/detail.aspx?FileName= DCYQ201212003&DbName=CJFQ2012.

Circuit design of a high scanning frequence and fast response laser methane remote sensor

Yuanjin Yang[1,4], Qi Wu[1], Yang Xu[1], Wenlong Wang[2], Chao men[2] and Bingxiong Yang[1,3]

[1]Nanjing University of Aeronautics and Astronautics, Nanjing, China
[2]Nanjing Puguang Chip Technology Co., Ltd, Nanjing, China
[3]Jiangsu Industrial Technology Research Institute, Nanjing, China

[4]Corresponding author's e-mail: yyj0213@nuaa.edu.cn

Abstract. To solve the problem of slow detection of laser methane sensors based on tunable diode laser absorption spectroscopy (TDLAS), a methane detection system with a high detection rate was proposed. Focusing on the hardware circuit design of the system which consists of a laser driver circuit, laser temperature control circuit, optical amplifier and filter circuit and main control circuit based on FPGA+ARM, this circuit is designed to achieve a laser scanning frequency of 50 kHz. A voltage-controlled current source is designed to drive the laser diode (LD). A TEC controller is designed to keep the operating temperature of the laser diode stable. Since the photocurrent can be very weak, to ensure the quality of signals from the photodiode (PD), a signal collection circuit consisting of the transimpedance amplifier, gain-controlled amplifier and low-pass filter was designed. An experiment is carried out to test system measurement capabilities. The results show the circuit can reach a response time of 2.5 ms while ensuring a measurement error of less than 4.3%.

1. Introduction

As an energy gas, methane is widely used in various fields, but it can be harmful as a flammable and explosive gas and one of the major greenhouse gases, which make it important to monitor methane in a certain situation [1]. For now, methane sensors mainly employ catalyst combustion, electrochemical, or semiconductor, which has some disadvantages such as slow response speed, short lifespan and low measuring range [2]. As a new type of methane sensor, tunable diode laser absorption spectroscopy (TDLAS) is an effective approach for the concentration of gases based on optics, avoiding the drawbacks of traditional methane sensors. TDLAS has the characteristics of high measurement accuracy, fast response speed, non-contact measurement and strong resistance to interference with non-target gases, which now is widely used in the fields of environmental protection and safety monitoring [3-4].

Due to the good propagation characteristics of the laser, TDLAS technology can achieve a good application in the remote measurement of methane. With the development of laser technology, much research on methane remote sensor systems has been proposed in recent years. A large number of devices based on TDLAS have come out. However, most of the research focuses on improving the measurement distance and accuracy. Therefore the existing detection systems still have the disadvantage of low scanning frequency which is generally below 10 KHz, which limits the detection rate of the system [5-8]. In practical applications, for example, on-vehicle sensors, the laser spot can move very fast on reflectors, but the existing systems make it difficult to meet the high-speed gas detection needs. In order to improve the detection speed of the system, a new circuit was designed. It can achieve a laser scanning frequency of 50 kHz, which greatly improves the detection speed and broadens the application scene of the laser methane detection system.

2. Principle of as detection

TDLAS technology is based on Lambert-Beer law. When a laser beam of a certain wavelength passes through a uniform gas, its intensity decreases due to the absorption of the laser light by the gas molecules. The Lambert-Beer law quantifies the change in intensity of monochromatic light before and after it passes through the gas [9-10].

$$I_0 = I_t e^{\left[-a(v)CL\right]} \tag{1}$$

where I_0 is the initial light intensity before passing through the light. It is the intensity of the laser after passing through the gas. v is the laser emission frequency. C is the concentration of the gas to be measured. L is the path length of the laser through the gas to be measured. a(v) is the absorption coefficient of the gas to be measured and it can be expressed as follows.

$$a(v) = PS(T)\varphi(v) \tag{2}$$

where P is the gas pressure; S(T) is the intensity of characteristic spectral lines, which is determined by gas temperature; $\phi(\upsilon)$ is a line function, which is related to the gas temperature and pressure. Its integral in the frequency domain is 1. The gas integral concentration can be expressed as:

$$CL = \frac{\dfrac{\ln I_0}{I_t}}{PS(T)\varphi(v)} \tag{3}$$

With temperature T, gas pressure P and the intensity of the gas absorption spectrum S (T), the gas integral concentration to be measured can be calculated. The schematic of the detection principles is shown in Figure 1.

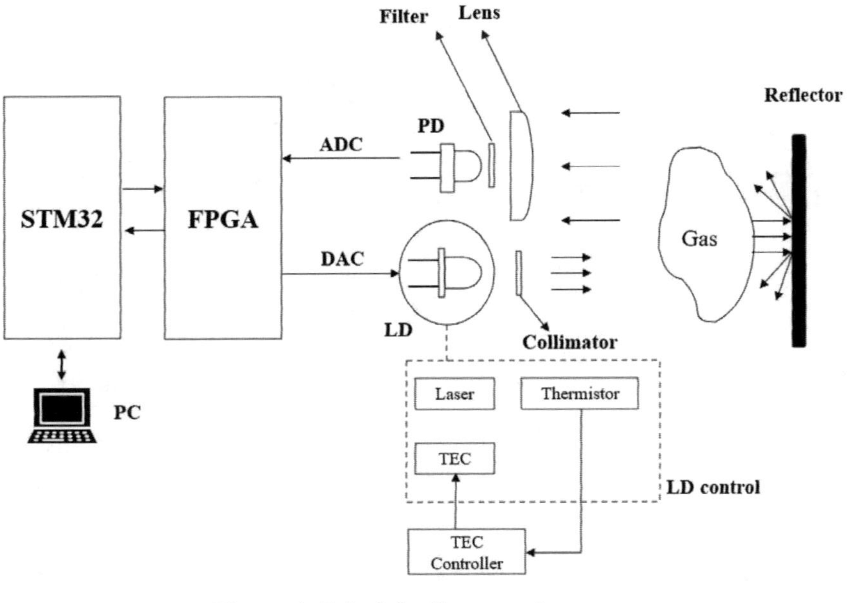

Figure 1. Principle diagram of sensor.

For laser detection of methane, the absorption intensity near the wavelength of 1653.7 nm is large and the rest of the interference gases such as H_2O and CO_2 in this band have much smaller absorption intensity than the methane, so the impact on methane detection can be ignored [11]. Therefore this circuit is designed based on a 1653 nm laser.

3. Hardware circuit design of the system

The main functions of the circuit are to drive the laser and control its current, amplify and filter the photocurrent signal from PD, collect it and transmit the concentration data to the host computer. The circuit needs to drive the laser to emit a sawtooth wave with a frequency of 50 kHz. High-speed DAC and ADC are used to complete the emission and collection of signals. Due to the requirement of a high detection rate, the circuit adopts FPGA (AG1280) as the main controller and ARM (STM32G071) as the assistant controller. FPGA is used to control the high-speed DAC and ADC and sends the raw data to ARM. The data is digitally filtered by ARM. FPGA and ARM are communicated via SPI.

3.1. Circuit design of power supply

As a measurement system, it requires a low power supply ripple and noise. According to the power supply requirements of the loads of the circuit, the system power supply structure is designed, as shown in Figure 2. An LDO regulator AP7361C is used to make a pretreatment to the power supply. The laser diode requires a large and stable driving current. A high-performance, 300mA LDO regulator VRH3601NTX is used. The operational amplifiers of the amplifying and filtering circuit require both positive and negative power supply, so a low noise power supply LTC3260 chip is used, which includes an inverting charge pump with both positive and negative LDO regulators. The ADC and DAC need a large operating current and are supplied by ADP123. A dual DC/DC converter TPS62410 which provides two independent output voltage rails is used to power the FPGA and MCU (channel 1) and TEC controller (channel 2). The core voltage for FPGA is supplied by a 1.2 V LDO LP3986.

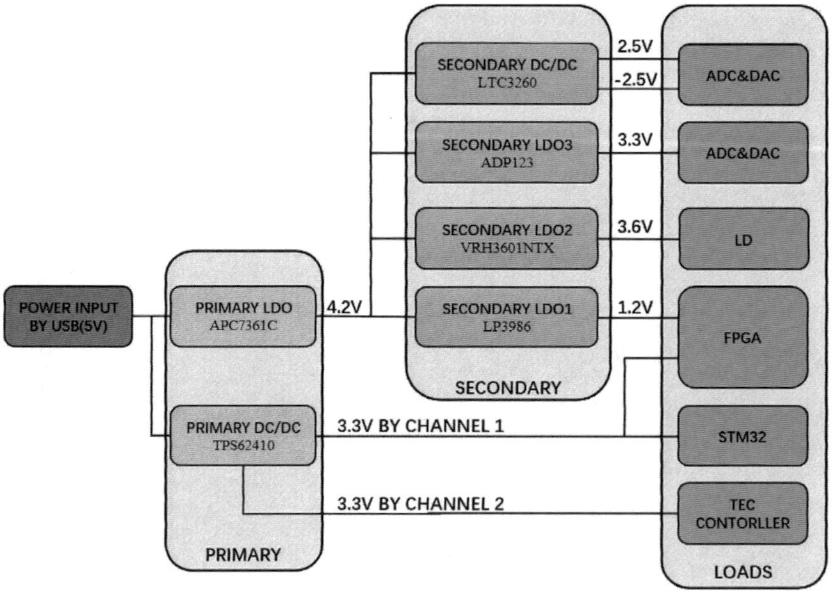

Figure 2. Power supply structure.

3.2. Circuit design of LD driving

The circuit drives the current of LD as a sawtooth waveform with a frequency of 50 kHz. The accuracy of the sawtooth waveform will directly affect the signal quality and detection accuracy. An FPGA is utilized to generate voltage waveforms through a high-performance DAC 3PD5651E, which has a 10-

bit resolution and a maximum sampling rate of up to 125 MSPS. The sawtooth generated by the DAC cannot directly drive the LD. Therefore, a voltage-controlled current source is used to drive the laser diode, which enables precise control of the current. The op-amp RS8751 with low input offset voltage is used. It has a low-input bias current and low noise, making it no influence impact on the signal. The schematic of the laser diode driving circuit is shown in Figure 3.
The driving current can be expressed as:

$$I_{LD} = \frac{V_{Set}}{R_7} \tag{4}$$

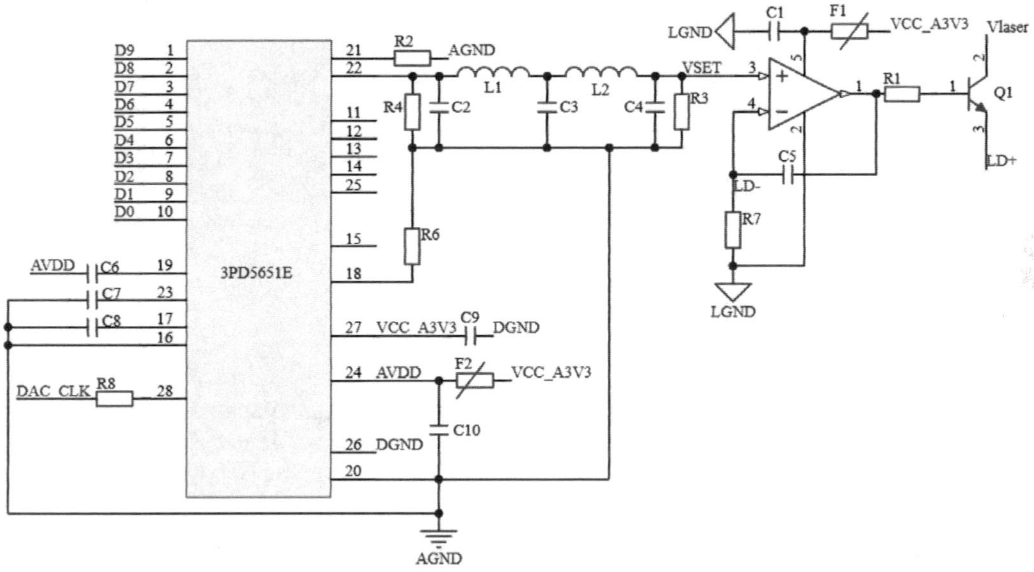

Figure 3. LD driving circuit.

3.3. *Circuit design of LD temperature control*

The wavelength of light emitted by the laser diode is highly affected by temperature, which can cause wavelength shifts. Therefore it is necessary to ensure that its operating temperature is stable. A Thermoelectric cooler (TEC) is used to regulate its temperature. By controlling the magnitude of current flowing through the TEC and controlling its cooling and heating power, the LD used in this circuit has an internal integrated thermistor and TEC. In order to ensure the precision of temperature control, a monolithic TEC controller ADN8834 is used, which has a small package, high-efficiency architecture, and two zero-drift, rail-to-rail op-amps integrated. ADN8834 thermal feedback loop can be constructed through the internal integration op-amps. The chopper 1 amplifier receives the feedback voltage from the thermistor, converts or regulates the input to a linear voltage output, and inputs it into the chopper 2 amplifier to compare with a temperature setpoint voltage. The Chopper 2 amplifier is used as the PID compensation amplifier. The ADN8834 achieves temperature stabilization through the TEC driving current control. The schematic of the TEC controller is shown in Figure 4.

3.4. *Circuit design of amplifier and filter*

The signal from PD is a current signal, which needs to be converted into a voltage signal that can be collected. Since the received photocurrent signal may be as low as tens of nanoamps, a large gain is required. The circuit needs to have a low noise to ensure signal accuracy. A double-stage amplifier circuit with a transimpedance amplifier and a controlled gain amplifier is used. A second-order low-pass filter is designed to eliminate high-frequency noise.

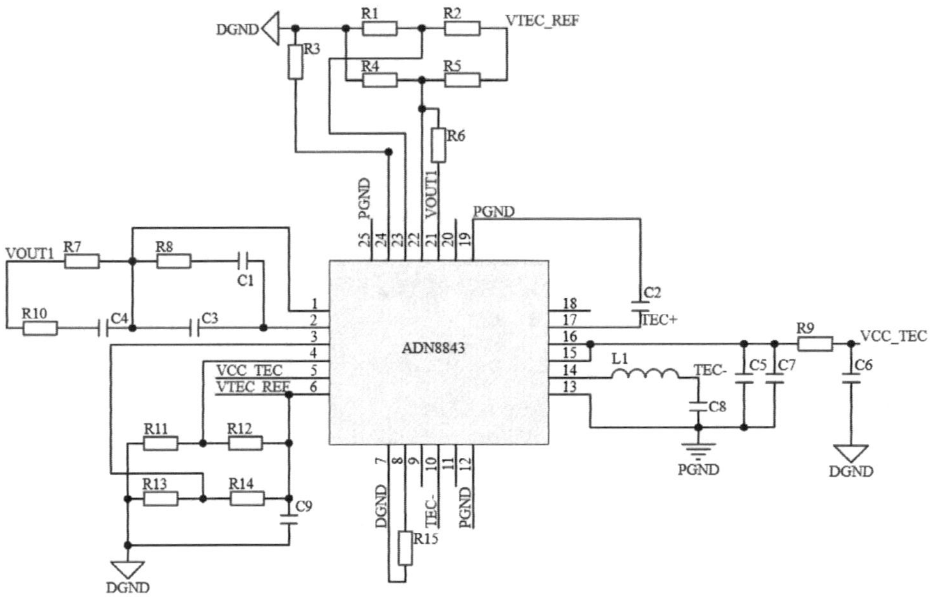

Figure 4. TEC controller circuit.

3.4.1. Transimpedance amplifier (TIA) circuit. TIA amplifies the photocurrent signal as primary amplification and completes the current-voltage conversion. According to the design requirements, the signal frequency is 50 kHz. The op-amp needs to have a high bandwidth. Besides, the signal can be extremely weak. The circuit noise is strictly required. A High-speed precision op-amp with a 500 MHz bandwidth is used as the TIA op-amp, which has a very low input bias current and noise. It is suitable for the amplification circuit of weak photocurrent signals. The gain of TIA is 10^6, the output voltage can be expressed as:

$$V_{PD\_Out1} = I_{PD} \bullet R_1 \tag{5}$$

The circuit schematic is shown in Figure 5 as PART A.

3.4.2. Gain-controlled amplifier circuit. In practical applications, due to the different types and distances of reflectors, the strength of the received signals may have a large difference, so it is necessary to adjust the gain of the signal in real-time. A digital potentiometer MCP4018 and an op-amp RS8752XM are used to build a gain-controlled amplifier circuit. The MCP4018 has a maximum resistance of 10 KΩ with 128 steps selectable. The RS8752XM combines low power, low noise, high speed and high precision. The schematic is shown in Figure 5 as PART B. The value of the digital potentiometer and resistor R_2 determines the gain. The ARM makes real-time adjustments on the digital potentiometer via I^2C according to the received signal. The gain can be expressed as:

$$V_{PD\_Out2} = V_{PD\_Out1}\left(\frac{128R_2}{n \bullet 10^4} + 1\right) \tag{6}$$

where n is 1-128 controlled by the program.

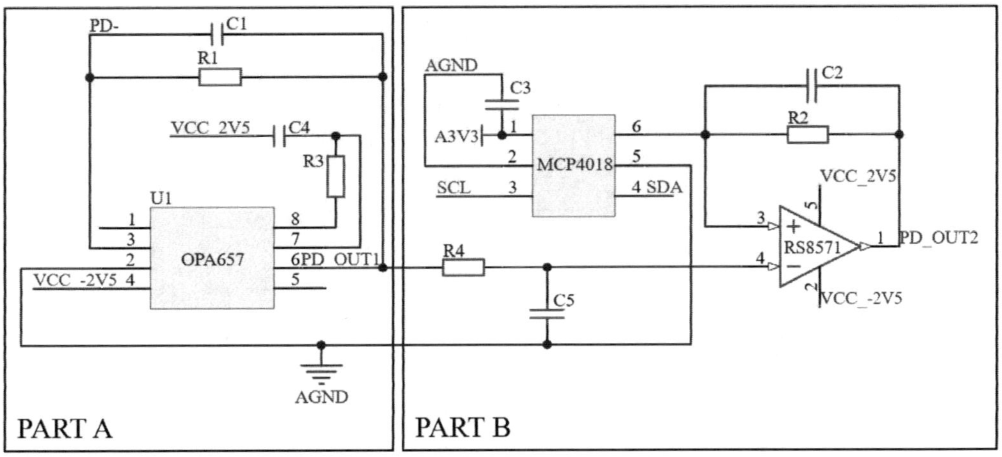

Figure 5. TIA and Gain-controlled amplifier circuit.

3.4.3. Filter circuit. The received signal will contain a variety of optical noise, mechanical noise and circuit noise. These noises generally have a high frequency, which is greatly harmful to the signal. Therefore, a second-order low-pass filter is designed to eliminate the effects of these high-frequency noises. The schematic is shown in Figure 6. The cut-off frequency is decided by R_8, R_9, C_2 and C_3. According to the signal frequency of 50 kHz, by setting R_8 = 1 K, R_9 = 9.2 K, and $C_2 = C_3$ = 100 pF, the cut-off frequency can be calculated.

$$f_c = \frac{1}{2\pi\sqrt{R_8 \bullet R_9 \bullet C_2 \bullet C_3}} = 524718Hz \tag{7}$$

The gain of the filter is

$$GAIN = \frac{R_4}{R_3} + 1 \tag{8}$$

Besides, an adder is to superimpose a 1V reference voltage to the signal to ensure the integrity of the converted signal.

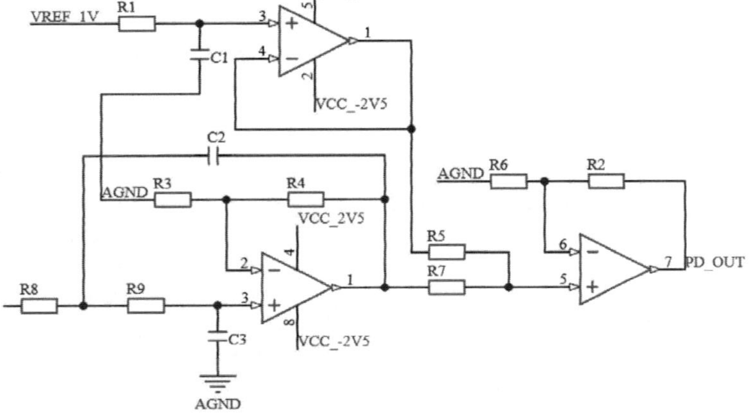

Figure 6. TIA and Gain-controlled amplifier circuit.

3.5. Circuit design of high-speed ADC sampling

In order to meet the requirements of high-speed and accurate acquisition of the circuit, a high-speed analog-to-digital converter MS9280 is used, which provides 10-bit accuracy at 50 MSPS data rates and guarantees no missing codes over the full operating temperature range. The voltage data collected by the MS9280 is transmitted in parallel to the FPGA, which receives and processes the data and transmits it to the ARM for concentration calculation. The circuit schematic is shown in Figure 7.

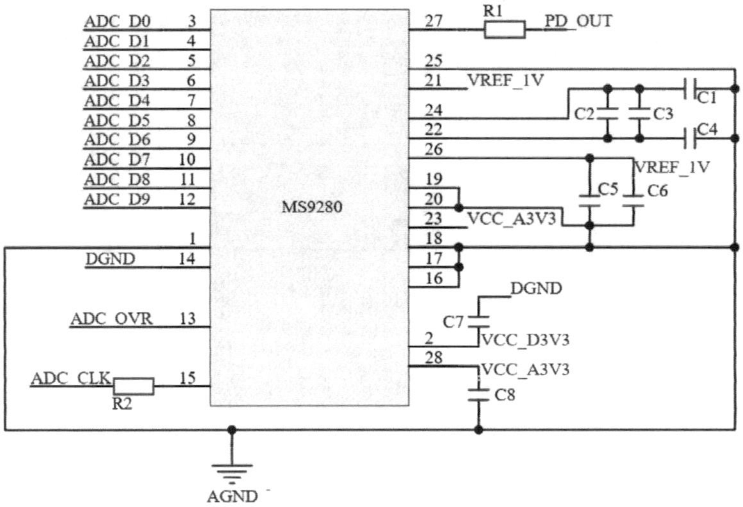

Figure 7. ADC sampling circuit.

4. Measurement results and processing

PD, LD and their optical components are connected to the circuit board for experimentation. The reflector is set as ordinary printing paper at a distance of 10 m. Gas clusters are modeled by using N_2 and CH_4 gas mixtures in gas bags including 2000 ppm·m, 4000 ppm·m and 6000 ppm·m. The gas bag is fixed. The detection system is rotated so that the light path can sweep through the gas bag at a uniform speed. The speed is 9.5 m/s. The width of the gas bag is 0.4 m. The time for the optical path to pass through the gas bag is 42 ms. The measured data are shown in Figure 8.

As the system measures the concentration in the area covered by the laser beam spot when the gas bag enters and leaves the optical path, the measured concentration will slowly rise and fall for the light spot cannot fully cover the airbag. Taking the data when the measured value is stable, it can be seen that the measurement errors of the system at 2000 ppm·m, 4000 ppm·m and 6000 ppm·m are 4.3%, 3.5% and 3.2%. The measurement points are set where the measured concentration is greater than 90% of the target concentration to be the valid points. It takes a response time of less than 2.5 ms to rise from 10% to 90% of the target concentration.

5. Conclusion

In this paper, given the slow detection rate of the current methane telemetry system, the circuit design of a methane detection system with a 50 kHz laser scanning frequency was completed. The use of a high-precision DAC and voltage-controlled current source achieves a high-precision laser current driving. The TIA circuit, controlled gain amplification and low-pass filter are designed. The sampling is completed by a high-precision ADC to ensure the accuracy of the signal. The use of FPGA and ARM meets the requirements of the system for high-speed acquisition and processing of signals. The circuit designed can achieve a high detection accuracy and response speed and also can be applied to situations with high-speed detection needs.

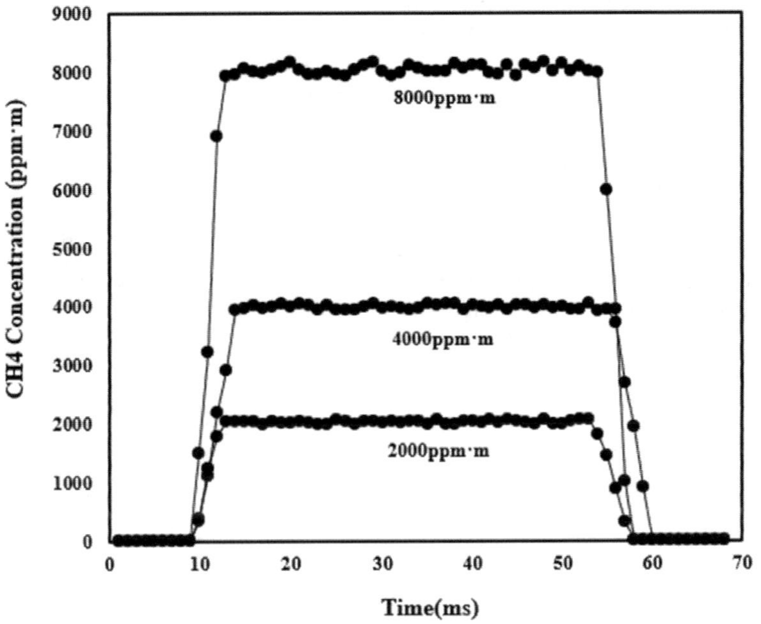

Figure 8. Measured data of response time.

References

[1] Wang, R., Peng, J., Cao, Y., Mei, J., Wang, G., Liu, K., Chen, W., & Gao, X. (2023). Double-enhanced multipass cell-based wavelength modulation spectroscopy CH4 sensor for ecological applications. Optics Express, 31(2), 3237-3248. https://doi.org/10.1364/OE.480496.

[2] Shuk, P., Mcguire, C.M., & Brosha, E.L. (2019). Methane Gas Sensing Technologies in Combustion: Comprehensive Review. Sensors & Transducers, 229(1), 1-10. https://www.osti.gov/servlets/purl/1503194.

[3] Dong, L., Tittel, F. K., Li, C., Sanchez, N. P., Wu, H., Zheng, C., Yu, Y., Sampaolo, A., & Griffin, R. J. (2016). Compact TDLAS-based sensor design using interband cascade lasers for mid-IR trace gas sensing. Optics express, 24(6), A528-A535. https://doi.org/10.1364/OE.24.00A528.

[4] Wang, B., Lu, H., Li, A., Chen, Y., Dai, T., Huang, S., & Lian, H. (2020). Research of TDLAS methane detection system using VCSEL laser as the light source. Hongwai Yu Jiguang Gongcheng/Infrared and Laser Engineering, 49(4), 0405002. https://doi.org/10.3788/IRLA202049.0405002.

[5] Zhang, T., Hu, J., Wei, Y., Li, Y., Wang, Z., & Liu, T. (2020). High-precision synchronous detection method for multi-gas using the least square fitting. In: Applied Optics and Photonics. Xiamen. 252013. https://doi.org/10.1117/12.2579816.

[6] Li, G., Ma, K., Jiao, Y., Jiang, Q., Zhang, X., Zhang, Z., Wu, Y., & Song, D. (2020). Performance enhancement of DFBL-based near-infrared CH4 telemetry system using a focus tunable lens. Microwave and Optical Technology Letters, 63, 1147 - 1151. https://doi.org/10.1002/mop.32733.

[7] Liang, W., Wei, G., He, A., & Shen, H. (2021). A novel wavelength modulation spectroscopy in TDLAS. Infrared Physics & Technology, 114, 103661. https://doi.org/10.1016/J.INFRARED.2021.103661.

[8] Li, G., Song, Y., Zhao, H., Dong, E., Li, J., Jia, L., Liu, Y., Zhang, S., & Zhao, F. (2023). A near-infrared multi-gas sensor based on IWTD-CEEMDAN and WOA-BiLSTM for detection of

CH4leaked in industrial production. Infrared Physics and Technology, 131, 104695. https://doi.org/10.1016/j.infrared.2023.104695.

[9] Liang, W., Wei, G., He, A., & Shen, H. (2021). A novel wavelength modulation spectroscopy in TDLAS. Infrared Physics & Technology, 114, 103661. https://doi.org/10.1016/J.INFRARED.2021.103661.

[10] Qixin, Liu, Huifang, Wang, Yiding, & Zheng, et al. (2016). A near-infrared acetylene detection system based on a 1.534 μm tunable diode laser and a miniature gas chamber. Infrared physics and technology, 75, 93-99. https://doi.org/10.1016/J.INFRARED.2021.103661.

[11] Li, G., Song, Y., Zhao, H., Dong, E., Li, J., Jia, L., Liu, Y., Zhang, S., & Zhao, F. (2023). A near-infrared multi-gas sensor based on IWTD-CEEMDAN and WOA-BiLSTM for detection of CH4leaked in industrial production. Infrared Physics and Technology, 131, 104695. https://doi.org/10.1016/j.infrared.2023.104695.

Investigate cosmic rays: Analysis of the air showers

Chenglong Wen[1, 4], Yiqian Li[2] and Guanghai Feng[3]

[1] Department of Physics, University of Toronto, Toronto, M5S 1A1, Canada
[2] Department of Physics, University of North Carolina at Chapel Hill, Chapel Hill, NC 27599, the United States
[3] Department of Astronomy, Xiamen University, Xiamen, 361000, China

[4] Corresponding author's e-mail: wcl960612@gmail.com

Abstract: In this lab report, the concept of cosmic rays and their components that can be detected on the earth's surface is introduced. In this experiment, four CosmicWatch muon detectors were used to detect the components of cosmic rays. This report describes the setup of the experiment as well as the method and algorithms used to filter the true signals of cosmic rays out of the background radioactive noises. Some process of the algorithm is visualized using histograms. The rate of air showers was found to be 0.151 mHz ± 0.022 mHz for a separation of 6.73 cm ± 0.08 cm, and 0.096 mHz ± 0.015 mHz for a separation of 13.5 cm ± 0.1 cm. Other sources of uncertainties are also analyzed in this report.

1. Introduction

The study of cosmic rays has been an important aspect of particle physics since the discovery of cosmic rays by Victor Hess in 1912 [1]. It is not yet possible to build observatories based on image-forming optics due to various detection problems. For example, the integration time of detectors can be too long to distinguish cosmic rays from noises [2]. Thus, there are several ways to observe cosmic rays from other aspects, such as scintillation detectors and Air Cherenkov detectors. For our experiment, we use what is called the Cosmic Watch Muon detector, which can be taken as a kind of scintillation detector.

Developed as a portable device, the Cosmic Watch muon detector has become a valuable tool in Astro / particle research applications [3]. In this case, it is used to detect cosmic rays, which are composed of high-energy relativistic particles that originate from the sun or outside the solar system. The two main instruments within Cosmic Watch detectors performing the function of detection are photomultipliers and scintillators. When charged particles pass through a scintillator, they deposit energy and thus produce photons. The photomultiplier then converts the photons captured into electrical signals, which can be recorded by the detector.

In this experiment, we tried to detect the muon component of cosmic rays using Cosmic Watch muon detectors. By setting up different configurations of the detector system, we can filter the background radioactive noises and observe particles with different incident angles. Another goal of this experiment is to develop a method to synchronize all detectors in the detector system, to provide a comprehensive measurement.

2. Theory

The main components of cosmic rays are ionized hydrogen and helium nuclei, with the former being the dominant component [4]. These particles being relativistic means that they have a speed close to the speed of light, hence their large kinetic energy. However, high-energy cosmic rays undergo a series of energy losses including GZK cutoff, which is caused by energy loss due to the interactions with the CMB (cosmic microwave background), and effects of the solar wind and the geomagnetic field of the earth, hence that the primary cosmic rays reaching the upper atmosphere cannot exceed the energy limit of 10^6 GeV [5][6]. As the primary cosmic rays collide with the atmosphere, nuclear collisions happen due to their high energy, producing mesons like pions and kaons [7]. While the neutral meson would decay into photons, the charged meson would decay into neutrinos and highly penetrative muons, which do not lose much energy besides their ionization when they pass through a variety of materials [8][9]. With a mean lifetime as short as 2.197 μs, muons would also decay into electrons, positrons, and neutrinos. Yet a muon with sufficiently high energy would travel fast enough to reach the sea level considering the time dilation before decaying. As calculated, the time for a muon to travel from where it is created (an altitude of 15 km) to sea level is about 5.5 μs in the frame of the muon. Thus, there is only a reduction by a factor of 10 in the sea-level flux of muons [10]. In addition, electrons and positrons, which are less penetrative, would be easily blocked by solid-like concrete walls in buildings, and thus cannot be the dominant particles we detected. In summary, when cosmic rays enter the Earth's atmosphere, they interact with air molecules, resulting in proton scattering. The protons scatter into pions, which then decay into muons [11].

By indirectly measuring the energy of photons released in ionization, only charged particles can be detected when passing through the scintillators inside Cosmic Watch detectors [12]. Heavy-charged particles are rarely observed due to their special interactions with various scintillators [13]. Thus, the most commonly detected particles are high-energy charged particles including electrons and muons in this experiment. The data was taken indoors, hence most of the particles detected were muons. The background radioactive noises which include the alpha, beta, and gamma decay are the most common signals that are detected. We need to find the actual signals from cosmic rays out of the background noises.

3. Methodology

The experimental apparatus consists of 4 Cosmic Watch Muon Detectors shown in Figure 1.

Figure 1: The 4 Cosmic Watch detectors used in this experiment.

The Cosmic Watch detector will record an event when it detects high-energy charged particles, which can be muons, electrons, and other radioactive background noises. One main function of the device is that two detectors can be connected using a cable. When a signal is detected by both detectors that are connected within a short time window, they will report a coincidence for that event. This could happen when one particle passes through two detectors, or it is a real coincidence between two different particles. The main data we need from the detectors are the index of the events, the timestamp, and the coincidence number, which indicates whether it coincides with the other detector.

The first step we did was to filter muons from the background noise. To achieve this, we used two detectors, one on top of the other, and connected them. Hence when a particle passes through both detectors in a short time, we consider it to be a muon. Also, the path of muons coming from the atmosphere is more likely to be vertical. The overall angular distribution of muons at sea level is approximately $I_0 \cos^2(\varphi)$, where φ is the zenith angle and $I_0 \approx 70\ m^{-2}\ s^{-1}\ sr^{-1}$ is the flux at $\theta = 0$ [8]. Hence a large percentage of muons would be able to pass through both detectors. That is, we can only look at the data with the presence of coincidence. The detector was run for 12 hours in this step.

Next, we wanted to filter the signals for air showers. We used four detectors that were built as in Figure 1. The top-left detector is connected to the top-right one. The bot-left detector is connected to the bot-right one. We expect that air showers will travel through the four detectors at the same time. However, we cannot connect more than two detectors. Hence, we need to find an algorithm to filter the events where four detectors coincide.

Known that two detectors on the top are connected, the same as the bottom two. Hence, we know that when a coincidence happens, it is either a muon traveling horizontally or an air shower. What we need to do now is to search for the air showers in the coincidence data. Consider the timestamp of coincidence on the top as t_i, at the on the bottom as b_j. Note that the index i and j are different. We can generate a new sequence d_{ij} by taking the difference: $d_{ij} = t_i - b_j$.

Note take the sequence has the length $i \times j$. Ideally, for d_{ij} that is small enough, t_i and b_j are possibly the timestamps for the same particle.

The main problem is, there is a time offset between two detectors. The two detectors on the top might record an air shower at 1000 ms, but the same air shower might be recorded at 9000 ms by the bottom detectors. This happens when the detectors at the bottom are turned on 8 seconds earlier than the top detectors. Also, the clock in the detector can deviate over a long time, hence the time offset is time-dependent. We can find the time offset by plotting a 2D histogram, where the y-axis is the difference d_{ij}, and the x-axis x_{ij} is generated by $x_{ij} = i$. Note that x_{ij} also has the length of $i \times j$.

Now we can find the function of the time offset $O(t)$ over time. Hence, we can find the adjusted t_i by using $t_i' = t_i - O(i)$. The new time difference is $d'_{ij} = t'_i - b_j$, hence for all d'_{ij} that is small enough, it is an air shower. This can be visualized using a histogram which is shown in the result section. The 4 detectors were run for 88 hours.

For the next setup. Based on the setting in Figure 1, we increased the distance between the left and right detector to the width of 1 detector. Other settings remained the same. We used the same method to find the air showers in this case. The detectors were run for 113 hours.

4. Result and Analysis

4.1 Result

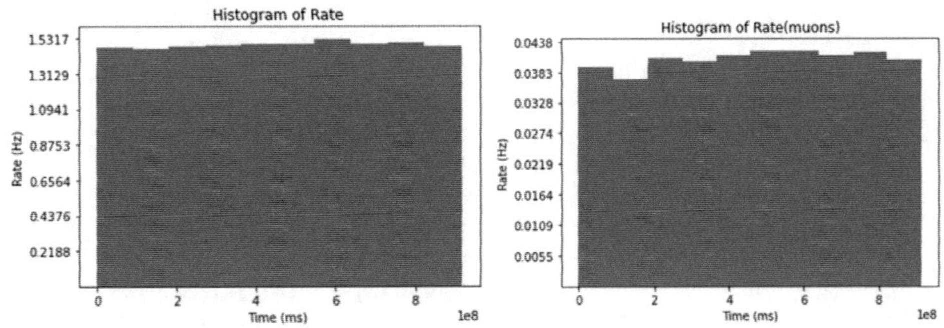

Figure 2: The rate histogram for all signals (left), for muons (right). The x-axis is time in ms. The y-axis is the rate in Hz.

We filtered the muons out of background noises first and made 2 histograms for comparison as in Figure 2. We observed that the detection rate of the detector is about constant, but the rate varied a little for muon detection, which means that the number of muons detected on the Earth's surface does not need to be constant.

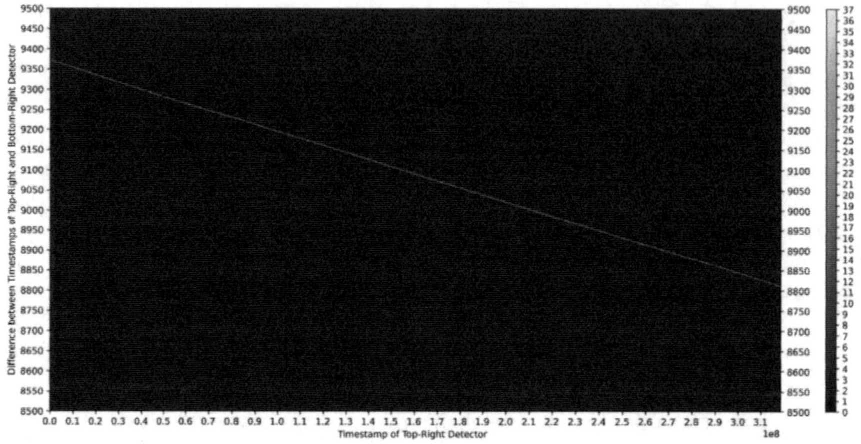

Figure 3: The 2D histogram of the top-right detector and the bottom-right detector in the first air shower detection. The x-axis is the timestamp x_{ij} of the top-right detector in ms. The y-axis is the time difference d_{ij} in ms. The green line is the function of time offset $O(t)$.

For the first air shower detection, we used the 2D histogram to find the time offset between the top and bottom detectors. Figure 3 is a sample 2D histogram for the top-right detector and bottom-right detector. We observed a straight line in the graph which means the function $O(t)$ is a linear function over time.

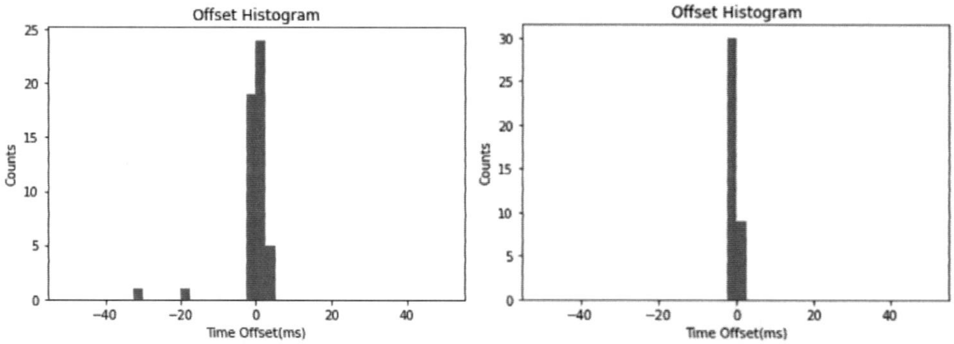

Figure 4: The histogram of the adjusted time difference d'_{ij} for the first air shower detection (left), and the second air shower detection (right). The x-axis is the time offset in ms. The y-axis is the number of counts. The bin width is 2.5 ms.

Note that for d'_{ij} that is small enough, it is an air shower. By Figure 4, we claim that for $d'_{ij} < 5\ ms$, it can be considered as an air shower. Hence in the first air shower detection, there were 48 air showers detected in 88.5 hours. The rate is 0.151 mHz ± 0.022 mHz. In the second air shower detection, there were 39 air showers detected in 113 hours. The rate is 0.096 mHz ± 0.015 mHz. Remind that in the first air shower detection, the detectors on the left and right are contacted. In the second air shower detection, they are separated by the width of one detector. The width of the detector is 6.73 cm ± 0.08 cm.

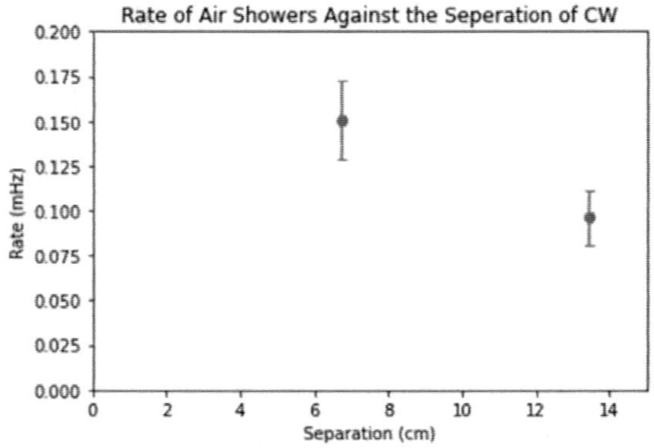

Figure 5: The rate of air showers against the separation of CosmicWatch detectors with error bars. The x-axis is the separation in cm. The y-axis is the rate of air showers in mHz.

Note that in Figure 5, we can see that the rate lies outside of each other's uncertainties. Hence, we conclude that the change in the separation of muon detectors will affect the rate of air showers.

4.2 Uncertainty analysis

Note that independent counting follows Poisson distribution, where the standard deviation is simply the square root of the number of counts. Hence the number of air showers with uncertainties should be 48.0 ± 6.9, and 39.0 ± 6.2.

The other uncertainty is that there might be a real coincidence instead of an air shower in the detection. We need to calculate the rate of coincidence, that is $R = 2N_1N_2\tau$. N_1 and N_2 are the rate of coincidence for the 2 detectors, and τ is the coincidence window, which we set to be 10 ms. Both N_1 and N_2 are at about 10^-9 Hz. Hence the rate is too low to be considered.

The insufficiency for this experiment is that in our setup, the detectors can only capture muons or air showers with an incident angle no greater than about 40 degrees. However, we can still analyze the data to observe the trend due to its generality.

5. Conclusion

In conclusion, the Cosmic Watch muon detector can be used to detect single muons and air showers. After analyzing the data given by the detector, we found out that when the horizontal separation of detectors is 6.73 cm ± 0.08 cm, the rate of air showers is 0.151 mHz ± 0.022 mHz. When the separation is 13.5 cm ± 0.1 cm, the rate of air showers is 0.096 mHz ± 0.015 mHz. The plus-minus sign here means 1 standard deviation. Through the observation, the rate of air shower decreases as the separation of Cosmic Watch detectors increases. Since we only have 2 samples, we cannot find the trend. However, we can conclude that the separation of detectors will affect the rate of air showers that are detected. Further analysis of the momentum and radius of air showers can be made with more experiments.

References

[1] Spencer, R., and Raply, C. A history of radio detection of cosmic rays. Vol. 59. Doi: https://doi.org/10.1093/astrogeo/aty149. 2018.

[2] Griffin et al. Searching for Fast Optical Transients using VERITAS Cherenkov Telescopes. arXiv:1206.6535. 2011.

[3] S. N. Axani, K. Frankiewicz, J. M. Conrad. The Cosmic Desktop Muon Detector: a self-contained, pocket-sized particle detector. Massachusetts Institute of Technology.

[4] K. O. et al., "Cosmic rays-particle data group," Nuclear Instruments and Methods in Physics Research Section A: Accelerators, Spectrometers, Detectors, and Associated Equipment, Vol. 38, no. 090001, p. 6, 2014.

[5] R. Abbasi et al., “IceCube high-energy starting event sample: Description and Flux Characterization with 7.5 years of Data,” Physical Review D, Vol. 104, no. 2, 2021. doi:10.1103/physrevd.104.022002.
[6] T. K. Gaisser, R. Engel, and E. Resconi, Cosmic rays, and particle physics. Cambridge University Press, 2016.
[7] R. Alves Batista et al., “Open questions in cosmic-ray research at ultrahigh energies,” Frontiers in Astronomy and Space Sciences, Vol. 6, 2019. doi:10.3389/fspas.2019.00023.
[8] Axani S N. The physics behind the CosmicWatch desktop muon detectors[J]. arXiv preprint arXiv:1908.00146, 2019.
[9] M. Aaboud et al., “Study of the rare decays of B0 s and B0 mesons into muon pairs using data collected during 2015 and 2016 with the ATLAS detector,” Journal of High Energy Physics, Vol. 2019, no. 4, Apr. 2019, doi: https://doi.org/10.1007/jhep04(2019)098.
[10] Liu, L. The Speed and Lifetime of Cosmic Ray Muons. Massachusetts Institute of Technology. 2007.
[11] Blanchard, J. A Measurement of the Angular Distribution of Cosmic Muons. 2012.
[12] Attix, F. H. Introduction to Radiological Physics and Radiation Dosimetry (4th ed.). ISBN: 978-3-527-61714-2. 2008.
[13] Leo, W. R. Techniques for Nuclear and Particle Physics Experiments (2nd ed.). Springer. doi:10.1007/978-3-642-57920-2. p. 173, 1994.

Journal of Physics: Conference Series **2724** (2024) 012053 doi:10.1088/1742-6596/2724/1/012053

Research on denoising the temperature monitoring signal of a distributed optical fiber downhole cable

Xiaowei Zhao[1], Dong Wang[1,2,5], Kang Liu[3], Jianqiao Guo[3], Hongfu Zhao[4] and Liangxiao Tan[4]

[1] College of Geodesy and Geomatics, Shandong University of Science and Technology, No. 579, Qianwangang Road, Qingdao, China
[2] National Demonstration Center for Experimental Surveying and Mapping Education, Shandong University of Science and Technology, No. 579, Qianwangang Road, Qingdao, China
[3] Yankuang Energy Group Co., LTD. Xinglongzhuang Coal Mine, No. 10, Xinglongzhuang Road, Jining, China
[4] Yankuang Energy Group Co., LTD. Dongtan Coal Mine, No. 1088, Gongjian Road, Jining, China

[5]Corresponding author's e-mail: 2215532673@qq.com

Abstract. Aiming at the noise mixed in the downhole cable skin temperature signal measured by the BOTDR distributed fiber optic sensing temperature measurement system, a new improved wavelet threshold function is proposed, and denoising experiments are completed based on the MATLAB platform for the measured cable skin temperature signal. The denoising results of several threshold function methods are compared and analyzed from the aspects of graphical intuitive comparison as well as quantitative evaluation indicators. The results show that the improved threshold function denoising fiber optic temperature monitoring signal has a higher SNR and lower RMSE, there is no obvious distortion phenomenon, and the denoising performance is significantly improved compared with the traditional threshold function, which has a high application value.

1. Introduction

With the increase in power supply demand in coal mines, the number and length of underground cables laid in the coal mine working face have been increasing, and their temperature changes are crucial to the normal work of the whole electrical distribution system. The humid environment of the mine tunnel and the gradual aging of the insulating skins in long-term working conditions can easily cause the cables to heat up, leading to the occurrence of fires, which puts the safety of people and property at serious risk [1]. Due to the harsh electrical environment in mines, it is very difficult to extinguish fires, and the existing means of fire prevention use expensive materials, which makes the cost of fire prevention extremely high [2]. Therefore, regular and continuous monitoring of cable temperatures and the necessary temperature control and maintenance are of great significance to maintain the safety order of coal mine production and the economic interests of enterprises.

BOTDR distributed fiber optic sensing temperature measurement technology uses the Brillouin scattering effect in the optical fiber, which can continuously and real-time acquire and transmit optical signals, independent of the influence of the harsh underground environment, to obtain high-resolution and high-precision information on the temperature change. However, the temperature signal measured by this technique contains a certain amount of noise, and how to denoise it is an important research area at present. Wang [3] denoised distributed fiber optic conduit temperature measurement signals by the EMD algorithm and proved that the algorithm has good denoising ability in the frequency domain, while the denoising effect in the time domain is average. Narasimhappa et al. [4] introduced Kalman filtering into the noise filtering of fiber optic gyro signals, which removes some of the interfering noise but is computationally complex, resulting in a long response time for the system.

The wavelet threshold is the most common method to deal with the noise of optical fiber signals. This method has the characteristics of less computation, simple implementation, and good denoising effect, and the key is how to select the threshold function [5]. The traditional threshold function denoising performance is poor, and it will lose some helpful information while removing noise [6]. It is necessary to seek a more efficient and practical threshold function denoising method. For this reason, with downhole cable temperature monitoring as the main objective, a new threshold function is established for the noise that exists in the system, and the effectiveness and excellence of the proposed improved threshold function in the application of distributed fiber optic downhole cable temperature monitoring signal denoising is verified by denoising the measured cable skin temperature signals, which is of high value for promotion.

2. Distributed fiber optic cable temperature measurement system and its noise signal characteristics

2.1. Relationship between temperature and cable damage

Cables are made up of cores, insulation, and protective layers, and there is a close relationship between the cable core temperature and the damage that occurs to the cable. Once the cable is overloaded, the core temperature will rise sharply, accelerate cable insulation aging and its expansion, and even cause thermal breakdown. A cable fire is not sudden, but the cable core is in the working process, the temperature continues to rise, and the insulation gradually ages, ultimately leading to accidents. The cable core temperature is difficult to measure directly, and the theory shows that the cable core temperature, cable skin temperature, and cable insulation layer temperature trends are synchronized. Therefore, in actual engineering, often through the measurement of cable skin temperature, the cable transient thermal circuit model [7] is used, and then the cable core temperature is indirectly projected.

2.2. BOTDR distributed fiber optic temperature measurement system

The BOTDR technique utilizes the linear relationship between the change in frequency of spontaneously Brillouin-scattered light and the change in temperature to achieve distributed sensing. BOTDR measures temperature by first injecting a broadband light pulse into the optical fiber. As the light pulse propagates through the fiber, it interacts with the acoustic waves in the fiber to produce Brillouin scattering. The difference between the frequency of scattered light and the frequency of the light pulse is proportional to the temperature. By measuring the frequency shift of the scattered light, the temperature at each location in the fiber can be determined, thus obtaining information about the temperature change along the fiber and completing the distributed measurement of temperature. The main components of the system are shown in Figure 1.

Since the technology requires only single-ended injection and the device structure is relatively simple, it is more suitable for cable skin temperature measurement in such a special environment as underground coal mines.

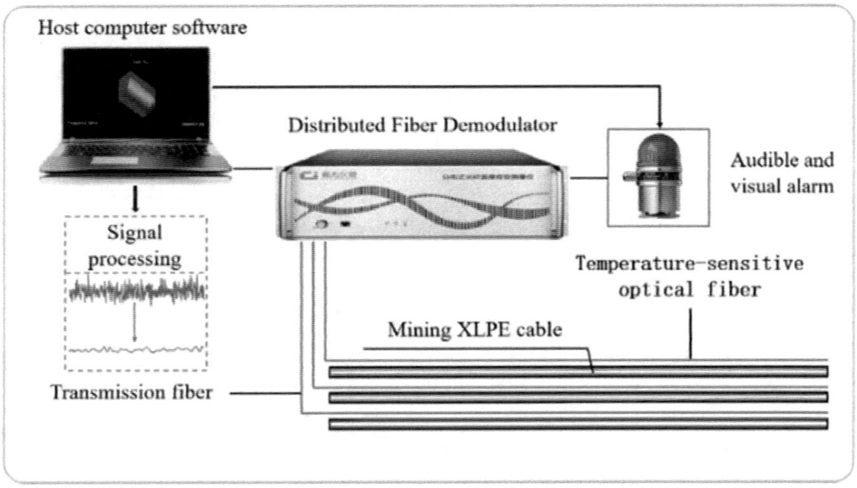

Figure 1. BOTDR distributed fiber optic temperature monitoring system.

2.3. Sources of noise in the system and their characterization

When the BOTDR distributed fiber optic temperature measurement system is used for temperature monitoring, factors such as the structure, quality, thermal diffusion, and temperature response of the fiber optic cable used itself will affect the accuracy of the measurement. At the same time, there is semiconductor thermal noise, voltage fluctuation noise, and scattered particle noise due to random electron action in the system. This noise has the following characteristics [8]:

- Consisting of a large number of short sharp pulses whose amplitude and phase are random, the signal waveforms perceived by different models of sensing fibers vary.
- The amplitudes of the noise are all much larger than the intensity of the measured signal, and most of the useful information is submerged in noise, which leads to a very low SNR of the signal.
- The noise of the system has a certain cumulative characteristic, which is not linear but nonlinearly related to the change in temperature.
- In the measured fiber optic signals, the useful temperature signals are low in frequency, while the noise has a wide spectrum with distribution in both low- and high-frequency bands, mainly white noise, and transient pulses.

Therefore, to achieve the required monitoring accuracy, this paper adopts the wavelet transform threshold denoising method, i.e., cable temperature signal noise reduction processing, to obtain more real and reliable cable skin temperature information.

3. Improved wavelet threshold function denoising algorithm

3.1. Wavelet thresholding denoising method

Wavelet threshold denoising is a signal processing technique designed to reduce or eliminate noise from a signal. The basic idea is that the signal is decomposed into several subbands by wavelet transform, which determines which coefficients are caused by noise based on the magnitude of the coefficients in the subbands, sets these coefficients to zero or shrinks them, and finally inverts the coefficients to obtain the denoised signal.

In wavelet thresholding denoising, the signal is first wavelet decomposed, and the energy in each subband is calculated. Then, the energy, in which subbands are due to noise, is determined by comparing it with a predetermined threshold. If the energy in a subband is less than the threshold, the signal in that subband is assumed to be noise and is zeroed or reduced to a very small value. Finally, the transformed subbands are merged, and the signal is restored to the time domain by inverse wavelet transform, i.e., the signal is reconstructed to remove the noise. The principle is shown in Figure 2.

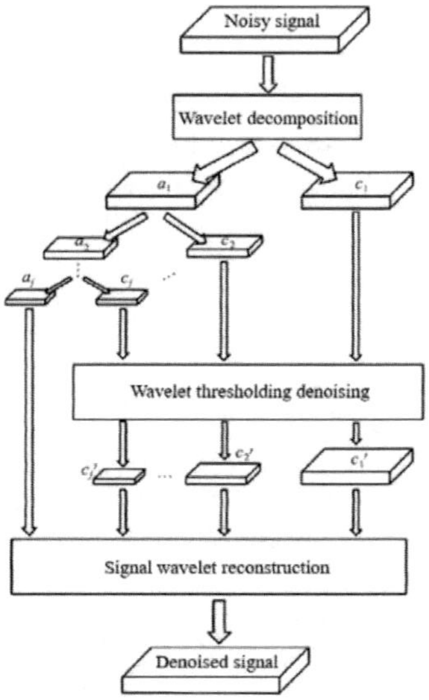

Figure 2. Principle of wavelet thresholding denoising.

The wavelet threshold denoising method needs to consider several factors in practical applications, including the selection of the wavelet base and decomposition level, the type of threshold function, and the setting of threshold rules, in which the merit of the threshold is especially critical. If the threshold value is too high, some real signals will be removed, while if the threshold value is too low, the denoising effect will be reduced. In this paper, the following improved threshold selection method, in which the threshold value accords with the noise distribution characteristics, is selected to calculate the threshold value [9]:

$$T = \frac{\sigma\sqrt{2\,ln\,N}}{ln(j+1)} \tag{1}$$

$$\sigma = \frac{median(|w_{j,k}|)}{0.6745} \tag{2}$$

where σ is the standard variance of noise; N is the length of signals; j is the number of decomposition levels; and $w_{j,k}$ is the wavelet coefficient.

3.2. *Improved wavelet threshold function*

In wavelet threshold denoising, hard and soft threshold functions [10] are traditional types of threshold functions. Equation (3) and equation (4) show the wavelet coefficients after noise reduction for hard and soft thresholding functions, respectively:

$$\widehat{w}_{j,k} = \begin{cases} w_{j,k}, & |w_{j,k}| \geq \lambda \\ 0, & |w_{j,k}| < \lambda \end{cases} \tag{3}$$

$$\widehat{w}_{j,k} = \begin{cases} sgn(w_{j,k})(|w_{j,k}| - \lambda), & |w_{j,k}| \geq \lambda \\ 0, & |w_{j,k}| < \lambda \end{cases} \tag{4}$$

where $w_{j,k}$ is the wavelet coefficient before denoising; λ is the threshold value; and $sgn(*)$ is a symbolic function.

In practical applications, the hard threshold function lacks continuity, which will result in the pseudo-Gibbs phenomenon in wavelet reconstruction. There is a constant error between the signal denoised by the soft threshold function and the real signal, and the details are lost, so a more reasonable threshold function needs to be constructed.

In this paper, for the above characteristics of the noise of the temperature monitoring signal of the BOTDR system, a new threshold function is proposed to calculate the wavelet coefficients after noise reduction by further improving the existing threshold function:

$$\widehat{w}_{j,k} = \begin{cases} (1-\mu)(w_{j,k}) + \mu sgn(w_{j,k})\left(|w_{j,k}| - \dfrac{M\lambda}{e^{\cos(\frac{\pi}{2}\times\frac{\lambda}{|w_{j,k}|})}}\right), & |w_{j,k}| \geq \lambda \\ (1-\mu)sgn(w_{j,k})\dfrac{|w_{j,k}|(1-M)}{1+\mu}, & |w_{j,k}| < \lambda \end{cases} \tag{5}$$

where $\mu = e^{-(|w_{j,k}|-\lambda)^2}$; $w_{j,k}$ is the wavelet coefficient before noise reduction; $sgn(*)$ is the sign function; λ is the threshold; and M is the adjustable parameter, $0 < M \ll 0.9$.

The new threshold function adopts the idea of a weighted average and introduces the cosine function and exponential function at the same time, utilizing the bounded symmetry of the cosine function and the rapid change in the exponential function in a fixed interval and is the neutralization of the traditional threshold function. The value of parameter M has a direct influence on the final denoising effect, which can be flexibly adjusted according to the characteristics of the temperature signals received by different models of sensing fibers. When M is 0.2, 0.5, and 0.9, the comparison between the function image and the hard and soft threshold function image is shown in Figure 3.

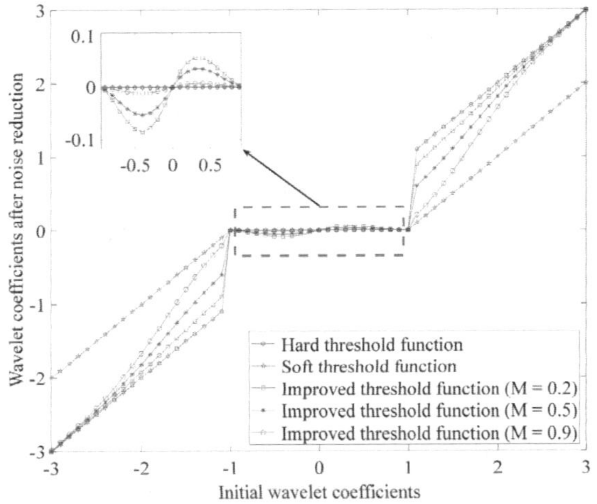

Figure 3. Threshold function comparison results.

3.3. Mathematical analysis of new threshold functions

3.3.1. Continuity analysis.

$$\lim_{w_{j,k}\to\lambda^+} \widehat{w}_{j,k} = \lim_{w_{j,k}\to\lambda^+}\left[(1-\mu)(w_{j,k}) + \mu sgn(w_{j,k})\left(|w_{j,k}| - \frac{M\lambda}{e^{\cos(\frac{\pi}{2}\times\frac{\lambda}{|w_{j,k}|})}}\right)\right] = 0 \tag{6}$$

$$\lim_{w_{j,k}\to\lambda^-} \widehat{w}_{j,k} = \lim_{w_{j,k}\to\lambda^-}\left[(1-\mu)sgn(w_{j,k})\frac{|w_{j,k}|(1-M)}{1+\mu}\right] = 0 \tag{7}$$

$$\lim_{w_{j,k}\to\lambda^+} \widehat{w}_{j,k} = \lim_{w_{j,k}\to\lambda^-} \widehat{w}_{j,k} = 0 \tag{8}$$

Thus, it follows that the function has continuity.

3.3.2. Deviation analysis. When $w_{j,k} \longrightarrow +\infty, \mu \longrightarrow 0$,

$$\lim_{w_{j,k}\to+\infty}\left(\widehat{w}_{j,k} - w_{j,k}\right) =$$

$$\lim_{w_{j,k}\to+\infty}\left[(1-\mu)(w_{j,k}) + \mu sgn(w_{j,k})\left(|w_{j,k}| - \frac{M\lambda}{e^{\cos\left(\frac{\pi}{2}\times\frac{\lambda}{|w_{j,k}|}\right)}}\right) - w_{j,k}\right] = 0 \tag{9}$$

Similarly, when $w_{j,k} \longrightarrow -\infty, \mu \longrightarrow 0$:

$$\lim_{w_{j,k}\to-\infty}\left(\widehat{w}_{j,k} - w_{j,k}\right) = \lim_{w_{j,k}\to-\infty}\left[(1-\mu)sgn(w_{j,k})\frac{|w_{j,k}|(1-M)}{1+\mu} - w_{j,k}\right] = 0 \tag{10}$$

$$\lim_{w_{j,k}\to+\infty}\left(\widehat{w}_{j,k} - w_{j,k}\right) = \lim_{w_{j,k}\to-\infty}\left(\widehat{w}_{j,k} - w_{j,k}\right) = 0 \tag{11}$$

Thus, it follows that the constant deviation between $\widehat{w}_{j,k}$ and $w_{j,k}$ of this function is 0.

Through the above theoretical analysis, the new threshold function overcomes the shortcomings of the traditional threshold function, and the wavelet detail coefficients are continuous after denoising, which solves the problem of constant deviation between the denoised value and the true value. The denoising effect is perfect, and it has a certain degree of superiority in the application of practical denoising.

4. Example analysis

4.1. Experimental data and evaluation indicators

Figure 4 shows the temperature monitoring signal of the skin of a section of distributed fiber optic cable with a length of 600 m in the tunnel of the internship mine of Shandong University of Science and Technology in Tai'an Campus. The sampling resolution of the system is 1 m, and the temperature measurement accuracy is ±1°C. It can be seen that the data contain much noise, and the signal fluctuates with more burrs.

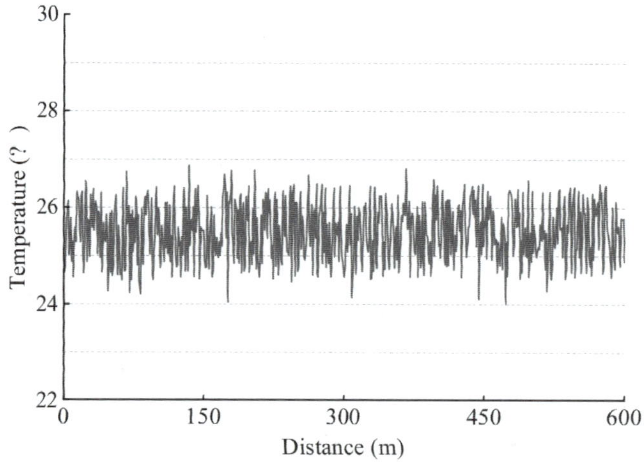

Figure 4. Cable skin raw temperature signal.

Before the formal experiment using control variables and other methods, the MATLAB platform was used to screen the optimal combination of wavelet base and decomposition layers suitable for the signal to denoise the result of the signal-to-noise ratio (SNR) and the root mean square error (RMSE) as a basis for judgment. The larger the SNR is, the smaller the RMSE is, proving a better denoising effect. SNR and RMSE of the signal after denoising are as follows:

$$SNR = 10\log\left[\frac{\sum_{i=1}^{N} x^2(i)}{\sum_{i=1}^{N}\left(x(i)-\hat{x}(i)\right)^2}\right] \tag{12}$$

$$RMSE = \sqrt{\frac{1}{N}\sum_{i=1}^{N}\left(x(i)-\hat{x}(i)\right)^2} \tag{13}$$

where $x(i)$ is the signal before noise reduction; $\hat{x}(i)$ is the reconstructed signal after denoising; and N is the sampling length.

Comparing the results of several pre-experiments, the denoising wavelet basis used is finally selected as sym3 (sym3 wavelet is a nearly symmetric tightly branched set of double orthogonal wavelets, whose waveforms are similar to this temperature signal), and the number of wavelet decomposition levels is 3.

4.2. Results and discussion

4.2.1. Comparison of denoising effects. The measured temperature signal is denoised using the hard and soft threshold functions and the improved threshold function (M = 0.5) in this paper, and the results before and after denoising are shown in Figure 5((a),(b),(c),(d)).

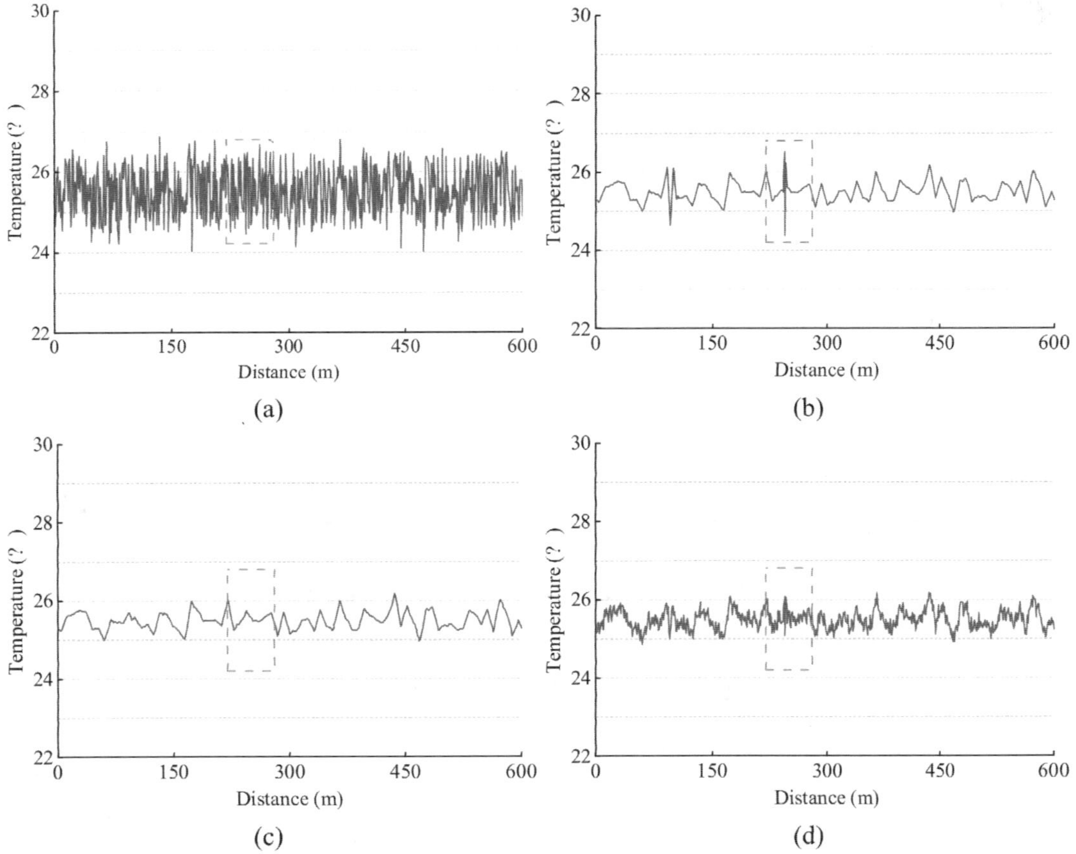

Figure 5. Denoising effect of each threshold function on the measured temperature signal.

In addition to comparing the denoised waveforms, the original temperature data within the distance from 234 to 253 of the temperature sensing fiber and the denoised temperature data with different threshold functions are extracted for a more detailed comparison, and the results are shown in Figure 6:

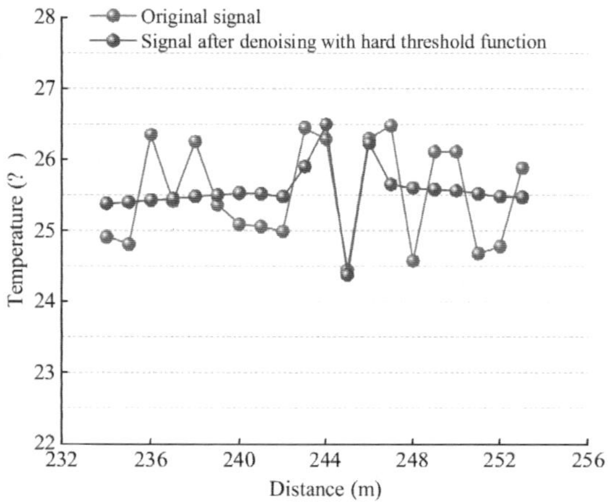

(a)

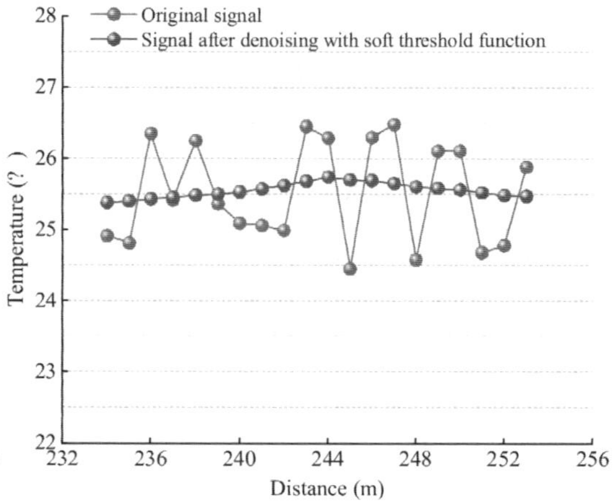

(b)

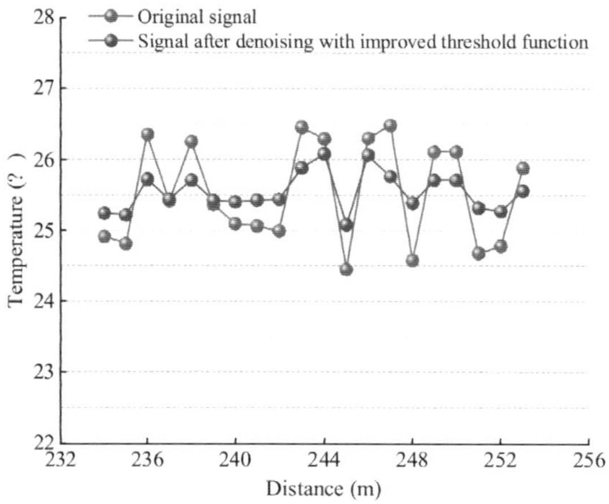

(c)

Figure 6. Comparison of signal values before and after denoising in the range from 234 to 253.

Combining the above two sets of graphical results, we can see that several thresholding functions remove the noise in the signal, but the traditional hard thresholding function denoises the signal in a few localized positions in the denoising is incomplete, resulting in the phenomenon of spikes and oscillations; the soft thresholding function denoises the signal as a whole is too smooth, and there is denoising excessively, which results in the distortion of denoised signals. In contrast, the signal processed by the improved threshold function in this paper no longer contains spikes and oscillations, and the shape of the signal after denoising is more consistent with the original signal. Intuitively, the denoising effect is better than that of the other two threshold functions. To more accurately determine the denoising performance of various threshold functions, it is necessary to quantitatively analyze the denoised signal. The SNR and RMSE of the signal after denoising are calculated by various threshold functions, as shown in Table 1.

Table 1. Indicator evaluation results of the denoising effect with different threshold functions.

Wavelet threshold function	SNR	RMSE
Hard threshold function	32.9480	0.444626
Soft threshold function	32.8377	0.450309
Improved threshold function (M = 0.5)	34.6684	0.364731

Comparing the three denoising methods, the improved threshold function of this paper has the largest SNR and the smallest RMSE after denoising the temperature signal measured by the distributed fiber, and the two evaluation indexes are better than the other two threshold functions, which indicates that the signal denoised by the method has the least residual noise, the highest degree of noise reduction, and the denoising result is the closest to the actual value. On the other hand, when the parameter M takes different values, the denoising effect is also different (Table 2), which further proves its applicability and flexibility.

Table 2. Indicator evaluation results of the denoising effect with different M values.

The value of M	SNR	RMSE
0.1	36.2426	0.304272
0.3	35.4503	0.333333
0.5	34.6684	0.364731
0.7	33.9121	0.397914
0.9	33.1887	0.432470

4.2.2. Reliability analysis. To validate the reliability of the distributed optical fiber temperature monitoring information after denoising by the improved threshold function, the temperature of the cable skin is measured using an infrared thermometer at the same location every 30 m from the monitoring point at 20 m from the starting position of the temperature-sensing fiber. With the distance as the abscissa and the temperature value as the ordinate, two temperature curves are obtained by polynomial fitting of the temperature scatter points measured by the two methods, and the comparison of the two curves is shown in Figure 7. According to Figure 7, it can be concluded that the cable skin temperature information obtained by the two monitoring methods is basically consistent, and the difference between the two groups of results is always within the accuracy range of the distributed fiber optic temperature measurement system, which proves that the fiber optic temperature monitoring information after denoising by the improved threshold function in this paper has high reliability.

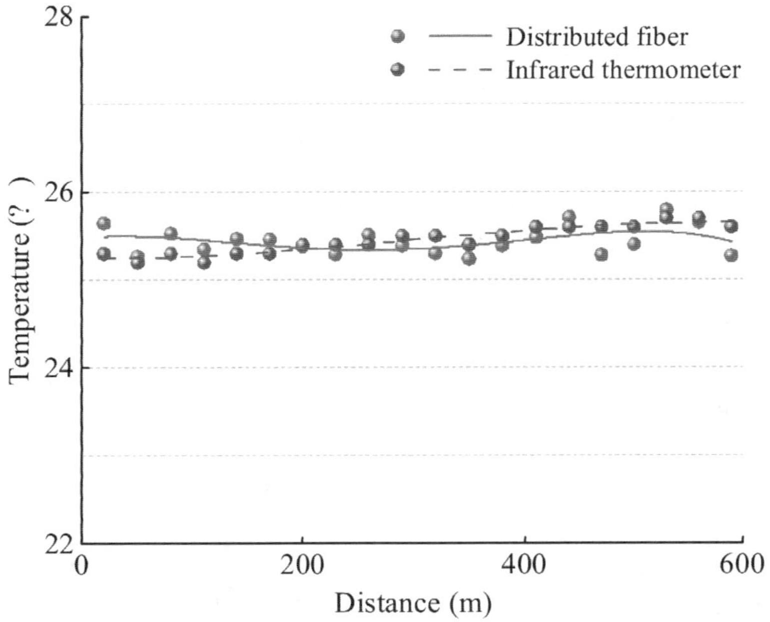

Figure 7. Polynomial fitting of monitoring results in two ways.

Comprehensive analysis of the above, it can be concluded that this paper improves the threshold function denoising the overall quality of the signal and that the traditional threshold function has been significantly improved. The denoising results more truly reflect the temperature information of the measured cable skin, which can provide an effective basis for the safety diagnosis of underground cables in mines.

5. Conclusions

(1) Aiming at the noise interference problem of the BOTDR distributed fiber optic cable temperature monitoring signal, a new improved wavelet threshold function is proposed for denoising, which effectively improves the pseudo-Gibbs phenomenon after denoising using the hard threshold function, solves the constant deviation problem of denoising with the soft threshold function, and strengthens the wavelet thresholding's ability to remove noise.

(2) Through example analysis, this paper improves the threshold function denoising signal after the SNR can be increased from 32.9480 to 36.2426, the RMSE can be reduced from 0.444626 to 0.304272, the denoising performance is significantly improved compared to the traditional threshold function, and the denoised results and real cable skin temperature information are consistent, indicating that the method has a better application value, which is conducive to the promotion of modern mine safety monitoring and monitoring. It is beneficial to promote the development of modern mine safety monitoring and evaluation technology systems.

(3) For the underground cable skin fiber optic temperature monitoring signal, in this paper, to improve the threshold function, the denoising algorithm has been able to filter most of the interference noise, but there are still a few residual noises that are difficult to remove. It is proposed that future studies can take into account a variety of factors and combine with other different denoising methods for systematic processing, to provide more accurate monitoring of the temperature state of the cable for the coal mine underground.

References

[1] Ren X W, Wang Z X, Wang R, Kang F R, Wang Y, Han D Y, Wang W F, Qi L H and Xu Y Y 2022 Study on fire hazard of cable in coal mine *Safety in Coal Mines* **53** 151-156.

[2] Liang Y T, Yang Y L, Guo S D, Tian F C and Wang S F 2022 Combustion mechanism and control approaches of underground coal fires: a review *International Journal of Coal Science & Technology* **10**.

[3] Wang X 2019 Denoising analysis of distributed optical fiber pipe temperature measurement signal based on EMD *Computer & Telecommunication* **15** 51-54.

[4] Narasimhappa M, Sabat S L and Nayak J 2016 Fiber-optic gyroscope signal denoising using an adaptive robust Kalman filter *IEEE Sensors Journal* **16** 3711-18.

[5] Zhang H, Yuan Z X, Li X Y, Wang S Z, Sun X H and Wang Z X 2021 Improved wavelet threshold denoising method for distributed optical fiber lightning location *Electric Power Construction* **42** 75-82.

[6] Xie B, Xiong Z Q, Wang Z J, Zhang L J, Zhang D Z and Li F S 2020 Gamma spectrum denoising method based on improved wavelet threshold *Nuclear Engineering and Technology* **52** 1771-76.

[7] Mazzanti D 2015 Life estimation of HVDC cables under the time-varying electrothermal stress associated with load cycles *IEEE Transactions on Power Delivery* **30** 931-939.

[8] Bai Q, Wang Q L, Wang D, Wang Y, Gao Y, Zhang H J, Zhang M J and Jin B Q 2019 Recent advances in brillouin optical time domain reflectometry *Sensors* **19**.

[9] Zhao R Z, Song G X and Wang H 2001 Improved model for wavelet coefficient threshold estimation *Journal of Northwestern Polytechnical University* **19** 625-628.

[10] Donoho D L 1995 Denoising by soft-thresholding *IEEE Trans. Information Theory* **41** 613-627.

Pointer instrument positioning and indication recognition algorithm based on YOLOV5s

Fei Li[1,2], Wenliang Zhang[1] and Bo He[1]

[1]Guangxi Beihai Power Supply Bureau of China Southern Power Grid, Beihai Guangxi 536000, China

[2]Corresponding author's e-mail: 648840893@qq.com

Abstract. The substation inspection robot has low accuracy in positioning and indication reading of the pointer instrument. An algorithm design for indicator recognition is proposed. The algorithm mainly includes dial area extraction, scale line extraction and dial center determination, zero scale line determination, and indicator recognition. An experimental system for pointer instrument positioning and indication recognition is constructed to verify the proposed algorithm. The results indicate that the positioning accuracy of the two instruments is 90.5% and 100% respectively. Ten positioned instruments are selected, and the reading recognition error is 0.126. It shows that the proposed pointer instrument positioning and indication recognition algorithm can improve the efficiency of substation instrument recognition.

1. Introduction

Based on autonomous navigation, the substation inspection robot integrates multiple sensors to realize instrument reading recognition and equipment thermal anomaly recognition. The pointer instrument has the advantages of low price, simple structure, convenient maintenance, and strong stability. Because of these advantages, the pointer instrument is widely used in substations. Currently, the pointer instrument has no data interface and cannot realize automatic indication acquisition. The indication records of the pointer instrument are mainly collected manually, and the working state of the instrument is checked according to the instrument value. This method has disadvantages such as a heavy workload and a big subjective influence on the staff, which may cause omission and misreading. Substation inspection robots carrying detection devices can automatically identify the pointer instrument reading accurately and quickly. The robots can reduce the inspection workload and improve instrument identification efficiency. Because of the environmental impact or instrument type, the instrument recognition accuracy cannot meet the requirements of application promotion. Therefore, it is very significant to study research automatic positioning and reading of pointer instruments for substation inspection robots.

The common detection methods of pointer instruments are mainly based on machine vision technology and artificial intelligence technology. Zhang et al. [1] used the visual saliency region of the pointer instrument to construct the instrument detection model, used the undirected graph sorting algorithm to suppress the interference of non-pointer images, and extracted the instrument pointer well. Liu et al. [2] preprocessed the instrument image with super-resolution, such as non-local denoising and depth denoising. The pointer was segmented and located using the example segmentation algorithm Mask-RCNN. Finally, the indication was read using the angle method. Li et al. [3] used the Faster R-

CNN algorithm to detect the pointer dial area and detect the pointer. The pointer was extracted according to the Hough transform on the obtained pointer area.

Although the research on the positioning and indication recognition of pointer instruments has made great progress, the positioning of pointer instruments in substation inspection still relies on manual work, which is very time-consuming and labor-intensive. The pointer instrument reading recognition based on artificial intelligence has limited reasoning speed due to the complexity of its network. At the same time, due to the lack of computing power, the existing model cannot be directly deployed to embedded development platforms. Aiming at solving these problems, this paper uses the YOLOv5s target detection model with the smallest model in YOLOV5 with superior performance as the basis to point the pointer instrument. Then it uses digital image processing technology to identify the instrument to achieve better reading results.

2. Design of pointer instrument detection algorithm

The overall process of the automatic positioning and reading recognition system is as follows: firstly, the data set is collected and established through the laboratory and the network, the YOLOv5s model is trained, and then the inspection robot reaches the specified detection point. Then the trained model performs real-time positioning of the pointer instrument position, extracts the instrument dial area according to the instrument position selected by the box, and finally reads the indication.

Redmon et al. proposed the YOLO algorithm by summarizing object detection as a regression problem [4]. The YOLO algorithm is based on a separate end-to-end network, completing the input of the original image to the output of the object location and category. With the deepening and improvement of the algorithm, the algorithm is widely used in target detection, and YOLOv5 was updated to version 5.0 in April 2021 [5]. The YOLOv5 algorithm has advantages such as high precision, small file memory usage, and high speed. YOLOv5 is easier to deploy on embedded development boards and is suitable for mobile terminals such as inspection robots. YOLOv5s has the smallest volume, and the weight of the latest 5.0 version is only 14 MB. The instrument detection algorithm in this paper adopts the YOLOv5s 5.0 version.

2.1. The Network model of YOLOv5s

The YOLOv5s network structure includes four parts, and the functions are introduced as follows.

Input: Random scaling, cropping, and arrangement using Mosai data enhancement can adapt well to small target detection, enrich the data set, and reduce the computational requirements of training GPUs. The optimal anchor frame value in different training sets is adaptively calculated. The filling of black edges at both ends and the computation are reduced using adaptive image scaling, and the target detection rate is improved.

Backbone: Focus structure and CSP structure are used. The slicing operation is used in the Focus structure. After slicing operation and convolution operation with 32 convolution kernels, the original image with a size of $608\times608\times3$ is transformed into a feature map with a size of $304\times304\times32$. Two CSP structures are designed in YOLOv5s. The CSP1_X structure is used in the Backbone network, and the CSP2_X structure is used in the Neck network.

Neck: FPN and PAN structures are adopted. Strong semantic features are conveyed from top to bottom in the FPN structure. A feature pyramid containing two PAN structures is added to the FPN structure to convey strong positioning features from bottom to top. The fusion ability of network features is enhanced by feature aggregation of different detection layers from different backbone layers [6].

Prediction: The error between the predicted value and the real value is evaluated by using GIOU_Loss as the bounding box's loss function to solve the problem of boundary mismatch. The prediction box with a lower score is discarded by Non-maximum suppression [7].

2.2. Introduction of Data set

A pointer instrument data set was made, and 1538 pointer instrument pictures (1076 training sets and 462 test sets) were collected from the laboratory and network. Partial test images are shown in Figure 1.

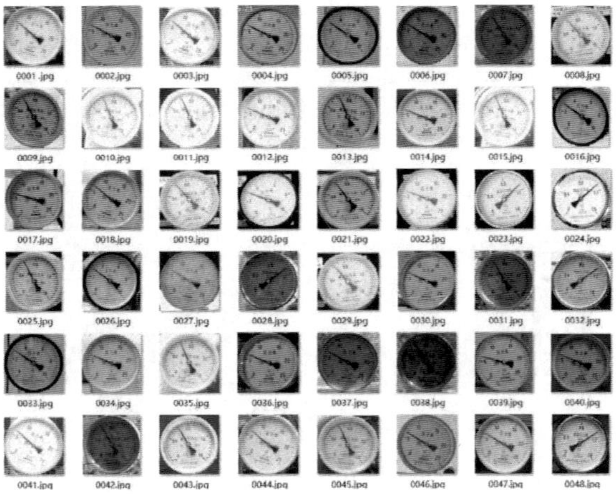

Figure 1. Instrument image set.

2.3. Settings of training parameter

The network training parameters are set as shown in Table 1.

Table 1. Parameter settings.

Parameters	Setting	Implication
Epochs	300	Number of training
Weights	Yolov5s.pt	Weight file
Batch Size	32	Number of samples selected for one training
Image Size	640	Size of input pictures
Workers	8	Number of threads used to process data
Patience	100	Number of overfitting iterations
Iterations	300	Number of iterations

2.4. Evaluation indicators

The evaluation indexes of the experiments are the precision rate P, the recall rate R, the average precision mean (mAP_0.5), and the harmonic mean (mAPP_0.5:0.95).

The precision rate is the proportion of the true instrument detected as an instrument, shown in equation (1). The recall rate is the proportion detected in all instrument samples, shown in Equation (2). The larger the values indicate the better instrument recognition effect.

$$P = \frac{Q_P}{Q_P + T_P} \tag{1}$$

$$R = \frac{Q_P}{Q_P + T_N} \tag{2}$$

where Q_P represents that both the target and the detection are instruments. T_P represents that both the target and the detection are not instruments. T_N represents that the target is the instrument, but the detection is not an instrument.

As shown in Figure 2, the precision rate and recall rate change with the iteration number. The accuracy and recall rate tend to be stable above 95% around 200 training rounds.

The precision mean is the curve area value surrounded by the precision and recall rates. The average precision mean is the mean value of the average precision of all learning categories [8]. It can be found from Figure 3 that mAP tends to be stable around 200 training rounds, in which the average precision mean is stable around 1.0, and the model training results are ideal.

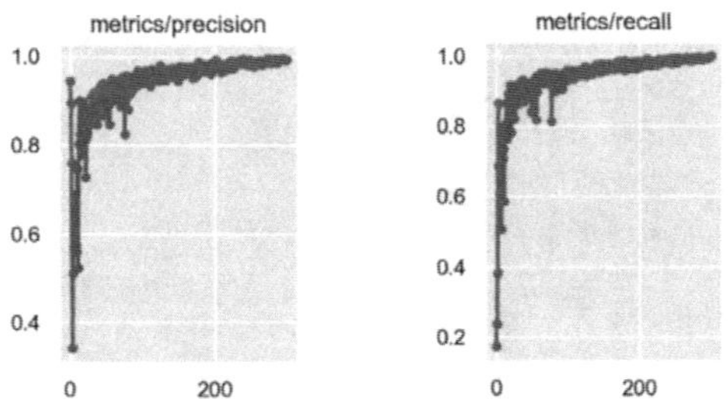

Figure 2. Precision rate and recall rate.

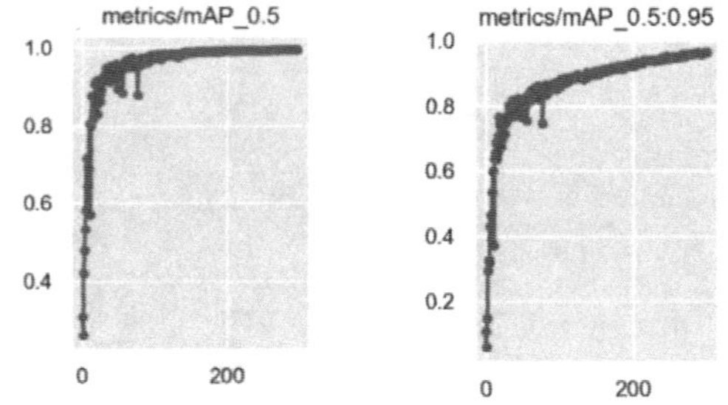

Figure 3. Precision means.

3. Design of pointer instrument number recognition algorithm

After the positioning of the pointer instrument is completed, the identification and display of the dial indicator are carried out, including the dial area extraction, the scale line extraction, the dial center determination, the zero scale line determination, and the indication identification.

3.1. Dashboard area extraction

After detecting the instrument type and instrument indication through the YOLOv5 algorithm, the instrument area of the original image is cut according to the detection result. After the target detection, part of the background has been removed. Hough circle extracts the dial area to read the instrument number more accurately [9]. The mean filtering and grayscale conversion of the instrument image is performed, and the instrument dial area is obtained after detecting the center position and circle radius by the probability Hough circle. Finally, the instrument dial area is extracted on the instrument cutting image to obtain the dial ROI area, and the process images of dial area extraction are shown in Figure 4.

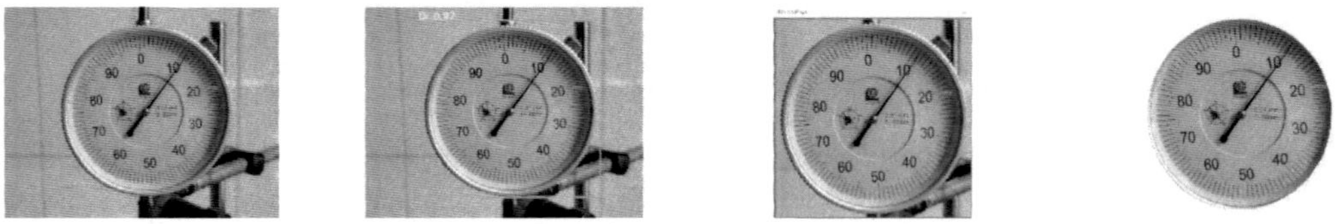

(a) Original figure (b) Test figure (c) Crop figure (d) ROI area

Figure 4. Screenshot of the dial area extraction process.

3.2. Scale line extraction and determination of the center of the instrument dial

Under the actual working conditions, there is a big deviation between the center position found by the Hough circle transformation and the center point of the instrument dial. The scale line contour is extracted using characteristics of the connected region of the scale line contour. The center point of the instrument dial is determined by fitting the scale line and finding the intersection point. The binary image is obtained by median filtering, graying, and adaptive threshold segmentation of the dial area image. The scale line contour is screened by the area, length-width ratio, and position characteristics of the scale line connected domain. The scale line contour is fitted into a straight line based on the least square method. Half of the line set is randomly divided into two parts, and the intersection point of the two-part line sets is obtained. The point set is stored in a two-dimensional 0 array with the image size. Then the position of the point in the two-dimensional array is added by 1. Finally, the center of the instrument dial is found, the point with the largest value. The instrument calibration line and the dial center test results are shown in Figure 5.

3.3. Scale line extraction and determination of the center of the instrument dial

After the binary image of the instrument dial is removed from the scale line area, the pointer contour still cannot be directly extracted from the dial text, shadow, and instrument pattern. The instrument pointer is extracted by Hough line detection. The pointer line is obtained by skeleton refinement and least square linear fitting. The pointer extraction is shown in Figure 6.

3.4. Scale line extraction and determination of the center of the instrument dial

There are manual selection methods and maximum-minimum scale line determination methods for the zero scale line determination of the pointer instrument [10]. The manual selection method wastes manpower, and the maximum and minimum scale line determination method cannot meet the circular full-scale line instrument. An instrument indication recognition algorithm is proposed based on zero-scale line marking. This method is suitable for reading fixed and a small number of easily marked instruments. In instrument image detection, the mouse marks the zero scale line. There are the angle and the distance methods in the instrument indication calculation, and the angle method is used in this paper. The instrument reading result is shown in Figure 7.

Figure 5. Instrument dial center.

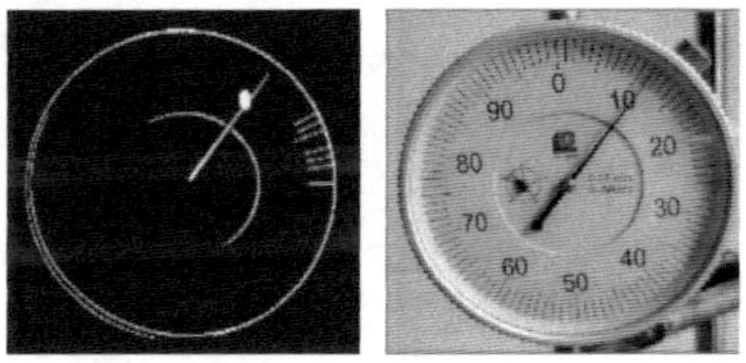

Figure 6. Instrument pointer.

Figure 7. Recognition result.

4. Experimental analysis

4.1. Test results analysis of instrument positioning

462 pointer instrument pictures are selected as the target detection set for experiments to test the model effect. The test results show that the recognition rate of single target detection of simple instruments and multi-type instruments are 100% and 90.5% respectively, and the overall recognition accuracy is 95.25%. The model detection accuracy is high, which lays a foundation for the recognition of instrument representation numbers and has a wide range of applications.

4.2. Instrument reading recognition error test

To test the accuracy of the algorithm, 10 instrument images with different numbers are collected for the instrument reading recognition experiment. The instrument reading recognition results are shown in Table 2. It can be found from the table that the proposed algorithm has a high recognition ability for clear pointer instrument images, and the average error of reading recognition is 0.126. The experiment results reveal that the recognition ability and accuracy of the proposed algorithm are strong.

5. Conclusion

According to the characteristics of current substation inspection and pointer instrument positioning and indicator recognition, a pointer instrument positioning and indicator recognition algorithm is proposed for substation inspection robots. The artificial intelligence-based YOLOv5s algorithm is adopted to locate the pointer instrument, greatly reducing the labor burden. Using digital image processing technology, this paper designs an algorithm to identify the indicator, which avoids the subjective errors of inspectors and is more accurate and efficient. Firstly, an automatic positioning and data acquisition algorithm of a pointer instrument based on YOLOv5s is proposed. Through a self-made instrument data set, the YOLOv5s network model is trained. The model can automatically identify instrument types and locate instrument positions in images. The model has been tested with high accuracy. Secondly, after the instrument is identified by the target detection algorithm, a pointer indicator recognition algorithm based on image processing technology is proposed, which mainly includes dial area extraction, scale line extraction, determination of dial center point, pointer extraction, and determination and reading of zero scale line. The test results show that the reading recognition error is 0.126, which indicates that the proposed algorithm has good recognition accuracy.

Table 2. Recognition results of the instrument reading recognition results.

No	Meter Reading	Manual reading 1	Manual reading 2	Manual reading 3	Average Manual reading	Automatic reading	Error
1		57.5	57.6	57.6	57.56	57.41	0.15
2		18.2	18.3	18.2	18.23	18.14	0.09
3		78.9	78.8	78.9	78.86	78.82	0.04
4		36.7	36.7	36.8	36.73	36.72	0.01
5		56	56.1	56.1	56.67	55.97	0.7
6		7.5	7.5	7.5	7.5	7.53	0.03
7		79.1	79.3	79.2	79.2	79.04	0.16
8		17	16.9	17	16.97	16.99	0.02
9		43.2	43.3	43.3	43.27	43.22	0.05
10		53.3	53.4	53.5	53.4	53.39	0.01

References

[1] Zhang W, Xiong Q, Zhang J, Wang Y, Jiang P, and Wang C. Pointer type meter reading recognition based on visual saliency (In Chinese). Journal of Computer-Aided Design & Computer Graphics, 2015,27(12):2282-2295.
[2] Liu J, Wu H, Chen Z. Automatic identification method of pointer meter under complex environment. Proceedings of the 2020 12th International Conference on Machine Learning and Computing. 2020: 276-282.
[3] Zhang S, Wan J, Wang H, Guan M, Yang B, and Li Fan. Convolutional neural network based on attention mechanism for reading recognition of pointer-type meter images (In Chinese). Electric Power Automation Equipment, 2022,42(04):218-224.
[4] Redmon J, Divvala S, Girshick R, et al. You only look once: Unified, real-time object detection. Proceedings of the IEEE conference on computer vision and pattern recognition. 2016: 779-788.
[5] Redmon J, Farhadi A. Yolov3: An incremental improvement. arXiv preprint arXiv:1804. 02767, 2018.
[6] Trinh H C, Le D H, Kwon Y K. PANET: a GPU-based tool for fast parallel analysis of robustness dynamics and feed-forward/feedback loop structures in large-scale biological networks. Plos one, 2014, 9(7): e103010.
[7] Rezatofighi H, Tsoi N, Gwak J Y, et al. Generalized intersection over union: A metric and a loss for bounding box regression. Proceedings of the IEEE/CVF conference on computer vision and pattern recognition. 2019: 658-666.
[8] Zhou Q, Zhang W, Li D, and Niu F. Ship classification and detection method for optical remote sensing images based on improved YOLOv5s (In Chinese). Laser & Optoelectronics Progress, 2022,59(16):476-483.
[9] Ballard D H. Generalizing the Hough transform to detect arbitrary shapes. Pattern recognition, 1981, 13(2): 111-122.
[10] Li N, Wang J, Dong X, and Shi S. Pointer meter recognition based on improved Hough transform (In Chinese). Chinese Journal of Liquid Crystals and Displays, 2021,36(08):1196-120.

Dynamic characteristics for 2 μm-self-similar pulse thulium-doped fiber laser

Weici Liu [1,2,3]**, Xingxing Li** [1] **and Zhenguo Lin** [1]

[1] School of Engineering, Guangzhou College of Technology and Business, Foshan, Guangdong, 528138, China
[2] Institute of New Generation Electronic Information Technology, Guangzhou College of Technology and Business, Foshan, Guangdong, 528138, China

[3]Corresponding author's e-mail: liuweici@m.scnu.edu.cn

Abstract. A passive mode-locked thulium-doped fiber laser is a generator for 2 μm ultra-short laser pulses. In addition, mode-locked self-similar pulses also exhibit rich nonlinear dynamic characteristics in thulium-doped fiber lasers. Therefore, this article studies the dynamic characteristics of self-similar pulses for the 2 μm passive mode-locked thulium-doped fiber laser. The research results also indicate that as the gain coefficient increases, the peak power of self-similar pulses increases, pulse width widens, and the linearity of chirp increases.

1. Introduction

The fiber lasers provide an ideal platform for exploring the complex nonlinear dynamic behavior of mode-locked pulses. They not only have the advantages of compact structure, good stability, and high beam quality but also have the characteristics of easy dispersion adjustment and high nonlinear coefficient in fibers [1].

Due to the 1 μm-band and 1.5 μm-band passive mode-locked fiber lasers, the developments are relatively mature. Many studies on mode-locked pulse dynamics have been carried out in these bands, and many important breakthroughs have been made [2]. In addition, mode-locked pulses also exhibit rich nonlinear dynamic processes in fiber amplifiers, making them an important method for studying optical phenomena such as high-energy pulse generation and self-similar pulses [3].

Self-similar pulses are typical mode-locked pulses formed in the near-zero positive dispersion region. This type of pulse avoids pulse splitting caused by nonlinear phase shift by generating linear chirp during transmission. Its time-domain characteristic is that the pulse contour is parabolic in shape and can self-evolve. Therefore, it can maintain the overall shape unchanged during transmission so that the pulse width and amplitude change proportionally [4]. This pulse that evolves in a self-similar manner is also known as a similariton. The frequency spectrum is steep on both sides, and the top is convex, which can tolerate a higher nonlinear phase shift, resulting in a higher corresponding pulse energy. When self-similar pulses are amplified outside the cavity, they can maintain their linear chirp and parabolic pulse shape unchanged, making it easy to achieve pulse compression by eliminating chirp. They are widely used to achieve high-energy femtosecond pulses.

2 μm-band laser has a wide range of applications in scientific research and civilian applications, which has aroused great research interest. The 2 μm-band laser can achieve laser generation in other wavelengths, including wavelength conversion based on energy level transitions of rare earth ions and

wavelength conversion based on nonlinear effects of the medium. It is widely used in laser surgery, optical communication, material processing, and so on.

In recent years, with thulium-doped fibers as gain media, the 2 μm-band laser pulse has also been deeply studied due to its wide application requirements. The related fiber device preparation technology and passive mode locking technology have become increasingly mature [5]. This makes the passive mode-locked fiber laser and fiber amplifier in 2 μm-band an ideal platform for studying pulse dynamics. At present, not only are various mode-locked pulse forms implemented in this band, but also, due to the different optical properties of fiber materials in this band, some unique pulse nonlinear dynamics have become the focus of research.

Resonant cavity structure is the basic way to achieve high average power output for fiber lasers. Passive mode-locked thulium-doped fiber laser has not only the advantages of compact structure, economic practicality, and high stability but also flexible and controllable resonant cavity structure and parameters, making it an ideal platform for exploring mode-locked nonlinear dynamics. Therefore, a 2 μm-self-similar pulse thulium-doped fiber laser, as a new high-performance laser technology, is a frontier research field worth exploring.

2. Theoretical model

The mode-locking fundamental equation of the self-similar pulsed thulium-doped fiber laser is as follows:

$$\frac{i\beta_2}{2}\frac{\partial^2 A}{\partial T^2}+\frac{g}{2}A=\frac{\partial A}{\partial z}-i\gamma\left|A\right|^2 A \tag{1}$$

Where β_2 is normal group-velocity-dispersion (GVD), and T is time scaled; A stands for the normalized-pulse-amplitude (NPA), and g represents the fiber's gain. γ is the nonlinearity coefficient.

3. The simulation and analysis

The evolution processes of a 2 μm-self-similar pulse in thulium-doped fiber are numerically simulated by the split-step Fourier method (SSFM). $\beta_2 = 4.2\times10^{-2}$ ps^2/m, $\gamma = 6\times10^{-3}$ W^{-1}m^{-1}, T_0=0.4 ps, and E_0=90 pj. The numerical simulation and analysis results are as follows.

3.1. The generation and evolution of 2 μm-self-similar pulse when g= 0.6 m^{-1}

The simulation result in Figure 1 shows that 2 μm-self-similar pulses are generated, which can maintain their respective waveforms unchanged during the propagation. As the pulse propagates, the pulse width and peak power gradually increase.

As shown in Figure 2, the 2 μm-self-similar pulse was obtained.

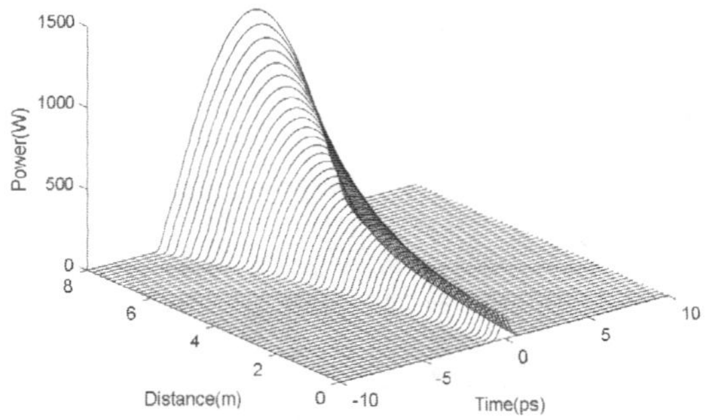

Figure 1. The evolution of the 2 μm-band-self-similar-pulse with g= 0.6 m^{-1}.

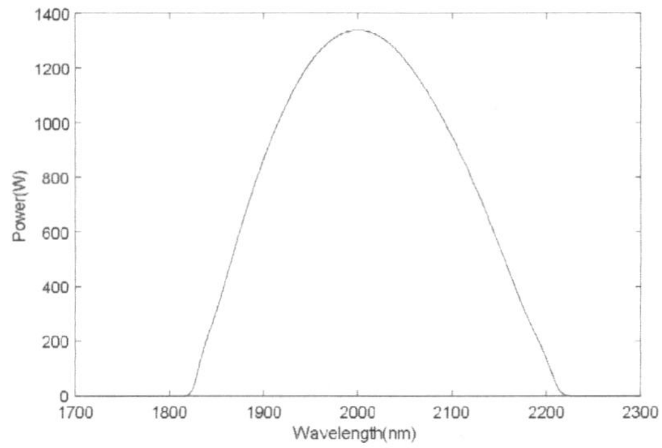

Figure 2. The spectrum of the 2 μm-band-self-similar pulse with g= 0.6 m^{-1}.

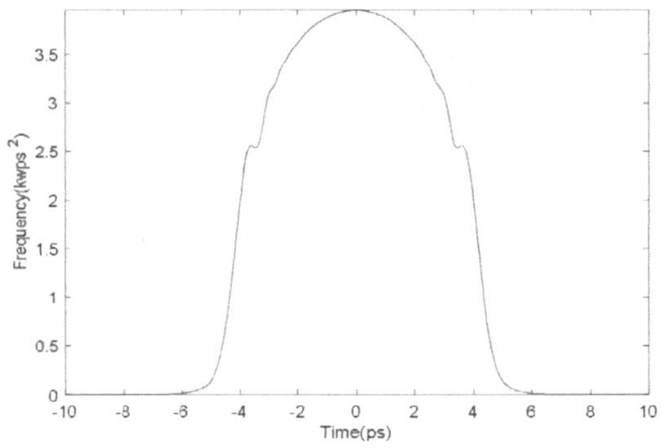

Figure 3. The frequency of the 2 μm-band self-similar pulse g= 0.6 m^{-1}.

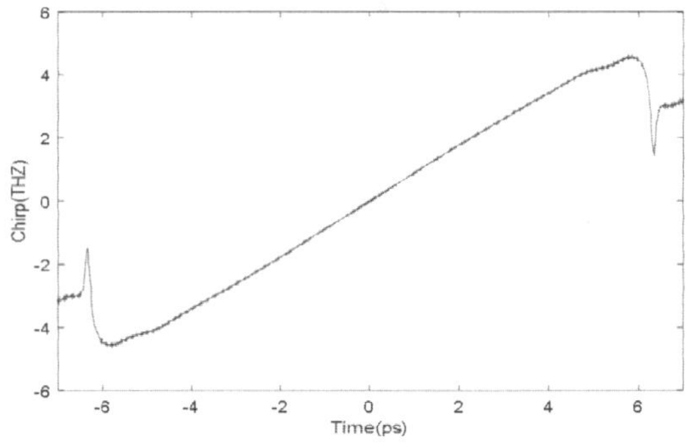

Figure 4. The chirp of the 2 μm-band self-similar pulse g= 0.6 m^{-1}.

The frequency of the 2 μm-band-self-similar pulse is shown in Figure 3. Figure 4 shows a good linear chirp achieved during the self-similar pulse evolution.

3.2. The generation and evolution of 2 μm-self-similar pulse when g= 0.7 m^{-1}

Passively mode-locked 2 μm-band-self-similar pulse fiber laser has advantages over 1 μm-band and 1.5 μm-band for the distinct optical properties of 2 μm-band. The large gain bandwidth of thulium-doped fibers results in fundamental differences in their nonlinear dynamic characteristics. After changing the gain coefficient, the generation and evolution characteristics of 2 μm-self-similar pulses are shown as follows.

Comparing Figures 1-4 with Figures 5-8, we can observe that as the gain coefficient increases, the 2 µm-self-similar pulse width widens, the peak power increases, and the linearity of the chirp improves.

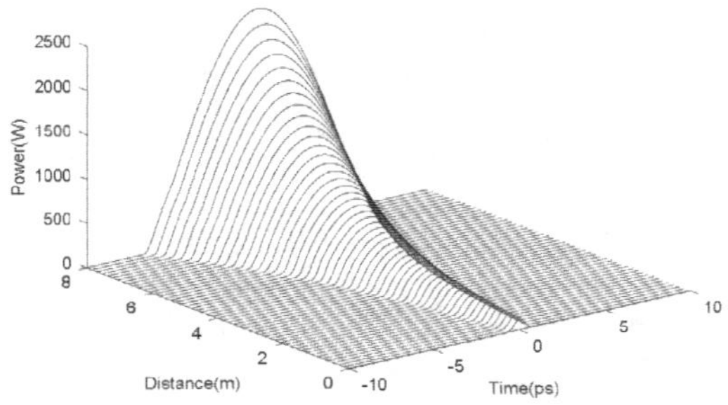

Figure 5. The evolution of the 2 µm-band-self-similar pulse with g= 0.7 m^{-1}.

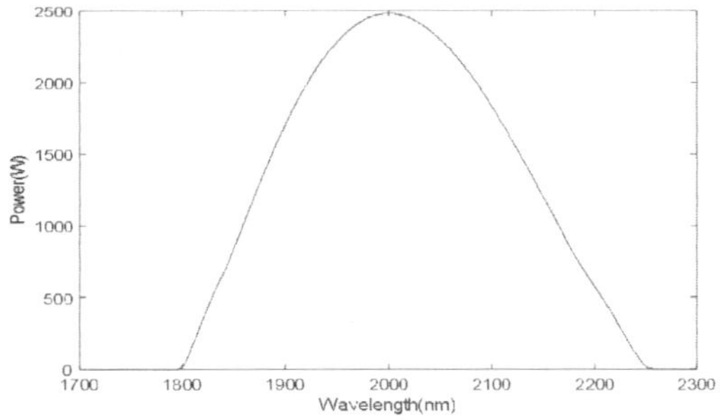

Figure 6. The spectrum of the 2 µm-band-self-similar pulse with g= 0.7 m^{-1}.

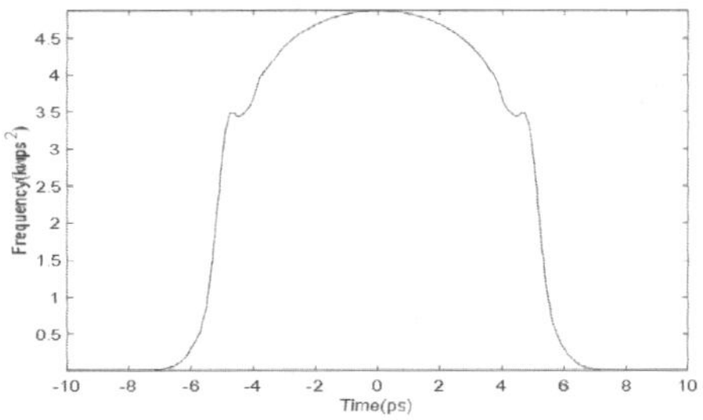

Figure 7. The frequency of the 2 µm-band-self-similar pulse g= 0.7 m^{-1}.

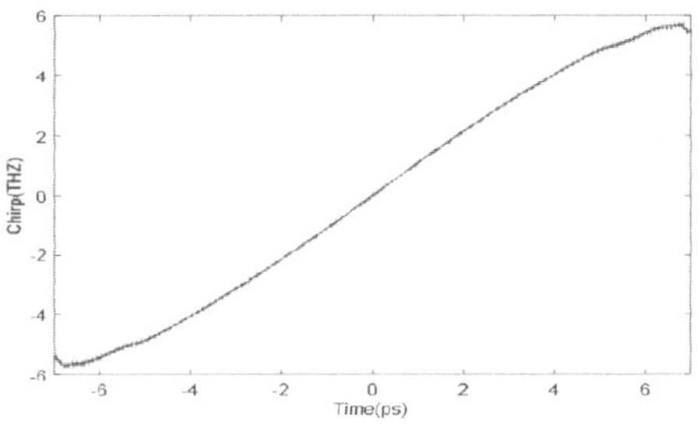

Figure 8. The chirp of the 2 µm-band-self-similar pulse g= 0.7 m^{-1}.

4. Conclusion

In summary, this article investigates 2 μm-band-self-similar pulses generated by a thulium-doped mode-locked fiber laser, and the evolution process of parabolic pulses generated by the laser in the cavity was analyzed. Passive mode-locked thulium-doped fiber lasers provide a good platform for studying 2 μm-band-self-similar dynamics. For passive mode-locked fiber lasers, their flexible resonant cavity parameters are conducive to achieving diverse mode-locking mechanisms. Studying the 2 μm-band-self-similar pulse dynamics characteristics is not only beneficial for revealing complex physical phenomena but also for better meeting practical needs by utilizing their pulse diversity.

Acknowledgment

The work was supported by grants from the Key Project of DEGP (Nos. 2022KTSCX166、2021KTSCX154), the Young Talents Project of DEGP (Nos. 2021KQNCX125, 2020KQNCX109), the Project of Guangzhou College of Technology and Business (Nos. KYPY2022039、KABS202108、KYZD202204), the Innovation and Entrepreneurship Training Program for College Students of Guangzhou College of Technology and Business (No. xj202313714088).

References

[1] Li, X., Wang, D. N., Hua, K., Chen, Q., Ge, Y., & Xia, Q. K. (2022). Saturable absorber based on graphene for a hybrid passive mode-locked erbium-doped fiber laser. Optical Fiber Technology, 70, 102867.

[2] Vladimir, Lazarev, Alexander, Krylov, Dmitriy, & Dvoretskiy, et al. (2016). Stable similariton generation in an all-fiber hybrid mode-locked ring laser for frequency metrology. IEEE transactions on ultrasonics, ferroelectrics, and frequency control.

[3] Sharafali., A., Dhasarathan, V., & Ung, B. (2021). Self-similar pulse compression by suspended core photonic crystal fiber with cubic-quintic nonlinearities. Results in Optics.

[4] Deng, Y. X., Tu, C. H., & F.-Y. Lü. (2009). Study of self-similar pulse nonlinear polarization rotation mode-locked fiber laser. Acta Physica Sinica, 58(5), 3173-3178.

[5] Zhang, N., Sun, H., & Luo, Y. (2021). Research on key technologies for 2 μm band of thulium-doped fiber laser. Journal of Physics: Conference Series, 1812(1), 012014 (6pp).

Study on the measuring device of medical centrifuges

Huijuan Wang[1, 2], Mengdie Xiang[1], Bowen Deng[1] and Yongdong Wu[1]

[1] Chongqing Academy of Metrology and Quality Inspection, Chongqing 401123, China

[2]Corresponding author's e-mail: 540063612@qq.com

Abstract: This paper designs a measuring and detecting device for medical centrifuges. The device integrates the measurement of rotational speed, temperature, and time. The photoelectric tachometer was transformed into a laser speed detection module to adapt to the rotational speed measurement of the closed centrifuge and a wireless temperature detection module could be inserted into the slot of the centrifuge tube to measure the temperature of the tube position when the centrifuge rotates at a high speed. This method achieved reliable and accurate measurement of the speed, temperature, and time of various centrifuges with low costs.

1. Introduction

Medical centrifuges are commonly used in medical laboratories or biochemical experiments. Medical centrifuges use rapid rotation to rapidly precipitate suspended particles in mixtures to separate various substances and components with different values of specific gravity. To separate samples, it is necessary to maintain the corresponding centrifugal rotational speed and set the centrifugation time when the medical centrifuge is used for the pre-treatment of a medical laboratory or a biochemical experiment. To prevent some of the high-temperature resistant substances from deteriorating due to friction heat, temperature control may also be used. Therefore, the accuracy of rotational speed, temperature, and time directly affects the centrifugal effect and thus test results. It is very important to conduct quantitative detection and control the accuracy of these three parameters for medical centrifuges.

At present, there are a variety of measurement methods for the rotational speed of medical centrifuges, mainly including photoelectric, magnetoelectric, and vibrating types. In Zhao et al.'s work [1], a photoelectric tachometer was proposed to measure the speed of medical centrifuges. A reflection mark is pasted on any rotating arm of the centrifuge and the photoelectric tachometer is used to measure the observation hole of the centrifuge. This method is accurate and reliable, and the detecting equipment has multiple models and low costs, but it is only applicable to centrifuges with observation holes. More than 30% of medical centrifuges without observation holes are closed in the market, so rotational speed measurement cannot be done by this method. An electromagnetic sensor is used to measure the rotational speed of the medical centrifuge at Hunan University. In this method, a magnet is attached to the inner wall of the medical centrifuge, and a Hall sensor is placed in the test tube of the rotating shaft of the medical centrifuge. The Hall sensor is used to detect the number and time of pulses, and the rotational speed of the medical centrifuge is calculated [2]. This method can be used to measure medical centrifuges with or without observation holes, but the installation is difficult, and because of the shielding effect of the metal rotor, the anti-interference ability and the accuracy of speed measurement are poor. Li et al. [3] from Shenyang University of Technology proposed a principle of vibration measurement. A vibration sensor is fixed on the outer shell of a medical centrifuge to measure rotational speed. It can effectively measure both types of medical centrifuges (with and without observation holes). However, this method has high equipment costs and low economic efficiency Moreover, it requires unbalanced objects placed in the medical centrifuge to achieve vibration, and the rotor of the medical centrifuge can be easily damaged by a slightly improper operation.

At present, the commonly used method for temperature measurement of medical centrifuges is to place the wired temperature sensor into the centrifuge chamber for measurement [4]. In order to avoid touching the temperature sensor during the high-speed rotation of the centrifuge, this method is generally used when the equipment is in a static state. However, in a non-centrifugal state, there is no internal airflow, and the temperature distribution is extremely uneven. In addition, the rotor of the centrifuge generates heat due to friction during rotation, so the actual temperatures in a static state and a centrifugal state differ enormously. In addition, the method for temperature measurement is to firmly paste the temperature sensor on the inside of the cavity wall of the medical centrifuge, so that the temperature is measured after the rotor of the medical centrifuge rotates. However, because the refrigeration system of the medical centrifuge is generally used for cavity wall refrigeration, the temperature of the inner wall of the medical centrifuge is significantly lower than that of the tube, and the measured temperature is not accurate [5].

2. Overall scheme design of medical centrifuge measuring device

In order to solve the difficulties and problems in the measurement and detection of centrifuges, this paper designs a measuring and detecting device for medical centrifuges. The device integrates the measurement of rotational speed, temperature, and time. The photoelectric tachometer was transformed into a laser speed detection module to adapt to the rotational speed measurement of the closed centrifuge and a wireless temperature detection module could be inserted into the slot of the centrifuge tube to measure the temperature of the tube position when the centrifuge rotates at a high speed. This method achieved reliable and accurate measurement of the speed, temperature, and time of various centrifuges with low costs.

This detecting device mainly includes a main control unit, a laser speed detection module, a wireless temperature detection module, etc., as shown in Figure 1.

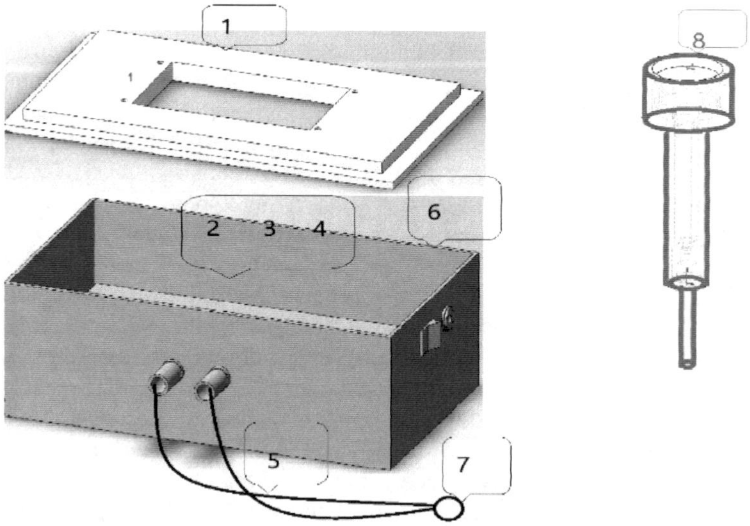

Figure 1. Three-dimensional diagram of the detecting device for a medical centrifuge.

In the three-dimensional diagram of the detecting device, Part 1 is the host screen; Part 2 is the photoelectric circuit and photovoltaic system; Part 3 is the temperature signal reception system; Part 4 is the battery; Part 5 is the optical fiber; Part 6 is the main control unit, containing Parts 2-4; Part 7 is the laser speed measurement module; Part 8 is the wireless temperature detection module, whose probe is small and can be put in the test tube.

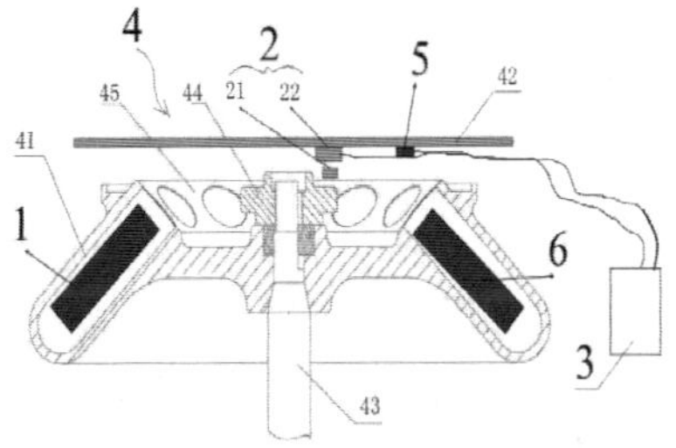

Figure 2. Graphical representation of the use of the calibrating device for medical centrifuge metrology.

The device is placed in a medical centrifuge for use, as shown in Figure 2, where Number 41 represents the tube position of the medical centrifuge, Number 42 represents the top cover of the medical centrifuge, and Numbers 43 and 44 represent the rotating shaft and the rotor of the medical centrifuge. Numbers 1, 2, 3, 5, and 6 represent the parts of the device. Number 3 represents the main control unit, which is placed outside the medical centrifuge, and Number 2 represents the laser speed measurement module. Numbers 2 and 3 are connected by the optical fiber. Number 1 represents a wireless temperature collector. Number 6 represents the temperature collect or weight unit, whose weight is consistent with Number 1. Number 5 represents the temperature signal connector, and the signal is transmitted to the main control unit (Number 3) through the signal line. The device can detect medical centrifuges with a rotating speed of 0~30000 r/min, a temperature control range of 0~40°C, and a rotation time of 0~24 h. It is suitable for various types of centrifuges with or without observation holes, operating at low temperatures or normal temperatures.

The laser reflection unit (Number 21) is attached to the rotation axis of the centrifuge, Number 22 represents the laser transmitting and receiving end of the device. One end of Number 22 is connected to the main control unit (Number 3), which is extended into the interior of the medical centrifuge through the optical fiber opening gap and attached to the upper edge of the centrifuge top cover (Number 42), thus constituting the laser speed measurement module. Number 1 represents a wireless temperature collector, and the data it collects is transmitted to the inside of the centrifuge through a wireless signal. In addition, the temperature acquisition unit is equipped with a weight unit (Number 6), which is consistent with the wireless temperature collector to ensure smooth rotation of the medical centrifuge. The main control unit (Number 3) is also connected to a clock module to display and calculate centrifuge running time; the collected speed, temperature, and time values are all displayed on the display screen simultaneously through the main control unit.

3. Design of each module of the device

3.1 Design of rotational speed measurement module

This rotational speed measurement module uses infrared laser feedback technology, as shown in Figures 1 and 2. Number 22 represents the laser transmitting and receiving unit, which is connected to the main control unit (Number 3) through an optical fiber. Number 21 represents the reflection unit, which is pasted and fixed on the rotating platform of the medical centrifuge rotor. Number 22 emits infrared light, and with every turn of the rotor, the reflection unit (Number 21) reflects an infrared light signal to Number 22. The reflected light is transmitted back to the main control unit (Number 3) through the optical fiber, and the main control unit (Number 3) calculates the speed of the medical centrifuge by calculating the frequency of the reflected light. Every time a laser signal is received, the rotating platform inside the medical centrifuge rotates for one round.

3.2 Design of temperature measurement module

As shown in Figures 1 and 2, the wireless temperature collector (Number 1) is inserted into the centrifuge tube slot of a medical centrifuge to measure the temperature in the centrifuge tube slot. The collector, which is shaped like a centrifuge tube, contains a high-precision temperature sensor, a battery, a signal transmitter, and a flexible circuit board. All the above components are integrated into a small unit with a diameter of 22 mm. In order to avoid the Doppler effect of signals and ensure the stable and continuous transmission of temperature signals during rapid rotation, this module adopts wireless transmission to output temperature data, and the wireless receiving unit (Number 5) is fixed on the internal top cover of the medical centrifuge, transmitting the data it obtains to the main control unit (Number 3) through the signal line and then displaying the temperature data in real time through the display screen.

Because the items placed in the rotor of the medical centrifuge need to be balanced, we designed a counterweight unit (Number 6). Its weight is equivalent to that of the wireless temperature detection unit (Number 1) and its centrifuge tube corresponds to the centrifuge tube slot of the wireless temperature detection unit (Number 1) during use to ensure balanced and stable high-speed rotation of the medical centrifuge rotor.

3.3 Design of the main control unit

The main control unit is the overall control platform and data summary platform of the whole medical centrifuge measuring and detecting device, and its structure is shown in Figure 3.

The outer shell of the main control unit is arranged with a display screen, operation keys, and a USB Type-C interface. The microprocessor is located in the shell, and the built-in clock module is used to display and calculate the running time of the centrifuge. The microprocessor is connected to the display screen, operation keys, wireless receiving module, laser sensor, and the outer end of the transmitting fiber and receiving fiber connection. The power module is powered by lithium batteries and charged through a Type-C interface.

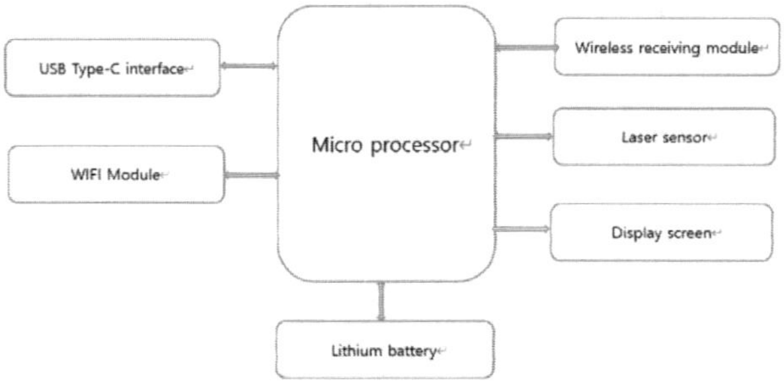

Figure 3. Diagram of main control unit structure.

4. Test and analysis of the device

4.1 Scheme of device test [7][8][9]

In order to verify the measurement accuracy of the device, the device, the photoelectric tachometer, and the vibrating type measuring instrument for centrifuge rotational speed was used to measure the rotational speed of the same centrifuge (with the same rotor installed) (for the convenience of comparison, a medical centrifuge with observation holes was selected for measurement); the device and the wired temperature inspecting instrument were used to measure the temperature of the same centrifuge (a low-temperature centrifuge was selected). Timing measurements were performed on the same centrifuge using this device and a stopwatch, respectively.

1) The device was used for the quantitative detection of medical centrifuges [10][11].

The medical centrifuge was placed on a stable horizontal table, the wireless temperature recording module and its counterweight were placed symmetrically in the rotor of the centrifuge, and a test tube filled with water was placed in one of the other free positions to load the medical centrifuge to a full load.

The centrifuge top cover was opened, and the reflection unit was pasted on the rotating platform of the rotor of the medical centrifuge. The laser transmitting and receiving module of the device was pasted and fixed on the bottom of the medical centrifuge top cover. After the top cover of the medical centrifuge was closed, the transmitting optical fiber, the receiving optical fiber, and the wire of the wireless receiving unit were extended through the opening gap of the medical centrifuge, and then connected to the main control unit.

Then the pre-cooling of the medical centrifuge started, and the temperature measurement value of the medical centrifuge in a static state was read on the device after the set temperature was reached.

After that, the medical centrifuge started to cool and rotate, and the set temperature and speed were reached. After the medical centrifuge reached stability, the measurement values of rotational speed, temperature, and timer time of the medical centrifuge were read on the display screen of the device.

2) The rotational speed was measured by the photoelectric tachometer and vibration speed measuring instrument.

The top cover of the medical centrifuge was opened, and then the tube was installed, fill with water, and placed in the tube position. After that, the medical centrifuge was loaded to a full load, the reflective strip was attached on the rotating platform of the rotor to the position of the observation hole, and the vibration probe of the vibration meter was stuck on the outer wall of the medical centrifuge side. After the top cover of the medical centrifuge was closed and the centrifuge was started to reach the set rotational speed, the rotational speed of the medical centrifuge was measured by pointing the observation hole with the photoelectric tachometer, and the value of the measured rotational speed was read at the terminal of the vibrating type speed measuring instrument.

3) The temperature was measured with a wired temperature inspecting instrument.

The temperature probes 1 and 2 of the wired temperature inspecting instrument were inserted through the bottom of the medical centrifuge and pasted on the inner and middle inner walls of the bottom of the medical centrifuge, respectively. Probe 3 was inserted through the gap of the upper cover and placed in the placement position of the centrifuge test tube. Then the power supply of the centrifuge was turned on, the temperature was set, and the centrifuge was pre-cooled. After the temperature of the medical centrifuge reached stability, the temperature measurement values of Probes 1, 2, and 3 were read. Then Probe 3 was removed, the centrifuge rotor was started to reach the set speed and stabilized, and the temperature measurement values of Probes 1 and 2 were read [6].

4) The timer time was measured with a stopwatch.

The centrifuge timer time was set, and the stopwatch measurement was started when the speed reached the set value.

4.2 Results of device test

According to the test method, the speed, temperature, and time of H2050R, LC-LX-H185C, Microfuge 20, Legend Micro 17, and TGL16M were measured (the five devices numbered 1, 2, 3, 4, and 5 respectively). To cover the measurement range of the five types of equipment, three measurement points—1000 r/min, 5000 r/min, and 15000 r/min—were selected for speed, 5°C for temperature, and 5 min for time. The data obtained are as follows.

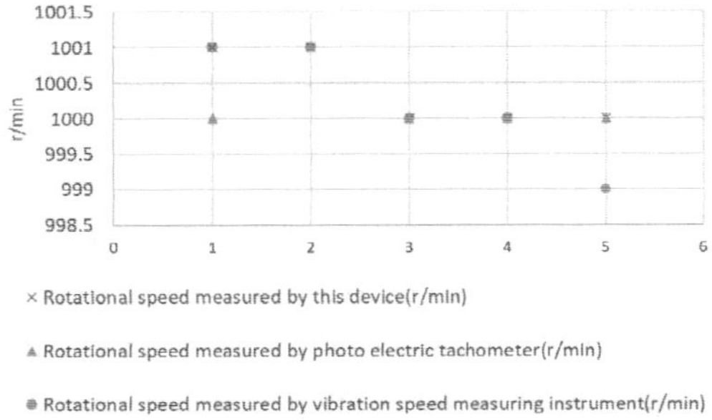

Figure 4. Rotational speed measured at the 1000 r/min measurement point.

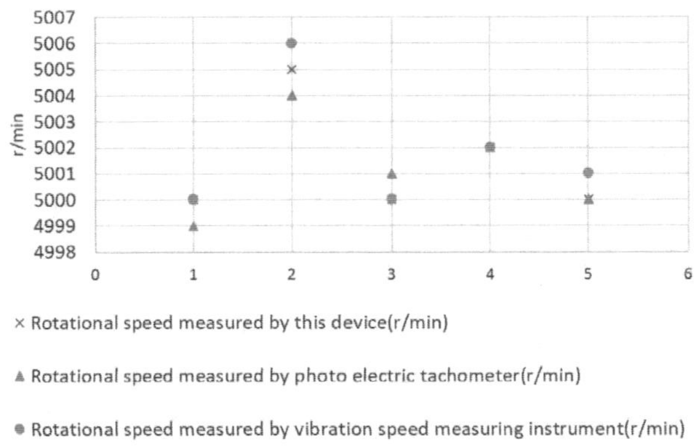

Figure 5. Rotational speed measured at the 5000 r/min measurement point.

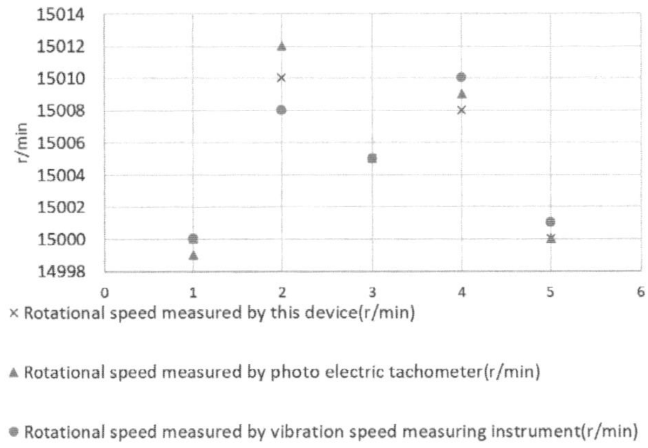

Figure 6. Rotational speed measured at the 15000 r/min measurement point.

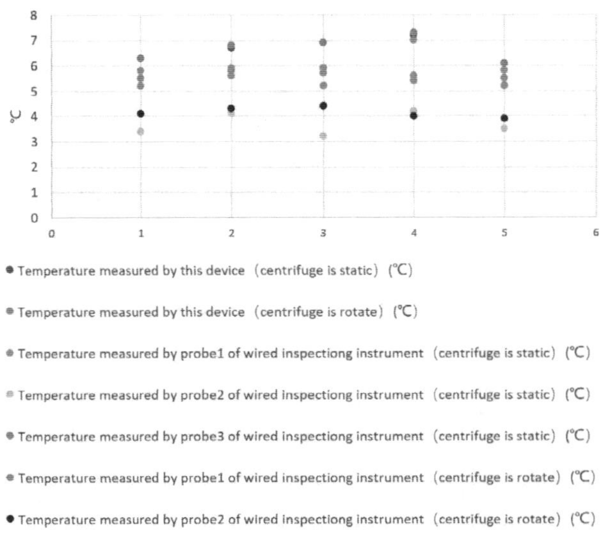

Figure 7. Temperature measured at the 5°C measurement point.

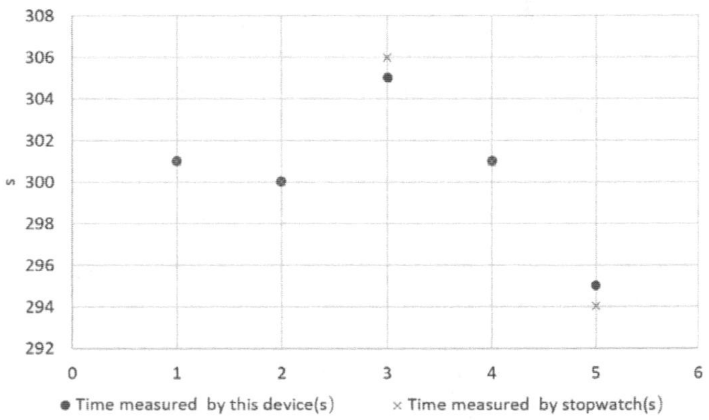

Figure 8. Time measured at the 5 min measurement point.

4.3 Statistical analysis

By comparing with other methods of centrifuge measurement, we obtained the following results from a statistical analysis:

1) From the Figure 4-6, we can see that the results of the rotational speed measurement experiment for this device, photoelectric tachometer, and rotational speed measuring instrument are basically the same, which proves that the rotational speed measurement of this device is accurate and reliable.

2) From the Figure 7, we can see the result of temperature measurement. In this experiment, when the medical centrifuge was not rotating, the temperature at the test tube position was also measured. The temperatures measured by this device are consistent with those measured by the wired temperature inspecting instrument, which proves that the temperature measurement of this device is accurate and reliable. The wired temperature inspecting instrument detected different temperatures at the bottom, side wall, and tube position of the centrifuge, and the device detected different temperatures when the centrifuge rotor was rotating and not rotating, which proves the necessity of the design and use of the device. Because the use of a low-temperature centrifuge is to ensure that the temperature of the material to be centrifuged is within the control range, it is more practical to measure the temperature of the tube position during the rotation of the rotor when testing the medical centrifuge.

3) From the Figure 8, we can see that in the time measurement experiment, the time measured by the device and the stopwatch are basically the same, which proves that the time measurement of the device is accurate and reliable. This device can automatically measure time and speed synchronously, which is more convenient for practical use.

5. Conclusion

This paper takes the measurement and detection of medical centrifuges as the research object. Starting from the detection requirements of medical centrifuges, this paper uses a laser reflection tachometer, a wireless temperature sensor, a timing counter, and other common, low-cost equipment to achieve accurate measurement of rotational speed, low-temperature centrifuge temperature and centrifugation time for medical centrifuges. Moreover, the proposed device has good portability, practicability, and operability with low costs and high safety standards.

References

[1] Zhao Yifeng, Li Weijia, Qian Jianguo. Measurement and Detection of Centrifuge Speed [J]. Chinese Medical Equipment, 2012, 27(8): 67-69.

[2] Zhang Huaixin. Study on the calibration method of laboratory centrifuge speed [J]. China Metrology, 2015(8): 104-105.

[3] Li Hong, Li Guang, et al. Measurement and detection system of speed of closed medical centrifuge [J]. Testing in China, 2017, 43(11): 70-73.

[4] Zheng Xiaoqing. Study on calibration method of medical refrigerated centrifuge [J]. Metrology and Testing Technology, 2021, 48(6): 1-6.
[5] Xue Xuemin, Tang Yaoxu, et al. Research on temperature uniformity of medical refrigerated centrifuge [J]. Metrology and Testing Technology, 2021, 48(12): 35-39.
[6] Huang Fan, Fu Lei, et al. Study on temperature distribution characteristics of medical cryogenic centrifuge in different radius area, 222, 49 (5): 10-16.
[7] YY/T0657, Medical Centrifuge, China Quality and Standards Publishing & Media Co., Ltd., Beijing, (2017).
[8] JJF (Hebei) 167, Calibration Specification for Medical Centrifuges, China Quality and Standards Publishing & Media Co., Ltd., Beijing, (2019).
[9] JJF (Nei meng) 039, Calibration Specification for Laboratory Centrifuges, China Quality and Standards Publishing & Media Co., Ltd., Beijing, (2019).
[10] Zou H, Bai J M, Chen M. Experimental Research on Sample Reagent Temperature in Cavity of Refrigerated Centrifuge Rotor for Medical Labs [J]. Fluid Machinery, 2010, 102(C1): 1091-1099.
[11] Zhou Wei, Sun Li, Ni Haifeng. Calibration method of medical refrigerated centrifuge [J]. Shanghai Metrology and Test, 2018, 45(06): 50-51.

AUTHOR INDEX